The aerodynamic principles that make flight possible were little known or barely understood as recently as a hundred years ago. Although their roots can be found in the fluid dynamics of ancient Greek science, it was not until the scientific breakthroughs at the beginning of the twentieth century that it became possible to design successful flying machines.

This book presents a history of aerodynamics, intertwined with a review of the aircraft that were developed as technology advanced. Beginning with the scientific theories and experiments of Aristotle and Archimedes, the book continues through the development of applied and theoretical aerodynamics in the early 1900s and concludes with modern hypersonic and computational aerodynamics. It is a thoroughly engrossing account of successes and failures, human competition, and the role of aerodynamics in the development of science and technology in this century.

The author, John D. Anderson, Jr., is a professor of aerospace engineering, faculty member of the Committee on the History and Philosophy of Science, and affiliate professor of history at the University of Maryland and a special assistant for aerodynamics at the National Air and Space Museum of the Smithsonian Institution. He has written extensively on various topics in the field of fluid dynamics.

Cambridge Aerospace Series 8

General Editors
MICHAEL J. RYCROFT AND ROBERT F. STENGEL

A History of Aerodynamics

Cambridge Aerospace Series

1. J. M. Rolfe and K. J. Staples (eds.): *Flight Simulation*
2. P. Berlin: *The Geostationary Applications Satellite*
3. M. J. T. Smith: *Aircraft Noise*
4. N. X. Vinh: *Flight Mechanics of High-Performance Aircraft*
5. W. A. Mair and D. L. Birdsall: *Aircraft Performance*
6. M. J. Abzug and E. E. Larrabee: *Airplane Stability and Control*
7. M. J. Sidi: *Spacecraft Dynamics and Control*
8. J. D. Anderson, Jr.: *A History of Aerodynamics*

A History of Aerodynamics

and Its Impact on Flying Machines

JOHN D. ANDERSON, JR.

University of Maryland

CAMBRIDGE
UNIVERSITY PRESS

PUBLISHED BY THE PRESS SYNDICATE OF THE UNIVERSITY OF CAMBRIDGE
The Pitt Building, Trumpington Street, Cambridge, United Kingdom

CAMBRIDGE UNIVERSITY PRESS
The Edinburgh Building, Cambridge CB2 2RU, UK
40 West 20th Street, New York, NY 10011-4211, USA
10 Stamford Road, Oakleigh, VIC 3166, Australia
Ruiz de Alarcón 13, 28014 Madrid, Spain
Dock House, The Waterfront, Cape Town 8001, South Africa

http://www.cambridge.org

First published 1997
First paperback edition 1998
Reprinted 1999, 2000, 2001

Printed in the United States of America

Typeset in Times

A catalog record for this book is available from the British Library

Library of Congress Cataloging in Publication Data is available

ISBN 0 521 45435 2 hardback
ISBN 0 521 66955 3 paperback

Contents

Appendixes

Foreword

John Anderson's book, *A History of Aerodynamics and Its Impact on Flying Machines*, represents a milestone in aviation literature. For the first time aviation enthusiasts – both specialists and popular readers alike – possess an authoritative history of aerodynamic theory. Not only is this study authoritative, it is also highly readable and linked to the actual (and more familiar) story of how the airplane evolved. Countless books exist on famous aviators, historic aircraft, air transportation, and the impact of air power on modern warfare; few books, however, touch upon how baseline theoretical work on aerodynamic theory made modern flying machines possible. A professor of aerospace engineering, John Anderson brings this complex story to life.

The dream of flight is indeed an ancient dream – one that is manifest in myth, in art, and in the earliest records of civilization. John Anderson begins with Aristotle and Archimedes, and then traces how this dream became transformed into a rudimentary science through the inspired work of Leonardo da Vinci, around 1500. The heirs of Leonardo da Vinci – George Cayley, Otto Lilienthal, Samuel Langley, and the Wright brothers – carried on, each establishing important benchmarks in the theory of aerodynamics. When the Wrights flew at Kitty Hawk on December 17, 1903, they inaugurated the modern air age, having demonstrated that a practical means of powered human flight had been achieved. The rapid advance of aeronautical design after the Wrights is remarkable; in fact, this story is one of the most compelling chapters in the history of technology. The saga of modern aerodynamics is told here for the first time in a comprehensive way. John Anderson links the theory of aerodynamics to the developmental history of flying machines. He touches on all the major theorists and their contributions and, most important, the historical context in which they worked to move the science of aerodynamics forward.

Dr. Anderson's historical reconstruction is fascinating and, from the standpoint of current aeronautical literature, an important new avenue for us to understand the history of human flight. How airplanes were designed and redesigned to fly "faster, farther, and higher," one could argue, is a critical chapter in the story of the twentieth century. The airplane – in the space of a few decades – evolved from a fragile, underpowered aircraft of wood and fabric into the sleek jet-powered aircraft of today. Around this transformation were related breakthroughs in powerplant design, in aircraft structures and materials, and in aircraft stability and control. Through his historical analysis John Anderson pulls the curtain back, to allow us to see how aerodynamics shaped the evolution of flying machines, past and present.

John Anderson has served for many years as a Special Assistant for Aerodynamics at the Smithsonian's National Air and Space Museum. In doing so, he has enriched our programs in research, exhibits, and public outreach. He is one of a small company of aerodynamicists who have assumed the role of historian; such a shift has greatly enlarged our comprehension

of the theoretical underpinnings of flight. I applaud his creative labors and warmly endorse
this important study.

Von Hardesty
Smithsonian Institution

Preface

My office at the University of Maryland is under the flight path for airplanes going to and from the College Park airport, usually single- or twin-engine general-aviation aircraft. Each one, as it flies overhead, typifies the laws of aerodynamics in action; each is a reminder of how humans have harnessed these laws and put them to practical use in the design of flying machines. However, it is easy to forget that only a little more than a century ago these laws were so little known or so misunderstood that no one had been able to build a machine that would fly.

How did we finally come to understand the basic laws of aerodynamics, at least to an extent sufficient to design successful flying machines? The story of that understanding reaches all the way back to ancient Greek science and the theories and studies of Aristotle and Archimedes. It has been an exciting quest, and the purpose of this book is to tell that story, to present the history of aerodynamics.

The unique aspects of this book are as follows:

(1) It is the first to be devoted exclusively to the *history* of aerodynamics.
(2) It provides an interpretation of the history of aerodynamics as seen through the eyes of a practicing aerodynamicist, and thus it is complementary to previous studies carried out by professional historians of science and technology.
(3) It presents new research findings, previously unpublished, that have answered some long-standing questions posed by historians of aeronautics. Previously unknown aspects of work in applied aerodynamics by such pioneers as George Cayley, Horatio Phillips, Otto Lilienthal, Samuel Langley, and the Wright brothers have come to light as a result of the research in preparation for this book. Also, several technical discrepancies and inconsistencies associated with the historical data have been clarified simply by carrying out proper aerodynamic analyses, combined with some new thinking. In this sense, parts of the book make new contributions to scholarship in the history of technology.
(4) In addition to examining the history of aerodynamics per se and assessing the state of the art of aerodynamics during various historical periods, the book seeks to answer an important question: How much of the contemporary state of the art in aerodynamics at any given time was incorporated into the actual design of flying machines at that time? That is, what was the impact of aerodynamic knowledge on contemporary designs of flying machines?

Finally, this book covers the developments in both theoretical and applied aerodynamics from ancient Greek science through the revolutionary breakthroughs achieved in the circulation theory of lift by Kutta and Joukowski and the boundary-layer theory of Prandtl, all within the first decade of the twentieth century. The history of aerodynamics divides very nicely at that juncture, with the early history of aerodynamics covering the period prior to the time of explosive growth and widely divergent developments in the field after 1910. The book covers that early history as well as twentieth-century aerodynamics to the present.

xi

I wish to acknowledge the invaluable help of some of my colleagues at the National Air and Space Museum of the Smithsonian Institution during the preparation of this manuscript, in particular Peter Jakab, Tom Crouch, and Von Hardesty of the Aeronautics Department. Their professional advice and encouragement made all the difference in the world to me. I hope that this book measures up in some way to their expectation. Also, the archive materials and artifacts in the National Air and Space Museum were valuable sources in my search for new information and new insights; of particular note was access to Samuel Langley's laboratory notebooks, on file in the Ramsey Room, the rare-book room of the museum. I also wish to acknowledge the invaluable support of my wife, Sarah-Allen, who has continually reminded me of the value of careful historical research. The majority of this manuscript was typed for me by Susan Cunningham, my longtime friend and colleague from the University of Maryland. I hope that my "word-smithing" in this book has half the quality of her "word-processing."

Finally, my ride through the history of aerodynamics has been pure, unadulterated fun. I wish the reader the same enjoyable trip through this book.

John D. Anderson, Jr.

PART I

The Incubation Phase

CHAPTER 1

Aerodynamics
What Is It?

Aerodynamics . . . the branch of dynamics that treats of the motion of air and other gaseous fluids and of the forces acting on bodies in motion relative to such fluids.

Webster's Third New International Dictionary

As you read these words, around the world there are thousands of airplanes in flight through the earth's atmosphere. We take such things for granted; the airplane is a fixture in our everyday life. However, it will be informative to pause for a moment to consider that these aircraft are miracles of modern engineering wherein many diverse fundamental laws of nature are applied and combined in a useful fashion so as to produce safe and efficient flying machines. Some of these fundamental laws involve the science of aerodynamics, without which modern flight would be impossible. Indeed, an airplane flying overhead typifies the laws of aerodynamics in action, and we often forget that only two centuries ago these laws were so little known or so misunderstood that no one had been able to build a flying machine that could lift off the ground, let alone fly long distances.

Why were the principles of aerodynamics considered so mysterious, and why was it so difficult to develop them into an applied science? What happened over the centuries in various fields of science that finally resulted in the discipline of aerodynamics? What were the driving forces that focused human thought on the idea of a flying machine? These and many other questions about the evolution of aerodynamics are addressed in this book. Aerodynamics is a beautifully *intellectual* discipline, incorporating elements from a millennium of human thought that finally coalesced during the nineteenth century to produce the exponential growth in powered flight that we see today.

The story of aerodynamics is itself dynamic, replete with smashing successes, abject failures, and intense human competitions that have had major social, economic, and political consequences. It is a story of human beings, some exceptional and some average, some admirable and some not so. Above all, it is the story of an intellectual quest to understand those laws of nature that have allowed the continuing development of aircraft that can fly faster, higher, and more efficiently than their predecessors. The purpose of this book is to tell that story. The approach will be scholarly, but the story is also entertaining. This book is directed to a cosmopolitan audience: to practicing aerodynamicists who want to know more about the legacy of their profession, to engineers and scientists who want to understand the place of aerodynamics within the much larger world of science and technology, to historians who want to reflect on the role of aerodynamics in the history of science and technology, and to those in the general public who simply are curious about the origins of flight and how the airplane was developed.

This book is a historical presentation of the evolutionary and sometimes revolutionary developments in aerodynamics, beginning with its roots in the theories about fluid dynamics during antiquity and moving forward to the modern hypersonic and computational aerodynamics of today. However, it is more than simply a presentation of the history. The book is also a critical evaluation and interpretation of the history as seen through the eyes of a practicing aerodynamicist.

How should a book on this topic be organized? There were various options that could have been chosen. For example, one might take a thematic approach, discussing separately and in turn such aspects as the understanding of aerodynamic drag, the evolution of theoretical aerodynamics, the development of experimental aerodynamics, and so forth. Alternatively, one might take a strictly chronological approach, detailing the century-by-century development of the discipline (for the twentieth century, the decade-by-decade development). The choice made here is a mixture. A quick examination of the table of contents will reveal an apparent chronological structure. However, there is a theme carried out within that structure in order to focus an understanding of the historical developments and to provide some interpretation of events. With that in mind, the organization is as follows:

(1) Each chapter covers a certain chronological period in the history of aerodynamics. Within each period, the evolution of the state of the art of aerodynamics is discussed from various points of view: experimental aerodynamics, theoretical aerodynamics, and the connections (whenever present) between the two.

(2) For each period, a particular representative flying machine is identified and discussed. Then this question is addressed: How much (if any) of the contemporary knowledge of aerodynamics was applied to the development of that flying machine? This question is used to focus the story of the historical developments in the science and art of aerodynamics in order to provide more context and clarity for the chronology. Also, the question is important in its own right. An irrepressible desire to design and build machines that could fly had provided much of the rationale and driving force behind the development of aerodynamics. However, it is clear that the existing state of the art during many historical periods was not directly reflected in flying-machine designs. Why not?

The continuing importance of that question was highlighted by the British aviator B. C. Huncks, who in July 1914 addressed the Royal Aeronautical Society in London: "I consider that the present day standard of flying is due far more to the improvement in piloting than to the improvement in machines." That was a clear expression of the frustration felt by many of the pioneers of flight, from the Middle Ages to well into the twentieth century, namely, that their contemporary flying machines (real or imagined) were neither as good as they would have desired nor as good as they could have been, given the existing state of the art. Therefore, by structuring this book according to the theme just described, I am striving to present a critical evaluation of the historical development of aerodynamics, rather than simply to relate the chronology of aerodynamics as it grew through the ages.

Before progressing further, we need to examine our use of the terms "state of the art" and "aerodynamics." The term "state of the art" appears to be of twentieth-century origin, as it does not appear in the nineteenth-century literature and earlier writings. Moreover, it did not appear in *Webster's Third New International Dictionary* until 1986. However, the fact that "state of the art" is a recently coined term does not preclude its use to describe the body of information and practice that existed at *any* time in history within a given discipline. Therefore, although eighteenth-century writers did not use the term, there was indeed in existence a collection of scientific and engineering knowledge and expertise that can be identified as the state of the art for that time. Second, the applied science of aerodynamics is considered to be a twentieth-century product. This is true if a strict interpretation is applied to the term, namely, that body of knowledge directly applicable to airplanes and modern missiles. However, consider the quotation at the beginning of this chapter. It defines aerodynamics as being concerned with interactions between the atmosphere and moving bodies (not just airplanes or missiles). Interest in the motions of objects in the atmosphere goes back at least as far as Aristotle. Hence, aerodynamics can be traced back to at least 350 B.C., although it was not called "aerodynamics" at that time. Indeed, the word "aerodynamics" was coined in the nineteenth century. It is defined in *A New English Dictionary on Historical Principles,* edited by James A. H. Murray, Oxford, at the Clarendon Press, 1888, as follows: "Aerodynamics [is the] branch of pneumatics which treats of air and other gases in motion, and of their mechanical efforts." Murray's *Dictionary* also gives a first historical reference to use of the word "aerodynamics" by quoting a statement from the *Popular Encyclopedia* (1837) as follows: "Aerodynamics; a branch of aerology, or the higher mechanics, which treats of the powers and motion of elastic fluids."

In addition to the foregoing considerations, aerodynamics is a branch of the more general subject of fluid dynamics, which encompasses the flows of fluids and gases in general, not just air. Although aerodynamics is sometimes viewed as a modern science, in many respects it is inseparable from the older science of fluid mechanics. Therefore, in a loose interpretation, the history of aerodynamics is rooted in the history of fluid dynamics, and that interpretation will be followed in this book. Thus, we shall consider the history of aerodynamics as going back to antiquity, with its early history being the same as the history of fluid dynamics. Therefore, we shall, for example, discuss Leonardo da Vinci's contributions to aerodynamics, even though the word "aerodynamics" was not used during his time.

Aerodynamics: Philosophical Antecedents of the Technology

The purpose of this chapter is to set the stage for the remainder of the book. An essential part of this staging is a basic understanding of the *technical* nature of aerodynamics. With such an understanding, we shall be better able to appreciate the historical significance of various contributions made over the centuries, in addition to being in a better position to follow the thought processes of the early researchers and pioneers of aerodynamics as they struggled to piece together the basic elements of this discipline.

Aerodynamics. What is it? A dictionary definition is given at the beginning of this chapter. To elaborate, aerodynamics is the science of the motions of gases (sometimes, more specifically, the motions of air) and the effects of those motions on various bodies or surfaces in the flow. The effects in which we are interested can be as diverse as the generation of lift and drag on an airplane, the force of the wind on a windmill, the dispersion of smoke from a smokestack, the aerodynamic heating of the space shuttle, and so forth. For our study of the history of aerodynamics, we shall be focusing primarily on an understanding of the generation of *lift* and *drag* on a flying object. This is a proper focus, because the development of a machine that could fly was the driving force behind the blossoming of modern aerodynamics (although, as we shall see, prior to the twentieth century there had been a major philosophical chasm between scientists working in theoretical fluid mechanics and inventors working to get into the air). The early inventors of flying machines were well aware of the importance of being able to predict aerodynamic *forces,* and equally aware of their lack of understanding of the generation of such forces. For example, in the third annual report of the Aeronautical Society, in 1868, we find the following statements:

> With respect to the abstruse question of mechanical flight, it may be stated that we are still
> ignorant of the rudimentary principles which should form the basis and rules for construc-
> tion. No one has yet ventured to give a correct experimental definition of the primary laws
> and amount of power consumed in the flight of birds; neither, on the other hand, has any
> tangible evidence been brought forward to show that mechanical flight is an impossibility
> for man .... We are equally ignorant of the force of the wind exerted on surfaces of various
> sizes, forms, and degrees of inclination: these are generally *assumed* on the mathematical
> laws of the resolution of forces, considered as the rigid impulse of inelastic weight and
> matter, and demonstrated by the aid of diagrams combined with a system of weights, cords,
> and pulleys, which convey but a very distant idea, relative to the conditions of the present in-
> quiry, where the elastic and yielding nature of the air is the cause of such unforeseen results,
> differing in according to the width, form, angle, and speed of the surface of the impact.

Indeed, the society was more emphatic about the importance of understanding aerodynamic
forces in its annual report two years later, in 1870, where it was categorically stated that
"the first great aim of the Society is the connecting of the velocity of the air with its pressure
on plane surfaces at various inclinations." Clearly, the prediction of aerodynamic force has
been a primary concern in aerodynamics.

Experimental measurements of the aerodynamic forces acting on a vehicle are carried
out primarily in wind tunnels. The wind tunnel itself is an aerodynamic device: It is a
machine designed to produce in the laboratory a well-defined, smooth, uniform flow of a
gas (usually air) in which a model of a flight vehicle is placed for the purpose of measuring
lift, drag, and other parameters. Over the years, the primary focus for many researchers was,
and in some cases still is, the aerodynamics of the wind tunnel itself. Thus, at appropriate
times throughout this book, some emphasis will be placed on the historical development of
wind tunnels. Also, as we shall see, often knowledge of the *details* of a flow field around a
flight vehicle is required if one is to calculate the aerodynamic forces. Although we shall
focus on aerodynamic forces as the main technical concern in this historical presentation,
we shall not ignore the importance of a detailed understanding of phenomena within the
flow field.

To begin this presentation of the technology of aerodynamics, consider the flow of air
over the body shown in Figure 1.1. The motion of the air over the body (an airfoil, i.e., a
section of an airplane wing) is shown by four *streamlines,* sketched above and below the
airfoil. These streamlines trace out the paths of small parcels of air (called *fluid elements*) as
they move through the flow field. Figure 1.2 shows a photograph of the streamlines over an
airfoil, made visible by injection of small jets of smoke into the airflow ahead of the airfoil.
The streamlines shown in Figures 1.1 and 1.2 can be thought of as the streamlines that
would be observed if a body were stationary and air were flowing over it (such as occurs
in a wind tunnel) or as the streamlines that would be observed if the body were moving
into stationary air, with the observer moving along with the body (such as occurs when

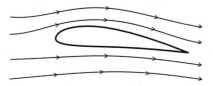

Figure 1.1   Streamlines over a body immersed in a flow.

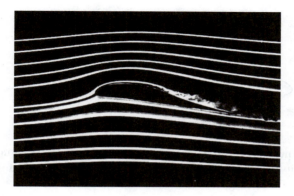

Figure 1.2   Smoke photograph of the low-speed flow over an airfoil taken in a smoke tunnel at Notre Dame University. (Courtesy of Dr. T. J. Mueller.)

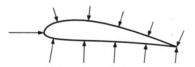

Figure 1.3   Pressures acting on the surfaces of a body.

   **Figure 1.4**   Shear stresses acting on the surfaces of a body.

an airplane flies through the quiescent atmosphere). These two points of view are totally equivalent, as was first recognized in the late fifteenth century by Leonardo da Vinci, and that provided the basis for the use of wind tunnels to simulate flight through the air.

The flow over the airfoil shown in Figures 1.1 and 1.2 exerts an aerodynamic force on the body, an aerodynamic force that derives from two natural sources:

(1)   There will be a distribution of *pressures* exerted over the surface (Figure 1.3). Those pressures will vary from point to point over the surface and will act *perpendicular* to the local surface. In general, there will be a net imbalance in the pressures acting over the surface, leading to exertion of a force on the body. For example, for the airfoil shown in Figure 1.3, the pressures on the bottom surface will be higher than the pressures on the top surface, generating an upward lifting force on the body.

(2)   Because of the friction generated between the moving air and the body surface, there will be a distribution of *shear stresses* over the surface, as sketched in Figure 1.4. Those shear stresses will vary from point to point over the surface and will act *tangentially* to the surface. As will be intuitively obvious from this sketch, shear stress is a major contribution to drag.

If you hold this book on the palm of your left hand and slide it across that palm by pushing it with your right hand, you will be exerting a force on the book; the book will experience the force via contact with your hands. Similarly, the pressure and shear-stress distributions exerted on the airfoil surface are the two hands of nature that grab hold of the body immersed in the flow and exert a force on the body – the *aerodynamic force*. This resultant

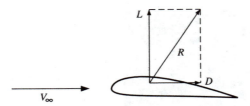

Figure 1.5 Resultant aerodynamic force and its resolution into lift and drag.

aerodynamic force, R, is shown in Figure 1.5. Also shown is the flow velocity far upstream of the airfoil, V_∞; this velocity is called the *free-stream velocity*. This leads to the following definitions of aerodynamic lift and drag:

(1) *Lift* is that component of R perpendicular to the free-stream velocity (the lift force is labeled L in Figure 1.5).
(2) *Drag* is that component of R parallel to the free-stream velocity (the drag force is labeled D in Figure 1.5).

With Figures 1.3–1.5 in mind, we can easily devise a strategy for prediction of aerodynamic lift and drag: First, develop a theory that will allow detailed prediction of the distribution of pressures over the body surface. Next, develop a separate theory that will allow detailed prediction of the distribution of shear stresses over the body surface. Finally, add (in mathematical terms, integrate) the detailed distributions of pressures and shear stresses over the body surface, thus obtaining the resultant force R, and hence the components L and D. This strategy dates back to the seventeenth century, when Isaac Newton devised theoretical models for prediction of pressure and shear-stress distributions over a surface immersed in a flowing fluid. The strategy is straightforward; actual calculations of pressure and shear-stress distributions are not. Even today, such a calculation for a real three-dimensional configuration is a complex task. For the early aerodynamicists, it was a task that was rarely accomplished.

Let us carry this thought process one step further. Assume that we have a complete theory with which we can predict the detailed flow-field distributions of pressure, velocity, and so forth, *everywhere* throughout the entire flow, not just at the body surface. That is, assume that we can predict the complete flow sketched in Figure 1.1. Then, if we can calculate the flow properties at any arbitrary point in space, we can apply the same theory to calculate the properties at the surface, which ultimately will give us the lift and drag. In reality, that is what usually happens. In order to predict the lift and drag on a body, aerodynamicists will first develop a general aerodynamic analysis that will allow predictions of pressures and velocities at all points in the flow. Then that theory will be applied at the surface to obtain the surface pressures and shear stresses. Finally, the surface pressures and shear stresses will be integrated over the entire body surface to yield the aerodynamic force on the body. Later we shall see that the major advances in the historical development of aerodynamics involved revolutionary theories that were directed at a solution for the *complete flow field*, or at least *certain regions* of the flow field away from the body surface.

One final philosophical aspect of aerodynamics must be discussed here. Most aerodynamic flows are quite complex, and whereas nature has had no difficulty in creating such complexities, over the centuries humans have had plenty of trouble in attempting to understand and decipher them. To help our thinking about aerodynamics and to aid in rational

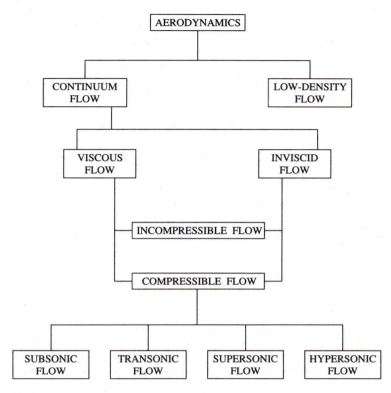

Figure 1.6   A paradigm of aerodynamic flow regimes.

analysis of problems in aerodynamics, various classifications of flows have been proposed. (I have heard a few aerodynamicists state that all of aerodynamics can be reduced to the study of some 15 to 20 different types of flow patterns. Unfortunately, that is an oversimplification.) For our purposes, it is useful to divide aerodynamics into several general types of flows, as diagrammed in Figure 1.6. There are two general categories in aerodynamics that depend on the density of the flow: continuum flow and low-density flow. It is easy to obtain a feel for continuum flow: Simply wave your hand through the air around you; it feels like a continuous medium. Virtually 99% of all aerodynamic problems can be analyzed by assuming that air is a continuous substance – thus defining a continuum flow. However, within the earth's atmosphere, as altitude increases, density decreases. At very high altitudes, the air is so thin that the assumption of a continuum begins to break down. For example, if you were to wave your hand through the air around you at an altitude of 400,000 ft, you would be sensing the impact of individual, widely spaced molecules, rather than feeling a continuous substance. When vehicles (such as the space shuttle) are flying at very high altitudes (e.g., 300,000 ft), some of their aerodynamic problems cannot be analyzed assuming continuum flow, but rather must be treated in the context of a low-density flow. Across the whole spectrum of aerodynamics, such low-density flows are specialized cases, accounting for less than 1% of all applications.

Within the category of continuum flow, aerodynamic problems naturally divide into two types: viscous flow and inviscid flow (Figure 1.6). "Inviscid flow" is an idealization that ignores the effects of friction and thermal conduction. There is a massive body of theoretical literature based on the assumption of inviscid flow, mainly because such a flow is easier to analyze. Such analyses are capable of predicting only lift and pressure drag; prediction of skin-friction drag is impossible for an inviscid analysis, because friction is ruled out from the beginning. In contrast, analyses of viscous flow must take into account the effects of friction, clearly a more realistic situation. One pays dearly for this realism, because viscous flows are much more difficult to analyze.

Both viscous flows and inviscid flows can be either incompressible or compressible. "Incompressible flow" is another idealization – this time the assumption being that the density of the flow is constant. The assumption of incompressible flow works well for liquids (e.g., the flow of water), but it yields much less realistic results for air. The exception to this is that for airflows *at low speeds* (e.g., velocities less than 300 mph), the density varies so little that the flow can be assumed to be incompressible. That was the speed range for almost all aircraft up to the beginning of World War II and is still the range for most light, general-aviation aircraft today. Almost all studies of aerodynamics in the first 40 years of the twentieth century concerned incompressible-flow aerodynamics. At higher speeds, the air density becomes an important variable throughout the flow field, and that requires compressible-flow analysis. Taking account of density variations adds considerable complexity to any analysis of a flow field.

Within the category of compressible flow, there are four distinct types of flows: subsonic, transonic, supersonic, and hypersonic. The distinction is made on the basis of Mach number, which is defined as the velocity of the flow divided by the speed of sound. For *subsonic flow,* the Mach number is less than 1 everywhere throughout the flow field. *Transonic flow* is that regime in which there is a mixture of regions of locally subsonic flow and supersonic flow (i.e., where the flow over a body has some regions of subsonic flow and other regions of supersonic flow). An F-15 Eagle cruising at a free-stream Mach number of 0.9 provides a case in point; parts of the flow will be subsonic (Mach numbers between 0.9 and 1.0), whereas the accelerated flow occurring in the expansion over the wing surface will be slightly supersonic (Mach numbers slightly above 1.0). A flow field in which the number is everywhere greater than 1 is called *supersonic*. Such flows almost always shock waves. Finally, at the high-speed end of the flight spectrum, where Mach become very large, we have *hypersonic flow*. As a rule of thumb, hypersonic flow flows with Mach numbers greater than 5.

s section we have described some philosophical antecedents and structure for the theoretical and experimental aerodynamics. This section provides a model or *n* for the concepts of aerodynamics, the structure on which we shall hang our nt discussions of the history of aerodynamics.

The Equations of Aerodynamics

Aerodynamics is a highly quantitative field of science.[1] The basic governing equations of theoretical aerodynamics, as we understand them today, compose a system of partial differential equations. For viscous flow, they are called the *Navier-Stokes equations*. For inviscid flow, they reduce to simpler forms, called the *Euler equations*. The statements of the Euler equations in the eighteenth century and the Navier-Stokes equations in the nineteenth century were historic milestones in the evolution of theoretical aerodynamics (these equations are discussed further in Appendixes A and B).

We note that a great deal of the early thought about aerodynamics was concerned with the following question: How do the lift and drag on an object immersed in a moving fluid depend on the *velocity* and the *density* of the flow? That was a straightforward question, and after a few centuries of intellectual effort the answer was obtained, as reflected in two straightforward equations:

$$L = \tfrac{1}{2}\rho V^2 S C_L$$

and

$$D = \tfrac{1}{2}\rho V^2 S C_D$$

where ρ is the free-stream density, V is the free-stream velocity, S is a reference area (e.g., wing area for an airplane), and C_L and C_D are the lift and drag coefficients, respectively. For low-speed flight, far below the speed of sound, the lift and drag coefficients have essentially fixed values for a given airplane at a given angle of attack. In turn, these equations state that lift and drag vary

(1) directly with density, and
(2) as the *square* of the velocity.

The equations for lift and drag in terms of the lift coefficient C_L and the drag coefficient C_D seem disarmingly simple. However, do not be misled. We have mentioned that the lift and drag on a flight vehicle are the integrated effects of the pressure and shear-stress distributions exerted over the surfaces of the vehicle (Figures 1.3–1.5). Prediction of these pressure and shear-stress distributions is not easy. This complexity, in relation to the equations for L and D, is buried in the predictions of C_L and C_D. Although the equations look simple, the question of the values for L and D now shifts to the question of the values for C_L and C_D. In turn, the values for C_L and C_D are obtained from detailed knowledge of the surface pressure and shear-stress distributions. In short, the equations are simple relations for L and D *provided that we know the values for C_L and C_D,* but theoretical or experimental values for C_L and C_D are not easily obtained.

Summary and Timeline

To understand the history of technology, it is useful to understand the technology itself. This chapter is intended to provide some understanding before we begin our journey through the winding roads of history.

The remainder of this book will take a chronological path, although it is not simply a chronology of the history of aerodynamics, but rather a critical and interpretive examination of the main themes involved in the development of aerodynamics. In the subsequent chapters it will be useful for us to consult a timeline of the history of aerodynamics (Table 1.1). Also, major connections will be made between the evolution of human

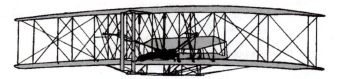

Figure 1.7   Wright Flyer, 1903. (Angelucci,[175] with permission.)

Table 1.1. *A timeline for the history of aerodynamics*

Date	Developments
350 B.C.	Aristotle describes a model for a continuum and suggests that a body moving through a continuum encounters a resistance.
250 B.C.	Archimedes suggests that a fluid is set into motion by the existence of a pressure difference exerted on the fluid.
1490	Leonardo da Vinci's contributions: 1. law of continuity: $AV =$ constant. 2. observations and sketches of various flow patterns, representing the first qualitative contribution to experimental aerodynamics. 3. statement of the "wind-tunnel principle." 4. statement that air resistance is *directly proportional* to the *area* of the body, i.e., $R \propto A$. 5. introduction of the concept of streamlining a body to reduce drag.
1600	Galileo is the first to understand that aerodynamic resistance varies *directly* as the *density* of the fluid, i.e., $R \propto \rho$.
1673	Edme Mariotte, in Paris, states that aerodynamic resistance varies as the *square* of the velocity, i.e., $R \propto V^2$.
1687	Isaac Newton introduces Newtonian mechanics, the beginning of rational mechanical analysis, leading to the Newtonian sine-squared law for aerodynamic force. *Theoretical* verification that $R \propto V^2$.
1690	Christiaan Huygens, in Paris, publishes experimental data that further substantiate $R \propto V^2$.
1732	Invention of the Pitot tube by Henri Pitot in France.
1738	Daniel Bernoulli's *Hydrodynamica* is published. Initial statement of the pressure–velocity relationship.
1742	Development of the ballistic pendulum and the whirling arm by Benjamin Robins in England.
1744	Statement of d'Alembert's paradox.
1752	Leonhard Euler publishes the Euler equations for fluid flow, the first proper mathematical modeling of an inviscid flow. Euler derives the "Bernoulli equation."
1759	John Smeaton studies the forces on flat planes normal to an airstream and introduces the Smeaton coefficient for calculation of air forces.
1763	Jean-Charles Borda, in France, is the first to observe the effects of aerodynamic interference between two closely spaced bodies.
1788	Joseph Lagrange introduces the concepts of the velocity potential and stream function, two fundamental ideas in theoretical aerodynamics.
1789	First appearance of Laplace's equation, the fundamental equation for analysis of incompressible, inviscid, irrotational flow. Also, Laplace is the first to properly calculate the speed of sound in air.
1799	George Cayley, in England, introduces the concept of the modern-configuration airplane via an etching on a silver disk.
1809–10	Publication of Cayley's "triple paper," the substantive beginning of applied aerodynamics.
1840	Publication of the equations of fluid dynamics for flow with friction, the Navier-Stokes equations, thus putting into place all the equations necessary to describe a real fluid flow.
1903	First powered flight by the Wright brothers.
1904	Ludwig Prandtl's boundary-layer concept.
1906	The circulation theory of lift is developed.
1908	Prandtl's supersonic-shock and expansion-wave theory.
1915	Prandtl's lifting-line theory.
1922	Thin-airfoil theory by Max Munk.

Table 1.1 *(cont.)*

1925	Development of linearized supersonic aerodynamic theory.
1928	Prandtl-Glauert compressibility correction.
1940	Early applications of supersonic-flow theory.
1950	Area rule.
1955	The emphasis on hypersonic aerodynamics begins.
1960	Supercritical wing.
1962	Beginning of computational fluid dynamics.

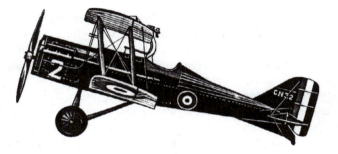

Figure 1.8   British S.E.5, 1917. (Angelucci,[175] with permission.)

Figure 1.9   Douglas DC-3, 1935. (Angelucci,[175] with permission.)

Figure 1.10   Boeing 707, 1958. (Angelucci,[175] with permission.)

knowledge of aerodynamics and the application of that knowledge to flying machines. We shall reflect somewhat on the evolution of the airplane itself. That evolution can be broadly divided into four periods: (1) the period from antiquity to the work of the Wright brothers at the beginning of the twentieth century, as exemplified by the Wright Flyer (Figure 1.7); (2) the era of the strut-and-wire biplane, as exemplified by the British S.E.5 (Figure 1.8); (3) the era of the mature propeller-driven airplane, symbolized by the Douglas DC-3 (Figure 1.9); (4) the age of the jet-propelled airplane, typified by the Boeing 707 (Figure 1.10).

Let us now start at the beginning.

CHAPTER 2

The Early History of Aerodynamics
From Antiquity to da Vinci

Dissect the bat, study it carefully, and on this model construct the machine.

Leonardo da Vinci, *Sul volo degli uccelli* (1505)

The flying-machine sketch shown in Figure 2.1 is one of more than 500 sketches concerning flight that survive from the notebooks of Leonardo da Vinci (1452–1519). Dating from the period 1486–90, it shows a device designed to be powered by a man, with the wings being flapped up and down and back and forth to provide lift and thrust simultaneously. Such machines are called *ornithopters*.

Da Vinci had a variety of such designs, with the pilot either lying prone (as in Figure 2.1) or dangling vertically under the machine as in a modern-day hang glider. In all cases the pilot would supply the motive power with arm, leg, or torso movements that would be mechanically translated into flapping of the wings. At first glance, the ornithopter in Figure 2.1 would appear to have no redeeming aerodynamic value – it is so contrary to our modern ideas about airplanes. On the other hand, it was the first serious design for a flying machine, and it is chosen as the seminal development from which to focus the historical discussion in this chapter.

Looking more closely at da Vinci's flying machine (Figure 2.1), we see an ornithopter in which the pilot lies prone and operates the wings by pushing and pulling a variety of levers. The wings themselves are essentially wooden spars; we can surmise that he intended some fabric covering for the wings in order to increase the area of the lifting surface. We know that he drew pictures of wings covered with net-like fabric. Figure 2.2 shows an ingenious da Vinci idea for independently testing an ornithopter wing: On one pan of a balance there would be a man and a wing, and weights would be placed on the other pan until the two were in balance. The wing would then be dropped downward under the combined forces of gravity and the man pushing on a lever. Lift would be generated by the wing movement, and the resulting imbalance of the two pans would be an indication of the lift. That da Vinci's interest in flying machines was serious is further reflected in his thoughts on a retractable landing gear and on the use of flap valves on the wings that would open and allow free flow of air through the wing surface during the upward movement of the wing. We can safely say that Leonardo was driven by a strong interest in flight, although there is no evidence that he ever attempted to construct or fly one of his machines. Moreover, although da Vinci and other scientists shared and explored numerous other engineering ideas during the Renaissance, his concepts for flying machines were uniquely his. From a modern perspective, many of da Vinci's concepts about the nature of flight were misguided; but his ornithopters embodied the first serious thoughts about flying machines. Therefore, we use this machine to pose the following questions:

(1) What was the aerodynamic state of the art in 1490?
(2) How much of that state of the art was reflected in the flying machine in Figure 2.1?

14

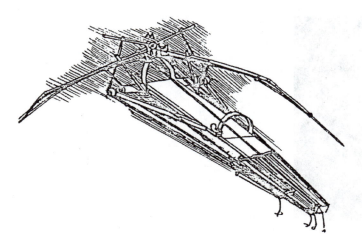

Figure 2.1   An ornithopter by da Vinci (1490).

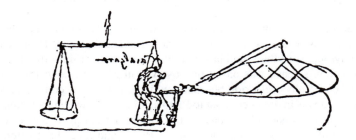

Figure 2.2   Balance mechanism for testing wings, a concept by da Vinci.

The first question prompts us to examine the evolution of aerodynamics from antiquity to 1490.

Aristotle and the Beginning of the Road to Aerodynamics

If we were to view modern physical science as a massive tapestry, then a few of its fundamental threads could be traced all the way back to the origins of Greek science in the five centuries before the birth of Christ. Indeed, the history of aerodynamics began there, in particular with the golden age of Greek culture and art that peaked at about 350 B.C. The most important Greek scientist of that time was Aristotle (384–322 B.C.).

Aristotle (Figure 2.3) was born in the Ionian colony of Stagira, on the northwest shore of the Aegean Sea. His father, Nicomachus, was personal physician to Amyntas II, king of Macedonia. At the age of 17, Aristotle was sent to study at Plato's academy in Athens (for its time, the equivalent of Harvard combined with the Massachusetts Institute of Technology). He remained there for 20 years, first as a student and later as a teacher. He was by far the most important figure to come out of the academy. Later, Aristotle left Athens to become a teacher and philosopher at the court of Hermias in Anatolia. He then moved on to Macedonia, where he was personal tutor to the teenage Alexander the Great. At about 335 B.C., Aristotle returned to Athens and established his own school at the

Figure 2.3   Aristotle.

Lyceum, where he began a collection of books and maps, surrounded by students and re-searchers. After his death in 322 B.C., that became known as the Peripatetic school of Greek philosophy.

Aristotle lived during the most intellectually stimulating time in Greek history. He went to the best school and was associated with some of the most influential people. He developed a corpus of philosophy, science, ethics, and law that influenced the world for the following 20 centuries. Aristotelian science was based on observation and was concerned not so much with the "why" of the physical world but rather with the results and consequences of various phenomena. That was in direct contrast to Plato's philosophy, which had attempted to construct a mathematical model of the physical sciences, not caring if the results of such reasoning were contrary to observed phenomena. Rouse and Ince[2] have offered an interesting comparison of the two: Plato's love of ideas earned him the title of a speculative mathematician, whereas Aristotle's love of facts marked him as a contemplative natural scientist.

Aristotelian science later led down a number of blind alleys during the further development of science, a situation exacerbated by the fact that Aristotle's writings were adopted as absolute truth by the Catholic church during the Middle Ages – an unchanging situation that Aristotle would have strongly opposed, because dynamic change became part of his philosophy in later life.

One of those blind alleys derived from Aristotle's concept that all matter was made up of four elements, earth, water, air, and fire, in the order of their ascending status: Fire rose above air, which rose above water, which in turn rose above earth. That idea was not original with Aristotle; it arose from earlier Greek philosophy and was strictly taught at Plato's academy. Aristotle simply reinforced the concept.

Another blind alley was Aristotle's concept of motion. He reasoned that a body in motion through the air had to experience a constant application of force to continue its motion. For a projectile in motion through the air, Aristotle reasoned that the air itself (the medium) sustained the motion by separating in front of the projectile and then rapidly filling in the region behind the projectile, thus constantly exerting a force on the back of the projectile. That reasoning became known as the *medium theory* of motion, and it was widely held to the time of Galileo, almost 20 centuries later. (Today, we know from Newtonian mechanics that a net force is necessary to *change* the motion of a body, but for a body in constant, unchanging motion, the net force is zero. This is the essence of Newton's first and second laws.)

On the other hand, Aristotle's scientific thought produced two concepts that bear on the development of aerodynamics. The first is the concept of a *continuum*. He wrote that "the continuous may be defined as that which is divisible into parts which are themselves divisible to infinity, as a body which is divisible in all ways. Magnitude divisible in one direction is a line, in three directions a body. And magnitudes which are divisible in this fashion are continuous" (Aristotle, *Treatise on the Heavens*, Book I). Return to Figure 1.6, which lists the various classifications of aerodynamic flows. Continuum flows account for about 99% of all aerodynamic applications. Today, students of aerodynamics intuitively accept the concept of continuum flow. It is not widely appreciated that this fundamental idea is one of Aristotle's important contributions to the science of aerodynamics. (It is interesting to note that the concept of low-density flow can also be traced to Greek science. The assumption of the gas or fluid as composed of distinct, individual atoms and molecules, *not* continuous, originated with the Greek philosopher Democritus, born about 465 B.C., who established the Atomic school of philosophy.)

Aristotle's second contribution to aerodynamics was the idea that a moving body passing through the air or another fluid encounters some aerodynamic "resistance." He wrote that "it is impossible to say why a body that has been set in motion in a vacuum should ever come to rest. Why, indeed, should it come to rest at one place rather than another? As a consequence, it will either necessarily stay at rest, or if in motion, will move indefinitely unless some obstacle comes into collision with it" (*Treatise on the Heavens*, Book I). A conclusion from this reasoning is that because a body will eventually come to rest in a fluid, there must be a *resistance* acting on the body. Today, we call this the aerodynamic *drag*.

Archimedes: the Founder of Fluid Statics

Fluid statics deals with the characteristics of fluids that are at rest, and aerodynamics deals with fluids (gases or liquids) that are in dynamic motion. So what does fluid statics have to do with aerodynamics, and in turn what does Archimedes (287–212 B.C.), the founder of fluid statics, have to do with the history of aerodynamics? Not very much. However, there are three aspects of the general mechanical principles advanced by Archimedes that have some bearing on aerodynamics.

Archimedes (Figure 2.4) was born in Syracuse, Sicily. Living about a century later than Aristotle, Archimedes was able to take advantage of the scholarly facilities that had been amassed in the relatively new city of Alexandria, Egypt, where he read the manuscripts collected in the great library and studied under the disciples of Euclid. During that time, he developed a great interest in mathematics and for the rest of his life would consider his major contributions to be in that field. However, the world remembers Archimedes more for his practical, engineering inventions, such as the Archimedean screw for lifting water to higher elevations, and the development of the mechanical lever. After returning to his home city of Syracuse, Archimedes spent a lifetime in the employ of Hieron II, king of Syracuse, for whom he devised all manner of mechanical inventions, including engines of war used to defend Syracuse from Roman conquest. By far, the most famous story about Archimedes is that concerning the crown of Hieron. The king was concerned that his crown, supposed to be of solid gold, had actually been debased by the addition of silver.

Figure 2.4   Archimedes.

He asked Archimedes to investigate the matter. Legend has it that the solution occurred to Archimedes while he was at the public baths. So excited was Archimedes that he ran home, forgetting to dress, shouting *"Eureka! Eureka!"* ("I have found it! I have found it!"). His solution involved immersing the crown, and then equal weights of gold and of silver, in water and measuring the displaced water in all three cases. From those relative displacements the actual amounts of gold and silver in the crown were calculated. In spite of his many practical accomplishments, Archimedes felt that engineering was beneath the dignity of pure mathematics, and he deliberately left no written record of his engineering endeavors. The mathematician in Archimedes was dominant to the very end; in 212 B.C. Syracuse was finally sacked by Roman soldiers under the command of Marcellus, and Archimedes was killed by a Roman sword as he was studying a geometric problem drawn in the sand. (The killing of Archimedes was a soldier's blunder. Marcellus had given orders that Archimedes and his house were to be spared, so great was Roman admiration for the great thinker. Marcellus subsequently buried Archimedes with honor and befriended his surviving relatives.)

Archimedes' place in the history of aerodynamics derives from three concepts:

(1)   He clearly stated that a fluid (gas or liquid) is a continuous substance and can be treated mathematically as a continuum. In that sense, he reinforced and actually used the continuum concept first stated by Aristotle a hundred years earlier.

(2)   Archimedes had some conception of pressure, and he understood that every point on the surface of a body immersed in a fluid was under some force due to the fluid. He stated that in a fluid, "each part is always pressed by the whole weight of the column perpendicularly above it." That was the first statement of the principle that, to use modern terms, the pressure at a point in a stationary fluid is due to the weight of the fluid above it and hence is linearly proportional to the depth of the fluid. It is interesting that Archimedes' concept of pressure held sway through the middle of the eighteenth century; even Daniel Bernoulli agreed with it. It would

finally fall to Leonhard Euler, in 1752, to state the proper conception of pressure as a *point property* defined as

$$p = \lim_{dA \to 0} (dF/dA)$$

where dA is a very small area centered at a given point in the fluid, and dF is the correspondingly small force on that area due to the fluid.

(3)   Today, we understand that in order to set a stagnant fluid into motion, a *difference* in pressure must be exerted across the fluid. We call this pressure difference over a unit length the *pressure gradient*. Archimedes had a vague understanding of this point when he wrote that "if fluid parts are continuous and uniformly distributed, then that of them which is the least compressed is driven along by that which is more compressed."

Liberally interpreted, this means that when a pressure gradient is imposed across a stagnant fluid, the fluid will start to move in the direction of decreasing pressure. That statement by Archimedes was a clear contribution of Greek science to basic aerodynamics.

A Leap across the Centuries to da Vinci

The period from the death of Archimedes to the time of da Vinci witnessed the zenith of the Roman Empire and its fall, the dearth of intellectual activity in western Europe during the Dark Ages, and the surge of new thought that characterized the Renaissance. In terms of the science of aerodynamics, the 17 centuries that separated Archimedes and da Vinci yielded no worthwhile contributions. Although the Romans excelled in creating highly organized civil, military, and political institutions and in carrying out large engineering feats such as building construction and distribution of water from reservoirs to cities via aqueducts, they contributed nothing of substance to scientific theory.

Therefore, the state of the art of aerodynamics inherited by da Vinci in 1490 was fragmentary and totally immature. Moreover, there is some question regarding how much of Greek science was known to da Vinci. The works of Aristotle and Archimedes were available in Latin, but da Vinci had no formal education and could not read Latin. That forced him to depend on the few translations into Italian that were available at that time. In da Vinci's notebooks there are indications that he was familiar with six of Archimedes' treatises. However, regarding the only treatise that concerned aerodynamics – that entitled *On Floating Bodies* – Clagett[3] observed that da Vinci's notes were not written in his usual left-to-right fashion, indicating that those particular notes had been written by someone else.

Leonardo da Vinci (1452–1519), one of the most powerful intellects and most creative geniuses of the Renaissance (Figure 2.5), was born in the small Tuscan village of Vinci, near Florence, on April 15, 1452, the illegitimate son of Ser Piero da Vinci, a young lawyer, and Caterina, a girl of peasant stock. He lived with his mother until she married in 1457, at which time Leonardo was taken into Ser Piero's family and raised in part by a kindly stepmother. At the age of 16, Leonardo moved to Florence, where he began an apprenticeship as a painter and sculptor. At the age of 30, he moved to Milan to serve the powerful family of Ludovico Sforza, designing military machines and weapons, in addition to painting, sculpting, and practicing architecture. He spent 20 years at the Sforza court, one of his most intense creative periods. It was there that he completed the famous wall painting, *The Last Supper,* in 1497 and formulated his ideas about mechanical flight, as described in copious notes. As

Figure 2.5   Leonardo da Vinci (1452–1519).

was so often the case during the Middle Ages, political power was extremely ephemeral, and in 1500, when the French king, Louis XII, took control of Milan, Leonardo moved back to Florence. In 1502–3 he served Cesare Borgia as a military engineer during campaigns in central Italy. Subsequently, Leonardo became relatively mobile, returning to Milan in 1506, going to Rome in 1513, where he produced art for the Vatican under the patronage of Giuliano de' Medici, brother of Pope Leo X, and finally residing in Cloux, France, where he died on May 2, 1519.

Although da Vinci served some of the most powerful families in Italy – Sforza, Borgia, and Medici – he had no specific professional position. Indeed, he was a failure as a "careerist," and his day-to-day life was not a great success. There is some feeling that Leonardo's accomplishments were compromised because he had so many varied interests, frequently having insufficient time to finish the tasks he began. On the other hand, the things he did produce earned him a solid reputation in his own day, mainly as a painter and a sculptor. Today we regard him as a genius, as evidenced in his art and in the thoughts and sketches recorded in his notebooks. His mind ranged across the diverse areas of painting, sculpture, basic science, mathematics, machine design, military engineering, flying machines, anatomy, and medicine.

Da Vinci's aerodynamic concepts were amazingly advanced and would have constituted a quantum jump in the state of the art of aerodynamics if they had been widely disseminated. However, that was not to be, for he recorded his ideas in a most haphazard fashion, and to make things worse, after his death his notes changed hands numerous times and became scattered, remaining unused for centuries. That situation was described by Hart: "Leonardo's notebooks, for so long lost and at last saved, can well be described as a vast scrap-heap, the study of which shows their writer to have displayed an ingenuity and a capacity for original thinking on a variety of matters far ahead of the days in which he lived. It is these notebooks, as well as the few paintings and the numerous cartoons and sketches that have survived, that place him in the forefront of the world's greatest intelligences" (p. 18).[4]

The disposition of da Vinci's notes after his death is an interesting study in human dedication, ignorance, apathy, and greed. After da Vinci died in southern France, a close friend and companion, Francesco Melzi, served as executor of his estate, which included all of da Vinci's notes. Melzi faithfully looked after the notes for 50 years, until his death in 1570, at which time supervision of the notes broke down. Francesco's son, Orazio, inherited the notes but had no understanding of their value. He allowed the family tutor, Lelio Gavardi di Asola, to take 13 volumes of the notes to Pisa, where he left them with Ambrosio Mazzenta, a law student at the University of Pisa. Later, Orazio allowed many others to help themselves to the remaining "spoils" of da Vinci's estate. In 1590, Mazzenta took religious vows with the Barnabite order and passed the notes in his possession to his brother Guido. Later, seven volumes were passed back to the unappreciative Orazio, who promptly sold them to Pompeo Leoni, a sculptor and friend of King Philip II of Spain. (The fact that these notes were sold indicates some growing appreciation of their value, especially on the part of the purchaser, Leoni.) We have to thank Pompeo Leoni for trying to consolidate the notes at that time. He acquired four more volumes over the next few years and began to try to rearrange them in some kind of useful order. He cut out numerous portions from all of da Vinci's papers and combined them, along with some loose drawings, into a single volume of 402 pages and 1,700 drawings; this volume is called the *Codex Atlanticus,* and for modern scholars it has become the most useful of all the remaining da Vinci papers. Leoni died in 1610. His heir, Polidoro Calihi, sold da Vinci's notes to Count Galezzo Arconati in 1625, who in turn presented them to the Ambrosian Library in Milan in 1636. That was not to prove as secure a repository as might be expected, because in 1796 Napoleon seized the *Codex Atlanticus* and 12 other volumes of da Vinci's notes and moved them to Paris. There the *Codex* was placed in the Bibliothèque Nationale, and the other 12 volumes were sent to the library of the Institute of Paris. It was at that point that a wider audience began to become aware of da Vinci's notes, appropriately through the efforts of the Italian physicist Giovanni Battista Venturi (who studied the flow of fluids in tubes, and for whom the Venturi tube is named). While in residence in Paris, Venturi carefully studied the 12 volumes of da Vinci's notes in the Institute of Paris, where he lettered each volume as manuscript A, manuscript B, and so forth, designations that continue to the present day. Moreover, Venturi presented a paper to the French Academy, "Essai sur les ouvrages physico-mathematics de Leonard de Vinci," wherein some of the intellectual power of da Vinci's mind was finally revealed to a wide audience. The lettered manuscripts remain in Paris; but in 1815 the *Codex Atlanticus* was returned to the Ambrosian Library in Milan. In recent years, other notes by da Vinci have been found. For example, two manuscripts were discovered in Madrid in 1965, labeled *Codex Madrid I* and *II*. In addition to these major manuscripts, others can be found in England (*Codex Leicester* and *Codex Arundel*) and in Turin (*Codex on the Flight of Birds*).

Da Vinci's Aerodynamics

Leonardo's enduring interest in flying machines reached its peak during the period from 1488 to 1514: "The conquest of the air was his dream and his obsession – and so were the birds whose freedom of the air he wanted to emulate" (p. 311).[4] Da Vinci was observed to buy birds in cages and then set them free, so strong were his feelings about the freedom of flight. In the *Codex on the Flight of Birds,* Leonardo stated that "a bird is an instrument working according to mathematical law, an instrument which is within the capacity of man to reproduce with all its movements, though not with a corresponding degree of strength, for it is deficient in the power of maintaining equilibrium. We may therefore say that such an instrument constructed by man is lacking in nothing except the life of the bird, and this life must be supplied from that of man." This statement leads us to the da Vinci ornithopter (Figure 2.1) and to da Vinci's ideas on aerodynamics.

Figure 2.6   Sketch by da Vinci showing cross section of flow of water in a river.

Leonardo had an interest in the characteristics of basic fluid flows. For example, one of the fundamental principles of modern fluid mechanics is that mass is conserved; in terms of a fluid moving in a tube, this means that the mass flow (e.g., the number of pounds per second) passing through any cross section of the tube does not vary. For an incompressible flow (flow of a fluid, or low-speed flow of a gas), this principle leads to the basic relation that

$$AV = \text{constant}$$

where A is the cross-sectional area of the tube at any location, and V is the velocity of the fluid at that same location. This relation is called the continuity equation, and it states that for fluid moving from one location in a tube to another location where the cross-sectional area is smaller, the velocity will become greater by just the right amount that the product of A times V will remain the same. Leonardo noted that effect in regard to the flow of water in a river, observing that in those locations where a river became constricted, the water velocity increased. He quantified his observation in the following statement, with an accompanying sketch (Figure 2.6): "Each movement of water of equal surface width will run the swifter the smaller the depth . . . and this motion will be of this quality: I say that in *mn* the water has more rapid movement than in *ab,* and as many times more as *mn* enters into *ab*; it enters 4 times, the motion will therefore be 4 times as rapid in *mn* as in *ab*."[2] Here we have, for the first time in history, a quantitative statement of the special form of the continuity equation that holds for low-speed flow. In addition to this quantitative contribution, da Vinci, being a consummate observer of nature, made many sketches of various flow fields. A particularly graphic example is shown in Figure 2.7, from the *Codex Atlanticus*. Here we see the vortex structure of the flow around a flat plate. At the top, the plate is perpendicular to the flow, and da Vinci accurately sketched the recirculating, separated flow at the back of the plate, along with the extensive wake that trails downstream. At the bottom, the plate is aligned with the flow, and we see the vortex that is created at the juncture of the plate surface and the water surface, as well as the bow wave that propagates at an angle away from the plate surface. These sketches by da Vinci are virtually identical with the photographs of such flows that can be taken in any modern fluid-dynamics laboratory, and they demonstrate the detail to which Leonardo observed various flow patterns.

Da Vinci's contributions to basic fluid dynamics were complemented by his thinking on the applied aerodynamics associated with flying machines. Much of his thought about aerodynamics was influenced by his study of bird flight. Also, Leonardo began to study the ancient writings of Euclid, and after 1496 he became intensely involved in studying geometry. The degree to which Leonardo became immersed in mathematics is reflected in this statement: "No knowledge can be certain, if it is not based upon mathematics or upon some other knowledge which is itself based upon the mathematical sciences. Instrumental, or mechanical science is the noblest and above all others, the most useful." Elsewhere he

Figure 2.7   Sketches by da Vinci showing complex flow fields over objects in a flowing stream.

was even more emphatic: "Let no man who is not a mathematician read the elements of my work."

To bring this to bear on aerodynamics and bird flight, we find this statement in the *Codex on the Flight of Birds:* "A bird is an instrument working according to mathematical law, an instrument which is within the capacity of man to reproduce with all its movements." Here we clearly see Leonardo's approach to the design of a flying machine going far beyond simply copying the physical structure of a bird; indeed, he was seeking the governing laws on which bird flight is based, so that those same laws could be used for the design of a machine.

As we examine Leonardo's thinking on aerodynamics in more detail, our discussion will revolve around three aspects: lift, drag, and general flow characteristics. There are entries in his notes, some extensive, that bear on all three categories.

In regard to the generation of lift, there is no evidence of any scientific thinking on this subject before da Vinci. R. Giacomelli, the well-known historian of mechanics at the University of Rome, has stated that in being first, da Vinci had the advantage of not being burdened with any preexisting notions about lift. Giacomelli notes that "lift . . . had not at all attracted the attention of anybody before Leonardo, so that no opinion or theory to be taken into account hampered the way to an examination of the question in full freedom of spirit" (p. 1018).[5] In spite of that apparent advantage, da Vinci's explanation of lift was flawed. He argued that when a surface struck the air (such as the downward movement of a bird's wing), the air would be compressed below the surface and would tend to support the surface. In the *Codex Trivultianus* we find the following statement: "When the force generates more velocity than the escape of the resisting air, the same air is compressed in the same way as bed feathers when compressed and crushed by a sleeper. And that object by which air was compressed, meeting resistance on it, rebounds in the same way as a ball striking against a wall."

To use more modern terms, da Vinci was saying that a high-pressure, high-density region of air is formed under a lifting surface, which in turn exerts an upward force on that surface. Today we know that lift is achieved primarily because of the low pressure (suction) over the top of a wing and that the pressure on the bottom of the wing, albeit higher than that on the top, is not much higher than the free-stream static pressure. Moreover, da Vinci attempted to quantify the aerodynamic force exerted on the bottom of the surface by stating, without proof, that it was proportional to the velocity. That, too, was incorrect; the aerodynamic force is proportional to the *square* of the velocity (see Chapter 1). Specifically, we find in the *Codex Atlanticus* the following statement: "Air which with great velocity is struck by a body compresses an amount of itself, in proportion with that velocity."

Leonardo's conception of the generation of lift was a natural assumption, when viewed in light of the absolute lack of rational physical science in his day. Imagine us standing in da Vinci's shoes in 1490: If we were to beat the air with some surface, such as a large piece of cardboard, we certainly would feel some resistance, which would appear to be a high pressure on the bottom of the cardboard. Moreover, experience would show us that that force would increase with velocity, and it would be easy to allow intuition to tell us that the force probably would vary directly with velocity. Of course, intuition in physical science is frequently wrong, as in this case.

Toward the end of his life, Leonardo recorded an observation on the flow field over a lifting object (in that case, a bird) that qualitatively was much closer to identifying the actual source of lift. In *Codex E,* written around the year 1513, we find the following statement:

> What quality of air surrounds birds in flying? The air surrounding birds is above thinner than the usual thinness of the other air, as below it is thicker than the same, and it is thinner behind than above in proportion to the velocity of the bird in its motion forwards, in comparison with the motion of its wings towards the ground; and in the same way the thickness of the air is thicker in front of the bird than below, in proportion to the said thinness of the two said airs.

For the sake of comparison, return to Figure 1.3, which is a qualitative sketch of the pressure distribution over a lifting airfoil. There are lower pressures over the top surface and higher pressures over the bottom surface, with the highest pressure at the leading edge (the stagnation-point pressure). Now reconsider the foregoing statement from *Codex E,* substituting the word "pressure" for "thinness" and "thickness," with "thinner" meaning "lower pressure" and "thicker" meaning "higher pressure." What results is a clear explanation of the actual pressure distribution over an airfoil. In short, we have Leonardo da Vinci, in the year 1513, giving a valid description, expressed in the technical language of the early sixteenth century, of the sources of lift as well as pressure drag (form drag) on an aerodynamic body. Thus, Leonardo was three centuries ahead of his time, because George Cayley, in 1809, was the next person to appreciate the actual source of lift.

There are two important corollaries to da Vinci's thinking on lift. One is his statement of what today we call the "wind-tunnel principle," namely, that the aerodynamic results are the same whether the body moves through a medium at a given velocity (the case of free flight through the atmosphere) or the medium flows past the stationary body at the same velocity (the case of a model mounted in a wind-tunnel flow). In the *Codex Atlanticus,* da Vinci makes the following statements: "As it is to move the object against the motionless air so it is to move the air against the motionless object." "The same force as is made by the thing against air, is made by air against the thing." Therefore, the basic principle that allows us to make wind-tunnel measurements and apply them to atmospheric flight was first conceived by da Vinci. Giacomelli has called this the "principle of aerodynamic reciprocity."[5]

The second corollary associated with da Vinci's thinking on lift concerns the production of lift on a *fixed* wing moving against the air (as opposed to the more prevalent concept, in his notes, of beating the air with the wings of an ornithopter). It is clear that Leonardo was the first person to understand the mechanics of bird flight. On the basis of his numerous careful observations of birds in flight he was able to deduce that the up-and-down flapping of a bird's wings did not contribute much to lift, but rather was the means by which the tip feathers produced a forward thrust for propulsion. The lift was produced as the forward motion of the bird caused air to flow across the wings. Thus, rather late in life, da Vinci concluded that a flying machine could have *fixed wings,* with a separate mechanism for propulsion. In the *Codex Atlanticus,* in 1505, da Vinci wrote the following: "Therefore if air moves against motionless wings the same air supports the heaviness of the bird through air." Hence, we see that in 1505 da Vinci was moving away from his flapping-wing ornithopter concepts and was beginning to think about fixed-wing aircraft. Once again, that was a precursor (by three centuries) to George Cayley's separation of lift and propulsion in a fixed-wing aircraft in 1799.

In contrast to his theory of lift, which was a unique and historic advance, da Vinci's concept of drag was at first patterned after the ancient "medium theory" of Aristotle. In that conception, the air would assist the motion of a body by filling in the space behind the body and exerting a thrust that would overcome any air resistance; it was reasoned that only in that way could an object maintain any motion at all. A theory opposed to the medium theory had been advanced by Joannes Philoponus (John the Grammarian) in the sixth century. Philoponus, a Christian who lived in Alexandria, advanced the "impetus theory," which held that a body acquired impetus from an original motive source (a rock acquires impetus from the arm of the person throwing it) and that the body would continue to move until all of its impetus had run out. In that theory, the action of the air would be purely that of a resistance (drag). Leonardo at first felt at ease with both the medium and impetus theories; he believed that a body would first move under the impetus applied and that the medium would subsequently assist its motion. However, later in life, Leonardo completely abandoned the medium theory. In the *Codex Leicester,* written between 1505 and 1508, Leonardo presented a long discussion attempting to prove that air does *not* push a body from behind and that its only effect is that of a resistance to motion. Leonardo later postulated, incorrectly, that that resistance, just as in the case of his consideration of lift, was proportional to the velocity. However, a second postulate was correct, namely, that the resistance due to air is proportional to the surface area of the body.

As a corollary to his new concept of drag, Leonardo pointed out the advantages of streamlined shapes to reduce drag. In *Codex G* there is a sketch of a fish and comparable hull shapes for ships (Figure 2.8). Leonardo argued that the resistance to the movement of a fish is small because the water flows smoothly over the afterbody of the fish, closing in on the afterbody and not creating a void as would be the case if the flow separated from the surface. Today we know that the function of streamlining a body is to prevent flow separation with its attendant high-pressure drag. Leonardo did not think in these terms, but he certainly was on the right track. He also applied the concept of streamlining to projectiles from cannons (Figure 2.9). This sketch, taken from the *Codex Arundel,* shows some aerodynamically advanced shapes, with fins for stability.

Given the void in scientific knowledge that existed before Leonardo, it seems clear that he made substantial contributions to the state of the art of aerodynamics. Or did he? The state of the art should represent what is *available for use by others,* as well as a certain state of affairs that will foster future developments and advancements. Da Vinci's work satisfied neither of those criteria. His work on aerodynamics was essentially bottled up in his notes, which were unavailable to others during his lifetime and for long afterward, and which were further masked by his reverse "mirror-like" handwriting. Leonardo's work on

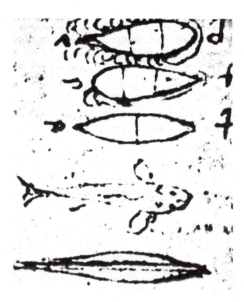

Figure 2.8   Sketches by da Vinci of streamlined shapes.

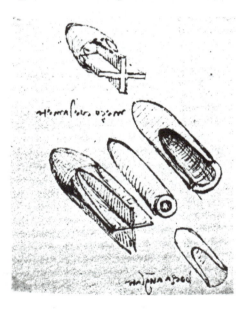

Figure 2.9   Sketches by da Vinci of streamlined projectiles.

aerodynamics really came to light only in the nineteenth and twentieth centuries, by which time the state of the art had advanced well beyond his thinking, and therefore his thinking is of only historical interest.

How much of that state of the art was reflected in the da Vinci ornithopter? The ornithopter was a product of Leonardo's attempt to emulate the basic mechanism of a bird, based on his extensive observations of bird flight and his detailed knowledge of the skeletal and muscular structures of birds. There is no evidence in his notes of any quantitative calculation associated with his ornithopter sketches.

Developments in Aviation from Antiquity to the Year 1500

Throughout history, lack of knowledge about aerodynamics and basic design principles never deterred people from attempting to build flying machines and fantasizing about flying through the air. Indeed, for any period during the evolution of aerodynamics there were actual events and accomplishments in the wider world of aviation. Hence, to provide additional understanding and appreciation for the historical development of aerodynamics, we shall be summarizing the relevant events in aviation. For this chapter, such discussion will be short, indicative of the virtual lack of substantive developments in aviation up to the year 1600. We can single out only the following aspects worth noting:

(1) The kite was first developed in China around 1000 B.C. It appeared much later (ca. A.D. 1300) in Europe in the shape of a windsock. The first known illustration of the more conventional plane-surface diamond shape dates back to 1618. We have to assume that Leonardo was familiar with kites, which may have provided some inspiration for his invention of the parachute (Figure 2.10).

(2) From about A.D. 800 to 1500 there were various stories about men who fashioned wings from wood, feathers, and cloth and jumped off roofs, trees, and other heights, desperately flapping away in a vain effort to fly. Called "tower jumpers" by the historian Gibbs-Smith,[25] all such people met with absolute failure.

(3) The windmill, with a horizontal spindle and blades rotating in a plane perpendicular to the ground, can be traced to Europe in about A.D. 1290. That was an early example of a machine utilizing the wind's energy to produce useful work.

Clearly, the technical void that faced Leonardo in his efforts to understand aerodynamics was accompanied by a similar void in aviation accomplishment.

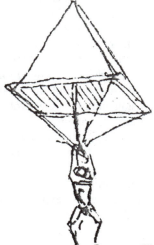

Figure 2.10   Sketch showing da Vinci's concept of a parachute.

CHAPTER 3

The Dawn of Aerodynamic Thought

To George Cayley and the Concept of the Modern-Configuration Airplane

> All the theories of resistance hitherto established are extremely defective, and that it is
> only by experiments analogous to those here recited that this important subject can ever be
> completed.
>
> Benjamin Robins, paper in *Philosophical Transactions* (1746)

Figure 3.1 shows a sketch of a hand-launched glider, approximately 1 m in length, designed by Sir George Cayley in 1804 (a full-scale model is on view at the British Science Museum in South Kensington, London). Today, such a glider may seem trivial, almost a child's toy, but in 1804 that glider represented a major technological breakthrough.[6] It was the first modern-configuration airplane. Here we see a heavier-than-air machine with a *fixed wing,* a *fuselage,* and *horizontal and vertical tail structures.* That was totally at variance with contemporary thought, which focused on ornithopter concepts. Although da Vinci had come to the conclusion late in life that a flying machine could be designed with fixed (rather than flapping) wings, that idea had not been made available to the general public. Therefore, in terms of the practical advancement of aeronautics, George Cayley, two centuries later, was responsible for the concept of the modern-configuration aircraft. He proposed a fixed wing to generate lift, a separate mode of propulsion to overcome the "resistance" (drag) to the machine's motion through the air, and both vertical and horizontal tail surfaces for directional and longitudinal static stability.

That concept was first illustrated by Cayley in a very unconventional manner. In 1799 he engraved on a silver disk an outline of a fixed-wing aircraft (Figure 3.2). On one side is a sketch of the aircraft, a machine with a fixed wing, a fuselage (occupied by a person), horizontal and vertical tail structures at the rear end of the fuselage, and a pair of "flappers" for propulsion. The means to achieve lift (the fixed wing) and propulsion (the flappers) were clearly separate, in contrast to the actions of ornithopter wings, which were intended to provide lift and propulsion all in the same motions. Cayley's concept was further emphasized on the flip side of the disk, which showed, for the first time, a lift-and-drag diagram for a lifting surface. The arrow shows flow from right to left, and the heavy diagonal line represents a wing cross section at a rather large angle of attack to the flow. In the right triangle above the wing, we see that the hypotenuse represents the resultant aerodynamic force, and the horizontal and vertical sides represent the drag and lift, respectively. Return to Figure 1.5, which shows an aerodynamic-force diagram for a modern airfoil; clearly, Cayley's diagram on the silver disk conveys exactly the same information. Today that silver disk, no larger than the size of an American quarter, is on display at the British Science Museum.

Cayley's 1804 glider was designed to test his fixed-wing concept. Launched by hand, it flew successfully, and it gave Cayley the incentive to pursue further experiments on the aerodynamic characteristics of aircraft.

Figure 3.1   George Cayley's sketch of his 1804 glider.

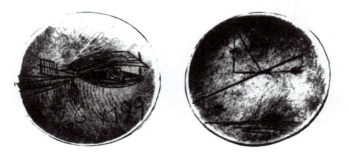

Figure 3.2   George Cayley's concept for the modern airplane, etched on a silver disk.

Galileo and the Beginnings of Rational Science

To understand the state of the art of aerodynamics in 1804, we must examine the evolution of aerodynamics from the time of da Vinci's death (1519) to Cayley's initial glider (1804). The roots of the mechanical science of aerodynamics were planted in the seventeenth century, when the elements of classical mechanics were being investigated by Galileo Galilei (1564–1642) at the beginning of the century, and when mechanics was being placed on a firm mathematical foundation by Isaac Newton (1642–1727) toward the end of the century. Galileo and Newton were giants in the history of science, and it is far beyond the scope of this book to relate all their contributions. Only their investigations that bear directly on the advancement of aerodynamics will be discussed here.

Galileo's contributions to aerodynamics were small compared with his work in astronomy and basic mechanics. However, there were several aspects of his research that relate directly to our aerodynamic model constructed in Chapter 1 and deserve some discussion here.

Galileo Galilei (Figure 3.3) was born in Pisa, Italy, February 15, 1564, the eldest of seven children. His father, Vincenzio Galilei, a musician and musical theorist, was the first to apply the mathematics of number theory to the analysis of musical harmony. (Interestingly enough, Vincenzio became disenchanted with mathematics as a practical tool, and at a later date he tried to discourage Galileo from studying mathematics at the university level.) First tutored as a young child in Pisa, Galileo later attended the well-respected school at the monastery of Santa Maria at Vallombrosa near Florence. In 1581 Galileo was sent to the University of Pisa as a medical student. However, during his studies at Pisa he became attracted to mathematics, and in spite of the opposition of his father, Galileo left the university without a degree in order to independently study the works of Euclid and Archimedes. That independent education led, in 1589, to his assignment to the chair of mathematics at the University of Pisa. There he soon became unpopular with his faculty colleagues because of his efforts to discredit the entrenched Aristotelian view of physics, and because of his disrespectful attitude toward the university administration. That, in

Figure 3.3   Galileo Galilei.

combination with the poor pay, prompted him to leave Pisa for the more lucrative chair of mathematics at Padua (his father died in 1591, leaving Galileo with financial responsibility for the family). The University of Padua attracted the best scholars and students in Italy and provided Galileo with a relatively free and tolerant atmosphere for his research in physics. During his residence at Padua, he took a Venetian mistress, Marina Gamba, with whom he had two daughters and one son. The eldest daughter, Virginia, born in 1600, was to become Galileo's main solace in later life. Galileo became aware of the telescope, invented by an obscure Dutch lens grinder, Hans Lippershey, in 1608, and improved it to a magnifying power of 30-fold. With that came the first real fame for Galileo, because he discovered mountains on the moon's surface, found the Milky Way to be composed of distinct stars, and spotted four satellites of Jupiter. One year later he left Padua to serve as a mathematician and philosopher to the grand duke of Tuscany in Florence. In 1611 he traveled to Rome, where he received the accolades of the church for his astronomical observations. However, after that his relations with the church changed for the worse. In 1616 he was criticized by the church for preaching the Copernican heliocentric philosophy (i.e., that the earth revolved around the sun), rather than the established geocentric theory, dating from Aristotle and Ptolemy, that the earth was the center of the universe and that all the heavenly bodies revolved around the earth. In March of 1632 the first of Galileo's two primary works was published: *Dialogue Concerning the Two Chief World Systems.* There Galileo made clear his logical arguments, supported by his astronomical observations, in favor of the heliocentric system. The church lost no time; within months he was tried by the Inquisition and sentenced to life imprisonment. His book was banned and placed on the Index (a list of books not to be read by the faithful). However, several influential members of the church who were supporters of Galileo were able to have his sentence commuted to permanent house arrest under surveillance. Galileo was allowed to move first to Siena, and then in 1634 to Arcetri, in the hills above Florence. It was there that he wrote his second influential volume: *Discourses and Mathematical Demonstrations Concerning Two New Sciences,* which dealt with the basic principles of the strength of materials and the mathematical aspects of kinematics. That book was published in 1638, by which time Galileo was totally blind. Four years later he died and was buried at Santa Croce in Florence.

Like da Vinci, Galileo moved in some of the best circles in Italy, his trouble with the church notwithstanding. Also like da Vinci, Galileo learned by observation. Much of the content of *Two New Sciences* was derived from his own physical experiments concerning the velocities, accelerations, and trajectories of moving bodies.

Galileo's contributions to mechanics advanced the mechanical science of aerodynamics. For example, he introduced the concepts of inertia and momentum. He observed that the effect of a force was to *change* a motion, in contrast to the earlier Aristotelian tenet that a force was needed simply to sustain motion. In that sense, Galileo's principle was a precursor to Newton's first and second laws of motion, about 50 years later, although Galileo never quantified his principle. Galileo observed that a body rolling down an inclined plane continuously increased its velocity under the force of gravity, but upon rolling out onto a horizontal surface, its velocity remained unchanged. Furthermore, when the body rolled off the end of the table, its horizontal component of velocity remained the same, whereas its vertical component toward the ground increased. From that, Galileo was able to prove mathematically that the body's trajectory through the air was a parabola. In that regard, Galileo was concerned with the aerodynamics of the falling body. He noted that "but the resistance of the air influences the path of the body." Hence, along with Aristotle and da Vinci before him, Galileo theorized that a body moving through the air experienced an aerodynamic force that retarded its motion – a "resistance" or drag. However, like his predecessors, Galileo was unable to quantify that resistance. Perhaps that was why, on another occasion, he remarked that "the air resistance is small enough to ignore it," possibly succumbing to the almost universal tendency of many scientists and engineers (even to the present day) to choose to ignore those phenomena that they cannot understand or quantify.

Although Galileo was unable to quantify the aerodynamic resistance, he made a partial contribution. In experiments involving observations of falling bodies and the motions of pendulums, he was able to deduce that the aerodynamic resistance was proportional to the air density. Here we have, for the first time in history, the appearance of a second variable factor in the lift-and-drag equations given in Chapter 1, namely, that L and D are proportional to the density ρ. Recall that the first variable – that L and D are proportional to a suitable reference area S – had been established by da Vinci. Thus, by 1638, with the publication of Galileo's *Two New Sciences,* it was known that

$$D = K\rho S f(V)$$

where K is a proportionality constant, and $f(V)$ is some function of velocity. Precisely what function of velocity was quite another question. We have already seen that da Vinci believed that resistance was directly proportional to velocity. By a completely different (and erroneous) investigation, Galileo also concluded that the aerodynamic force was proportional to velocity [i.e., in the foregoing equation, $f(V) = V$].

The period between Galileo's contributions to experimental science and Newton's contributions to theoretical science spanned the middle of the seventeenth century. During that time there were several minor advancements and one major development that bear on aerodynamics. First, let us consider the minor advancements.

In 1628, Benedetto Castelli (1577–1644), one of Galileo's pupils, and later a trusted colleague, published the book *Della misura delle acque correnti,* in which he stated the law of continuity for an incompressible flow, namely, that $AV = $ constant, where A is the cross-sectional area of the flow, and V is the velocity. We have already seen that da Vinci understood this law, but as with most of his work, it was not published in his time. Because Castelli was the first to put this law in print, it became known as Castelli's law in Italy.

Contemporary with Castelli was Evangelista Torricelli (1608–47), who was also influenced by the work of Galileo, and who became Galileo's close companion for the last three months of his life. Torricelli carried out numerous experiments on the velocities of liquids flowing from holes in the bottoms of filled containers, and he was the first to show that the velocity is directly proportional to the square root of the height of the fluid in the container. Torricelli's law, that $V = \sqrt{2gh}$, can be found in all modern introductory physics books. He also invented the barometer, at about 1644. In 1893, Ernst Mach[7] called Torricelli the founder of the theory of hydrodynamics; however, that is not generally accepted by most modern historians of science. For example, Tokaty[8] believes that there is not sufficient evidence of contributions by Torricelli to justify such a label.

Also active during that period was Blaise Pascal (1623–62) in France. Pascal advanced the discipline of fluid *statics* (which had begun with Archimedes) to a well-developed science. He understood that the action of the atmosphere was that of a weight impressed on the surface of a fluid. He was the first to advance the theory that the pressure at any point within a fluid is the same in all directions through that point, and that it depends only on the depth of the fluid at that point. (Recall from our basic description of aerodynamics in Chapter 1 that in a flow, pressure is a point property that can vary from one point to another in the flow field; indeed, one of the important tasks in theoretical aerodynamics is to predict the variations in pressure for all points in a flow.) In Pascal's time, the fact that pressure is truly a point property was not fully understood – in the world of fluid *statics,* pressure was thought of, at best, as a property that was constant across a horizontal layer of fluid at a given depth. Hence, even in Bernoulli's time, in the eighteenth century, pressure was considered as an equivalent "height of fluid." As we shall see, Leonhard Euler, in 1754, was the first person to recognize that pressure in a *moving* fluid is strictly a point property.

The Velocity-Squared Law: The First Aerodynamic Breakthrough

Toward the end of the seventeenth century there was a major development in the advancement of aerodynamics. Until the middle of the seventeenth century the prevailing theory had been the incorrect notion that force was directly proportional to the flow velocity. However, within the space of 17 years at the end of the century, that situation changed dramatically. Between 1673 and 1690, two independent sets of experiments, conducted by Edme Mariotte (1620–84) in France and Christiaan Huygens (1629–95) in Holland, along with the theoretical fundamentals published by Isaac Newton (1642–1727) in England, clearly established that the force on an object varies as the *square* of the flow velocity (i.e., if the velocity doubles, the force goes up by a factor of 4). After centuries of halting, minimal progress in aerodynamics, the rather sudden realization of the velocity-squared law for aerodynamic force represented the first *major* scientific breakthrough in the evolution of aerodynamics.

Credit for the velocity-squared law rests with Edme Mariotte (Figure 3.4), who first published it in 1673. Mariotte lived in obscurity for the first 40 years of his life. We have no information concerning his personal life, his education, or his work until 1666, when suddenly he was made a charter member of the newly formed Paris Academy of Sciences. Most likely he was self-taught in the sciences. He came to the attention of

Figure 3.4 Some members of the Paris Academy of Sciences; Mariotte is second from left.

the Paris Academy because of his pioneering theory that sap circulated through plants in a manner analogous to blood circulating through animals. Controversial at that time, his theory was confirmed within four years by numerous experimental investigators. Mariotte quickly proved to be an active member and contributor to the academy. His interests were diverse: experimental physics, hydraulics, optics, plant physiology, meteorology, surveying, and general scientific and mathematical methods. Mariotte is credited with bringing *experimental* science to France, inspiring the same interest in experimentation that flourished during the Italian Renaissance in the work of da Vinci and Galileo. Indeed, Mariotte was a prolific experimentalist who took pains to examine the linkages between existing theory and experiment – a novel approach for that time. The Paris Academy essentially consumed Mariotte's later life; he remained in Paris until his death, on May 12, 1684.

Mariotte was particularly interested in the forces produced by various bodies impacting on other bodies or surfaces. One of those "bodies" was a fluid; Mariotte studied and measured the forces created by a moving fluid impacting on a flat surface. The device he used for those experiments is shown in Figure 3.5. Here we see a beam dynamometer wherein a stream of water impinges on one end of a beam, and the force exerted by that stream is balanced and measured by a weight on the other end of the beam. The water jet flows from the bottom of a filled vertical tube, and its velocity is known, from Torricelli's law, as a function of the height of the column of water in the tube. With that experimental apparatus, Mariotte was able to prove that the force of impact of the water on the beam varied as the square of the flow velocity. He presented those findings in a paper read to the Paris Academy of Science in 1673: *Traité de la percussion ou choc des corps.*

The esteem in which Mariotte was held by some of his colleagues was reflected in the words of J. B. du Hamel: "The mind of this man was highly capable of all learning, and the works published by him attest to the highest erudition. In 1667, on the strength of a singular doctrine, he was elected to the Academy. In him, sharp inventiveness always shone forth combined with the industry to carry through, as the works referred to in the course of this treatise will testify. His cleverness in the design of experiments was almost incredible, and he carried them out with minimal expense" (p. 120).[9]

There was at least one colleague who was not so admiring of Mariotte, one who represented another side of the historical debate on the credit for the velocity-squared law:

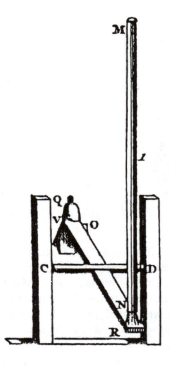

Figure 3.5 Mariotte's mechanism for measuring the force exerted on an object in a fluid flow.

Figure 3.6 Christiaan Huygens.

Christiaan Huygens (1629–95). Indeed, Giacomelli and Pistolesi[10] give Huygens credit for the first proof of the velocity-squared law.

Christiaan Huygens (Figure 3.6) was born April 14, 1629, in The Hague, The Netherlands, to a family prominent in Dutch society.[11] Several members of the family served as diplomats during the reign of the house of Orange. Christiaan was well educated; he was tutored by his father until the age of 16, after which he studied law and mathematics at the University of Leiden. Devoting himself to physics and mathematics, Huygens made substantial contributions, including improvements in existing scientific methods, development of new techniques in optics, and invention of the pendulum clock. Huygens, like Mariotte, was made a charter member of the Paris Academy of Science in 1666. Huygens moved to Paris to be closer the activities of the academy, living there until 1681. During that period, Mariotte and Huygens worked, conversed, and argued together as colleagues in the academy. In 1681, Huygens moved back to The Hague, where he died on July 8, 1695. During his life, Huygens was recognized as Europe's greatest mathematician. However, he was somewhat solitary and did not attract a following of young students, and he was reluctant to publish, mainly because of his exceedingly high personal standards. For those reasons, Huygens did not greatly influence the scientists of the next century; indeed, his work was not widely known during the eighteenth century.

In 1668, Huygens began to study the fall of projectiles in resisting media. Following da Vinci and Galileo, he started out with the belief that resistance (drag) was proportional to velocity. Within a year, his analysis of the experimental data convinced him that resistance was proportional to the square of the velocity. That was four years before Mariotte published the same findings in 1673, but Huygens did not publish his data and conclusions until 1690. That somewhat complicates the question of the credit for the velocity-squared law. The picture was further blurred by Huygens himself, who accused Mariotte of plagiarism.

Mariotte and Huygens were colleagues, and Huygens clearly stated that they discussed and shared thoughts. In such situations, sometimes the exact attribution of the credit for new ideas is not clear. Ideas frequently evolve as a result of discussions among groups and individuals. What is clear is this: Mariotte published the velocity-squared law in a paper presented to the academy in 1673; Huygens published the same conclusion 17 years later.

Using the published scientific literature as the measure of proprietorship, we have to conclude that Mariotte deserves first credit for this law. However, it is clear that Huygens's experiments, which were carried out before Mariotte's publication, also proved the velocity-squared law. Thus, by the end of the seventeenth century, there was direct experimental proof from two independent investigations that aerodynamic force varies as the square of the velocity. Of even greater importance was that the same law had been derived *theoretically* on the basis of the rational, mathematical laws of mechanics advanced by Newton in his *Principia,* published in 1687.

Newton and the Flowering of Rational Science

The early years of the seventeenth century brought a series of rapid developments in experimental physics, fueled in part by the work of Galileo. It is fitting that the end of the century saw the natural fruition of that experimental work in the development of a rational mathematical theory by Isaac Newton (1642–1727) (Figure 3.7). Newton's contributions to physics and mathematics were pivotal. The publication of his *Principia* in 1687 represented the first complete, rational, theoretical approach to the study of mechanical phenomena. However, Newton did not operate in an intellectual vacuum; we have already discussed several precursors to Newton's laws. Newton benefited from the work of others who preceded him, principally Galileo, Descartes, and Huygens. Newton wrote that "if I have seen further, it is because I have stood on the shoulders of giants." A thorough discussion

Figure 3.7   Sir Isaac Newton.

of Newton's contributions to science would be beyond the scope of this book. We shall consider only those aspects of his work that had an impact on aerodynamics.

Isaac Newton was born on December 25, 1642, near the small English town of Woolsthorpe. He was raised by his mother; his father had died five months before his birth. As a child, he showed an interest in mechanical diagrams, which he scratched on the walls and window edges of his house in Woolsthorpe. His mother had intended a farmer's life for her son, but the young Newton was absentminded and lackadaisical. Instead, with the encouragement of an uncle, Newton set his sights on college. He entered Trinity College at Cambridge in 1661 and received his B.A. degree in 1665. For the next two years Newton retreated to the country in Lincolnshire to avoid the plague that was running rampant in Europe and had closed the university. It was during that two-year period that he conceived many of his basic ideas on mathematics, optics, and mechanics that were later to appear in print. Newton said of those two years that "I was in the prime of my age of invention and minded mathematics and philosophy more than at any time since."

In 1667, Newton returned to Cambridge and became a minor fellow at Trinity. He earned an M.A. degree in 1668 and was appointed Lucasian professor in 1669. Newton remained at Cambridge for the next 27 years. The holder of the Lucasian professorship was required to give one lecture each week during the school year. In that capacity, from 1669 to 1687 Newton presented a long series of lectures that evolved into his famous *Philosophiae naturalis principia mathematica,* first published in 1687. During those years at Cambridge, Newton spent a great deal of time in his chambers, except for the time spent lecturing. His lectures were poorly attended, and few understood what he was saying. Many times he had no audience. Newton ate very little, sometimes forgetting to eat at all, and he presented a somewhat unkempt appearance. However, that did not keep him from becoming a fellow of the Royal Society in 1672 and a member of Parliament in 1689. The Royal Society had intended to publish his *Principia,* which was the starting point for what today we call classical physics, but sufficient funds were not available, and Newton's friend, the noted astronomer Edmund Halley, financed the publication of the *Principia*. In his last decade, Newton became bored with his professorship and with Cambridge in general. In 1696 he was appointed warden of the mint, where he applied his knowledge of chemistry and laboratory techniques to assaying. In his later years, Newton was a principal force in the Royal Society, not always to the betterment of science. Indeed, Newton was frequently critical of the work performed by younger scientists, and that attitude tended to inhibit scientific advancement in England at the beginning of the eighteenth century. Newton also became embroiled in a lengthy quarrel with the German scientist and mathematician Gottfried Leibniz over which of them conceived and developed the elements of the calculus. That dispute drove a wedge between some scientists in Britain and continental Europe. On March 20, 1727, Newton died of complications associated with gout and inflamed lungs. He was buried with distinction in Westminster Abbey in London, arguably the most important English scientist.

Newton's contributions to aerodynamics appear in Book II of the *Principia*, subtitled "The Motions of Bodies (in Resisting Mediums)." Book II deals exclusively with fluid dynamics and hydrostatics.[12] During the last part of the seventeenth century, practical interest in fluid dynamics was driven by problems in naval architecture, particularly the need to understand and predict the drag on a ship's hull, an important concern for a country that was ruling large portions of the world through the superior performance of its powerful navy. Newton's interest in fluid mechanics may have derived partly from such a practical problem, but he had a much more compelling reason for calculating the resistance of a body moving through a fluid. There was a prevailing theory, advanced by René Descartes, that interplanetary space was filled with matter that moved in vortex-like motions around the planets.[13] However, astronomical observations, such as the definitive work of Johannes Kepler in his *Rudolphine Tables*, published in 1627, indicated that the motions of the heavenly bodies through space were not dissipated, but rather that those bodies executed regular, repeatable patterns. The only explanation for that, if the bodies were moving through space filled with a continuous medium as Descartes had theorized, would be for the aerodynamic drag on each body to be zero. The central purpose of Newton's studies in fluid mechanics was to prove that there was a finite drag on a body (including the heavenly bodies) moving through a continuous medium. If that could be shown to be true, then the theory of Descartes would be disproved. Indeed, in Proposition 23 of the *Principia*, Newton calculated *finite* resistances on bodies moving through a fluid and showed that such resistances were "in a ratio compounded of the squared ratio of their velocities, and the squared ratio of their diameters, and the simple ratio of the density of the parts of the system." That is, Newton derived the velocity-squared law, while at the same time showing that resistance varies with the cross-sectional area of the body (the "squared ratio of their diameters") and the first power of the density (the "simple ratio of the density"). In so doing, Newton presented the first theoretical derivation of the essence of the drag equation

$$D \propto \rho S V^2$$

However, in Newton's mind, his contribution was simply to refute the theory of Descartes. That was stated specifically by Newton in the scholium accompanying Proposition 40, dealing with *experimental* measurement of the resistance of a sphere moving through a continuous medium. Because such spheres had been shown both theoretically and experimentally to exhibit *finite* resistances while moving through a fluid, "the celestial spaces, through which the globes of the planets and comets are continually passing towards all parts, with the utmost freedom, and without the least sensible diminution of their motion, must be utterly void of any corporeal fluid, excepting, perhaps, some extremely rare vapors and the rays of light." For Newton, that was the crowning accomplishment from his study of fluid dynamics. For modern aerodynamicists, in relation to our aerodynamic model constructed in Chapter 1, the meaningful accomplishment was the theoretical proof that aerodynamic force varies (1) with the first power of the fluid density, (2) with the first power of the body reference area, and (3) with the second power of the velocity. Of course, that was simply the theoretical justification, for by Newton's time such variations had already been established by experimental evidence.

In regard to aerodynamics, Newton's work in Book II of the *Principia* contributed a second fundamental finding, namely, a relationship for the shear stress at any point in a fluid in terms of the velocity gradient existing at that same point. The shear stress τ exerted on a surface immersed in a moving fluid is directly proportional to the velocity gradient normal to the surface:

$$\tau = \mu(dV/dn)$$

where the proportionality constant is μ, the viscosity coefficient. Integration of the effects of the shear stresses over the entire surface of the body gives rise to the skin-friction drag exerted on the body. For slender, streamlined bodies, skin friction is by far the largest contributor to the total drag on the body; hence Newton's relationship for shear stress was of vital importance for aerodynamics. The foregoing equation does not appear explicitly in the *Principia;* rather, Newton advances the following hypothesis in Section IX of Book II: "The resistance arising from the want of lubricity in the parts of a fluid is, other things being equal, proportional to the velocity with which the parts of the fluid are separated from one another." In modern terms, the "want of lubricity" is the action of friction in the fluid, namely, the shear stress τ. The "velocity with which the parts of the fluid are separated from one another" is the rate of strain experienced by a fluid element in the flow, which in turn can be mathematically represented by the velocity gradient, dV/dn. A mathematical statement of Newton's hypothesis is simply

$$\tau \propto dV/dn$$

With the proportionality constant defined as the coefficient of viscosity μ, that becomes

$$\tau = \mu(dV/dn)$$

as written earlier. This equation is called the Newtonian shear-stress law, and all fluids and gases that obey the law are called "Newtonian fluids." Virtually all gases, including air, are Newtonian fluids. Hence the Newtonian stress law, as first hypothesized in the *Principia,* represented a major contribution to the state of the art of aerodynamics at the end of the seventeenth century.

Angle-of-Attack Effects and the Newtonian Sine-Squared Law

Consider again the equations for lift-and-drag:

$$L = \tfrac{1}{2}\rho V^2 S C_L$$

and

$$D = \tfrac{1}{2}\rho V^2 S C_D$$

By the end of the seventeenth century, the essence of these equations had been experimentally and theoretically justified (i.e., that aerodynamic force was directly proportional to the fluid density, the reference area, and the square of the velocity).

Not so well established was the effect of the angle of incidence (as we say today, the angle of attack) on the aerodynamic force. In the lift-and-drag equations, the lift-and-drag coefficients C_L and C_D are functions of the angle of attack that a given body takes with respect to the free stream. The question of how aerodynamic force varied with regard to the angular orientation of a body in the flow was never seriously addressed prior to Newton's time.

In an indirect sense, Isaac Newton was responsible for the first technical contribution toward analysis of angle-of-incidence effects on aerodynamic force. Proposition 34 in Book II of the *Principia* is a proof that the resistance of a sphere moving through a fluid is half that of a circular cylinder of equal radius with its axis oriented in the direction of its motion. The fluid itself is postulated as a collection of individual particles that impact directly on the surface of the body, subsequently giving up their components of momentum normal to the surface, and then traveling downstream tangentially along the body surface. That fluid model was simply a hypothesis on the part of Newton; it did not accurately model the action of a real fluid, as Newton readily acknowledged. However, consistent with that mathematical model, buried deep in the proof of Proposition 34 is the result that the impact force exerted by the fluid on a segment of a curved surface is proportional to $\sin^2 \theta$, where θ is the angle between a tangent to the surface and the free-stream direction. That result, when applied to a flat surface (e.g., a flat plate) oriented at an angle of attack α to the free stream (Figure 3.8), gives the resultant aerodynamic force on the plate:

$$R = \rho V^2 S \sin^2 \alpha$$

This equation is called Newton's sine-squared law; it does not appear explicitly in this form in the *Principia,* although it follows directly from the derivation given by Newton in Proposition 34. The first application of Newton's sine-squared law to a flat plate at an angle of attack is historically obscure. It is such a simple extrapolation of the work in Proposition 34 that we have to assume that it followed shortly after publication of the *Principia*.

It is important to note one of the aerodynamic consequences of the sine-squared law. Returning to Figure 3.8, from the relationship involving L, R, and D we can write

$$L = R \cos \alpha = \rho V^2 S \sin^2 \alpha \cos \alpha$$

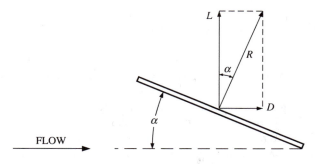

Figure 3.8 Aerodynamic force on a flat plate at an angle of attack α.

and

$$D = R \sin \alpha = \rho V^2 S \sin^3 \alpha$$

Hence

$$L/D = \cos \alpha / \sin\alpha = \cot \alpha$$

In these relations, the angle α is in *radians*. Recall that one radian is equal to 57.3°. Now consider a flat lifting surface, such as that shown in Figure 3.8. Assume that the flat surface is at a small angle of attack, say 3°. That is equal to 0.052 radian, a very small number. The sine of 0.052 is smaller yet, a number much less than unity. In turn, the square of the sine is particularly small. Examining the foregoing equation for lift, we note that for a flying machine at a given velocity and with a given wing area, the sine-squared variation predicts very small lift at small angles of attack. However, for steady, level flight, the lift must equal the weight. If we were to accept the Newtonian sine-squared law as correct (as we shall see shortly, it is not), then we would have only two options to counter the small value of $\sin^2\alpha$ and to increase the lift so that it would equal the weight of the flying machine:

(1) Increase the wing area S. That could lead to enormous wing areas, which would make the flying machine totally impractical.
(2) Increase the angle of attack α. Unfortunately, that would lead to greater drag along with greater lift. Indeed, the drag will increase faster than the lift when α is increased. That would result in a lift-to-drag ratio L/D that would decrease as α was increased. (Note from the foregoing equations that $L/D = \cot \alpha$, which constantly decreases as α increases.) Because L/D is a measure of aerodynamic efficiency, flying at large angles of attack would be undesirable, to say the least.

For those reasons, calculations using the Newtonian sine-squared law led to very pessimistic predictions regarding the aerodynamic characteristics of flying machines – so pessimistic that the foregoing results were used during the nineteenth century to argue against the practical feasibility of heavier-than-air flight.

Newton's Contributions to Aerodynamics

It is ironic that Newton's sine-squared law has had a rebirth in modern aerodynamics, namely, for the prediction of pressure distributions on the surfaces of hypersonic vehicles. The physical nature of hypersonic flow, where the bow shock wave lies very close to the vehicle surface, closely approximates the fluid model used by Newton – a stream of particles in rectilinear motion colliding with the surface and then moving tangentially over the surface. Hence, the sine-squared law leads to reasonable predictions for the pressure distributions over blunt-nosed hypersonic vehicles, an application that Newton could not have foreseen.

Perhaps Newton's most important contribution to aerodynamics was embodied in his famous laws concerning motion – three laws that provide the basis for all classical mechanics. Of greatest importance for aerodynamics was Newton's second law, which relates force to the time rate of change of momentum for a moving body (i.e., $F = ma$). Newton's second law, as applied to fluid flow, is one of the basic equations on which all theoretical aerodynamics is based. Newton's second law is one of the governing three fundamental physical principles used to obtain both the Euler equations and the Navier-Stokes equations (see Appendixes A and B).

Figure 3.9   Daniel Bernoulli.

The Sunrise of Hydrodynamics: Daniel Bernoulli and the Pressure–Velocity Concept

The fundamental advances in aerodynamics in the eighteenth century began with the work of Daniel Bernoulli (1700–82) (Figure 3.9). Newtonian mechanics had unlocked, but not opened, the door to modern hydrodynamics. Bernoulli was the first to open that door, though just by a crack. Leonhard Euler and others who would follow would fling the door wide open.

Daniel Bernoulli was a member of a prestigious family that dominated European mathematics and physics during the early part of the eighteenth century. He was born in Groningen, The Netherlands, February 8, 1700. At the University of Basel, Bernoulli received a master's degree in philosophy and logic in 1716. He went on to study medicine in Basel, Heidelberg, and Strasbourg, obtaining his Ph.D. in anatomy and botany in 1721. During those studies, he maintained an interest in mathematics, moving briefly to Venice, where in 1724 he published an important work entitled *Exercitationes mathematicae.* That earned him much attention and a prize from the Paris Academy of Sciences, the first of 10 he would receive. In 1725, Bernoulli joined the St. Petersburg Academy in Russia, which had earned a substantial reputation for scholarship and intellectual accomplishment at that time. The next eight years proved to be Bernoulli's most creative period. While at St. Petersburg, he wrote his famous book *Hydrodynamica,* completed in 1734, but not published until 1738. In 1733, Daniel returned to Basel to occupy the chair of anatomy and botany, and in 1750 he moved to the chair of physics, created specifically for him. He continued to write, to give well-attended lectures in physics, and to make contributions to mathematics and physics until his death in Basel on March 17, 1782.

Daniel Bernoulli was famous in his own time. He was a member of all the important learned societies and academies (e.g., Bologna, St. Petersburg, Berlin, Paris, London, Bern, Turin, Zurich, and Mannheim). His importance for fluid dynamics derives from *Hydrodynamica* (1738), wherein he coined the term "hydrodynamics." That book ranged over such topics as jet propulsion, manometers, and flow in pipes. Of most importance, however, he attempted to find a relationship between the variation of pressure with velocity in a fluid flow. He used Newtonian mechanics, along with the concept of *vis viva* (living force), introduced by Leibniz in 1695. That was actually an energy concept; *vis viva* was defined by Leibniz as the product of mass times velocity squared, mV^2; today, we recognize this as twice the kinetic energy of a moving object of mass m. Also, Bernoulli treated pressure in terms of the height of a column of fluid, much as Archimedes had done 20 centuries earlier. The concept that pressure is a *point* property that can vary from one point to another in a flow cannot be found in Bernoulli's work.

Let us critically examine Bernoulli's contributions to aerodynamics. In modern aerodynamics we find the "Bernoulli principle":

In a flowing fluid, as the velocity increases, the pressure decreases.

This is an absolute fact that is frequently used to explain the generation of lift on an airplane wing: As the flow speeds up in moving over the top surface of the wing, the pressure there decreases, and that lower pressure exerts "suction" on the top of the wing, thus generating lift. A quantitative statement of the Bernoulli principle is *Bernoulli's equation:* If points 1 and 2 are two different points in a fluid flow, then

$$p_1 + \tfrac{1}{2}\rho V_1^2 = p_2 + \tfrac{1}{2}\rho V_2^2$$

The Bernoulli equation is perhaps the most famous equation in all of fluid dynamics. Clearly, if V_2 is larger than V_1, then p_2 is smaller than p_1; that is, as V increases, p decreases. Question: How much of this did Bernoulli ever state? Answer: Not much. In *Hydrodynamica*, the primary source for all who have investigated his contributions, Bernoulli did attempt to derive the relationship between pressure and velocity. Using the concept of *vis viva*, Bernoulli applied an energy-conservation principle to the apparatus sketched in Figure 3.10. Here we see a large tank, *ABGC*, filled with water, to which has been attached a horizontal pipe, *EFDG*. The end of the pipe is partially closed; it contains a small orifice through which the water escapes. Stating that the sum of the potential and kinetic energies of the fluid in the pipe was constant (an incorrect statement, because in a flowing fluid there is work done by the pressure, in addition to the existence of kinetic and potential energies – such "flow work" was not understood by Bernoulli), he obtained the following differential equation for the change in velocity, dV, over a small distance, dx:

$$\frac{V\,dV}{dx} = \frac{a - V^2}{2c}$$

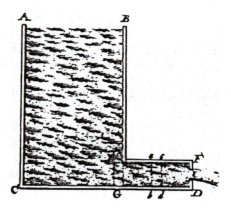

Figure 3.10   Sketch from Bernoulli's *Hydrodynamica* showing water flowing from a tank.

where a is the height of the water in the tank, and c is the length of the horizontal pipe. That equation is a far cry from the Bernoulli equation we use today.

$$p_1 + \tfrac{1}{2}\rho V_1^2 = p_2 + \tfrac{1}{2}\rho V_2^2$$

However, Bernoulli went on to interpret the term $V(dV/dx)$ as the pressure, which allows us to interpret the relation in *Hydrodynamica* as the form

$$p = \frac{a - V^2}{2c}$$

Because a and c are constants, this relation says, qualitatively, that as velocity increases, pressure decreases. From that we are led to conclude the following:

(1) The principle that pressure decreases as velocity increases is indeed presented in Bernoulli's book, though in a slightly obscure form. Hence it is clearly justified to call this the Bernoulli principle, as is done today. However, nowhere in his book does Bernoulli emphasize the importance of this principle, showing a certain lack of appreciation for its significance.

(2) Bernoulli's equation does not appear in his book, nor elsewhere in his work. *It is quite clear that Bernoulli never derived nor used Bernoulli's equation.*

This is not to denigrate Bernoulli's contributions to aerodynamics. His work provided a starting point for other investigators in the eighteenth century. He was the first to examine the relationship between pressure and velocity in a flow using the new scientific principles of the eighteenth century. As far as I can determine, he was the first to use the elements of the calculus to analyze a fluid flow, as illustrated in the foregoing differential equation from *Hydrodynamica*. His findings stimulated work by other investigators, including Euler, d'Alembert, and Lagrange. An English translation of *Hydrodynamica* is available.[14]

Jean le Rond d'Alembert and His Paradox

Although not monumental, Bernoulli's work served as a catalyst for other researchers in the eighteenth century. Particularly motivated by Bernoulli's findings was Jean le Rond d'Alembert (1717–83), whose work provided a bridge between the physical concepts of Bernoulli and the elegant mathematical modeling of Euler, as discussed in the next section.

Born illegitimately in Paris on November 17, 1717, to Claudine de Tencin, who would become a famous salon hostess of that time, and Chevalier Destouches, a cavalry officer, d'Alembert (Figure 3.11) was immediately abandoned by his mother (she had escaped after 16 reluctant years in a nunnery and was afraid of being forcibly returned to the convent). However, the father quickly arranged for the child to live with a family of modest means named Rousseau, and d'Alembert lived with that family for the next 47 years. With the support of his father, d'Alembert was educated at the Collège des Quatre-Nations, where he studied law and medicine and later turned to mathematics. For the remainder of his life, d'Alembert would consider himself a mathematician. By a program of self-study, d'Alembert learned the works of Newton and the Bernoulli mathematicians. His early efforts in mathematics caught the attention of the Paris Academy of Sciences, of which he became a member in 1741. He published frequently and sometimes rather hastily, in

Figure 3.11   Jean le Rond d'Alembert.

order to be in print before his competition. However, he made substantial contributions to the sciences of his time. For example, he was (1) the first to formulate the wave equation of classical physics, (2) the first to express the concept of a partial differential equation, (3) the first to solve a partial differential equation (he used the method of separation of variables), and (4) the first to express the differential equations of fluid dynamics in terms of a field.

During the course of his life, d'Alembert became interested in many scientific and mathematical subjects, including vibrations, wave motion, and celestial mechanics. In the 1750s he held the honored position of science editor for Diderot's *Encyclopédie*, a major French intellectual endeavor of the eighteenth century that attempted to compile all existing knowledge into a large series of books. As he grew older, he also wrote papers on nonscientific subjects, mainly musical structure, law, and religion.

In 1765, d'Alembert became very ill. He was helped to recover by the nursing of Mlle. Julie de Lespinasse, the woman who was d'Alembert's only love throughout his life. Although he never married, d'Alembert lived with Julie de Lespinasse until she died in 1776. He had always been a charming gentleman, renowned for his intelligence, gaiety, and considerable conversational ability, but after Mlle. de Lespinasse's death, he became frustrated and morose, living a life of despair. He died in that condition on October 29, 1783, in Paris.

One of the great mathematicians and physicists of the eighteenth century, d'Alembert maintained ongoing communications and dialogue with both Bernoulli and Euler and ranks with them among the founders of modern fluid dynamics. In terms of the state of the art of aerodynamics, d'Alembert was responsible for the following contributions: He introduced a model for flows and an equation for the principle of mass conservation that were much more sophisticated than the earlier suggestions by da Vinci. He introduced the concept of a moving fluid element of fixed mass to model the flow of a fluid, wherein the volume of the fluid element could change if the flow was compressible, but would remain constant if the flow was incompressible. That was published in the form of a differential equation in his paper *Traité de l'équilibre et des mouvements des fluides pour servir de suite au traité de dynamique* (1744). That was the first time that the continuity equation was expressed in terms of a differential equation applying locally in a flow field. In regard to the moving-fluid-element model, d'Alembert introduced the ideas of local velocity components and acceleration that could vary from one point to another in the flow. That was a major step toward the modern conception of aerodynamics.

Today, d'Alembert's name is not usually associated with the contributions to aerodynamics just cited, but rather is known primarily in regard to *d'Alembert's paradox*. It had been known since the time of Aristotle that a body moving through a fluid experiences a resistance (aerodynamic drag). Indeed, branching out from the work of Newton, one of the main pursuits of eighteenth century hydrodynamicists was the attempt to calculate that resistance. Thus, one can well imagine the frustration felt by d'Alembert when in his 1744 paper he obtained the result of *zero drag* for the inviscid incompressible flow over a closed two-dimensional body. That result, although contradictory to observations in real life, is a correct theoretical result for the assumption of an inviscid (frictionless) low-speed flow. Figure 3.12 illustrates the nature of d'Alembert's paradox. The flow over a circular cylinder is shown. In Figure 3.12a, the flow is assumed to be inviscid. For such a flow, the streamlines are symmetrical about the body; indeed, they close behind the body, with a streamline pattern identical with that in front of the body. In turn, the pressure distribution over the back surface of the body is the same as that over the front surface. Consequently, there is no net pressure drag on the body. Moreover, because the flow is frictionless, there is no skin-friction drag. Conclusion: The drag on the body is zero. That was the conclusion

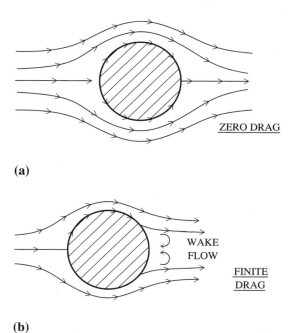

(a)

(b)

Figure 3.12   Flow over a circular cylinder: (a) theoretical inviscid low-speed flow field; (b) actual viscous flow, with separation of the flow at the back surface, and a wake that trails downstream.

reached by d'Alembert in 1744, based on his calculations of the flow fields over various two-dimensional body shapes. He carried out a different analysis in 1752, described in his paper *Essai d'une nouvelle théorie de la résistance des fluides.* The result was the same: zero drag. Finally, the same result was reported in yet a third paper published in 1768 in Volume V of *Opuscules mathématiques.* In that work, d'Alembert was frustrated to the point of making the following statement: "I do not see then, I admit, how one can explain the resistance of fluids by the theory in a satisfactory manner. It seems to me on the contrary that this theory, dealt with and studied with profound attention, gives at least in most cases, resistance absolutely zero: a singular paradox which I leave to geometricians to explain." Thus was born the famous "d'Alembert's paradox," which is referenced even today in all modern textbooks on aerodynamics.

Of course, d'Alembert's paradox is not a paradox at all, for his seemingly paradoxical findings were simply the correct results that naturally followed from the neglect of friction. As d'Alembert knew, there is a finite drag on any object moving through a fluid, and today we recognize that the neglect of friction was the reason that d'Alembert was not able to calculate that drag. For example, the actual flow over a circular cylinder is shown in Figure 3.12b; here, because of the mechanism of friction, the flow separates from the back of the body and forms a relatively large wake that trails downstream from the body. The recirculating, separated flow adjacent to the back surface is a low-energy flow – an almost "deadwater" region. In turn, because the flow does not close smoothly behind the body (as in the inviscid case), the pressures on the back face are lower than those on the front face, and hence a finite pressure drag is exerted on the body. Thus, d'Alembert's studies gave the theoretically proper results for an inviscid flow, and the seeming paradox arose because the original assumption of an inviscid flow was not valid.

There were experimental as well as theoretical contributions to aerodynamics in d'Alembert's work, and in that regard he had some impact on the continuing debate on the validity of the Newtonian sine-squared law. In 1777 he participated in a series of experimental measurements of the drag on ships' hulls moving in canals, work that was sponsored by the French government. As part of that work, the fundamental problem of the drag on a flat surface inclined at an angle to the flow was examined. In the publication resulting from that work, *Nouvelles expériences sur la résistance des fluides,* d'Alembert stated that the sine-squared law gave reasonably accurate results, as compared with experiment, for inclination angles of 50° to 90°, but that it was inaccurate for smaller angles, thus presenting major experimental data that called into question the validity of the sine-squared law. (That was not the first such instance, however; Borda, in 1763, was the first to disprove the sine-squared law, by measuring the aerodynamic force on a flat plate using a whirling-arm apparatus.) However, other researchers continued to use the sine-squared law for another century.

The High Noon of Eighteenth-Century Hydrodynamics: Leonhard Euler and the Governing Equations of Fluid Motion

Today there are thousands of aerodynamicists busy solving the governing equations of fluid motion for inviscid flows (flows without friction). Such methods are adequate to deal with many aspects of practical aerodynamic problems as long as drag is not being considered. These solutions may involve closed-form theoretical mathematics or, more likely today, may involve direct numerical solutions on a high-speed digital

Figure 3.13   Leonhard Euler.

computer. The governing equations that are being solved in such a "high-tech" fashion are more than two centuries old; they are called the *Euler equations* (Appendix A). The Euler equations represent a contribution to aerodynamics of much greater magnitude than any we have discussed thus far. They represent, for all practical purposes, the true beginning of theoretical aerodynamics. For that reason, practicing aerodynamicists often credit Leonhard Euler (1707–83) as the founder of fluid mechanics. That is perhaps excessive. As is almost always the case in the physical sciences, Euler benefited from earlier work, especially that of d'Alembert. On the other hand, Euler was indeed a giant in the history of aerodynamics, and his contributions tended to be revolutionary, rather than evolutionary.

Leonhard Euler (Figure 3.13) was born April 15, 1707, in Basel, Switzerland. His father was a Protestant minister who enjoyed mathematics as a pastime, and Euler grew up in a family atmosphere that encouraged intellectual activity. At the age of 13, Euler entered the University of Basel, which at that time had about 100 students and 19 professors. One of those professors was Johann Bernoulli, who tutored Euler in mathematics. Three years later, Euler received his master's degree in philosophy. Thus, three of the most influential figures in the early development of theoretical fluid dynamics (Johann and Daniel Bernoulli and Euler) lived in Basel, were associated with the University of Basel, and were contemporaries. Indeed, Euler and the Bernoullis were close friends. When Daniel Bernoulli joined the St. Petersburg Academy in Russia in 1725, he convinced the academy authorities to hire Euler as well. Euler never returned to Switzerland, although he retained Swiss citizenship throughout his life.

Euler's collaboration with Daniel Bernoulli in the development of fluid mechanics was productive during the years at St. Petersburg. It was there that Euler conceived of pressure as a point property that could vary from point to point throughout a fluid, and he derived a differential equation relating pressure and velocity. Then Euler integrated the differential equation, yielding the first derivation of what we call Bernoulli's equation. Thus the credit for Bernoulli's equation should legitimately be shared by Euler.

When Daniel Bernoulli returned to Basel in 1733, Euler succeeded him at St. Petersburg as a professor of physics. Euler was a dynamic and prolific scientist; by 1741 he had prepared 90 papers for publication and written the two-volume book *Mechanica*. The atmosphere in St. Petersburg was conducive to such achievement. Euler wrote in 1749 that "I and all others who had the good fortune to be for some time with the Russian Imperial Academy cannot but acknowledge that we owe everything which we are and possess to the favorable conditions which we had there."

However, in 1741, political unrest in St. Petersburg caused Euler to leave for the Berlin Society of Sciences, recently founded by Frederick the Great. Euler lived in Berlin for the next 25 years, where he transformed the society into a major academy. In Berlin, Euler continued his dynamic mode of working, preparing at least 380 papers for publication. There, in competition with d'Alembert, Euler formulated the basis for mathematical physics.

In 1766, after a major disagreement with Frederick the Great over some financial aspects of the academy, Euler moved back to St. Petersburg. That second stay in Russia became a time of physical suffering. In that same year, he became blind in one eye after a short illness. An operation in 1771 resulted in restoration of his sight, but only for a few days. He did not take proper precautions after the operation, and within a few days he was completely blind. However, with secretarial help he continued his work. His mind was as sharp as ever, and his spirit did not diminish. His literary output even increased, with about half of his papers being written after 1765! On September 18, 1783, Euler conducted business as usual – giving a mathematics lesson, making calculations of the motions of balloons, and discussing with friends the planet Uranus, which had recently been discovered. At about 5 p.m. he suffered a brain hemorrhage. His only words before losing consciousness were "I am dying." By 11 p.m., one of the greatest minds in history had ceased to exist.

Euler's contributions to theoretical aerodynamics were monumental. Whereas Bernoulli and d'Alembert made contributions toward physical understanding and the formulation of principles, Euler was responsible for the proper mathematical formulation of those principles, thus opening the door for future quantitative analyses of aerodynamic problems. The governing equations for an inviscid flow, incompressible or compressible, were presented by Euler in a set of three papers: *Principles of the Motion of Fluids* (1752), *General Principles of the State of Equilibrium of Fluids* (1753), and *General Principles of the Motion of Fluids* (1755). The derivation of those equations had depended on two vital concepts that Euler borrowed in total or in part from previous researchers:

(1)   The modeling of a fluid as a continuous collection of infinitesimally small fluid elements moving with the flow, where each fluid element can continually change its shape and size as it moves with the flow, but at the same time all the fluid elements taken as a whole constitute an overall picture of the flow as a continuum. That was somewhat in contrast to the individual and distinct particles in Newton's impact-theory model. The modeling of a flow by means of small fluid elements of finite size had been suggested by da Vinci,[8] though science and mathematics in da Vinci's time were not sufficiently advanced for him to capitalize on that model. Later, Bernoulli suggested that a flow could be modeled as a series of thin slabs perpendicular to the flow.[14] That was not unreasonable for the flow through a duct such as the horizontal pipe at the bottom of Figure 3.10. However, the thin-slab model lacked the degree of mobility that characterized a small fluid element that could move along a streamline in three dimensions. A major advance in flow modeling came

in 1744 with d'Alembert's moving fluid element, to which he applied the principle of mass conservation. Building on all those ideas, Euler refined the fluid-element model by considering an infinitesimally small fluid element, to which he directly applied Newton's second law expressed in a form that utilized differential calculus. Indeed, that led to the second point.

(2) The application of Newton's second law in the form of the following differential equation, a statement that force equals mass times acceleration:

$$F = M \frac{d^2 x}{dt^2}$$

In this differential equation, F is the force, M is the mass, and $d^2 x / dt^2$ is the linear acceleration (i.e., the second derivative of the linear distance x). Today, that is the most familiar form of Newton's second law; it was first expressed in that form by Euler in his paper *Discovery of a New Principle of Mechanics* (1750).

Using the foregoing two concepts, Euler derived the equations that today carry his name and provide the foundation for large numbers of modern aerodynamic analyses. Appendix A shows the set of partial differential equations that are known as the *Euler equations*. The first two sets of equations (the continuity and momentum equations) were derived in this form by Euler in his 1753 paper, considered one of his finest works.[15] The energy equation (Appendix A) was not treated by Euler; it came later, in the nineteenth century, with the development of the science of thermodynamics. However, for analysis of an inviscid, incompressible flow, the continuity and momentum equations are sufficient; for such a flow, the energy equation is redundant. And, of course, until the development of high-speed aircraft in the middle of the twentieth century, all problems in aerodynamics were basically treated as incompressible.

Some Polishing Touches: Lagrange and Laplace

Euler created the mathematical tools for theoretical aerodynamics, but using those tools was quite another matter. From a mathematical point of view, the Euler equations (Appendix A) compose a coupled system of nonlinear partial differential equations for which, to this very day, no one has found a general analytical solution. On the other hand, it was recognized as early as 1788 by Joseph-Louis Lagrange (1736–1813) that if the Euler equations should prove to be integrable (i.e., if the partial differential equations could be solved for algebraic formulas involving velocity, pressure, and density as functions of space and time throughout the flow field), then the behavior of a moving fluid could be determined "for all circumstances." Although Euler derived the governing equations of motion for an inviscid fluid, *he never solved those equations for any meaningful application*. However, close on Euler's heels came two mathematicians who took a slightly different tack, that of simplifying Euler's equations to obtain *approximate* equations that could be solved. One of those men was Lagrange.

Joseph Lagrange was born in Turin, Italy, on January 25, 1736. Self-taught in mathematics, he began to correspond with Euler at the age of 19. By the time he was 30, Lagrange had taken the post vacated by Euler in Frederick the Great's Berlin Academy. After Euler's death, Lagrange was considered the world's leading mathematician. In 1787 he moved to Paris, where he spent the rest of his life. He commanded such respect that he was virtually unaffected by the political turmoil of the French Revolution.

Lagrange made two contributions to the theory of fluid flow, both introduced in his book *Mécanique analytique* (1788):

(1)   A new model for fluid flow. Lagrange concentrated on a moving fluid element and wrote the governing equations in such a way as to calculate the pressure and velocity for that element as a function of time as it moved through space. The solutions appear in the form of these timewise variations for the fluid element being examined (i.e., each element is "tagged," and its instantaneous location in space is calculated). That approach is called the *Lagrangian method* in fluid dynamics. It is in contrast to the *Eulerian method,* which applies the physical principles to a fluid element, but with the equations in a form that allows the calculation of p and V directly as functions of x–y–z space and time. In modern aerodynamics, the Eulerian method is almost always employed; however, the Lagrangian method is somewhat more popular with physicists for calculating the properties of wave propagations from nuclear explosions.

(2)   Introduction of the *velocity potential* ϕ and the *stream function* ψ. Both ϕ and ψ are specially defined functions from which the velocity can be obtained by differentiation using differential calculus. When the Euler equations are combined and reworked such that ϕ or ψ is introduced and becomes the variable to be calculated, certain simplifications are obtained for some (but not all) types of flows. Both ϕ and ψ are used frequently in modern theoretical aerodynamics, and their introduction by Lagrange was a notable contribution to theoretical aerodynamics. They ultimately allowed the calculation of V (by differentiating ϕ or ψ with respect to distance) and, for a low-speed flow, the calculation of p through Bernoulli's equation.

Although the contributions by Lagrange were not of the same magnitude as those by Euler, they serve the purpose of "polishing" Euler's work, of providing new concepts in order to make solutions for the Euler equations more easily obtainable.

In that same category was the work of Pierre-Simon Laplace (1749–1827). Laplace's contributions to the state of the art of aerodynamics were quite oblique, but nonetheless important. The importance of Laplace's work for general progress in the sciences was summed up by Gillispie,[16] who stated that "Laplace was among the most influential scientists in all history." Laplace made substantial scientific contributions in regard to celestial mechanics, the motions of comets, the theory of heat, and the analysis of tides. His most important work was in mathematics, especially in the solution of partial differential equations. Much of his work is still taught today in courses dealing with advanced calculus.

Laplace was born in Normandy, France, on March 23, 1749. His family was of moderate means, their income being derived from the father's cider business. In 1766, Laplace became a student at the University of Caen, taking only two years to graduate. There he developed an intense interest in mathematics, a field in which he proved to be a genius. After his graduation, he was introduced to d'Alembert in Paris. Challenged by d'Alembert to solve a difficult mathematics problem within a week, Laplace returned the next day with the solution. Suitably impressed, d'Alembert was instrumental in Laplace's appointment as a professor of mathematics at the Ecole Militaire in Paris (Laplace was only 19 at the time). Five years later he was elected to the Academy of Sciences. His prolific output on various important problems in mathematics and physics ensured his fame; he was so highly respected that, like Lagrange, he was virtually untouched by the French Revolution. In 1788 he married Marie-Charlotte de Courly de Romanges, a woman 20 years younger than he. They had two children, a son who became a general in the French army, and a daughter who married a marquis and died in childbirth. Laplace's fame attracted the notice of Napoleon, who made him minister of the interior. He lasted only six weeks. Regarding Laplace's administrative abilities, Napoleon wrote that "he could never get a grasp on any question in its true significance; he sought everywhere for subtleties, had only problematic ideas, and in short carried the spirit of the infinitesimal into administration."[16] An obvious case of a square peg in a round hole, that was one of many examples in which men of scientific and mathematical genius have floundered when taken outside their environment. Using Laplace more for public relations than for substance, Napoleon appointed him to the Senate and made him chancellor of that body in 1803. In his later years, many of Laplace's theories in physics were challenged by his contemporaries, and his reputation began to suffer. By the time of his death in Paris, March 5, 1827, his circle of disciples had shrunk to a small fraction of what it had been during his earlier years.

Laplace's contributions to aerodynamics were only two – one general and one specific:

(1) His development of a theory for solution of partial differential equations helped to move the field closer to solutions for the Euler equations, although Laplace never addressed the Euler equations directly. He was responsible for one of the most famous equations in mathematical physics:

$$\frac{\partial^2 G}{\partial x^2} + \frac{\partial^2 G}{\partial y^2} + \frac{\partial^2 G}{\partial z^2} = 0$$

This equation first appeared in 1789 in a paper presented to the Paris Academy of Sciences in which Laplace reported his studies of the rings of Saturn by modeling them as infinitely thin layers of fluid in equilibrium under gravitational forces, wherein G denoted what later was recognized as a gravitational potential. However, for the flow of an inviscid (frictionless), incompressible fluid, the stream function ψ, as defined by Lagrange, satisfied Laplace's equation:

$$\frac{\partial^2 \psi}{\partial x^2} + \frac{\partial^2 \psi}{\partial y^2} = 0$$

Moreover, if the flow was irrotational (i.e., if the fluid elements of the moving fluid were only translating through space and were *not* rotating at the same time), the velocity potential ϕ, also defined by Lagrange, satisfied Laplace's equation:

$$\frac{\partial^2 \phi}{\partial x^2} + \frac{\partial^2 \phi}{\partial y^2} + \frac{\partial^2 \phi}{\partial z^2} = 0$$

Suddenly, Laplace's equation became a governing equation for inviscid, incompressible flow. Because mathematically it was a linear equation, it was much easier to solve than the complete nonlinear system of Euler's equations. However, once again a major impediment arose: Solution of Laplace's equation for practical aerodynamic shapes (airfoils, wings, etc.) was still difficult. Moreover, any solution for even a simple shape, such as a circular cylinder, continued to predict zero drag on the body – d'Alembert's paradox seemed to persist.

(2) Laplace made an important contribution to aerodynamics by being the first person to properly calculate the speed of sound in a gas, an important parameter in

calculations concerning high-speed flows (compressible flows) of gases. All calculations in supersonic aerodynamics today depend on correct knowledge of the speed of sound. The speed of sound in air had been measured in Isaac Newton's time: A cannon was fired, and the time required for the sound to be heard a long distance away was measured. With the distance between the observer and the cannon being known, the speed of sound could then be calculated. Newton calculated the speed of sound assuming that the air temperature was constant throughout the sound wave (an isothermal process). That was not a correct assumption, and as a result his calculations were wrong. That situation prevailed for almost a century, until Laplace made the correct assumption that the sound wave created a change in the gas temperature, but that the overall energy remained the same (an adiabatic process). Laplace was able to calculate a value for the speed of sound that agreed with the experimental data. For those readers interested in the equations for the speed of sound (denoted by a), Newton recognized that

$$a^2 = \frac{dp}{d\rho}$$

where $dp/d\rho$ is the change in pressure within the wave per unit change in density. For Newton's assumption of an isothermal process, that equation becomes

$$a = \sqrt{\frac{p}{\rho}}$$

which is incorrect. With Laplace's assumption of an adiabatic process within the sound wave, we have

$$a = \sqrt{\gamma \frac{p}{\rho}}$$

where γ is the ratio of the specific heat at constant pressure, c_p, to the specific heat at constant volume, c_v:

$$\gamma = \frac{c_p}{c_v}$$

This is the correct formulation to determine the speed of sound. An experimental measurement by J. L. Gay-Lussac and J. J. Welter in 1822 produced the value $\gamma = 1.3748$, very close to today's value of $\gamma = 1.4$.

The Rise of Experimental Aerodynamics

Quantitative experiments in aerodynamics did not begin until the late seventeenth century. The flow patterns observed and recorded by da Vinci (e.g., Figure 2.7) had represented a qualitative contribution of a sort. Over the ensuing centuries, however, no progress was made until the important force measurements of Mariotte in 1673, wherein the dependence of the aerodynamic force on the square of the velocity was first proved.

Progress in experimental aerodynamics resumed on November 12, 1732, at the Academy of Sciences in Paris, when Henri Pitot announced a new invention by which he could directly measure the local flow velocity at a point in a fluid. Later called the Pitot tube, this device has become the most commonplace instrument in modern aerodynamics laboratories.

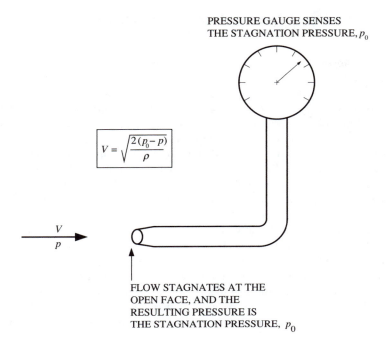

PRESSURE GAUGE SENSES
THE STAGNATION PRESSURE, p_0

$$V = \sqrt{\frac{2(p_0 - p)}{\rho}}$$

V

p

FLOW STAGNATES AT THE
OPEN FACE, AND THE
RESULTING PRESSURE IS
THE STAGNATION PRESSURE, p_0

Figure 3.14 Schematic of a Pitot tube.

A schematic of a Pitot tube is shown in Figure 3.14. It consists of a hollow tube, with its open face oriented perpendicular to the flow. The back of the tube is bent at a right angle; it then passes out of the flow and is connected to a pressure gauge; hence, most Pitot tubes are L-shaped. When a Pitot tube is first placed in a flow, during the first few milliseconds the fluid will surge into the hollow tube under the inertia effect of its velocity V. However, because the tube is closed off at its other end by a pressure gauge, the surging fluid soon will have no place to go. It then will stagnate and become "impacted" inside the tube. Additional fluid trying to enter the open face of the tube will encounter the impacted fluid inside the tube as an obstruction; the fluid outside the face of the tube will have to slow down as it approaches the open end, its velocity falling to zero right at the open face, just as if the tube were a solid rod. As fluid approaches the open end of the tube and slows down, its pressure will increase; the maximum pressure will be reached when the fluid velocity stagnates to zero at the open face of the tube. That pressure is called the *stagnation pressure* (or sometimes the total pressure). The stagnation pressure is transmitted through the fluid inside the tube up to the pressure gauge. Hence the pressure gauge on a Pitot tube registers the stagnation pressure of the fluid. Recall Bernoulli's equation for two points in a flow:

$$p_1 + \tfrac{1}{2}\rho V_1^2 = p_2 + \tfrac{1}{2}\rho V_2^2$$

If we let point 1 denote the flow properties ahead of the tube, and point 2 denote the properties after the fluid has stagnated at the face of the tube, then $v_2 = 0$ and $p_2 = p_0$, where p_0 denotes the stagnation pressure. Then Bernoulli's equation, when solved for V_1, yields

$$V_1 = \sqrt{\frac{2(p_0 - p_1)}{\rho}}$$

If we know the values for p_1 and ρ by other means, then the stagnation pressure p_0, as measured by the Pitot tube, allows us to calculate the flow velocity V_1. Thus the Pitot tube is a highly practical device for measurement of flow velocity.

In 1732, Henri Pitot did not have the benefit of Bernoulli's equation, which would be obtained by Euler 20 years later. Pitot's reasoning about the operation of his tube was purely intuitive, and he was able to correlate by *empirical means* the flow velocity corresponding to the measured difference between the stagnation pressure p_0, as measured by his Pitot tube, and the flow static pressure p_1, as measured by a straight tube inserted vertically in the fluid, with its open tube face parallel to the flow (not perpendicular as in the case of the Pitot tube). As discussed in Section 4.21 of *Introduction for Flight*[17] the proper application of Bernoulli's equation to determine a velocity from a Pitot measurement of stagnation pressure was not presented until 1913, when John Airey[18] published an exhaustive experimental study of the behavior of Pitot tubes and a rational theory for their operation, based on Bernoulli's equation. The Pitot tube, invented in the early part of the eighteenth century, was not properly incorporated into aerodynamics as a viable experimental diagnostic tool for two centuries.

The invention of the Pitot tube was just one event in the reasonably productive life of Henri Pitot (1695–1771) (Figure 3.15). Born in Aramon, France, on May 3, 1695, his youth was undistinguished; indeed, he intensely disliked academic studies.[19] While serving a brief time in the military, Pitot become interested in a geometry text purchased in a bookstore and subsequently spent three years at home studying mathematics and astronomy. In 1718 Pitot moved to Paris, and by 1723 he had become an assistant in the chemistry laboratory of

Figure 3.15   Henri Pitot.

the Academy of Sciences. It was before the academy that he announced his new device for measuring flow velocity: the Pitot tube. That invention was motivated by his dissatisfaction with the existing technique for measuring the flow velocity of water, which was to observe the progress of a floating object on the surface of the water. So he devised an instrument consisting of two tubes; one was simply a straight tube, open at one end, that was inserted vertically into the water (to measure the static pressure p), and the other was a tube with one end bent at a right angle, with the open end facing directly into the flow (to measure the total pressure p_0). From a bridge over the Seine River in Paris, he used the instrument to measure the flow velocity at different depths within the river. His presentation to the academy later that year had importance beyond the Pitot tube itself. Contemporary theory, based on the experience of some Italian engineers, held that the flow velocity at a given depth in a river was proportional to the mass above it; hence the velocity was thought to increase with depth. Pitot reported his stunning (and correct) findings that in reality the flow velocity decreased as the depth increased, thus introducing his new invention with flair. In his old age, Pitot retired to his birthplace, and he died at Aramon on December 27, 1771.

Experimental aerodynamics in the eighteenth century was driven by the work of four principal contributors: Henri Pitot and Jean-Charles Borda in France and Benjamin Robins and John Smeaton in England. The work of Pitot has already been examined; let us evaluate the impact of the others in more or less chronological order.

In 1746, the current understanding of the aerodynamic forces on bodies in flight was summarized by an English military engineer, Benjamin Robins (1707–51), who stated in a paper presented to the Royal Society (as quoted from Pritchard)[20] that "all the theories of resistance hitherto established are extremely defective, and that it is only by experiments analogous to those here recited, that this important subject can even be completed."

Robins was reacting to the situation that the advances being made in theoretical aerodynamics were not immediately contributing to practical calculations of aerodynamic forces to any reasonable accuracy. Important theoretical tools were being developed, but no one knew how to use them properly. The Newtonian sine-squared law was essentially the only practical tool available for calculating pressure distributions over the surface of a body, and its validity had been called into question. It was time for some serious experimental advances to fill the void, and Benjamin Robins took the lead in that respect.

Robins is virtually unknown to most practicing aerodynamicists. He was born in Bath, England, in 1707 to Quaker parents, but he never espoused the Quaker pacifist philosophy. He never married. He studied to be a teacher, but soon gave that up to pursue a scientific career as a military engineer. He was intensely interested in mathematics, and in 1727 he published a paper in the *Philosophical Transactions of the Royal Society* regarding demonstration of the eleventh proposition of Newton's treatise on quadratures (*De Quadrature*). That was the year of Newton's death. Robins became a strong supporter of the argument that Newton deserved priority over Leibniz for development of the calculus; he attacked, in writing, Leibniz and several of the Bernoullis who were perceived to be enemies of Newton in the calculus controversy. In addition to his interest in mathematics, Robins was an experimentalist, and it was in that respect that he made his lasting contributions. Toward the end of his relatively short life, Robins became interested in the use of rockets for military signaling. Robins traveled to India for the British East India Company to renovate some fortifications and died at St. David, India, on July 29, 1751.

Robins's contributions to experimental aerodynamics centered around two testing devices he invented: (1) a whirling arm for measuring aerodynamic forces at low speeds

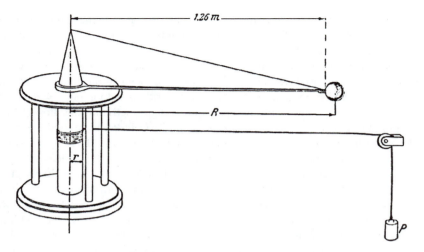

Figure 3.16   Whirling-arm mechanism designed by Benjamin Robins.

and (2) a ballistic pendulum for studying the aerodynamic characteristics of bodies at high speeds. Figure 3.16 shows a device in which an aerodynamic shape (a sphere) is mounted at the end of a long arm. The other end of the arm is attached to a shaft, which is rotated by a falling weight attached to the shaft via a cable-and-pulley system. As the arm rotates, the aerodynamic body moves through the air at some relative velocity and experiences an aerodynamic resistance that can be measured. However, the whirling arm had one distinct drawback: After some period of operation, the air in the vicinity of the whirling arm would start to rotate in the same direction as the arm, and it became difficult to determine the relative velocity between the moving body and the moving air. That diminished the accuracy of any force measurement as a function of relative velocity. Whirling arms are no longer used in aerodynamics, their function having been taken over by wind tunnels, which were developed in the late nineteenth century. Nevertheless, Robins's whirling arm was the only device for direct measurement of aerodynamic force in the eighteenth century. His invention of the ballistic pendulum was equally novel: A projectile was fired into a massive pendulum, the deflection of which was a measure of the momentum (hence velocity) of the projectile.

Using those two devices, Robins conducted extensive aerodynamic testing that contributed to the state of the art as follows:

(1) Robins verified Mariotte's seventeenth-century finding that aerodynamic force varies with the square of the relative velocity between a body and the airstream (at speeds less than the speed of sound).

(2) Robins was the first to show that two aerodynamic bodies with different shapes, but with the same frontal area, have *different* drag values. In Robins's time, the prevailing intuition was that drag was determined mainly by the frontal area of a body, with overall shape being secondary. Today we know better: The total shape of the body is critical for determining whether or not the flow will separate from the surface, and flow separation can dominate the production of drag on a body. (Realization of the role of flow separation did not come until later in the

eighteenth century, with the work of Borda.) In particular, Robins tested pyramid shapes, first with the apex forward toward the incoming flow, and then with the flat base forward. The latter case produced more drag. He also tested oblong flat plates at a 45° angle of attack, first with the long side as the leading edge, and then with the short side as the leading edge. He found that the drag values were quite different, with the latter case producing more drag. (Appreciation of aspect-ratio effects on wing aerodynamics began with the work of Wenham in the late nineteenth century.)

(3)   Robins observed that projectiles spinning on their axes of travel experienced side forces that made them deviate from a straight-line path. Today we know this as the Magnus effect; it is the reason that a spinning baseball will curve.

(4)   Robins was the first to report the large increase in drag associated with speeds near the speed of sound. Today this is called the *transonic drag rise*. During his ballistic-pendulum measurements, Robins observed that when projectiles were moving at speeds near the speed of sound, the aerodynamic force began to vary as the velocity *cubed* (not velocity squared as in the lower-speed cases). It is indeed amazing that the transonic drag rise was observed as early as the eighteenth century. For that discovery alone, Robins deserves a prominent place in the history of aerodynamics.

The courteous interactions among various researchers in the eighteenth century, such as the amicable relations among Daniel Bernoulli, Euler, and d'Alembert, suggest an unusually calm scientific landscape. That attitude toward interaction carried over to Robins, though it was not always to his benefit. Robins's aerodynamic experiments were reported in only two publications: *New Principles of Gunnery Containing the Determination of the Force of Gunpowder and Investigation of the Difference in the Resisting Power of the Air to Swift and Slow Motions* (1742); "Resistance of the Air and Experiments Relating to Air Resistance," in *Philosophical Transactions of the Royal Society* (1746). Those works were widely read. Indeed, Euler was so excited about Robins's book that he personally translated it into German in 1745, adding some commentary. In 1751, the year of Robins's death, it was translated into French. Euler's interest in Robins's work was both a hindrance and a help. The hindrance concerned Robins's observation of the side force exerted on a spinning projectile moving through the air. Euler considered that to be a spurious finding, due to manufacturing irregularities in the projectile. Recognized as the dominant hydrodynamicist of the eighteenth century, Euler far overshadowed Robins, and thus Robins's finding was not taken seriously for another century, until Gustav Magnus (1802–70) verified the phenomenon as a real aerodynamic effect. On the other hand, Robins's discovery of the transonic drag rise was praised by Euler, who agreed completely with Robins's data. Both Euler and Robins suggested that the use of bodies with small frontal areas could reduce the drag rise, thus presaging our modern understanding that the use of sharp, slender bodies tends to delay and reduce this drag in the transonic regime.

Eight years after Robins's death, another Englishman made a contribution to aerodynamics that was to have an important but controversial impact lasting into the twentieth century. John Smeaton (1724–92), a professional civil engineer, was largely responsible for making engineering a respected endeavor in the eyes of British society. Although Smeaton was responsible for only one contribution to experimental aerodynamics, it was an important development. John Smeaton was born in Austhorpe, England, June 8, 1724, a descendant of a long line of Scots. His father was a lawyer, and Smeaton began his education in law. However, he quickly found that his talents were for mechanical matters, and with the approval of his family he became a successful maker of scientific instruments. The Industrial Revolution was beginning in England, creating a demand for massive civil engineering projects. Smeaton took advantage of the new opportunities; he designed and constructed several harbors in England and established a reputation as a structural engineer. His outstanding accomplishment in that area was the rebuilding of the Eddystone lighthouse after the failure of two previous contractors – an achievement that earned him general fame in England. He became a fellow of the Royal Society and a charter member of the first professional engineering society, the Society of Civil Engineers, which after his death became known as the Smeatonian Society. Late in his life he became interested in matters more scientific ("natural philosophy," as science was denoted in those days). He studied the significance of the product of the mass times the velocity of a body (defined in modern terms as momentum) vis-à-vis the product of the mass times the square of the velocity (called *vis viva* in the eighteenth century, and today known as twice the kinetic energy).

In 1759, Smeaton was awarded the Copley Medal by the Royal Society, the same prize won by Robins 12 years earlier. The work for which he received that medal also earned Smeaton a place in the history of aerodynamics. There were more than 10,000 windmills in England at that time, as well as numerous mills driven by water power, and Smeaton carried out experiments on the forces exerted by air and water on the vanes of such devices. For those experiments, Smeaton adapted Robins's invention of the whirling arm, such that the windmill blades at the end of the arm not only were moved through space by the movement of the arm but also rotated, thus simulating the actual operation of a windmill (Figure 3.17). The windmill blades at the end of the whirling arm were spun by a cable-and-pulley mechanism activated by a falling weight. Smeaton published his experimental findings in the *Philosophical Transactions of the Royal Society* (1759). An English historian of aeronautics, J. L. Pritchard,[20] suggested that Smeaton may have been the first to observe the effect of camber, that is, to observe that lift is greater for flow over a curved surface than for flow over a flat surface at the same angle of attack. In his 1759 paper, Smeaton wrote that "when wind falls upon a concave surface, it is an advantage to the power as a whole." That observation apparently was lost on his contemporaries, for the same phenomenon was reported in more detail by George Cayley in 1809, without reference to the earlier work of Smeaton.

In addition to his observation of the effect of camber, in that 1759 paper Smeaton included a table of measurements of aerodynamic force on a flat surface perpendicular to the flow, tabulated versus wind velocity. The correlation for those measurements was given by Smeaton as

$$F = kSV^2$$

where F is the force in pounds exerted on the perpendicular surface, S is the surface area in square feet, V is the wind velocity in miles per hour, and k is a constant of proportionality. The numerical value reported by Smeaton was $k = 0.005$. The constant k became known as *Smeaton's coefficient,* and the value of 0.005 was used by many workers through the end of the nineteenth century. However, the accuracy of that value was questioned as early as 1809 by George Cayley. Indeed, the inaccuracy of Smeaton's coefficient had a major adverse impact on the early work of the Wright brothers.

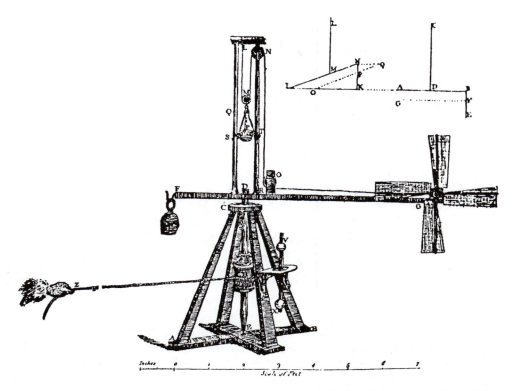

Figure 3.17 Smeaton's whirling-arm device with windmill blades being tested at the end of the arm.

Let us examine Smeaton's coefficient from a modern perspective. A sketch of a flat plate perpendicular to a flow is shown in Figure 3.18. This figure illustrates two situations. In Figure 3.18a we make the idealized assumption that the free-stream flow will stagnate over the entire surface of the plate; hence the pressure on the front surface of the plate will be the stagnation pressure. The flow will move around the edges of the plate, and there will be a large region of low-energy, recirculating, separated flow behind the plate. As a result, the pressure exerted on the rear surface of the plate will be much lower than the stagnation pressure on the front surface; indeed, the rear pressure will be on the order of the original free-stream pressure, p_∞. We see that the high pressure on the front of the plate (p_0) and the low pressure on the rear of the plate (p_∞) will result in a drag force on the plate. If the surface area of the plate is S, then the force F on the plate is

$$F = (p_0 - p_\infty)S$$

On the other hand, from Smeaton's paper we have

$$F = kSV_\infty^2$$

Combining the two equations, we have

$$(p_0 - p_\infty)S = kSV_\infty^2$$

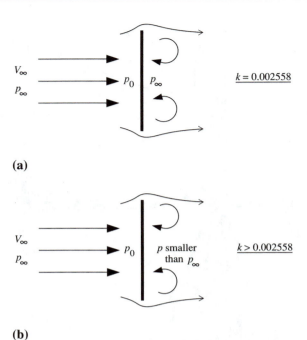

(a)

(b)

Figure 3.18   Flow over a flat plate oriented perpendicular to the flow: (a) The pressure on the back face is equal to the free-stream pressure. (b) The pressure on the back face is less than the free-stream pressure.

or

$$k = \frac{p_0 - p_\infty}{V_\infty^2}$$

Now let us return to Bernoulli's equation, which is valid for the low-speed, incompressible-flow conditions that prevailed for Smeaton's work. From Bernoulli's equation, the stagnation pressure is

$$p_0 = p_\infty + \tfrac{1}{2}\rho V_\infty^2$$

or

$$p_0 - p_\infty = \tfrac{1}{2}\rho V_\infty^2$$

Putting this result into the foregoing expression for k, we have

$$k = \frac{p_0 - p_\infty}{V_\infty^2} = \frac{\tfrac{1}{2}\rho_\infty V_\infty^2}{V_\infty^2} = \frac{1}{2}\rho_\infty$$

In other words, k is simply half the free-stream density. In Bernoulli's equation, a consistent set of units must be used; that is, for the English engineering system of units, V must be in feet per second (not miles per hour as used by Smeaton), and density in slugs per cubic foot. In those units, the standard sea-level value for ρ_∞ is 0.002377 slug per cubic foot. That will give a value of $k = 0.001189$ for that system of units. However, if V is in miles per hour (as Smeaton had it), then k must be converted. A convenient rule of thumb is that 60 mph is the same as 88 ft/s. Thus, converting k to pertain to miles per hour, we have

$$k = 0.001189 \left(\frac{88}{60}\right)^2 = 0.002558$$

where the squared term derives from the velocity-squared term in the original equation given by Smeaton. This value for k is about half that originally given by Smeaton – a glaring discrepancy. However, that is not the whole story. Returning to Figure 3.18a, recall that we made the assumption that the pressure on the rear surface was the free-stream pressure p_∞. Modern aerodynamic experiments for low-speed flows have shown that the pressure in a separated region is slightly *less* than p_∞ (Figure 3.18b). The actual pressure on the rear surface is a function of the Reynolds number (a parameter to be discussed later), which is defined as $\rho_\infty V_\infty h / \mu_\infty$, where h is the height of the plate, and μ_∞ is the viscosity coefficient. Different experimental conditions will involve different values for the Reynolds number and hence will yield different values for the pressure on the rear surface of the plate. That, in turn, will result in different measured values for Smeaton's coefficient, and many values for Smeaton's coefficient have been reported over the past two centuries. Pritchard,[20] secretary of the Royal Aeronautical Society during the middle of the twentieth century, reported a value for Smeaton's coefficient based on modern experimental data, formally agreed upon by the Royal Aeronautical Society:

$$k = 0.00289$$

In modern aerodynamics, problems are no longer formulated explicitly in terms of Smeaton's coefficient, so the controversy is now moot. But for workers such as Cayley, Lilienthal, and Langley in the nineteenth century and the Wright brothers at the beginning of the twentieth century, the use of Smeaton's coefficient was routine, and its value was a matter of serious concern.

The table of aerodynamic forces and velocities in Smeaton's 1759 paper was not compiled by Smeaton. Rather, it was sent to him by a friend, a Mr. Rouse, who had carried out his own experiments using a whirling arm. Smeaton clearly acknowledged Rouse's input, but users of those tables always referred to them as Smeaton's tables.

Another user of the whirling arm as an aerodynamic testing device was Jean-Charles Borda (1733–99). Although Borda's name is virtually unknown to modern aerodynamicists, his work yielded several fundamental advances in the understanding of applied and basic aerodynamics. Borda was born May 4, 1733, in Dax, France, into a family of nobility, the tenth child in a family of 16 children. He studied at the Collège des Barnabites at Dax, continued at the Jesuit Collège de la Flèche, and entered the Ecole du Génie de Mezieres in 1758, finishing a two-year course in one year. Schooled in mathematics, Borda scorned religion, and he never married.[21] He balanced productive technical and scientific activities with a career in the French navy. He was elected a member of the Paris Academy of Sciences in 1756. He participated in the American Revolutionary War and later was taken prisoner by the English while commanding a flotilla of ships in the Antilles in 1782. After the French Revolution, he played a leading role in developing the metric system of units and was responsible for coining the word "meter." He contributed to the calculus of variations and was recognized as a skillful synthesizer of calculus and experiment. Borda died in Paris on February 19, 1799.

Borda's contributions to experimental aerodynamics derived primarily from his rotating-arm experiments, as reported in a paper presented to the Paris Academy of Sciences in 1763:

(1)   He verified, once again, that aerodynamic force varies as the square of the velocity, thus adding to the body of evidence that began accumulating with the work of Mariotte in the seventeenth century.

(2)   Regarding the question of the variation of aerodynamic force with the angle of attack, Borda's experiments with a flat plate showed that such force varied as the sine of the angle of attack (*not* the sine squared as indicated by the Newtonian calculation). That is, Borda demonstrated experimentally that

$$R \propto V^2 \sin \alpha$$

where R is the aerodynamic force, and α is the angle of attack. That was the first published experiment to disprove the Newtonian sine-squared law as applied to a lifting surface. Within 15 years, additional experimental work directed by d'Alembert and sponsored by the French government reinforced the findings of Borda. Experiments in aerodynamics were beginning to raise real doubt about the validity of the sine-squared law.

(3)   Borda was the first to show the effect of aerodynamic interference. Previously, intuition had led to the belief that the combined drag on two closely spaced bodies would be the sum of the two individual drag values. Borda put two spheres near each other on his whirling arm and found that the total drag was different from the sum of the individual drag values for the two spheres tested separately. That became known as the Borda effect.[8]

(4)   Borda studied flows through ducts of various shapes, in an attempt to examine the validity of the continuity equation first established by da Vinci:

$$AV = \text{constant}$$

In particular, he found that in some cases in which the flow experienced a *sudden* increase in area, the continuity equation did not accurately describe the change in velocity. He noted that the flow did lose some of its *vis viva* (i.e., kinetic energy), but not to the extent predicted by the continuity equation. Today we recognize that the flow was separating from the walls of the duct in the region of the sudden area increase, and as a result the effective *stream-tube area* of the flow experienced a smaller area change. In modern aerodynamic applications, separated flow sometimes plays a dominant (and usually detrimental) role. Borda was the first to describe the *effect* of flow separation, and for that reason his finding became known as Borda's theorem (the existence of flow separation had been known qualitatively from early times, as shown by the sketches of da Vinci, Figure 2.7).

Interim Summary: The State of the Art of Aerodynamics Inherited by George Cayley

George Cayley's 1804 hand-launched glider (Figure 3.1) was presented as the symbolic aircraft for this chapter. We are now in a position to assess the state of the art inherited by Cayley as he began the major part of his work. In regard to theoretical aerodynamics, Cayley had available to him the following equations and concepts:

(1)   He had the continuity and momentum equations for an inviscid flow – the Euler equations – but at that time no one had been able to solve them. They were in the form of partial differential equations, precisely the form we use in modern aerodynamics. Those equations are all that one needs for analysis of low-speed flows. (For high-speed flows, one must also have the energy equation, which was developed in the late nineteenth century.) Although the Euler equations do not account for the effects of friction, solution of those equations will provide accurate pressure distributions over aerodynamic bodies and hence allow accurate calculations of lift. Unfortunately, Cayley could not take advantage of the Euler equations in that regard, because in 1804 there had been no solutions for the equations. The theoretical tools were there, but nobody knew how to use them.

(2)   The Newtonian sine-squared law was well known by Cayley. That law allowed for a simple calculation of the pressure distribution over a body and therefore provided Cayley the means to make predictions of both lift and pressure drag. However, even Newton had some doubt about the validity of the sine-squared law, and such doubt was reinforced by the whirling-arm experiments of Borda in 1763 and by the resistance measurements on ships' hulls conducted by d'Alembert in 1777.

(3)   In regard to the calculation of skin-friction drag (the major contribution to the total drag on a streamlined body), there were no theoretical tools at all, only a complete void, and that was a source of great frustration for Cayley and subsequent nineteenth-century investigators.

If the situation in the theoretical realm was less than promising, that in experimental aerodynamics was not much better. However, certain fundamental empirical findings were available to Cayley:

(1)   By 1804 it had become established fact that aerodynamic force varies directly as the air density, the cross-sectional area of the body, and the square of the velocity. Various pieces of those relationships had begun to fall into place as far back as da Vinci's time, with his statement that force varied with area, and that articulation continued with Galileo's experimental finding that force varied with the air density, culminating with the pioneering studies of Mariotte showing that the force varied as the square of the velocity.

(2)   The main concern of both theoretical and experimental researchers at that time was aerodynamic *drag*, rather than lift. People wanted to understand and be able to predict the *resistance* experienced by a moving body (a projectile through the air, or a ship's hull through the water). Only a few investigators were interested in examining the total aerodynamic force on a flat surface inclined at a small angle to a flow, and even there the primary interest was in the drag component. In terms of the angle-of-attack effects on the resultant aerodynamic force R, from which both drag and lift can easily be obtained, experimental data focused on the validation (or refutation) of the Newtonian sine-squared law. Borda's data in 1763 led him to conclude (correctly) that for small angles of attack, R varied as the sine of the angle of attack, not the sine squared, and 14 years later d'Alembert confirmed that experimentally. Therefore, by 1804, as Cayley was immersed in his basic aeronautical experiments, there were empirical data that could be used for prediction of angle-of-attack effects on aerodynamic lift, although much skepticism prevailed.

(3) Developments in experimental aerodynamics began to become more complex in
     the eighteenth century: the invention of the Pitot tube for velocity measurements,
     Robins's invention of the whirling arm and the ballistic pendulum for measure-
     ments of aerodynamic force, Robins's recognition of the transonic drag rise, and
     Borda's discovery of the aerodynamic interference between two bodies. However,
     in the eighteenth century, such increasing sophistication was not directly translated
     into the design of flying machines.

The thinking and the work of George Cayley, beginning with his conception of the modern-
configuration airplane in 1799 (Figure 3.2), were carried out in an intellectual atmosphere
of dawning enlightenment in regard to aerodynamics, but that atmosphere was still too
rarefied to have provided Cayley with many useful, practical tools. Therefore, in regard to
many of the ideas associated with the design of his 1804 glider (Figure 3.1), Cayley was on
his own. Biographies of Cayley are available,[22] and the definitive biography by Pritchard[23]
is required reading for any serious student of the history of aeronautics.

George Cayley was born December 27, 1773, in Scarborough, England. His mother,
Isabella Seton Cayley, was from a well-known Scottish family descended from Robert
Bruce. His father, Sir Thomas Cayley, was descended from the Normans who invaded
England in 1066. Because of the chronic ill health of his father, Cayley's parents spent
much time abroad, and Cayley's early days were spent at the family's house at Helmsley,
where he enjoyed much freedom. There he acquired an early interest in mechanical devices,
and he was frequently to be found in the company of the village watchmaker. After the death
of his grandfather (the first Sir George Cayley), the extensive family estate at Brompton
passed to George's father, who lived for only another 18 months, and then to the young
George. By 1792, Sir George Cayley had become the sixth baronet at Brompton Hall, at the
age of 19 years. He was to spend the rest of his life as a moderately prosperous Yorkshire
country squire.

As was not unusual in the eighteenth century, Cayley had virtually no formal education.
There is some evidence that he went to school briefly in York, but his main education came
from two influential tutors: (1) George Walker, a mathematician of high reputation, a fellow
of the Royal Society, and a man of considerable intellect; (2) George Morgan, a Unitarian
minister, scientist, and lecturer on electricity. Both tutors were freethinkers, and they had
a major impact on the wide breadth of education and open-mindedness acquired by Cayley
during the first 25 years of his life. Cayley never lost his enthusiasm for knowledge and
invention; by the early 1800s he was recognized as one of England's leading scholars in matters of science, technology, and social ethics. Captivated by the beauty and intelligence of
his first tutor's daughter, Sarah Walker, Cayley fell deeply in love. In 1795 the young couple
entered into a marriage that was to last for 62 years, ended only by Cayley's death in 1857.

George Cayley's Aerodynamics

Cayley's contributions to aeronautics also began early in his life. At the age of
19 years he designed and tested a simple helicopter-like model made of wood and feathers
and powered by a string-and-bow mechanism. That was the harbinger of a considerable
output of aeronautical thought, research, and invention to follow over the next 18 years. The
concept of the modern airplane – a machine with fixed wings, a fuselage, and horizontal and
vertical tail structures – was first advanced by Cayley in 1799, etched for posterity on his
silver disk. Putting thought into practice, in 1804 Cayley designed and successfully tested

the hand-launched glider shown in Figure 3.1, the first model of the modern-configuration airplane to fly. Of even more importance for the field of aerodynamics, Cayley carried out extensive aeronautical experiments, many of which used a whirling-arm apparatus. That work, during the first decade of the nineteenth century, was the first true research on airplane aerodynamics. The findings from his investigations were published in his "triple paper," "on Aerial Navigation," which appeared in three issues of *Nicholson's Journal of Natural Philosophy, Chemistry, and the Arts* (the November 1809, February 1810, and March 1810 issues). Cayley's triple paper was the highlight in the history of aerodynamics at the beginning of the nineteenth century.

In concert with his aeronautical interests, Cayley carried out extensive work on the design of internal-combustion engines. He recognized that the existing steam engines, with their huge external boilers, were much too heavy in relation to their power output to be of any use for aeronautics. To improve on that situation, Cayley invented the hot-air engine in 1799 and spent the next 58 years of his life trying to perfect the idea, along with a host of other mechanical designers of that day. The success of the gas-fueled engine in France in the mid-1800s finally ended Cayley's work on the hot-air engine.

For reasons not totally understood, Cayley directed his aeronautical efforts toward lighter-than-air balloons and airships during the period from 1810 to 1843, making contributions to the understanding of such devices and inventing several designs for steerable airships. Then, from 1843 until his death in 1857, he returned to the airplane, designing and testing several full-scale aircraft. One was a machine with triple wings (a triplane) and human-actuated flappers for propulsion (Figure 3.19). In 1849 that machine made a brief, floating flight off the ground, carrying a 10-year-old boy for several yards down a hill at Brompton. Another was a single-wing (monoplane) glider (Figure 3.20), which in 1853 flew across a small valley (no farther than 500 yards), with Cayley's coachman aboard as the pilot. At the end of the flight, the coachman was quoted as saying, "Please, Sir George, I wish to give notice. I was hired to drive and not to fly."

Although Cayley's place in history is based on his aeronautical contributions, that broadly educated and liberal-thinking man accomplished much more during his 84 years.

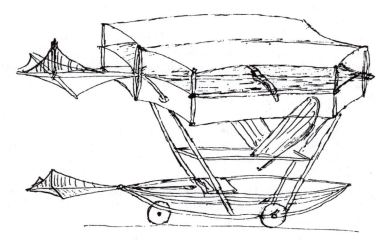

Figure 3.19 Triplane by Cayley (ca. 1849).

Figure 3.20   Monoplane glider by Cayley (ca. 1853).

Figure 3.21   Sir George Cayley.

Of particular note is his invention in 1825 of a tracted land vehicle, a forerunner of the Caterpillar tractor and the military tanks of the twentieth century. In 1847 he invented an artificial hand, a breakthrough in that field, replacing the simple hook that had been in use for centuries. Cayley's interests were purely humane; he expected and received virtually no financial compensation for that invention.

The other nonaeronautical interests of Cayley included concern for parliamentary reform, leading to a publication on that subject in 1818 and election to the chair of the powerful Whig Club at York in 1820; his work as a member of Parliament for Scarborough (starting in 1832); his founding of the Polytechnic Institution in Regent Street, London, in 1839, for scientific and technical exhibitions and education (now the Regent Street Polytechnic, an institution for continuing education); and his inventions to promote railway safety. His social conscience was never more in evidence than in his efforts to help unemployed laborers from the York area in 1842; he wrote an appeal for help in the newspapers and contributed a large sum of his own money (although he was not a rich man) to help relieve the social and economic distress in the area.

Cayley (Figure 3.21) was well liked and respected by his friends as a kind, thoughtful, and humorous country squire, and by his scientific and technical colleagues as one of the most innovative, knowledgeable, and well-read people in England at the time.

As for Cayley's contributions to the state of the art of aerodynamics, he was basically an experimentalist, what today might be called an "idea" person. He used both flight testing and

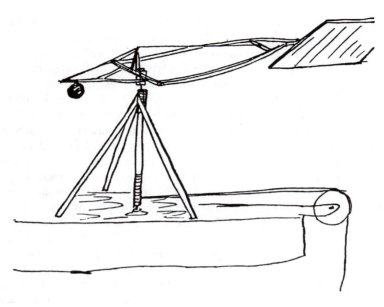

Figure 3.22   Whirling-arm device used by Cayley.

laboratory experiments to gain an understanding of basic aerodynamics that was advanced beyond that of any other person of his time. His flight tests involved both models and full-scale vehicles that glided through the air. (Cayley equipped some of his vehicles with propulsion mechanisms – flappers – that were ill-conceived and totally ineffective. For all practical purposes, those "powered" vehicles were essentially gliders.) The ground testing facilities used for his laboratory experiments comprised a stable of whirling-arm devices based on Robins's invention (Figure 3.22). Numerous such sketches can be found in his various notebooks, some in the archives of the Royal Aeronautical Society and others still in the hands of his heirs. Cayley was active in aeronautical research until two years before his death in 1857, but his major intellectual contributions to aerodynamics came early in his life, and were published in his triple paper of 1809–10. Cayley was well read in the technical literature of his day and knew the state of the art of aerodynamics. His paper referred to the Newtonian sine-squared law and to the work of Robins and Smeaton. He did not attempt to use the mathematical framework of fluid dynamics as developed by Euler, d'Alembert, Lagrange, and Laplace, for good reason: That framework was still not in a practical form in Cayley's time. Cayley's major contributions to basic and applied aerodynamics appeared in parts I and III of his triple paper (November 1809 and March 1810), where we find the first scientific statements of several phenomena that are today essential aspects of modern aerodynamics.

A problem of particular importance to Cayley was how the aerodynamic force on a lifting surface would vary as the angle of attack of the surface was changed, with special emphasis on the variation in the lift component of that force. That question had not been of primary interest to earlier investigators, whose efforts had been dominated by concern over the drag on various objects moving through a fluid. However, because his pioneering concept was of a fixed-wing aircraft that would be sustained in flight simply by the effect of the airflow over the wing, with the mode of propulsion being entirely separate from that of lift, Cayley had to be vitally concerned with the aerodynamic properties of a surface inclined to the flow, with particular emphasis on the variation of lift with the angle of attack. Unfortunately, the only practical means available for *theoretical* prediction of that variation was the Newtonian sine-squared law, and Cayley was well aware of its deficiencies. Indeed, early in part I of his paper, Cayley stated that "theory which gives the resistance of a surface opposed to the same current in different angles, to be as the squares of the sine of the angle of incidence, is of no use in this case; as it appears from the experiments of the French Academy, that in acute angles, the resistance varies much more nearly in the direct ratio of the sines, than as squares of the sines of the angles of incidence." Thus, Cayley knew the Newtonian sine-squared law was defective and was well aware of the experimental findings reported 30 years earlier by the "learned committee" chaired by d'Alembert. Cayley was frustrated by the absence of definitive data, theoretical or experimental, on the variation of lift with the angle of attack. For example, in discussing the flight of birds, he makes indirect reference to the fact that lift increases with the angle of attack in a "complicate ratio (useless at present to enter into)," indicating his feeling that the variation was rather complex and bewildering, insufficiently known to be worth discussing at that point. Cayley was being somewhat modest, because that same lack of data was what had driven him to design, construct, and operate his first whirling-arm device (1804), with which he had carried out the first meaningful aerodynamic experiments on lifting surfaces. At the end of the arm he had mounted a flat surface made of paper stretched tightly over a frame. That "flat-plate" lifting surface had then been tested at various angles of attack, ranging from $-3°$ ($3°$ below the horizontal) to $18°$. The area of the surface, for convenience of reducing the data, had been exactly 1 ft^2. In an entry in his notebook dated December 1, 1804, Cayley had listed in tabular form his findings at various angles of attack. A typical entry showed a measured aerodynamic lift force of 1 ounce for an angle of attack of $3°$ at a velocity through the air of 21.8 ft/s. It will be useful to look at that particular measurement by Cayley from our modern aerodynamic perspective. To lay the groundwork, we shall examine the low-speed aerodynamics of a flat plate as understood today.

A thin flat plate at some angle of attack in a low-speed flow is by no means an optimum lifting surface, for there is a tendency for the flow to separate from the top surface of the plate even at very low angles of attack. Figure 3.23 illustrates the streamline pattern over a flat plate at an angle of attack of $15°$. Thus sketch is drawn to scale from smoke-flow photographs. Such flow separation over the top surface will occur at the leading edge for even a small angle of attack, and the size of the separated region will grow as the angle of attack is increased. Recall that the lift on an airfoil is generated by the difference between

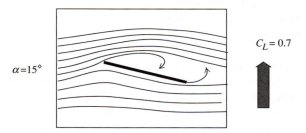

Figure 3.23   Flow streamlines over a flat plate at a 15° angle of attack, with the arrow on the right indicating the relative values for lift.

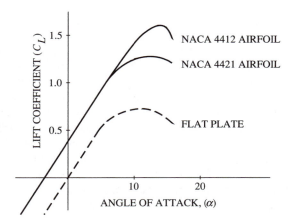

Figure 3.24 Lift coefficient versus angle of attack, comparison of a flat plate with two NACA airfoil sections.

the high pressures acting on the lower surface and the low pressures acting on the upper surface. Flow separation results in a higher pressure being exerted on the top surface than would be the case if the flow had remained attached, thus reducing the net difference in pressures between the lower and upper surfaces, and therefore reducing the lift. Thus, for many applications in aerodynamics, flow separation means a decrease in performance. To be more precise, the variation of the lift coefficient with the angle of attack is shown in Figure 3.24, where the flat plate is compared with two modern airfoil shapes. Clearly, the flat plate is an inferior lifting surface. The concept of flow separation, with its detrimental effect on lift, was not understood in Cayley's time. Indeed, it was one of those instances in which the progress of aerodynamics would have been advanced if Leonardo da Vinci's notes had been readily available to researchers in the eighteenth and nineteenth centuries. Leonardo's sketches of flow separation and the resulting vortex formation (Figure 2.7) could have greatly accelerated the work of Cayley and other nineteenth-century investigators. The fact was that da Vinci's notes were not available to these people – one of the classic "disconnects" that sometimes retard the advancement of technology.

There is another important dimensionless parameter used in modern aerodynamics that was not known in Cayley's time: the Reynolds number (Re).

$$\text{Re} = \frac{\rho V x}{\mu}$$

where ρ is the free-stream density, V is the free-stream velocity, x is the length along a flow, and μ is the coefficient of viscosity – a measure of the magnitude of the frictional effects in the flow. The Reynolds number is a controlling parameter for all viscous flows, and the Re value will also have an effect on flow separation. When comparing two or more sets of aerodynamic data obtained under different running conditions, it is desirable that all the data sets have the same Reynolds number. When the distance x is the complete chord length of an airplane's airfoil, the Re values for real airplanes in flight through the atmosphere can be well into the millions. In contrast, because of the much smaller sizes of the models used in wind tunnels, as well as the frequently lower velocities in wind tunnels than in free flight, the Re values for such laboratory tests are much lower than for full-scale flight.

Hence the laboratory findings often are somewhat compromised, especially in regard to flow separation, which is in part determined by the size of the Reynolds number. To be specific, let us estimate the Reynolds number for one of Cayley's experiments using his whirling arm. In Figure 3.22, at the end of the whirling arm is a lifting surface that is essentially a square; because the area is $1\,\text{ft}^2$, we take the chord length to be 1 ft. For the velocity of 21.8 ft/s given by Cayley, that corresponds to a Reynolds number of 139,000, based on the 1-ft chord length – a much smaller number than would be encountered in full-scale flight. *Hence Cayley's aerodynamic measurements were compromised because of that inordinately low Reynolds number.* Of course, Cayley had no way of appreciating his situation, because the understanding of the Reynolds number effect is a development of twentieth-century aerodynamics. Thus Cayley's lift measurements resulted in lift coefficients that were low. For example, on the basis of Cayley's whirling-arm data mentioned earlier (a lift force of 1 ounce on a flat surface at a 3° angle of attack in a flow at a velocity of 21.8 ft/s), we can calculate the lift coefficient as

$$C_L = \frac{L}{\frac{1}{2}\rho V_\infty^2 S} = 0.11$$

Figure 3.24, based on modern laboratory data at much higher Re values, shows that C_L for a flat plate at a 3° angle of attack is about 0.33 (higher than Cayley's measurement by a factor of 3).

There was another aerodynamic influence that greatly affected Cayley's measurements: the effect of the aspect ratio of the lifting surface. An *aspect ratio* is a geometric parameter of a wing defined as the square of the wingspan divided by the platform area of the wing (Figure 3.25). For Cayley's square lifting surface, the aspect ratio was 1. Today we know that at a constant angle of attack, reducing the wing aspect ratio will reduce the lift coefficient. Short stubby wings have low aspect ratios and hence low lift coefficients. The aerodynamic data shown in Figure 3.24 were taken from a flat-plate wing of relatively large aspect ratio; on the basis of modern data reported by Prandtl,[1] we know that the lift coefficient of 0.33 at a 3° angle of attack, read from Figure 3.24, should be reduced by a factor of about 0.44 to account for the aspect ratio of 1 in Cayley's experiments. That will give us a lift coefficient of 0.15 based on modern data, still higher than Cayley's measurement by about a factor of 1.4. However, in addition, we need to account for the Reynolds-number effect that was described earlier. For Reynolds numbers as low as 100,000, as in the case of Cayley's experiments, the lift coefficient is smaller than it would be at much higher Reynolds numbers; experimental data have shown that a lift coefficient at these low Reynolds numbers is about 68% of its value at high Reynolds numbers.[1] Using this factor, the aspect ratio 1, and a 3° angle of attack, the flat-plate lift coefficient of 0.15 noted earlier is reduced to $0.15 \times 0.68 = 0.10$, a value that is remarkably close to Cayley's measurement of 0.11. Cayley's original measurement was within 10% of the modern value

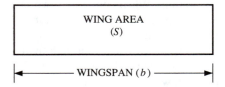

Figure 3.25   Definition of aspect ratio: $\text{AR} = b^2/S$.

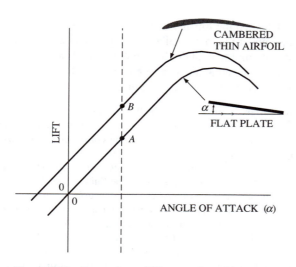

Figure 3.26 Comparison of lift curves for a thin, cambered airfoil and a flat plate.

– a most remarkable degree of accuracy considering the obvious drawback of having to use a whirling arm as a testing device and the lack of sophisticated instrumentation at that time. However, keep in mind that Cayley was testing a flat plate, which we know to be an inefficient lifting surface.

A second major contribution by Cayley was the recognition that a curved (cambered) airfoil shape will produce more lift than a flat plate at the same angle of attack (Figure 3.26). Note that the lift for a flat plate is zero at a zero angle of attack, whereas the cambered airfoil has positive lift at a zero angle of attack. Indeed, the lift curve for the cambered airfoil looks as if it is simply shifted to the left relative to that for the flat plate. Consequently, at a given angle (the dashed vertical line), the lift for the cambered airfoil (point B) is higher than that for a flat plate (point A). Cayley was the first to recognize this fact, and he quantified the lift characteristics of a cambered surface in the following statement from part I of his paper, based on his experimental measurements: "I may safely state, that every foot of such curved surface, as will be used in aerial navigation, will receive a resistance of one pound, perpendicular to itself, when carried through the air in an angle of six degrees with the line of its path, at a velocity of about 34 or 35 feet per second." Cayley used the word "resistance" to denote the total aerodynamic force, a convention followed throughout much of the early eighteenth- and nineteenth-century literature. At a low incidence angle, such as $6°$, that "resistance" is essentially *lift*. Cayley was quite accurate in his estimate; by comparison, the lift force on a modern-day NACA 2412 cambered airfoil will be about 1.17 pounds under the same conditions.

Why does a cambered airfoil produce lift at a zero angle of attack? That question perplexed the aerodynamics investigators of the nineteenth century. There was no mathematically based theory that would explain it. On a qualitative basis, today we know that the flow moves faster over the top of the curved surface, hence reducing the pressure on the top surface. That reduced pressure causes a pressure *difference* between the lower and upper surfaces, hence producing lift at a zero angle of attack. In contrast, for a flat plate aligned with the flow, the flow velocities over the top and bottom surfaces are the same;

there is no pressure difference between the top and bottom surfaces, and hence no lift. Of course, this simple picture from modern aerodynamics was not available in Cayley's time. Nevertheless, he made a halting attempt to explain the aerodynamic flow over a cambered airfoil and to suggest why lift is produced at a zero angle of attack. In part I of his triple paper, Cayley theorized as follows (with modern-day clarifications in brackets):

> A slender filament of the current [a streamline of the flow] is constantly received under the anterior edge of the surface [beneath the leading edge], and directed upward into the cavity, by the filament above it being obliged to mount along the convexity of the surface [the streamline adjacent to the surface must always flow tangentially along the surface], having created a light vacuity immediately behind the point of separation [the first streamline is being sucked into the concave space under the airfoil]. The fluid accumulated thus within the cavity has to make its escape at the posterior edge of the surface, where it is directed considerably downward; and therefore has to overcome and displace a portion of the direct current passing with its full velocity immediately below it [the downward flow coming from the trailing edge is inducing the flow in the whole region downstream of the leading edge to be directed downward relative to the free stream]; hence, whatever elasticity this effort requires operates upon the whole concavity of the surface, excepting a small portion of the anterior edge. This may or may not be the true theory, but it appears to me to be the most probable account of a phenomenon, which the flight of birds proves to exist.

It would seem that in the last part of this statement, Cayley was implying that the downward motion of the air at the trailing edge would push against the free-stream flow, and that push would be transmitted to the bottom surface of the airfoil by "whatever elasticity this effort requires." The general picture described by Cayley had some merit, when compared with our modern understanding of the generation of lift, but it was highly flawed by the insinuation that the flow would "push" on other parts of the flow, thus creating a lift force. (This is my interpretation of Cayley's statement; note that Cayley did not explicitly use the word "push.") Cayley was missing the essential point. He was saying that lift was produced on a cambered airfoil at a zero angle of attack because of the action of the air on the bottom surface, whereas in reality the dominant "action" is on the top surface. We know today that the major factor in the generation of lift on an airfoil is the expansion of air over the top surface, resulting in lower pressure on the top surface, hence creating a "suction" effect that forces the airfoil in the upward (lift) direction. On the other hand, Cayley's statement contained a dramatic insight to an alternative explanation for the generation of lift on a more global scale. Recall from our discussion in Chapter 1 that the aerodynamic force on a body is due to the net integration of the pressure distributions and the shear-stress distributions over the surface; indeed, the surface pressure and shear-stress distributions are nature's way of figuratively grabbing hold of a body in a flow and exerting a force on it. In modern aerodynamics, this is the most fundamental way of visualizing the generation of an aerodynamic force on a body. However, there is an alternative explanation,[17] as follows: The air flowing over an airfoil at a given angle of attack exerts an upward force on the airfoil. In turn, by Newton's third law (for every action there is an equal and opposite reaction), *the airfoil exerts a force on the air in the downward direction.* As a result, the general direction of the airflow downstream of a lifting body is inclined slightly downward. Cayley clearly was making that observation when he said that "the fluid . . . has to make its escape at the posterior edge of the surface, where it is directed considerably downward." Cayley's connection of the generation of lift with that downward motion of the air behind the airfoil was quite correct, and it was the first time that such an explanation had been suggested.

Thus, we can state that Cayley clearly was the first to offer a *scientific* examination of the generation of lift on an aerodynamic surface, with emphasis placed on lift at small angles of attack. His experimental findings were remarkably accurate, though his efforts to describe the physical reasons for the generation of lift, especially for a cambered airfoil, fell short of perfection. However, considering the total lock of understanding of such matters in Cayley's time, we have to conclude that Cayley's thinking was seminal. He was beginning to set applied aerodynamics on the right track.

Cayley was also concerned with the *drag* on a flying machine – a natural concern for an engineer who was designing for actual flight. There were several important insights into the problem of drag in Cayley's triple paper. However, before examining those insights, let us review the modern understanding of the drag on an airplane. Drag is fundamentally due to the pressure distributions and the shear-stress distributions integrated over the surface of the airplane, the drag component of the resulting aerodynamic force being in the direction of the free-stream flow (the drag direction). In aerodynamics as applied to subsonic airplanes, the drag is usually divided into two parts:

(1) *Parasite drag:* the drag due to skin friction over all surfaces of the airplane, plus the drag caused by any flow separation from the surface. The drag due to flow separation is a pressure effect: In a subsonic flow over a nonlifting body, if there is no flow separation anywhere on the surface, then the force in the drag direction due to the pressure distribution over the forward part of the body will exactly cancel the force in the direction opposite to the drag direction due to the pressure distribution over the rearward part of the body – hence no drag due to the pressure distribution. That effect was well known in Cayley's time, first having been shown in 1744 as d'Alembert's paradox. However, if flow separation occurs, the pressure distribution will be altered in such a fashion as to produce a net force in the drag direction, a contribution to drag that is called *form drag.* Hence, parasite drag is the sum of skin-friction drag and form drag (drag due to flow separation).

(2) *Induced drag:* the drag due to the generation of vortices at the tips of an airplane's wings, with the vortices trailing downstream of the wings. This is also a pressure effect. The presence of the wing-tip vortices (which are essentially small tornadoes at the ends of the wings) alters the pressure distribution over the wings in such a fashion as to produce an increased force in the drag direction. These wing-tip vortices are due to the fact that a wing that generates lift will have a higher pressure on the bottom surface of the wing and a lower pressure on the upper surface; hence the airflow in the vicinity of a wing tip will tend to curl around the tip from bottom to top, creating a circulatory motion that results in a trailing vortex. Hence the wing-tip vortices are intimately associated with the generation of lift: The greater the lift (hence the larger the pressure difference between the upper and lower wing surfaces), the stronger the vortices. Similarly, if the lift is zero, the vortices will have essentially zero strength, and the induced drag will be zero. Thus, we can view induced drag as *the cost of producing lift;* it is frequently called the "drag due to lift."

In summary, for a subsonic airplane we can write

Total drag = Parasite drag + Induced drag

where

Parasite drag = Skin-friction drag + Drag due to flow separation (form drag)

Now let us return to Cayley's work in light of the foregoing discussion of the various types of drag. First, at the fundamental level, Cayley did not have the picture of flow separation; however, he certainly appreciated its ramifications. At the end of part III of his triple paper, Cayley stated, in conjunction with a discussion of the shapes of low-drag projectiles, that "it has been found by experiment, that the shape of the hinder part of the spindle is of as much importance as that of the front, in diminishing resistance. This arises from the partial vacuity created behind the obstructing body. If there be no solid to fill up this space, a deficiency of hydrostatic pressure exists within it, and is transferred to the spindle." That was correct. A projectile with a large, blunt base will have a massive region of flow separation behind the base. The pressure in the region of separated flow will be less than it would be for attached flow (Cayley's "partial vacuity"). That low pressure existing at the base in turn results in greater drag (greater *form drag,* in the terms of our earlier discussion). On the other hand, if the projectile is smoothly tapered at the rear (a "solid filling the space," in Cayley's terms), there will be less form drag. Cayley clearly recognized that feature. (Unknown to Cayley, da Vinci had made the same observation.) Cayley continued in the same paragraph: "I fear however, that the whole of this subject is of so dark a nature, as to be more usefully investigated by experiment than by reasoning; and in the absence of any conclusive evidence from either, the only way that presents itself is to copy nature; accordingly I shall instance the spindles of the trout and the woodcock." Quite prophetic. Even today, flow separation is a phenomenon difficult to predict theoretically or computationally – still a state-of-the-art research problem in aerodynamics. To this day, experiment remains the primary means to obtain reliable data on flow separation.

Cayley was the first researcher to recognize the advantage of resolving the net aerodynamic force on a body into two components: (1) lift, the component perpendicular to the free-stream direction, and (2) drag, the component parallel to the free-stream direction, as shown on his silver disk (Figure 3.2). Ten years later, in part I of his triple paper, a similar lift-and-drag diagram was shown in the context of his discussion of soaring birds (Figure 3.27). Relative to that diagram, Cayley made a statement that is a fundamental contribution to applied aerodynamics:

> Draw *de* perpendicular to the plane of the wings, ..., and from the point *e*, assumed at pleasure in the line *de* let fall *ef* perpendicular to *df*. Then *de* will represent the whole force of the air under the wing; which being resolved into the two forces *ef* and *fd*, the former represents the force that sustains the weight of the bird, the latter the retarding force by which the velocity of the motion, producing the current *cd*, will continually be diminished.

Cayley was assuming that the resultant aerodynamic force was perpendicular to the chord of the wing. Today we know that that is not precisely true; the resultant aerodynamic force is usually inclined slightly behind the perpendicular. However, Cayley's fundamental contribution was resolution of the force vector into lift ("force that sustains the weight of the bird") and drag ("retarding force"). He went on to say that "in addition to the retarding force thus received [on the wing] is the direct resistance, which the bulk of the bird opposes to the current." That was a first reference to the component of drag that we now call *parasite drag.* He ends that thought by stating that "that is a matter to be entered into separately from the principle now under consideration; and for the present may be wholly neglected, under the supposition of its being balanced by a force precisely equal and opposite to

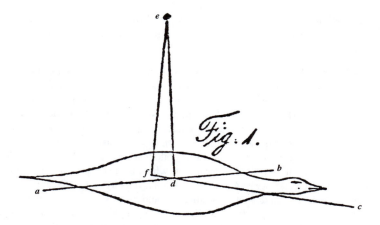

Figure 3.27   Resolution of the aerodynamic force on a bird, from Cayley's triple paper (1809).

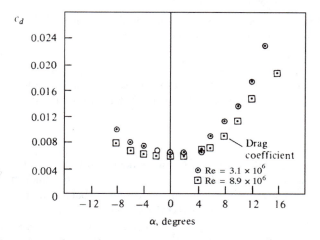

Figure 3.28   Drag coefficients for the NACA 2412 airfoil.

itself." There Cayley was directly influenced by d'Alembert's paradox, saying that the net force acting on the "bulk" of the bird was essentially zero (i.e., the parasite drag was negligible).

There were three more passages in Cayley's paper that represented substantial contributions to the understanding of aerodynamic drag. Of particular note is a short statement near the end of part II: "It has been before suggested, and I believe upon good grounds, that very acute angles vary little in the degree of resistance they make under a similar velocity of current." In that statement, Cayley is saying that for a lifting surface at small angles of attack, the drag will not vary much as the angle varies. Jumping to the modern world of aerodynamics, the variation of the drag coefficient with the angle of attack for an NACA 2412 airfoil is shown in Figure 3.28. Note that the drag coefficient changes very little with variation of the angle of attack as long as the angle of attack is small (near zero). Cayley's observation was therefore quite valid.

Another aspect of drag studied by Cayley involved his repetition of John Smeaton's experiments on flat plates mounted perpendicular to a flow, as described in the middle of part I of his triple paper:

> The results of Mr. Smeaton's experiments and observations was, that a surface of a square foot met with a resistance of one pound, when it travelled perpendicularly to itself through air at a velocity of 21 ft. per second. I have tried many experiments upon a large scale to ascertain this point. The instrument was similar to that used by Mr. Robins, but the surface used was larger, being an exact square foot, moving round upon an arm about five feet long, and turned by weights over a pulley. The time was measured by a stop watch, and the distance travelled over in each experiment was 600 feet. I shall for the present only give the result of many carefully repeated experiments, which is, that a velocity of 11.538 feet per second generated a resistance of 4 ounces; and that a velocity of 17.16 feet per second gave 8 ounces resistance. This delicate instrument would have been strained by the additional weight necessary to have tried the velocity generating a pressure of one pound per square foot; but if the resistance be taken to vary as the square of the velocity, the former will give the velocity necessary for this purpose at 23.1 feet, the latter 24.28 [feet] per second. I shall therefore take 23.6 feet as somewhat approaching the truth.

Let us use Cayley's data and calculate the Smeaton coefficient. Recall that Smeaton's expression for the force on a flat plate perpendicular to a flow is

$$F = kSV^2$$

where k is the Smeaton coefficient, S is the area in square feet, and V is the velocity in miles per hour. Using Cayley's datum of an 8-ounce force measured on an area of 1 ft^2 moving at a velocity of 17.16 ft/s (11.7 mph), we have

$$k = 0.0037$$

whereas Smeaton's value was 0.005. Hence, Cayley's data lead to a *modified* Smeaton coefficient that is less than that measured by Smeaton, but still higher than the modern value of 0.00289. Indeed, Cayley was one of the first to recognize that Smeaton's value was in error. In spite of Cayley's finding, the value of 0.005 was used for Smeaton's coefficient by many investigators over the nineteenth century. The controversy about Smeaton's coefficient spilled over to the early twentieth century; for example, the Wright brothers' early work on aerodynamics was compromised by the use of 0.005 for Smeaton's coefficient, until they independently recognized that it was in error.[24]

In regard to drag, we note one other observation by Cayley having to do with the effect of drag on the overall lift-to-drag ratio for an airplane. First, some additional background: The expression for the lift-to-drag ratio for a flat plate, based on Newtonian theory, is

$$\frac{L}{D} = \cot \alpha$$

where α is the angle of attack. Mathematically, this expression indicates that L/D will increase as the angle of attack decreases; indeed, for a zero angle of attack, this expression yields an infinitely large value for L/D – an impossible situation. This is illustrated by the solid curve in Figure 3.29. However, the Newtonian theory does not take account of friction drag. The effect of friction drag is to cause L/D to peak at some small angle of attack (the *maximum* lift-to-drag ratio), and then to go to zero at a zero angle of attack. This is illustrated by the dashed curve in Figure 3.29. Such trends were understood by Cayley.

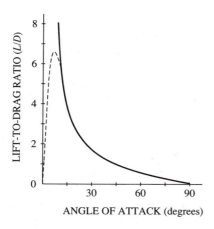

Figure 3.29   Sketch of the variation of the lift-to-drag ratio with the angle of attack for a flat plate.

For example, consider this discussion of the power required for sustained flight:

> As the acuteness of the angle between the plane and current increases [i.e., as the angle of attack *decreases*], the propelling power required is less and less. The principle is similar to that of the inclined plane, in which theoretically one pound may be made to sustain all but an infinite quantity; for in this case, if the magnitude of the surface be increased ad infinitum, the angle with the current may be diminished, and consequently the propelling force, in the same ratio.

Cayley was saying that if lift is to be generated by a larger wing area, the angle of attack can be decreased in order to sustain the weight of "one pound." He then recognized, from the Newtonian theory, that the ratio of lift to drag, L/D, theoretically goes to infinity at a zero angle of attack. Hence, in sustaining 1 lb of weight, the drag becomes smaller as the angle of attack becomes smaller, hence reducing the amount of "propelling force" needed to overcome the drag. That showed complete command of the basic principles of flight, along with good reasoning based on the contemporary aerodynamic theory. Cayley continued:

> In practice, the extra resistance of the car and other parts of the machine, which consume a considerable portion of power, will regulate the limits to which this principle, which is the true basis of aerial navigation, can be carried; and the perfect ease with which some birds are suspended in long horizontal flights, without one waft of their wings, encourages the idea, that a slight power only is necessary.

Thus, Cayley recognized the role played by parasite drag; a finite parasite drag prevents the L/D ratio from going to infinity, and indeed dictates a maximum L/D at some small angle of attack. He was also saying that the parasite drag must be kept small, so "that a slight power only is necessary." That remains an important goal for modern airplane designers: the smallest possible parasite drag.

As a corollary to discussion of the L/D ratio, we pose this question: How aerodynamically efficient were Cayley's flying machines? Because L/D is a direct index of the aerodynamic efficiency of an aircraft, we would like to know the values of L/D for Cayley's machines. We can estimate those values from a statement made in part I of his triple paper. In discussing one of his gliders, Cayley stated that it sailed "majestically" from the top of a hill, descending along an angle of about 18° with the horizon. A calculation based on that glide angle shows that the L/D ratio for the glider was 3.08, not an impressive value. Typical values of L/D for modern airplanes are 15–20, and values for modern gliders are above 40. Cayley did not have an efficient airplane. However, Cayley knew nothing about aspect-ratio effects. Today we know that low-aspect-ratio wings, such as those used by Cayley (aspect ratio about 1), entail large amounts of induced drag and hence are very inefficient.

Let us now summarize Cayley's major contributions to the state of the art of aero-dynamics:

(1)   Cayley performed the first serious measurements of the variation of lift with angle of attack, concentrating on the range of *small* angles germane to a flying machine. His measurements were remarkably accurate for his day, deviating by only 10% from estimates based on modern technology. Cayley's work was the first to show the importance of aerodynamic lift, as contrasted with earlier studies emphasizing only drag.

(2)   Cayley's basic concept of the mechanism of lift (a fixed wing) being totally separate from the mechanism of propulsion (with the related concept that lift would sustain the weight of the vehicle, and the sole purpose of propulsion would be to overcome the drag on the vehicle) was a quantum leap in the advancement of technical aeronautics. Late in his life, da Vinci had embraced the idea of a fixed-wing vehicle, in contrast to the flapping wings of an ornithopter, though he neither explored nor developed the idea in any technical way, and his notes were unavailable in Cayley's time. Thus, credit for the concept of the fixed-wing airplane must go to Cayley, because all subsequent aircraft configurations can be traced directly back to Cayley's work, not to da Vinci's thoughts.

(3)   As to the effect of camber on the lifting properties of a wing, Cayley was the first to appreciate the *aeronautical* implications of such camber. In 1759, Smeaton had noted that the wind effect on windmill blades was enhanced if the blades were curved – an obvious indication of the benefits of camber. However, Smeaton did not elaborate, nor was any technical effort made at that time to understand the effect of camber. Smeaton's observation was totally lost, unknown or ignored by subsequent investigators, having no impact on the state of the art of aerodynamics. In contrast, Cayley's work had substantial impact, earning Cayley the credit for appreciating the importance of camber and its role in fixed-wing flight.

(4)   Cayley's ideas concerning the drag on various components of a vehicle (drag on the wings, drag on the fuselage, etc.) were somewhat consistent with our modern conceptions of parasite drag and induced drag:

(a)   He reinforced the earlier work of da Vinci and Robins showing that the shape of the rear portion of a body has an important effect in determining the drag. Cayley was expressing the concept of *streamlining* a body, although that word had not yet entered the technical vocabulary. Although the *mechanism* of flow separation was not known to Robins and Cayley, the *result* of such flow separation, namely, increased drag, was certainly observed.

(b)   Cayley showed that the variation of drag with the angle of attack for a lifting surface is small as long as the angle of attack itself is small.

(c)   Cayley's measurements of the drag on a flat plate oriented perpendicular to a flow indicated that the accepted value for Smeaton's coefficient ($k = 0.005$)

was not correct; rather, the value $k = 0.0037$ was consistent with Cayley's data. In his triple paper, Cayley was very diplomatic. He did not explicitly state that Smeaton's findings might be in error. Rather, he stated that "I shall for the present only give the result of many carefully repeated experiments," leaving it to the reader to draw conclusions. In a sense, it is regrettable that Cayley did not take a stronger stance, for nineteenth-century investigators continued to use the erroneous value.

(d) Cayley recognized the role played by parasite drag in limiting the aerodynamic efficiency of a flying machine, emphasizing that parasite drag must be kept small, so that little power would be required.

In total, Cayley's contributions to the field of experimental aerodynamics were momentous – by comparison, the works of all previous investigators pale. However, Cayley made no contributions to *theoretical* aerodynamics. The reason was simple: At the beginning of the nineteenth century, the experimental tools available to Cayley, although crude by present-day standards, were considerably more useful than the theoretical tools available. Cayley simply chose the most productive approach.

The State of the Art as Reflected in Cayley's 1804 Glider

We have reviewed the fragmentary knowledge about aerodynamics that was available to Cayley in 1804. Although Cayley's triple paper was published in 1809–10, most of the work reported therein had been carried out prior to or during 1804 – his whirling-arm data and his basic conceptual thinking about the principles of heavier-than-air flight. Now we address the question whether or not Cayley's understanding of aerodynamics, as revealed in his triple paper, was reflected in the design of his 1804 glider (Figure 3.1).

First, we note that *nothing* of theoretical aerodynamics was reflected in that design. The basic partial differential equations that describe a frictionless (inviscid) flow – the Euler equations – were available in Cayley's time, but no one knew how to solve those equations for any reasonable aerodynamic applications – they were useless to Cayley for the design of his 1804 glider. Cayley was not a mathematician. He knew of the failure of the great eighteenth-century mathematicians (d'Alembert, Euler, Laplace, Lagrange) to solve those equations, and there is no evidence that Cayley even considered attempting their solutions. Cayley did take notice of the Newtonian sine-squared law, but only to the extent of recording his belief, based on the French empirical results, that the law yielded inaccurate data for small angles of attack. Cayley left no calculations involving that law and certainly based no design considerations on it. In short, theoretical aerodynamics was of no help whatsoever in the design of Cayley's 1804 glider.

In contrast, the existing body of knowledge in experimental aerodynamics was of some help to Cayley, especially because Cayley himself was responsible for much of it. His whirling-arm measurements of the lift on a flat plate at a small angle of attack had been remarkably accurate, and we see that the 1804 glider (Figure 3.1) had a wing surface inclined at a similar small angle relative to the fuselage. Moreover, the aspect ratios for the wing of the glider and for the experimental plate at the end of the whirling arm were essentially the same, and the wing sizes and velocities through the air were close in the two cases. Hence, Cayley's data from his whirling arm gave him a direct and reasonably accurate estimate of the lifting properties of his 1804 glider. Also, note that the wing of the glider was a flat surface, braced by cross-struts; Cayley made no effort to use camber in the wing.

Figure 3.30   A Cayley glider (ca. 1853).

However, note that 49 years later, in the design of a similar glider (Figure 3.30), Cayley used a cambered wing constructed of fabric. Cayley called it a "sail." The camber would be created by the aerodynamic pressures acting on the sail during flight. Thus, in 1853, Cayley was able to apply his knowledge of the effects of camber, whereas he had not been able to do so in 1804.

Cayley did the best he could with the knowledge and data available to him. However, the 1804 glider fell far short of incorporating the existing body of knowledge in aerodynamics, largely because the field of aerodynamics was still too immature and fragmented to allow meaningful applications. Some of the elementary tools were there, but nobody knew how to use them properly.

Developments in Aviation between 1600 and 1804

As for any meaningful advances in aviation to accompany developments in aerodynamics, the seventeenth and eighteenth centuries had little to offer. The field of aviation continued as a collection of fanciful ideas, and little else:

(1)   Tower jumpers continued in abundance – all singularly unsuccessful. Witness the celebrated attempt by the Marquis de Bacqueville to fly across the Seine River in Paris in 1742 with wings fixed to his arms and legs (Figure 3.31). The marquis suffered two broken legs as he plummeted down the riverbank. That was despite the publication of *De motu animalium* in 1680 by the mathematician Giovanni Alphonso Borelli, in which he demonstrated that a human being did not have the muscle strength to fly. That attempted flight occurred about the same time that Bernoulli was trying to develop the pressure–velocity relationship for a flowing fluid, and about 10 years before Euler published his famous equations for inviscid fluid flow. Clearly, the field of aviation was much less mature than the science of aerodynamics at that time.

(2)   Fanciful ideas for machines were being put forth during that time. Of particular note was the concept of a Jesuit priest, Father Francesco de Lana de Terzi, in 1670 for an "aerial ship" sustained by four copper spheres that would be evacuated (Figure 3.32). Today, we know that the differential in air pressure across those spheres would have crushed them. Much later, in 1781, a flapping-wing glider was proposed by Karl Friedrich Meerwein (Figure 3.33). Architect to the prince of Baden, Meerwein calculated that 126 ft^2 of wing surface would be necessary for his machine. On the basis of Cayley's later findings that about 1 lb of lift per square foot could be expected for such a machine, we see that Meerwein's glider would not have been large enough to generate the necessary lift as a fixed-wing machine; the weight of the machine plus the pilot would have been greater than the lift. Perhaps Meerwein was counting on additional lift from the flapping motion of the wings. He was reported to have made one or two short glides of no real

Figure 3.31 Depiction of the Marquis de Bacqueville attempting to fly across the Seine River in Paris (1742).

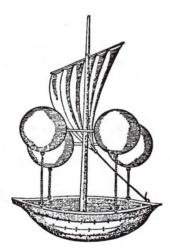

Figure 3.32 Concept by Francesco de Lana for an "aerial ship" sustained by four evacuated copper spheres (1670).

consequence.[25] Meerwein's glider was conceived after the primary theoretical and experimental advances of the eighteenth century – after the work of Pitot, Bernoulli, Euler, d'Alembert, Laplace, and others. Indeed, Newtonian physics was almost a century old at the time of Meerwein's attempt. Clearly, there was virtually no connection between experimental and theoretical aerodynamics and the concurrent aviation efforts.

(3) There were no connections between the scientists working to develop aerodynamics and the individuals attempting to fly. They were two different communities. With one exception (George Cayley), none of the mathematicians and scientists discussed in this chapter expressed the slightest interest in aviation. Robert Hooke, a colleague of Newton, was convinced that humans were too weak to fly on muscle

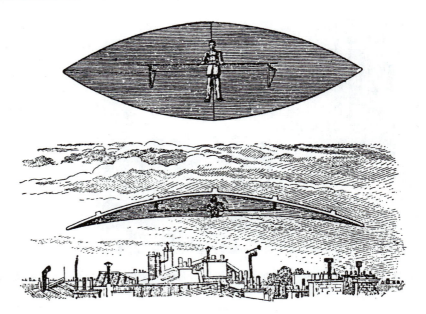

Figure 3.33   Flapping-wing glider conceived by Karl Meerwein (1781).

power alone. There is some evidence that Hooke made ornithopter models powered by springs and may even have designed a machine large enough to carry a person. Thus it is probable that Newton was exposed to the idea of aviation, albeit indirectly.

All of that changed with the work of George Cayley. Cayley was history's second example of an individual whose interests combined research in aerodynamics and the design of flying machines (da Vinci's earlier work had been lost). Cayley's flying machines embodied a quantum leap in aviation (Figures 3.1, 3.2, 3.19, 3.20, and 3.30). They were the first true airplanes – machines with features like fixed wings, a fuselage, and a tail, with a separate device for propulsion. Finally, at the beginning of the nineteenth century, aviation was on its way.

Reflections on the Historical Timeline

The first part of Table 1.1, ranging from 350 B.C. to A.D. 1809–10, lists the familiar milestones discussed in Chapters 2 and 3, from which we draw the following observations:

(1)   If we compare the development of the science of aerodynamics to the construction of a brick wall, we see from the timeline that only a few bricks were laid, almost at random, during the Greek and Roman periods, and likewise for the Middle Ages. However, the few bricks that were laid were important; they began to establish the architectural design for the entire wall. Of special note was the work of da Vinci, although we should consider his contributions to be like phantom bricks, because his notes become unavailable for centuries.

(2)   The pace of the bricklaying increased greatly at the end of the seventeenth century with the experimental work of Mariotte and Huygens and the theoretical

contributions of Newton. Indeed, the eighteenth century was a time of vast expansion in our brick wall, accompanied by strategic placement of key bricks within the wall.

(3) With the work of Cayley in the early 1800s, the basic framework for the brick wall of aerodynamics was completed. Certainly considerable numbers of bricks were missing, and there were a few very large, gaping holes. Nevertheless, the essence of the brick wall was in place. It remained for the developments of the nineteenth and twentieth centuries to fill in the gaps and complete the structure.

The entire portion of the timeline from 350 B.C. to A.D. 1809–10 represents an identifiable first phase in the history of aerodynamics, a phase we can call the incubation phase. That was a period when various basic, fundamental ways of looking at the physical world were attempted, and some rejected. During that period, a certain mind-set was being established. The basic intellectual models seeking to explain nature's ways were beginning to flower, and the definitions of terms used to describe those models were evolving. After 1810, the field of aerodynamics entered an adolescent period of rapid development during the nineteenth century, followed by a period of exponentially increasing sophistication to reach maturity in the twentieth century.

PART II

The Infancy of Aerodynamics, and Some Growing Pains

CHAPTER 4

The Infancy of Aerodynamics
To Lilienthal and Langley

The first great aim of the Society is the connecting of the velocity of the air with its pressure on plane surfaces to various inclinations. There seems to be no prospect of obtaining this relation otherwise than by a careful series of experiments. But little can be expected from the mathematical theory; it is a hundred and forty years since the general differential equations of fluid motion were given to the world by D'Alembert; but although many of the greatest mathematicians have attempted to adduce from them results of a practical value, it cannot be said that any great success has attended their efforts. The progress made has been very slight in the case of water, where the analysis is much simpler than for an elastic fluid like air; and the theory of resistance, which is part of hydromechanics, which has most direct bearing on aerial navigation is, perhaps, the part of the subject about which least is known.

Annual report, Aeronautical Society of Great Britain (1876)

In 1891, the sky over Germany hosted the first successful manned, heavier-than-air flying machine – the hang glider of Otto Lilienthal (Figure 4.1). If we compare it with Cayley's 1804 glider (Figure 3.1), we can detect little fundamental difference between the two flying machines. Both have fixed wings and horizontal and vertical tail structures. Both designers were knowledgeable and concerned about the static stability of their machines, recognizing that the center of gravity should be ahead of the aerodynamically generated aerodynamic center of the vehicle (notice the ballast weight hanging from the nose of Cayley's glider and the position of Lilienthal's body). Both machines were without any mechanical means of flight control, and both were unpowered gliders. However, there was one fundamental difference: The wing of Lilienthal's glider was a rigidly curved (cambered) airfoil shape, whereas the wing of Cayley's 1804 glider was simply a flat surface. To a casual observer that might seem a small difference, but to an aerodynamicist it represented a substantial change. The cambered airfoil used by Lilienthal reflected the progress in gaining an understanding of aerodynamics in the nineteenth century.

On August 10, 1896, Otto Lilienthal died, having been injured in the crash of one of his gliders the previous day. His death brought a temporary halt to the research necessary for successful heavier-than-air, powered flight in Europe. At the time of his death, Lilienthal had been working on a form of prime mover (an engine) to eventually be mated with his gliders. No similar disaster befell aeronautical research in the United States. Quite to the contrary, Samuel Langley, secretary of the Smithsonian Institution in Washington, D.C., had designed and built a relatively small, steam-powered, heavier-than-air flying machine and was actually testing that machine over the Potomac River in the summer of 1896. Figure 4.2 shows Langley's machine (which he called an *aerodrome*) winging its way over the river on May 6, 1896. That machine was too small to carry a man, but its flight on May 6 was the first sustained flight by an engine-powered, heavier-than-air flying machine.

Lilienthal's glider and Langley's 1896 aerodrome are the representative flying machines for this chapter, and again we shall evaluate the state of the art of aerodynamics during the nineteenth century and to what extent it was reflected in the design of those machines.

Figure 4.1   Otto Lilienthal flying one of his gliders (1894).

Figure 4.2   Langley's steam-powered aerodrome (1896).

The Inclusion of Friction in Theoretical Aerodynamics: The Works of Navier and Stokes

At the beginning of the nineteenth century, the equations for fluid motion, as derived by Euler, were well known. However, those equations did not take account of an important physical phenomenon, a phenomenon that was appreciated by scientists in the eighteenth and nineteenth centuries but was not sufficiently understood to be included in any theoretical analysis – namely, friction. Friction plays two dominant roles:

(1)   Extra terms must be added to the Euler equations to account for the effect of friction at any local point in the flow field. The physical reason for these extra terms is easily seen by visualizing a small element of fluid moving along a streamline (Figure 4.3). The fluid element is moving with a velocity V. Now visualize the fluid that is *outside* the moving element (i.e., the surrounding fluid around the

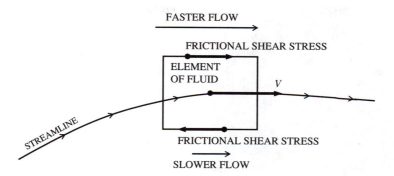

Figure 4.3 Sketch of the frictional shear stresses acting on a moving fluid element.

element). If the velocity of the surrounding fluid is the same as that of the element, then no friction effects are present (i.e., the surrounding fluid exerts no frictional force on the walls of the element). However, imagine that the fluid above the element is moving faster than V. The faster-moving fluid will "rub against" the upper surface of the element, exerting a frictional force on the upper surface acting to the right. In addition, imagine that the fluid below the element is moving slower than V. That slower-moving fluid will exert a "dragging" force on the bottom surface of the element acting toward the left. To account for those effects, the Euler equations must be modified by addition of extra terms to account for friction.

(2) The effect of friction between a fluid flow and a solid surface (such as the surface of an airplane wing) is to cause the flow velocity right at the surface to be zero (relative to the surface). This is called the no-slip condition in modern terminology, and in aerodynamic theory it represents a "boundary condition" that must be accounted for in conjunction with the solution of the governing flow equations. The no-slip condition is fully understood in modern fluid dynamics, but it was by no means clear to nineteenth-century scientists. The debate whether or not there was a finite relative velocity between a solid surface and the flow immediately adjacent to that surface continued into the second decade of the twentieth century.

The equations that describe fluid flows and contain terms to account for friction are called the *Navier-Stokes equations,* derived independently by Louis Navier in France and George Stokes in England in the nineteenth century, and they remain today the fundamental equations used to analyze viscous fluid flows. Moreover, they continue to be the focus of much basic research, with many applications in the field of computational fluid dynamics. The importance of the Navier-Stokes equations in modern aerodynamics cannot be overstated.

The first accurate representation of the effects of friction in the general partial differential equations for fluid flow was provided by Navier in 1822, as described in a paper presented to the Paris Academy of Sciences.[26] Although Navier's equations were in the correct form, his theoretical reasoning was greatly flawed, and it was almost a fluke that he arrived at the correct terms. Moreover, he did not fully understand the physical significance of what he had derived.

Claude-Louis-Marie-Henri Navier (1785–1836) (Figure 4.4) was born in Dijon, France, on February 10, 1785. His early childhood was spent in Paris, where his father was a lawyer to the National Assembly during the French Revolution. After the death of his father in

Figure 4.4   Claude-Louis-Marie-Henri Navier.

1793, Navier was left in the care of his mother's uncle, Emiland Gauthey (at the time of his death in 1806, Gauthey was considered France's leading civil engineer). As a result of his granduncle's influence, Navier entered the Ecole Polytechnique in 1802, barely meeting the school's admission standards. However, within a year his talents began to flower. In 1804 he entered the Ecole des Ponts et Chaussées, graduating in 1806 near the top of his class. There he studied with the famous French mathematician Jean Baptiste Fourier, a professor of analysis. Fourier's impact on Navier was immediate and lasting, with Navier becoming Fourier's protégé and lifetime friend. Over the next 13 years, Navier became recognized as a scholar of engineering science. He edited the works of his granduncle, which represented the traditional empirical approach to numerous applications in civil engineering. In that process, on the basis of his own research in theoretical mechanics, Navier added a somewhat analytical flavor to the works of Gauthey. That, in combination with textbooks that Navier wrote for practicing engineers, introduced the basic principles of engineering science to a field that previously had been almost completely empirical. Navier was responsible for precisely defining the concept of mechanical work in the analysis of machines. ("Work" is simply the product of force times the distance through which that force acts – today a fundamental concept in the analysis of mechanical systems. Navier called that product the "quantity of action.") In 1819, Navier took a teaching position at the Ecole des Ponts et Chaussées, where he permanently changed the style of teaching in engineering with his emphasis on physics and analysis. In 1831 he replaced the famous mathematician Augustin-Louis Cauchy at the Ecole Polytechnique. For the rest of his life, Navier lectured at the university, wrote books, and at times practiced his profession of civil engineering, particularly the design of bridges. It is ironic that the bridge design that brought him the most public notice was never completed. Toward the end of construction of a suspension bridge over the Seine, a sewer near one pier ruptured, flooding the area, weakening the foundation of the pier, and causing the bridge to sag. The damage could easily have been repaired. However, for various political and economic reasons, the Municipal Council of Paris, which had been opposed to the building of Navier's bridge from the beginning, took that opportunity to halt the project. The bridge was torn down, to Navier's great disappointment – one of the many examples throughout history in which engineering competence has been no match for fate and politics.

Bridges notwithstanding, Navier is recognized as the first to have derived the equations to describe fluid flow with friction. However, there was the irony that Navier had no conception of shear stress in a flow (i.e., the frictional shear stresses acting on the top and bottom surfaces of the fluid element in Figure 4.3). Rather, he was attempting to adapt Euler's equations of motion to take into account the forces acting between the molecules in a fluid. He assumed those intermolecular forces to be repulsive at close distances between molecules and attractive at larger distances. Thus, for a fluid that was stationary, the spacing between molecules would be determined by the equilibrium between the repulsive and attractive forces. By an elaborate derivation using that model, Navier produced a system of equations that were the same as Euler's equations of motion except for the addition of terms to account for intermolecular forces. Those terms involved second derivatives of the velocity multiplied by a constant, where the constant was a function of the spacing between molecules. As seen in the Navier-Stokes equations in Appendix B, that is indeed the proper form for the terms involving frictional shear stress, namely, a second derivative of velocity multiplied by a coefficient called the viscosity coefficient. The irony is that although Navier had no conception of shear stress and did not set out to obtain equations that would describe motion involving friction, he nevertheless arrived at the proper form for such equations.

It should be noted that Navier's finding was not totally a fluke. Our understanding of the physical significance of the viscosity coefficient derives from study of the kinetic theory of gases, and we can show that the viscosity coefficient is directly proportional to the molecular mean free path – the mean distance that a molecule moves between successive collisions with other molecules. Hence Navier's approach, wherein he was accounting for the spacing between molecules due to the balance between attractive and repulsive intermolecular forces, was not totally off the mark, although the mean free path and the mean spacing between molecules are different values. Right church, but wrong pew.

Although Navier did not appreciate the physical significance of his equations for fluid flow, one of his contemporaries did: Jean-Claude Barré de Saint-Venant (1797–1886). Saint-Venant was educated at the Ecole Polytechnique, graduating 12 years after Navier. After some 27 years as a municipal civil engineer, he retired to a life of teaching and research, dying at age 92 after a long and productive life. Saint-Venant was one generation younger than Navier and one rank lower in professional stature. Navier was elected to the Paris Academy of Sciences in 1824; Saint-Venant became a member in 1868. Saint-Venant was quite familiar with Navier's work, as reflected in his book *Mécanique appliquée de Navier, annotée par Saint-Venant* (1858). Seven years after Navier's death, Saint-Venant re-derived Navier's equations for a viscous flow, considering the internal viscous stresses (Figure 4.3), and eschewing completely Navier's molecular approach.[27] That 1843 paper was the first to properly identify the coefficient of viscosity and its role as a multiplying factor for the velocity gradients in the flow. He further identified those products as viscous stresses acting within the fluid because of friction. Saint-Venant got it right and recorded it. Why his name never became associated with those equations is a mystery. Certainly it was a miscarriage of technical attribution.

Sir George Gabriel Stokes (1819–1903) (Figure 4.5) was only a few hundred miles from Navier and Saint-Venant across the English Channel, but in the early 1840s he did not know of their work. Stokes was born at Skreen, Ireland, on August 13, 1819. The hallmark of

Figure 4.5   George Gabriel Stokes.

his family was religious vocation – his father was rector of the Skreen parish, his mother
was the daughter of a rector, and ultimately all three of his brothers became ministers in the
church. Stokes remained strongly religious and toward the end of his life became interested
in relationships between science and religion. He was president of the Victoria Institute of
London, a society for examining the relationship between Christianity and contemporary
thought, with emphasis on science. Stokes's education began with tutoring by his father,
which led to his admission to Bristol College in England, where he prepared for university
studies, entering Pembroke College, Cambridge, at the age of 18. Upon graduation, he
was immediately elected to a fellowship in Pembroke College. Eight years later, Stokes
occupied the Lucasian chair at Cambridge, the same position held by Newton almost two
centuries earlier. Because the Lucasian endowment was small, Stokes took an additional
position in the 1850s, teaching at the School of Mines in London. He held the Lucasian
chair until his death at Cambridge on February 1, 1903 (a scant 10 months before the Wright
brothers' successful first flight on December 17, 1903).

Stokes made a fundamental contribution to fluid dynamics by his derivation and application of the Navier-Stokes equations, the fundamental descriptors of a general three-
dimensional, unsteady, viscous fluid flow – thus providing the foundation for modern
theoretical and computational fluid dynamics. However, Stokes most likely would have
preferred to be remembered as a physicist and to some small degree a mathematician who
had made substantial contributions in the field of optics. Beginning about 1845, he studied
the propagation of light and how it interacted with the "ether" – a continuous substance
surrounding the earth, according to the prevailing theory of that day.

Stokes analyzed the properties of the hypothetical ether by analogy with his fluid-
dynamics equations of motion. He concluded that if the earth moved through a stationary
ether, the ether must be a very rarefied fluid. In a contradictory opinion, he also concluded
that the propagation of light required the ether to be much like a very elastic solid. Thus,
one of the first theoretical consequences of the Navier-Stokes equations was not a definitive flow-field calculation (as they are used today), but rather an inconclusive study of the

properties of the ether. To make things even more confusing, in 1846 Stokes showed that the laws of reflection and refraction applied equally well whether or not an ether existed. Of much greater importance for an understanding of the physics of light was Stokes's work on fluorescence, the phenomenon wherein a substance absorbs electromagnetic waves of one wavelength and emits waves of another wavelength. In particular, he reported observations of the blue light emitted from the surface of an otherwise transparent and colorless solution of sulfate of quinine when the solution was irradiated by invisible ultraviolet rays. His explanation for that phenomenon, wherein he coined the word "fluorescence," won him the Rumford Medal of the Royal Society in 1852. Later, he suggested the use of fluorescence to study the properties of molecules, and he was credited with developing the principles of spectrum analysis. Thus there is irony in the fact that today his name is heard more frequently in the field of fluid dynamics than in other fields of science and engineering.

As for Stokes's contributions in fluid dynamics, he was initially unaware of the work of Navier and Saint-Venant in France in deriving the equations of motion for a fluid with friction. Quite independently, he began with the concept of internal shear stresses in a moving fluid (Figure 4.3) and derived the governing equations for a viscous fluid (a fluid with internal friction), much as they would be derived today. In that process, he properly identified the dynamic viscosity coefficient μ as it appears in the Navier-Stokes equations in Appendix B. Stokes's work was published in 1845, two years after Saint-Venant's similar derivation.[28] Like most scientists studying fluid dynamics in the nineteenth century, Stokes dealt with incompressible flows. For such flows, the energy equation (as shown in Appendix B) is not essential. With that one exception, the situation as described by Stokes remains unchanged to the present day. Thus in modern fluid dynamics, where we deal with the Navier-Stokes equations on an almost daily basis, especially at the cutting edge of modern computational fluid dynamics, we are using ultramodern supercomputers to solve equations more than 150 years old that have weathered the test of time.

Interim Summary: The State of Theoretical Aerodynamics in 1850

In the context of our historical survey thus far (to the middle of the nineteenth century), the Navier-Stokes equations were the most powerful theoretical tools available. Prior to that time, the Euler equations, derived in the middle of the eighteenth century, had held that distinction. The Euler equations (valid only for inviscid flow) and the Navier-Stokes equations (valid for the more general case of viscous flow) were partial differential equations that, in principle, could allow direct determinations of pressure and velocity for an incompressible flow (and, in more recent times, also determinations of density and temperature for a compressible flow). However, the Euler equations had been sufficiently complicated that no mathematical solutions of any practicality had been obtained, and then along came the Navier-Stokes equations, which theoretically were an order of magnitude more complex than the Euler equations. Hence, by the middle of the nineteenth century, the tools necessary for theoretical aerodynamic analysis were in place, but nobody knew how to use them.

That dilemma led to an important development that would influence theoretical aerodynamics for the next 150 years: Because no general solutions could be obtained for either the Euler equations or the Navier-Stokes equations, after 1850 the research effort began to be directed toward *approximate* solutions. The idea was straightforward: One began by considering a specific type of flow or a specific application. Were there certain *physical* aspects of that flow that would justify dropping some of the mathematical terms that appeared in the Navier-Stokes equations, thus *simplifying* those equations? The physical argument would be that such terms were small in magnitude in comparison with other terms and therefore could be neglected. In some cases there might be terms that would be precisely zero because of the physical and/or geometric nature of a given application. Once those terms were neglected, were the resulting *approximate* forms of the Navier-Stokes equations (in such cases, no longer identified as the Navier-Stokes equations per se) simple enough to yield a mathematical solution? If the answer to both of the foregoing questions was yes, then some progress might be made. That philosophy of pursuing *approximate solutions* pervades aerodynamics to the present day. We shall be encountering it repeatedly, and it is important to appreciate the intellectual role that it plays in aerodynamics. Finally, we note that solutions for the *complete* Navier-Stokes equations are possible today only because of the modern techniques of *computational fluid dynamics,* and such solutions are purely numerical, rather than analytical. Such numerical solutions are feasible only with the use of modern high-speed digital computers (usually large supercomputers, such as a CRAY).

The Concepts of Vorticity and Vortex Filaments: Theoretical Advances Due to Helmholtz

The impasse presented by the extreme difficulty of finding analytical solutions to either the Navier-Stokes equations or the Euler equations for most practical fluid-dynamics problems seemed insoluble, but in the middle of the nineteenth century an answer was found – a major example of an "end run" around a seemingly intractable problem. That had to do with the introduction of a new way of describing the characteristics of a flow field, replete with new definitions and new conceptions. Central to that new approach were the concepts of *vorticity* and *vortex filaments* in the flow – concepts that were applied by the German scientist Hermann von Helmholtz.

In the sketch of the moving fluid element in Figure 4.3, the element is moving through space at velocity V. But there is a frictional shear stress on the top surface of the element pointing to the right, and a similar shear stress on the bottom surface pointing toward the left, with the result that the fluid element will have a tendency to *rotate* around some pivot point in the center of the element. In any real flow, all such fluid elements will have superimposed on their translational motions continuous changes in orientation that can be construed as rotational components of motion. Such a rotational component is sketched in Figure 4.6a and is labeled ω, the *angular velocity* of the fluid element. Flows wherein the fluid elements have a finite angular velocity are called *rotational flows*. Virtually all viscous flows are rotational, simply because of the existence of frictional shear stress within the fluid. In an idealized sense, flows with no friction (inviscid flows) may or may not be rotational. An inviscid flow that begins as a simple, uniform free stream far ahead of a body will not be rotational and will continue without rotation as it flows over the body, unless some extraneous mechanism is imposed on the flow and introduces rotation. An example of such an extraneous mechanism is a *curved* shock wave in supersonic flow; fluid elements flowing across a curved shock wave will be given a kind of "kick" that will introduce a rotational component into their motion. As stated earlier, flows with such rotational effects on the fluid elements are called rotational flows; in contrast, flows wherein the fluid elements have no rotational component of motion are called *irrotational flows*. A concept that is related to the preceding discussion is *vorticity,* defined as twice the angular velocity of a fluid element. In

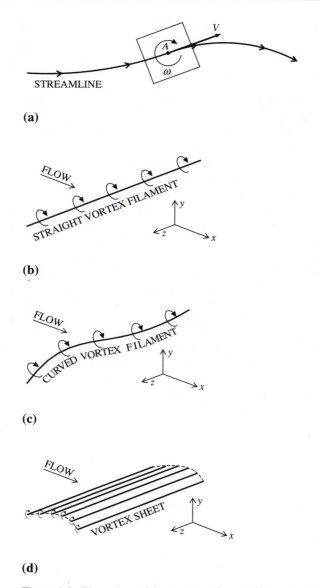

Figure 4.6   Illustrations of the concepts of vortex filaments and vortex sheets.

Figure 4.6a, the vorticity is equal to 2ω. Like angular velocity, vorticity is a point property of a flow – it can vary in magnitude from one point to another in a given flow field.

For an irrotational flow – one without vorticity – a special theoretical quantity can be defined, called the *velocity potential*, usually denoted by the symbol Φ. As discussed earlier, Joseph Lagrange introduced the concept of a velocity potential in his book *Mécanique analytique* (1787). The great advantage of the velocity potential is that its introduction into the Euler equations greatly simplifies those equations, thus allowing solution approaches that are sometimes tractable. Once Φ is obtained, the flow velocity can be extracted by differentiating Φ (in the sense of differential calculus). Indeed, mathematically Φ is *defined* as a scalar quantity from which the velocity components can be extracted by differentiation.

Once the idea of vorticity had entered the world of fluid dynamics, it was quickly expanded into two related concepts: the vortex filament and the vortex sheet. Figure 4.6a shows a fluid element moving in such a fashion that it has rotational motion around the point A at the same time that it is moving at velocity V. Imagine a straight line through point A, perpendicular to the page, such that the straight line connects other fluid elements that are also in rotational motion. Such a line is called a *straight vortex filament* (Figure 4.6b). The flow is in the x direction, and the vortex filament connecting adjacent fluid elements is shown along the z direction. A vortex filament does not have to be straight; it can be a curved line (Figure 4.6c). If we have a large number of straight vortex filaments, such as that in Figure 4.6b, lying side by side, we have a *vortex sheet* (Figure 4.6d).

If the concepts of vorticity, vortex filaments, and vortex sheets seem abstract and esoteric, to some extent that is correct. But they serve as conceptual tools that allow mathematical analyses of some types of flows – analyses that are essentially end runs around the Euler and Navier-Stokes equations. As is the case with most mathematical models designed for solution of physical problems, there is some physical relevance to the concepts of vortex filaments and vortex sheets. For example, consider two jets of fluid injected in parallel, but at different velocities (Figure 4.7a). There will be a thin mixing zone downstream of the jets within which large viscous (frictional) shear stresses will prevail, creating a thin zone of intense vorticity, as in Figure 4.7b, where the rotational motions of the fluid elements within the mixing zone are emphasized. This mixing zone is also called a *shear layer*. This shear layer is a physical counterpart of the mathematical model of a vortex sheet. Another example is the thin boundary layer along the surface of a body in a flow. Consider the airfoil sketched in Figure 4.8a, with a thin viscous boundary layer adjacent to the surface. That boundary

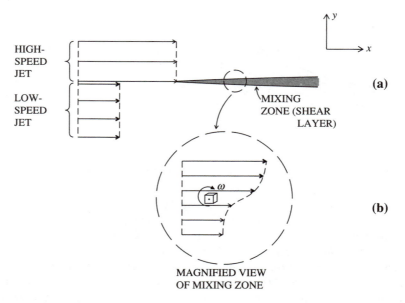

Figure 4.7 Mixing zone between two streams moving at different velocities.

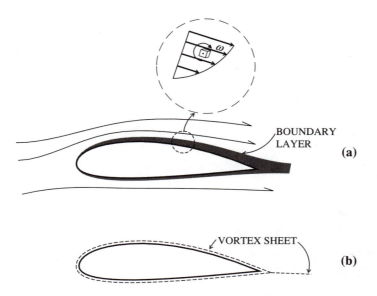

Figure 4.8 Analogy between the viscous boundary layer on a body and a vortex sheet on the surface.

layer is, by definition, the region where strong viscous shear stresses act near the surface and where large changes in the flow velocity occur (Figure 4.8a, inset). We shall examine the nature of boundary layers in more detail in later chapters; suffice it here to say that the thin boundary layer, because of the strong effects of viscous shear stresses, is a region of high vorticity, as reflected by the rotational motions of the fluid elements within the boundary layer (Figure 4.8a, inset). Thus the boundary layer is a type of shear layer that is a real-life counterpart of our conceptual vortex sheet. Indeed, we can imagine that there exists a vortex sheet wrapped around the surface of the airfoil (Figure 4.8b). In short, the mathematical models of vortex filaments and vortex sheets have some justification in real life.

The concept of a vortex sheet embodies an important characteristic. Imagine that the shear layer in Figure 4.7 is replaced by a vortex sheet (Figure 4.9). Note that the flow velocity changes *discontinuously* across this vortex sheet (i.e., changes discontinuously from V_1 to V_2 across the sheet). It is an important conceptual characteristic of a vortex sheet that the flow velocity changes discontinuously across the sheet, and it was that characteristic on which researchers in fluid dynamics in the middle of the nineteenth century attempted to capitalize.

Finally, we note that the conceptual models of vortex filaments and vortex sheets are applied to *inviscid* flows. These models represent mathematical discontinuities and singularities in such flows. However, in real-life flows, the mechanism of friction produces shear layers and boundary layers that, if thin enough, take on the trappings of the theoretical vortex sheet. Thus it is easy to see why some workers in the middle of the nineteenth century began to use the vortex sheet as a means to simulate some aspects of a flow that were inherently viscous in origin – again, an end run around their inability to solve the Navier-Stokes equations. The man most responsible for development of the theory of vortex filaments and vortex sheets was Hermann von Helmholtz.

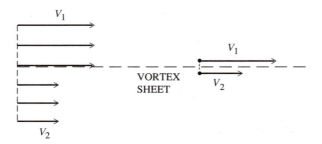

Figure 4.9   Across a vortex sheet, the flow velocity changes discontinuously.

Figure 4.10   Hermann Ludwig Ferdinand von Helmholtz.

Hermann von Helmholtz (1821–94) (Figure 4.10) was born at Potsdam, on the outskirts of Berlin, on August 31, 1821. Thwarted in his desire to attend a university to study physics, because of the family's financial status, he followed his father's advice to study medicine, for which he could obtain a government stipend in return for an eight-year commitment as an army surgeon. At the age of 17, Helmholtz entered the medical school of the Friedrich Wilhelm Institute in Berlin, receiving his M.D. degree four years later in November of 1842. During his medical studies, Helmholtz indulged his interest in physics by privately studying the works of Laplace, Biot, and Daniel Bernoulli. Assigned as a surgeon to the Potsdam regiment, he continued his interest in physics, publishing a paper on the mathematical principles of the conservation of energy in 1847. He maintained connections with scientific circles in Berlin, thereby gaining early release from his military obligation and appointment to the chair of physiology at Königsberg University in 1849. From that time on, Helmholtz pursued a life of intensive research, the first phase of which combined his interest in physics with his training in physiology to study how humans sense sound and how the eye processes light. In 1855 he accepted the chair of anatomy and physiology at the University of Bonn, where the dichotomy between his role as physician and his personal interest in physics grew more severe. That was exacerbated by his father, who continued to urge him to emphasize his medical activities and forget physics. But Helmholtz was rapidly gaining a reputation throughout Germany as a first-rate physicist. His father died in 1858, and Helmholtz turned increasingly toward physics. Never comfortable teaching anatomy, he left Bonn in 1858 to accept a chair at Heidelberg, then famous as a center for scientific activity. The 13 years at Heidelberg were the most productive in his professional career, leading to an appointment at the University of Berlin in 1871, where he would direct a massive new scientific laboratory. By 1885, Helmholtz was the most respected scientist in Germany, serving as the state's top scientific advisor. His status was enhanced by his marriage to the daughter of a Heidelberg professor, a beautiful, sophisticated woman much younger than Helmholtz who opened a broad range of new social contacts for the naturally reserved and shy physicist. Helmholtz remained at Berlin for the rest of his life. His health began to fail after 1885; the migraine headaches that had plagued him throughout life became much worse, causing long periods of professional inactivity, and in old age he suffered from major bouts of depression. On September 8, 1894, he died from complications following a stroke, and the world lost one of the most famous classical physicists of the nineteenth century.

Helmholtz's work was a turning point in the development of theoretical aerodynamics. He was the first to suggest the role of vorticity and to use the concepts of vortex filaments and vortex sheets in the analysis of invisid flow fields. Later, those would be the essential tools with which Kutta and Joukowski, and especially Prandtl, at the beginning of the twentieth century, would develop and implement the very powerful and important circulation theory of lift. Helmholtz's most important ideas relating to the advancement of aerodynamics were the following:

(1) After Cauchy[29] introduced the concept that fluid elements moving throughout a flow field could have rotational motion superimposed on their translational motion and Stokes[28] incorporated the idea of fluid-element rotational motion in his discussion of the flow of a viscous fluid, Helmholtz was the first to apply that concept to inviscid (frictionless) flow. Helmholtz coined the term *vorticity* and showed that the useful concept of a *velocity potential,* introduced many years earlier by Lagrange, was not valid for a rotational flow. In other words, the existence of a velocity potential presupposes an *irrotational* flow. That was an important restriction on the use of the velocity-potential concept and a major contribution by Helmholtz: "A research of the forms of motion in which no potential of the velocity exists seems to me of great interest. This research leads to the result that in the cases in which a potential of the velocity exists the smallest fluid particles do not possess rotatory motions, whereas when no such potential exists, at least a portion of these particles is found in rotatory motion."[30]

What was the advantage of defining a velocity potential in fluid dynamics, and what were its practical consequences? In the Euler equations (Appendix A) we have a number of partial differential equations with lots of unknowns (density, velocity, etc.). When a flow can be described in terms of the velocity potential, the Euler equations reduce to only one equation with one unknown: Φ. Lagrange knew that and used it, but he never appreciated that it should be restricted to only irrotational flows. Helmholtz made that important distinction and coined the term "velocity potential" for Φ.

(2) After Helmholtz made the distinction between irrotational and rotational flows, he went on to study the consequences of such flows, especially the latter, formulating and studying the vortex sheets sketched in Figure 4.6. Several fundamental mathematical properties of those vortex models, known in modern fluid dynamics

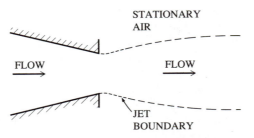

Figure 4.11   Model of a jet of air exhausting into the surroundings.

as *Helmholtz's vortex theorems*, were derived in his 1858 paper.[30] A key aspect of those vortex models is that the flow everywhere away from the vortex filaments and vortex sheets is irrotational (i.e., the vorticity is zero), but a filament or sheet itself is a singular line or sheet where the vorticity is infinite.

(3)   The singular mathematical nature of those vortex filaments and sheets led Helmholtz to the idea of *surfaces of discontinuity* in a real flow. In a paper published in 1868, Helmholtz reported his observation that a jet of water streaming into a reservoir of water had a well-defined boundary; it did not immediately diffuse into the surrounding water in all directions.[31] He also noted that the same phenomenon was seen when a jet of air was exhausted into the surroundings; when made visible by addition of smoke, a well-defined jet boundary appeared. This is sketched in Figure 4.11, where the jet boundary is shown as the dashed curve. Comparing Figures 4.9 and 4.11, we can make the intuitive leap of concluding that the physical jet boundary shown in Figure 4.11 can be modeled mathematically by the vortex sheet in Figure 4.9. In this case, there is a finite jet velocity on one side of the vortex sheet, and a zero velocity (the stationary air outside the jet) on the other side of the vortex sheet. In his 1868 paper Helmholtz made such an intuitive leap, thus introducing the concept of a *surface of discontinuity* (the vortex sheet) in an inviscid flow. What is discontinuous across this surface is the component of flow velocity tangent to the sheet.

In light of the foregoing discussion we cannot claim that Helmholtz's work led immediately to some quantum leap in the *practice* of aerodynamics – in fact, far from it. Helmholtz expressed little interest in the problem of powered flight, and none of his contemporaries was motivated to convert his esoteric and mathematical ideas to practical aerodynamic theory. However, those ideas were seeds that lay dormant for 40 years before flowering in the early twentieth century in the work of Kutta, Joukowski, and especially Prandtl. Without Helmholtz's pioneering concepts, theoretical aerodynamics most likely would have followed a slower and more difficult course.

Surfaces of Discontinuity: A Blind Alley for Drag Predictions

Figure 2.7 shows da Vinci's sketch of water flowing around a flat slab oriented first perpendicular and then parallel to the flow. The regions of separated flow are immediately apparent, bounded by what look like distinct surfaces across which rapid changes take place. Those surfaces, when viewed from afar, take on the appearance of "surfaces of discontinuity." Such phenomena are everyday observations in the world around us. Moreover, such

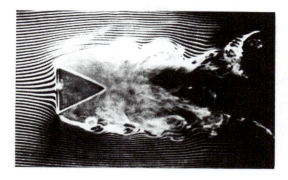

Figure 4.12   Smoke-flow photographs showing the streamlines over different bodies.

surfaces of discontinuity appear in controlled laboratory experiments (e.g., Helmholtz's observations of a jet of water streaming into a reservoir, and an air jet, made visible by smoke, exhausting into the surroundings). Figure 4.12 shows smoke-flow photographs taken by the French physiologist Etienne-Jules Marey in 1899. The upper photograph shows the flow over a wedge whose base is facing directly into the flow, and the lower photograph shows the flow over slender, airfoil-like shapes at a large angle of attack to the flow. Marey built the first smoke tunnel,[32] a device to make an airflow visible by injection of a number of small smoke filaments that subsequently flow over a model and give a reasonable visualization of the streamline pattern.[33] Those photographs clearly show regions of separated flow and what appear to be surfaces of discontinuity. Marey's photographs were taken 31 years after Helmholtz published the theoretical concept of such surfaces of discontinuity and obviously were not a factor in Helmholtz's thinking. However, those photographs helped to establish at a fairly early date that such surfaces do occur in real life.

It is no surprise that soon after Helmholtz's 1868 paper there were efforts to use the concept of surfaces of discontinuity as the means to predict aerodynamic drag on bodies. A century earlier, d'Alembert had shown that solutions for inviscid flows over two-dimensional body shapes always yielded the finding of zero drag – d'Alembert's paradox. With Helmholtz's introduction of the concept of surfaces of discontinuity, suddenly there was the prospect of resolving d'Alembert's paradox. The qualitative physical basis for that thinking is shown in Figure 4.13, with a flat plate at some angle of attack to the flow. According to d'Alembert's data, the aerodynamic drag on that plate should be zero. However, if

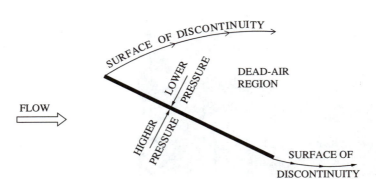

Figure 4.13 Hypothetical concept of a surface of discontinuity generated at the leading edge of a sharp flat plate.

Figure 4.14 Gustav Robert Kirchhoff.

there are two surfaces of discontinuity that trail downstream from the plate, one originating at the leading edge and the other originating at the trailing edge, then theoretically a region of low-energy, low-velocity flow – essentially a dead-air region – can exist between the two surfaces. In that low-energy, dead-air region, the static pressure will be low. Thus the pressure exerted over the top surface of the plate will have a low value compared with the pressure exerted over the bottom surface. As a result of that model of the flow, there clearly would be a net pressure drag on the plate, effectively nullifying d'Alembert's paradox. That qualitative flow picture, with its potential for prediction of aerodynamic drag, was particularly alluring to two contemporaries of Helmholtz: the German scientist Gustav Robert Kirchhoff (1824–87) (Figure 4.14) and the noted English scientist John William Strutt, more commonly known as Lord Rayleigh (1842–1919) (Figure 4.15). They worked independently on the problem, each unaware of the progress of the other. In 1869, only a year after Helmholtz had published the idea of surfaces of discontinuity, Kirchhoff

Figure 4.15   John William Strutt, Lord Rayleigh.

published a paper in which he considered the force exerted on a flat plate (a "lamina"), both perpendicular to the flow and inclined at an oblique angle to the flow, utilizing the model of surfaces of discontinuity.[34] For some unknown reason, Kirchhoff did not quantitatively evaluate the force for the case of the inclined plate. In 1876, Rayleigh published a paper on the same problem,[35] deriving equations for both the "mean pressure" and the center of pressure for a flat plate inclined obliquely to the flow. Referring to those two equations, Rayleigh commented in a footnote as follows: "Formulae (3) and (4) were given at the Glasgow Meeting of the British Association. I was then only acquainted with Kirchhoff's *Vorlesungen über mathematische Physik,* and was not aware that the case of an oblique stream had been considered by him (*Crelle,* Bd. LXX, 1869). However, Kirchhoff has not calculated the forces; so that the formulae are new." Rayleigh was making it clear that although he had been unaware of Kirchhoff's work, equations (3) and (4) in Rayleigh's paper still represented a new contribution to the state of the art.

In his 1876 paper, Rayleigh used a potential-flow solution (a solution for Laplace's equation for inviscid, incompressible flow) in combination with the flow model sketched in Figure 4.13 to derive a formula for the normal force on a flat plate at various angles of attack to the flow. The normal force, by definition, is the aerodynamic force perpendicular to the plate (Figure 4.16). By analogy to the lift and drag equations in Chapter 1, we can write the normal force as

$$N = \tfrac{1}{2}\rho V^2 S C_N$$

where C_N is the normal-force coefficient, and S is the area of the plate. In this equation, C_N is a function of the angle of attack α. The resulting equation for normal force obtained by Rayleigh is

$$N = \frac{\pi \sin \alpha}{4 + \pi \sin \alpha}\rho V^2 S$$

and hence

$$C_N = \frac{2\pi \sin \alpha}{4 + \pi \sin \alpha} \qquad \text{(from Rayleigh's paper)}$$

It is interesting to compare that with the Newtonian result, which is

$$N = (\sin^2 \alpha)\rho V^2 S$$

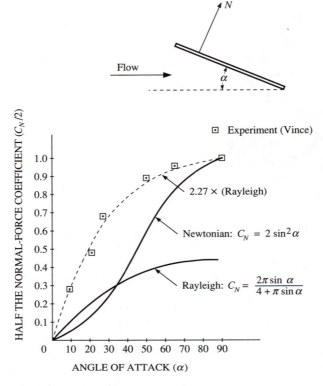

Figure 4.16   Normal-force coefficient for an inclined flat plate, comparison between theory and experiment.

and hence

$$C_N = 2\sin^2\alpha \qquad \text{(from Newtonian theory)}$$

The values obtained from Rayleigh and from Newtonian theory are shown in Table 4.1, compared with some experimental values obtained by Vince in 1798 and quoted by Rayleigh.

The first column is the angle of attack for the flat plate. The remaining columns tabulate the normal-force coefficient divided by 2. As discussed in Chapter 3, Newtonian theory gives very poor results for inclined bodies. That is reinforced by comparing the second column with the fourth column, which lists the experimental values obtained by Vince and published in 1798 in the *Philosophical Transactions*. Vince made his measurements on a flat plate in water, using a whirling-arm apparatus. Clearly, the Newtonian values do not even come close to agreeing with the experimental data, but neither do the values obtained by Rayleigh, tabulated in column three. Even with the plate perpendicular to the flow (an angle of attack of 90°), Rayleigh's value is smaller than the experimental value by more than a factor of 2. That did not phase Rayleigh. He simply *adjusted* his findings to agree with experiment at a 90° angle of attack. To accomplish that, he had to multiply his finding by a factor of 2.27. Then he ratioed *all* his remaining values at all angles of attack by that same factor. Rayleigh's *modified* values are tabulated in the fifth column. Remarkably,

Table 4.1. *Various values obtained for the normal-force coefficient*

α (degrees)	$C_N/2 = \sin^2\alpha$ (Newtonian)	$C_N/2 = \dfrac{\pi \sin\alpha}{4 + \pi \sin\alpha}$ (Rayleigh)	$C_N/2$ (experimental) (Vince)	$(2.27)\dfrac{\pi \sin\alpha}{4 + \pi \sin\alpha}$ (Rayleigh, modified)
90	1.000	0.440	1.000	1.000
70	0.883	0.425	0.974	0.965
50	0.587	0.376	0.873	0.854
30	0.250	0.282	0.663	0.640
20	0.117	0.212	0.458	0.481
10	0.030	0.120	0.278	0.272

after that adjustment, Rayleigh obtained excellent agreement over the whole range of angles of attack, the maximum discrepancy being less than 5%. Rayleigh made no apology for such an adjustment; he simply stated that "the result of Vince's experiments agrees with theory remarkably well." He did not elaborate at all on the fact that he had had to adjust his theoretical values by a factor of 2.27.

Rayleigh, a respected physical scientist, was operating in the mode of an engineer. Walter Vincenti[36] has drawn a clear distinction between the fundamental learning and thought processes of engineers and those of physical scientists. Vincenti makes the point that the end product of scientific thought is an "increment in knowledge," whereas the end product of engineering thought is an "artifact" – an engineering design or an operating system of some sort – with the generation of more knowledge being only secondary. Engineers construct physical models, using science and mathematics as accurately as is practical; however, there are many situations in which practical engineering solutions can be obtained only by resorting to gross simplifying approximations about the physics of the problem. Indeed, in some extreme cases, an engineering analysis is known to be based on an incorrect model, and yet it produces, for whatever reason, findings that are close enough to the true situation to be useful: "Such analytical assumptions are often realized to be wrong. They again are used for practical reasons and because they are known from experience to give conservative or otherwise acceptable results. Without them, a great deal of everyday design would not get done" (p. 215).[36] That kind of thinking is acceptable in engineering, and it was precisely that approach that Rayleigh took when he adjusted his theoretical findings, making no apology for the fact that his adjustment factor (2.27) was large.

To facilitate a further comparison of the results, Figure 4.16 shows a graph of the normal-force coefficient (divided by 2) as a function of the angle of attack. The small squares denote the experimental data obtained by Vince. The curve labeled Newtonian shows the results from Newton's sine-squared law, discussed in Chapter 3. Clearly, the Newtonian theory does not agree with experiment. Rayleigh's theoretical findings fare even worse; the curve labeled Rayleigh does not come close to the experimental data. But when Rayleigh's theoretical findings are multiplied by the factor of 2.27, the dashed curve is obtained, closely matching the experimental data, and prompting Rayleigh to claim that "the result of Vince's experiments agrees with theory remarkably well."

What was the problem with Rayleigh's theory? Why were his findings so far from the experimental values? From the vantage point of modern aerodynamics, the answer is straightforward. Return to Figure 4.13, which shows a flat plate at an arbitrary angle of attack, with the two surfaces of discontinuity bounding a separated-flow region over the back of the plate. Rayleigh treated that separated region as a dead-air region, and he assumed that the pressure in that region (and hence the pressure exerted on the top surface of the plate) was equal to the free-stream pressure far ahead of the plate. That is, if the flow far ahead of the inclined plate had a pressure of 1 atm, then the pressure in the dead-air region, and consequently the pressure exerted over the top surface of the plate, was assumed by Rayleigh also to be 1 atm. Today we know that there will indeed be a separated-flow region over the top surface of the plate, but the pressure in that region will be less than the free-stream pressure. Because Rayleigh assumed too high a pressure over the top surface, and that pressure was acting downward, tending to counteract the high pressure over the bottom surface, his prediction of the normal force was too small by the factor of 2.27. Modern data for the pressure over the back surface of a flat plate oriented at 90° to the flow can be found in *Fluid-Dynamic Lift*,[37] a principal source for aerodynamic data on lift, and companion to Hoerner's earlier book on drag.[38] The pressure over the back surface of a plate at a 90° angle of attack is given by

$$p = p_\infty - (1.1)\tfrac{1}{2}\rho V_\infty^2$$

where p_∞ denotes the free-stream pressure, ρ is the density of the fluid, and V_∞ is the free-stream velocity (implying a pressure coefficient of -1.1 over the back surface). Clearly, from this equation, the pressure over the top surface of the plate is less than the free-stream pressure p_∞. More important, with this more realistic upper-surface pressure, Rayleigh's original theoretical value of $C_N/2 = 0.44$ at a 90° angle of attack is modified by adding $\tfrac{1}{2}(1.1) = 0.55$. Thus, on the basis of modern data, Rayleigh's theory, properly modified for the correct pressure over the top of the plate, gives, for a 90° angle of attack $C_N/2 = 0.44 + 0.55 = 0.99$, essentially in agreement with the measured value of 1.0. Furthermore, at lower angles of attack, experience has shown that the pressure coefficient in the separated flow over an airfoil just beyond the occurrence of stall is about -0.6 at an angle of attack of about 20°. Taking Rayleigh's theoretical value of $C_N/2 = 0.212$ at a 20° angle of attack, we modify that for the more realistic pressure over the top surface, yielding $C_N/2 = 0.212 + \tfrac{1}{2}(0.6) = 0.512$, not too far from the measured value of 0.458.

Rayleigh's model of the flow over a flat plate at some angle of attack, with surfaces of discontinuity emanating from the leading and trailing edges, as sketched in Figure 4.13, was a good idea – it reflected the real, separated flow that actually occurs in nature. His error was in assuming free-stream pressure in that separated region. Today we know that the pressure in that region is less than the free-stream pressure. Because of that error, Rayleigh's theory did not directly advance the state of the art for prediction of aerodynamic drag and therefore was not used to any extent by designers of flying machines in subsequent years. From that point of view, the flow model using surfaces of discontinuity was a blind alley for drag predictions. On the other hand, his theoretical value for the mean pressure over the bottom of the flat plate was reasonably correct. His equation for that mean pressure,

$$\text{Mean pressure} = \frac{\pi \sin\alpha}{4 + \pi \sin\alpha}\rho V_\infty^2$$

is still used in the modern literature.[37,38]

A Note about Lord Rayleigh and His Contributions to Aerodynamics

We have discussed how Rayleigh's theoretical analysis, using the model of surfaces of discontinuity, led to a blind alley for drag prediction. That was not to his discredit. Indeed, his innovative use of Helmholtz's original work on surfaces of discontinuity showed a substantial degree of insight and originality. In fact, if he had had access to the modern aerodynamic findings regarding the pressure in the separated region over the top of a flat plate at various angles of attack, he would have been able to make a drag prediction that would have been right on the mark. In that sense, Rayleigh's method of analysis, considering the time at which he was conducting his research, was really quite exemplary. Moreover, his work on the prediction of forces on a flat plate was only part of his overall contribution to aerodynamics.

Lord Rayleigh was born John William Strutt, November 12, 1842, in Essex, England. He was educated at Cambridge University, where he was a pupil of the mathematician E. J. Routh and carefully followed the lectures of Sir George Stokes. In 1866, Strutt became a fellow of Trinity College at Cambridge; he was at the beginning of what promised to be a distinguished career in science. After graduation, Strutt visited the United States – not the usual travel choice for Europeans at that time, especially considering that America was just beginning to recover from the Civil War. On returning to England in 1868, he set up an experimental laboratory at the family seat in Terling Place that was to become the focal point for the rest of his life, as well as a famous scientific facility. In 1871 he married Evelyn Balfour, sister of Arthur James Balfour, well-known scholar and statesman. By that time, Strutt had begun work on his famous two-volume treatise *The Theory of Sound,* which is still the "bible" for researchers in the field of acoustics. In 1873 he succeeded to his inherited title and became Lord Rayleigh.

The decade of the 1870s was a time of major productivity for Rayleigh. He worked in the areas of radiation, acoustics, optics, and fluid dynamics, carrying out his theoretical work on the aerodynamic forces on flat plates. He also provided the scientific explanation for the blue appearance of the sky, based on his derivation of a law for light scattering by small particles, showing that the scattering is inversely proportional to the fourth power of the wavelength of the incident radiation. Visible radiation of the shortest wavelength is in the blue part of the light spectrum, and the shortest wavelengths experience the most scattering of all the visible wavelengths. As a result, when the sun's rays are incident on the earth's atmosphere, the sky appears blue.

In 1879 Rayleigh accepted the Cavendish professorship at Cambridge, vacated by the death of James Clerk Maxwell. He had an enduring effect on scientific education in England and in the United States, elevating the laboratory component of instruction in universities to a more respected and important role. Indeed, the emphasis he placed on experimental work, as well as the growing importance of establishing standards in physical measurements, eventually led, in no small measure due to his influence, to the establishment of the National Physical Laboratory in 1900 at Teddington in Middlesex. Rayleigh's reputation and influence continued their upward spiral; he became a part-time professor at the Royal Institution of Great Britain in London, served as the president of the British Association for the Advancement of Science, and was an advisor to many scholarly and governmental committees. He was elected to the Royal Society in 1873 and served as its secretary for 11 years.

During the period 1892–5 Rayleigh conducted a series of experiments in his laboratory at Terling Place that resulted in the discovery and isolation of argon, earning him the Nobel Prize in physics in 1904.

In 1905, Rayleigh published a definitive paper on shock waves in supersonic flow (as discussed further in a later chapter). In the 1876 paper[35] in which he reported his work on drag predictions for a flat plate, Rayleigh briefly discussed an equation for calculation of the stagnation pressure in a high-speed compressible flow, an equation still used today in the teaching of compressible flow. Rayleigh simply mentioned the compressibility effect in passing, making no use of it in his force predictions for a flat plate. In that same paper he noted that the compression process that creates a high stagnation pressure on a high-velocity body also results in a correspondingly large increase in temperature. In particular, he commented on the flow-field characteristics of a meteor entering the earth's atmosphere: "The resistance to a meteor moving at speeds comparable with 20 miles per second must be enormous, as also the rise of temperature due to the compression of the air. In fact it seems quite unnecessary to appeal to friction in order to explain the phenomena of light and heat attending the entrance of a meteor into the earth's atmosphere."[35] Note that 20 miles per second is equal to a velocity of 105,600 ft/s. Because the speed of sound in the earth's atmosphere is approximately 1,000 ft/s (it varies with temperature through the atmosphere), Rayleigh was talking about a Mach number greater than 100 (the highest velocity achieved in manned flight is 36,000 ft/s, or about Mach 36, the entry velocity of the Apollo command module returning from the moon). That range of Mach numbers is termed *hypersonic* in today's nomenclature. Rayleigh's statement, quoted earlier, is believed to be the first mention in the scientific literature of the implications of hypersonic aerodynamic heating.

Unlike most of the influential scientists and mathematicians of the eighteenth and nineteenth centuries who took absolutely no interest in heavier-than-air, manned flying machines, Rayleigh was an interested proponent of the idea. For example, on January 19, 1900, in his capacity as professor of natural philosophy at the Royal Institution of Great Britain, he delivered a discourse entitled "Flight" to the Royal Aeronautical Society. In the extended description of his talk subsequently published in the *Aeronautical Journal*, we find *no mention whatsoever of the application of theoretical fluid mechanics to flight,* a graphic example of the situation highlighted earlier: Basic advances in theoretical fluid dynamics, even as late as the end of the nineteenth century, did not find their way into the design of flying machines. There were two primary reasons for that situation: (1) The basic equations of fluid dynamics did not lend themselves to exact, analytical solutions and hence were difficult to apply to the practical problems of flight. (2) The quest for powered flight still carried the stigma of being somewhat irrational, and hence it was not a popular cause within most scientific circles. Rayleigh was somewhat of an exception – witness his interaction with the Royal Aeronautical Society and his appointment in 1909 as chairman of Britain's Advisory Committee for Aeronautics. (The Advisory Committee for Aeronautics in Britain later became the role model for the creation of the National Advisory Committee for Aeronautics in the United States in 1915.) Some historians have called Rayleigh the last great "polymath," the last scientist who might qualify as a Renaissance man. Perhaps his interest in powered flight reflected such an outlook. In any event, we can readily single out Rayleigh as the *first scientist of great stature to embrace the idea of powered flight.* By way of contrast, Lord Kelvin, an equally prestigious colleague, was quoted as saying "I have not the smallest molecule of faith in aerial navigation other than ballooning." Kelvin was expressing the prevailing opinion on the matter.

Rayleigh died at Terling Place on June 30, 1919. He had witnessed the beginning of powered flight, and at the time of his death, Prandtl's research group at Göttingen University was unraveling the secrets of the governing flow equations, using applied mathematical techniques to obtain meaningful lift and drag predictions for airfoils and wings. Rayleigh could not participate in that stage, but his work in fluid dynamics and the lending of his prestige to aeronautical pursuits provided momentum for the evolution of aerodynamics.

Osborne Reynolds and His Studies of Turbulent Flow

There are two types of viscous flows: *laminar flow,* in which the fluid elements move in a regular, ordered fashion and adjacent streamlines move smoothly over each other as if they were part of a medium made up of different will-ordered laminae, and *turbulent flow,* in which the fluid elements move in a disordered fashion and the streamlines form a tortuous, mixed-up, irregular pattern. The viscous shear stresses that cause skin-friction drag on a body (Chapter 1) are higher for turbulent flow than for laminar flow. Hence, it is vital to know whether the flow is laminar or turbulent. In reality, the viscous flow over a body starts out at the leading edge as laminar and then undergoes a transition to turbulent flow somewhere downstream of the leading edge. For accurate prediction of skin-friction drag, it is vital to have knowledge of where on the body surface this transition takes place (called the transition point). An understanding of the fundamental nature of turbulent flows sufficient to allow prediction of their properties is still today one of the unsolved problems of classic physics. In turn, accurate prediction of the transition point is one of the most challenging problems in modern aerodynamics. An important first step in the study of the transition from laminar flow to turbulent flow was taken in the latter part of the nineteenth century by Osborne Reynolds.

Osborne Reynolds (1842–1912) (Figure 4.17) was born October 23, 1842, in Belfast, Ireland. He was raised in an intellectual family atmosphere; his father had been a fellow of Queens' College, Cambridge, a principal of Belfast Collegiate School, headmaster of Dedham Grammar School in Essex, and finally reactor at Debach. Already in his teens he showed intense interest in the study of mechanics and appeared to have a natural aptitude. At the age of 19, he served a short apprenticeship in mechanical engineering before entering Cambridge University a year later. Reynolds was a highly successful student, graduating with highest honors in mathematics. In 1867 he was elected a fellow of Queens' College, Cambridge, only a year after Rayleigh had been elected a fellow of Trinity College (Reynolds

Figure 4.17   Osborne Reynolds.

and Rayleigh had been classmates at Cambridge). Reynolds went on to spend a year as a practicing civil engineer in London.

In 1868, Owens College (later the University of Manchester) established its chair of engineering, the second such chair in an English university (the first had been the chair of civil engineering at University College, London, in 1865). Reynolds applied for the Owens chair, writing in his application that "from my earliest recollection I have had an irresistible liking for mechanics and the physical laws on which mechanics as a science is based. In my boyhood I had the advantage of the constant guidance of my father, also a lover of mechanics and a man of no mean attainment in mathematics and their application to physics" (obituary, by Horace Lamb, *Proceedings of the Royal Society,* ser. A, vol. 88, February 24, 1913). Despite his youth and relative lack of experience, Reynolds was appointed to the chair at Manchester, where he remained until his retirement in 1905.

During his 37 years at Manchester, Reynolds distinguished himself as one of the leading practitioners of classical mechanics. He worked on problems involving electricity, magnetism, and the electromagnetic properties of solar and cometary phenomena. After 1873, he focused on fluid mechanics – the area in which he would make his most important contributions.

Reynolds was a scholarly man, with high standards. Engineering education was new to English universities at that time, and Reynolds had definite ideas about its proper form. He believed that all engineering students, no matter what their specialty, should have a common background based in mathematics, physics, and particularly the fundamentals of classical mechanics. He organized a systematic engineering curriculum at Manchester covering the basics of civil and mechanical engineering. Despite his intense interest in education, he was not a great lecturer. His lectures were difficult to follow, and he frequently wandered among topics with little or no connection. He was known to stumble upon new ideas during the course of a lecture and to spend the remainder of the time working out those ideas at the blackboard, oblivious to his students. He did not spoon-feed his students, and many did not pass his course, but the best students enjoyed his lectures and found them stimulating, such as J. J. Thomson, who in 1906 received the Nobel Prize in physics for demonstrating the existence of the electron.

In regard to Reynolds's emphasis on a research approach, his student and colleague, Professor A. H. Gibson, commented as follows in his biography of Reynolds, written for the British Council in 1946: "Reynolds's approach to a problem was essentially individualistic. He never began by reading what others thought about the matter, but first thought this out for himself. The novelty of his approach to some problems made some of his papers difficult to follow, [but his] more descriptive physical papers . . . make fascinating reading, and when addressing a popular audience, his talks were models of clear exposition."

At the turn of the century, Reynolds's health began to fail, considerably diminishing his physical and mental capabilities, a particularly sad state for such a brilliant scholar. He died at Somerset, England, in 1912. Sir Horace Lamb, noted researcher in fluid dynamics and a longtime colleague of Reynolds, commented as follows:

> The character of Reynolds was like his writings, strongly individual. He was conscious of the value of his work, but was content to leave it to the mature judgement of the scientific world. For advertisement he had no taste, and undue pretension on the part of others only elicited a tolerant smile. To his pupils he was most generous in the opportunities for valuable work which he put in their way, and in the share of cooperation. Somewhat reserved in serious or personal matters and occasionally combative and tenacious in debate, he was in

the ordinary relations of life the most kindly and genial of companions [obituary, by Horace Lamb, *Proceedings of the Royal Society*, ser. A, vol. 88, February 24, 1913].

Reynolds's three contributions to fluid mechanics were pivotal and seminal. The first was his study of the transition from laminar flow to turbulent flow in pipes. To put that contribution in perspective, we must fall back two decades and examine the work of a German hydraulics engineer, Gotthilf Heinrich Ludwig Hagen (1797–1884). Hagen was the first to report that two distinct types of flow could exist inside pipes, hinting of that situation in a closing remark in a paper published in 1839.[39] Concerning the flow of water in pipes, he referred to "strong movement" that the water demonstrated under certain flow conditions. He went on to express a certain degree of frustration: "The exact investigation of the results produced in this case appears hence to offer great difficulties; at least I have not yet succeeded in clarifying sufficiently the peculiarities which are then evidenced."[39] The strong movements observed by Hagen were associated with what today we call a turbulent flow. A more graphic description was given by Hagen in a paper published in 1855,[40] discussing the effects of heating a tube through which water was flowing. The tubes were made of glass to enable him to observe the nature of the flow:

> Since I invariably had the efflux jet before my eyes, I noticed that its appearance was not always the same. At small temperatures it remained immovable, as though it was a solid glass rod. On the other hand, as soon as the water was more strongly heated, very noticeable fluctuations of short period were established, which with further heating were reduced but nevertheless even at the highest temperatures did not wholly disappear.... With each repetition of the experiment the same phenomenon occurred, and when I finally made the graphic summary, I found that the strongest fluctuations always took place in that portion of the curve where the velocity decreased with increasing temperature....
> Special [observations] that I made with glass tubes showed both types of movement very clearly. When I let sawdust be carried through with the water, I noticed that at low pressure it moved only in the axial direction, whereas at high pressure it was accelerated from one side to the other and often came into whirling motion [p. 159].[2]

From the perspective of modern aerodynamics, we understand what happened in Hagen's experiment: Laminar flow was destabilized by addition of heat to the flow. In Hagen's experiment, at low temperature the water flow through the small glass tube was a laminar flow that was stable. Because heat was added, thus increasing the flow temperature, the laminar flow was shifted from a stable regime to an unstable regime. Given even a slight disturbance, that heated, unstable laminar flow easily made the transition to turbulent flow, just as Hagen described.

Hagen did not determine *quantitative* criteria for the conditions at which the transition from laminar flow to turbulent flow would occur. That was where Reynolds's contribution became so important. In 1883, Reynolds reported his findings from a series of fundamental experiments that would have lasting effects for analyses of where the transition from laminar flow to turbulent flow would occur.[41] His work, like that of Hagen, showed that there could be two distinct types of viscous flow – laminar and turbulent – but Reynolds's experiments were better controlled and better designed for quantitation than those of Hagen. Reynolds's experimental apparatus is shown in Figure 4.18 (an example of the elegant sketches of experimental apparatus seen in some technical papers before the development of modern photographic techniques). Reynolds filled a large reservoir with water, which fed into a glass pipe through a larger bell-mouth entrance. As the water flowed through the pipe, he introduced dye into the middle of the stream at the entrance of the bell mouth. Figure 4.19

Figure 4.18   Reynolds's experimental apparatus for studying transition (1883).

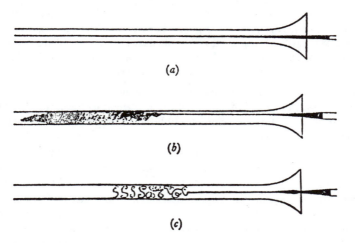

Figure 4.19   Reynolds's sketches of the transition phenomena for flow in a pipe.

(also from Reynolds's original paper) shows what happened to that thin filament of dye as it flowed through the pipe. The flow was from right to left. If the flow velocity was low, the thin dye filament would travel downstream in a smooth, neat, orderly fashion, with clear demarcation between the dye and the rest of the water (Figure 4.19a). If the flow velocity was increased beyond a certain value, the dye filament would suddenly become unstable and fill the entire pipe with color (Figure 4.19b). Reynolds clearly pointed out that the smooth dye filament corresponded to laminar flow in the pipe, whereas the agitated and totally diffused dye filament was due to turbulent flow in the pipe. Furthermore, he studied the details of that turbulent flow by visually observing the pipe flow illuminated by a momentary electric spark, much as we would use a strobe light today. He saw that the turbulent flow consisted of a large number of distinct eddies (Figure 4.19c). The transition from laminar flow to turbulent flow occurred when the parameter defined by $\rho V D/\mu$ exceeded a certain critical value, where ρ was the density of the water, V was the mean flow velocity, μ was the viscosity coefficient, and D was the diameter of the pipe. That dimensionless parameter, first introduced by Reynolds, would become known as the Reynolds number. Reynolds determined that the critical value for that parameter, the value above which turbulent flow would occur, was 2,300. That was indeed a fundamental finding: It indicated that the transition phenomenon did not depend simply on velocity by itself, nor on density by itself, nor on the size of the flow by itself, but rather on the particular *combination* of the variables defined earlier that make up the Reynolds number. No matter what the velocity or density or viscosity of the flow, and no matter what the size of the pipe through which the flow was moving, transition would occur, according to Reynolds's calculations, at a value of 2,300 for the combination $\rho V D/\mu$. That was a stunning discovery. Accurate determination of where on a surface the transition from laminar flow to turbulent flow will occur is perhaps the highest priority in modern aerodynamics, and the use of Reynolds numbers for that determination is still the approach used today.

Reynolds's second major contribution was the conception and implementation of a theoretical model for analysis of a turbulent flow – for detailed calculation of the velocity, density, and temperature distributions throughout a turbulent flow field. The governing equations for the flow-field variables in a viscous flow are the Navier-Stokes equations (Appendix B), partial differential equations written in terms of pressure p, density ρ, the x and y components of velocity (u and v, respectively), and temperature T. Solution of these equations will give, in principle, the variations of those properties over the whole x–y–z space as functions of time. For simplicity, let us consider a *steady flow,* where the flow-field variables at all points in the flow are independent of time, meaning that if we have a steady flow and we lock our attention onto a particular point moving in the flow, p, ρ, u, v, and T will not change – they will retain their same values, independent of time. This presupposes a steady, *laminar* flow; for such a flow there are no fluctuations at any given point. The situation is quite different for a turbulent flow. The experiments of Reynolds (and those of Hagen before him) showed that a turbulent flow is characterized by continually recurring turbulent eddies (some small, some large) that cause timewise fluctuations in the flow at any given point. Reynolds sketched turbulent, fluctuating flow in Figure 4.19c. No matter how small the turbulent eddies, a turbulent flow locally at any given point is an *unsteady* flow. In a turbulent flow, if we lock our attention onto a given point, we will see that the local p, ρ, u, v, and T values are changing as a function of time. However, Reynolds theorized that if one took a suitable *time average* for each flow property in a turbulent flow, that time average would be a steady value. Taking a hint from some methods in the kinetic theory of gases, Reynolds specifically assumed that each variable in a turbulent flow was locally composed of its time mean, say $\bar{u}$, and its timewise fluctuating component, u', such that the actual local value at any instant in time would be expressed as $u = \bar{u} + u'$. Moreover, the Navier-Stokes equations can be assumed to hold if the dependent variables (p, ρ, u, v, T, etc.) that appear in those equations are interpreted as their *time-averaged* values. However, when the time averaging of those equations is done mathematically, some extra terms appear in the equations that can be interpreted as a "turbulent viscosity" μ_T and a "turbulent thermal conductivity" k_T. Therefore, when the Navier-Stokes equations are to be used to study a turbulent flow, according to Reynolds the flow properties are to be used as their *time averages,* and the viscosity coefficient and the thermal conductivity are to be replaced by the sums $(\mu + \mu_T)$ and $(k + k_T)$, respectively, where μ_T and k_T are the apparent *increases* in viscosity and thermal conductivity due to the fluctuating, turbulent eddies. With this formalism, the Navier-Stokes equations become the *Reynolds-averaged Navier-Stokes equations for turbulent flow* – a set of equations used in what is today by far the most frequently employed theoretical approach to engineering analyses of turbulent flows. That method of treating a turbulent flow locally as the sum of a time-averaged mean and a fluctuating component was the most substantial and pivotal of Reynolds's contributions to fluid dynamics,[42] and its impact on aerodynamics has been historic. The vast majority of theoretical predictions of skin-friction drag on aerodynamic shapes have used, in one form or another, the time-averaged model of Reynolds.

Reynolds's theoretical model, important as it was, did not "solve" the problem of turbulence. The Reynolds-averaged Navier-Stokes equations introduced the turbulent viscosity μ_T and turbulent thermal conductivity k_T. In any analysis of turbulent flow, we need appropriate numbers for μ_T and k_T, and that can be a *big* problem, for such values will depend on the nature of the flow itself. In direct contrast, the values for μ and k (the molecular viscosity and molecular thermal conductivity) are known properties of the fluid that can be looked up in standard reference sources. Finding the proper values for μ_T and k_T for a given turbulent flow is called *turbulence modeling.* Reynolds introduced his time-averaged equations in 1894,[42] but today, 100 years later, research to find the best, most appropriate turbulence models to calculate μ_T and k_T is one of the highest priorities in aerodynamics.

Reynolds's third contribution of significance for aerodynamics, though of lesser importance than the two discussed earlier, was in determining the connection between skin friction and heat transfer. Today, there is an approximate relation used by engineers that relates the local skin-friction coefficient C_F to the local heat-transfer coefficient C_H via the *Reynolds analogy,* which can be written

$$\frac{C_H}{C_F} = f(\mathrm{Pr})$$

where $f(\mathrm{Pr})$ denotes a function of the Prandtl number ($\mathrm{Pr} = \mu C_P / k$, with C_P being the specific heat at constant pressure). First introduced by Reynolds in 1874,[43] the Reynolds analogy has come into its own since the middle of the twentieth century, when aeronautical engineers had to begin to cope with the problems of aerodynamic heating associated with supersonic and hypersonic flight.

Many a contribution in the physical sciences has had a certain half-life, with diminishing importance as the years have gone by, but Reynolds's contributions, viewed in the light of modern aerodynamic applications, have actually increased in significance. The entire field of modern turbulence modeling and even our basic views of the nature of turbulence and transition have derived from the ideas of Reynolds.

Applied Aerodynamics in the Nineteenth Century: The Fog Thickens

By the end of the nineteenth century, the basic principles underlying classical fluid dynamics were well established. There had been a complete formulation and understanding of the detailed equations of motion for a viscous fluid flow (the Navier-Stokes equations), as well as the beginnings of a quantitative, experimental data base on basic fluid phenomena, including the transition from laminar flow to turbulent flow. In essence, fluid dynamics was in step with the rest of classical physics at the end of the century – a science that was perceived at that time as being well known, almost mature, with nothing much more to be learned. That view prevailed despite the fact that the Navier-Stokes equations were intractable from the point of view of obtaining analytical solutions for practical problems, and the analysis of a turbulent flow was (and still is) virtually a branch of black magic.

On the other hand, transfers of technology from that advanced state of the art in fluid dynamics to the investigation of powered flight were virtually nonexistent. The idea of powered flight was still considered fanciful by the established scientific community – an idea that was not appropriate for serious intellectual pursuit. Even Lord Rayleigh, the only one of the scientific giants of the nineteenth century to show any interest in powered flight, contributed nothing tangible to applied aerodynamics. That situation could not have been stated more emphatically than in the following paragraph from the 5th annual report of the Aeronautical Society of Great Britain in 1870:

> Now let us consider the nature of the mud in which I have said we are stuck. The cause of our standstill, briefly stated, seems to be this: men do not consider the subject of 'aerostation' or 'aviation' to be a real science, but bring forward wild, impracticable, unmechanical, and unmathematical schemes, wasting the time of the Society, and causing us to be looked upon as a laughing stock by an incredulous and skeptical public.

Clearly they had a technology-transfer problem between the science of fluid dynamics and those who were attempting applications to powered flight. Those impracticable dreamers who insisted on pursuing *applied aerodynamics* in the nineteenth century were forced to go their own way. Their path, and the reasons for it, will be highlighted in the remaining sections of this chapter. In many cases that path became shrouded in a fog of misunderstanding and ill-conceived ideas.

The Aeronautical Society of Great Britain: A Ray of Hope

In the development of any scientific discipline, *technical credibility* is essential. Today, such credibility is established through intricate mechanisms of peer evaluation, widespread publishing in technical journals and books, and direct interactions among scientists and engineers (via the many technical conferences around the world, the telephone, video, etc.). Such a mechanism was first formalized in 1662, when the Royal Society of London was incorporated under a royal charter, becoming the first formal learned society in the field of science. The Royal Society was formed for the reading and publishing of scientific papers, for the interchange of ideas among scientists, and for the recognition of scientific achievement (e.g., the awarding of the Copley Medal).

It is no surprise that the increasing interest in powered flight eventually led to the founding of technical societies for the purpose of formally exchanging ideas and publishing papers on aeronautics. The first of those was the Société Aerostatique et Météorologique de France, founded in Paris in 1852. The second, and by far the most important, was the Aeronautical Society of Great Britain, founded in London in 1866. The first meeting of the council of the society was held at the residence of the duke of Argyll on January 12, 1866. The purpose of the society was stated by its first honorary secretary: "Before proceeding to the reading of papers, which it is the object of this meeting to encourage and discuss, it is necessary to claim on behalf of the Council perfect immunity in respect of any complicity with the views of their respective authors. It cannot be too often repeated that the Council, as a body, *has no theories of its own* .... The Aeronautical Society of Great Britain has been formed to encourage, to observe, to record, and to aid, in proportion as its ability is strengthened by the support of its members." Thus a formal mechanism for the establishment of technical credibility in aeronautics was finally in place – proof that practical applications in aeronautics were beginning to coalesce into an accepted field of endeavor. However, the desired technical prestige that would be associated with investigations in aeronautics in general, and aerodynamics in particular, was slow in coming.

One of the most interesting of the early papers to be published by the society was "on Aerial Locomotion and the Laws by Which Heavy Bodies Impelled through Air Are Sustained," by Francis H. Wenham. Delivered on June 27, 1866, and published in the first annual report, Wenham's paper focused on sustainment (i.e., the generation of an aerodynamic force, principally aerodynamic lift, on a object moving through air). It quickly revealed the confusion existing at that time regarding the basic sources of aerodynamic force: "The resistance against a surface of a defined area, passing rapidly through yielding media, may be divided into two opposing forces. One arising from the cohesion of the separated particles; and the other from their weight and inertia." Today we know that the aerodynamic force exerted on any object is the net result of pressure and frictional shear stress acting over the surface. One might attempt a very liberal interpretation of Wenham's "cohesion of the separated particles" as a reference to frictional shear stress. As we know from modern kinetic theory, the action of viscosity, indeed the quantitative value of the viscosity coefficient, depends in part on the strength of the force field between molecules – the intermolecular force. For two molecules in close proximity, the intermolecular force is large and repulsive, but that force becomes much weaker and changes to an attractive force at large separation distances. Was that the "cohesion of the separated particles" mentioned by Wenham? Not likely, for that would have required a more advanced understanding of molecular kinetic theory and its relation to viscosity than existed in 1866. Wenham went on: "In plastic substances, the first condition, that of cohesion, will give rise to the greatest resistance. In water this has very little retarding effect, but in air, from its extreme fluidity, the cohesive force becomes inappreciable, and all resistances are caused by its weight alone." That statement indicated a lack of understanding of frictional shear stress, which is *not* "inappreciable" in air nor in water. For slender, streamlined bodies in a gas or a liquid, the major contribution to drag is frictional shear stress.

As for Wenham's statement of the other cause of aerodynamic force, namely, the "weight and inertia" of the particles, one might try to interpret that as a reference to the pressure acting on the surface. From modern kinetic theory we know that the pressure on a surface is due to molecules and atoms in purely random motions impacting a surface, being reflected from it, and consequently experiencing changes in momentum. The cumulative effect of the "time rate of change" in momentum (the rate of change in momentum with respect to time) for those particles as they impact a surface is the resulting force on the surface, and that force, taken per unit area, is the pressure. In modern mechanical terms, the time rate of change in momentum is called inertia, and of course the momentum of each particle is proportional to its mass (*not*, strictly speaking, its weight). Wenham's statement that the second cause of the aerodynamic force on a surface was the "weight and inertia" of the particles of the fluid seems more an intuitive statement than a science-based opinion. It was essentially on the right track, because pressure is involved with the inertia (hence the mass) of the fluid particles, but Wenham's statement revealed an incomplete understanding of the actual mechanism. For practical purposes, Wenham's statement of the two sources of aerodynamic force on an object was not correct (nor were the opinions of other researchers at that time). The more nearly correct ideas of Cayley, 60 years earlier, would seem to have been overlooked.

On the positive side, Wenham's paper contained the first statement of an important principle in applied aerodynamics. On the basis of extensive studies of the flight of birds, he noted that "the swiftest-flying birds possess extremely long and *narrow* wings and the slow, heavy flyers short and wide ones" (emphasis in original). He went on to suggest that the wings of flying machines should be long and narrow – the first recognition of the advantage of a high-aspect-ratio wing. An aspect ratio is a geometric quantity defined as the square of the wingspan divided by the planform area (i.e., top-view area) of the wing. For a rectangular wing, the aspect ratio is simply the span divided by the chord. High-aspect-ratio wings are long and narrow, and low-aspect-ratio wings are short and stubby. Today we understand the scientific reasons for the aerodynamic advantages of high-aspect-ratio wings for subsonic flight. In 1866, Wenham, like everyone else, had no clue. The advantage of such wings is related to the induced drag, a concept that was not understood until 1918. In any event, Wenham theorized correctly that most of the lift for a wing at a moderate angle of attack would be produced by the front portion of the wing. It followed, he said, that the most efficient wing configuration would be a number of long, narrow wings stacked vertically – a multiwing concept. (That design – a large number of venetian-blind-shaped wings stacked in four tandem decks – was used by Horatio Phillips in an airplane in 1908. The flight was successful; Phillips was airborne for approximately 500 ft. But by that time the Wright brothers dominated the headlines, and Phillips's airplane became only a footnote to history.) Wenham's paper was an indication of the lack of technical sophistication in aerodynamics at that time – there were no equations in the paper. Aerodynamics, as a quantitative science, had a long way to go.

The society did more than present and discuss papers. In 1868, just two years after its founding, the society organized the first aeronautical exhibition – a display of flying machines and balloons at the Crystal Palace in London. Outside of a steam-powered model by John Stringfellow, the exhibition was an odd collection of unsuccessful artifacts. But the exhibition was an indication that work toward the development of flying machines was becoming more respectable. For the next three decades, as reflected in the publications of the society, applied aerodynamics bumped along, sometimes momentarily getting off the ground, but never achieving sustained flight.

An interesting comment on the role of advanced engineering research appeared in the third annual report of the society in 1868. The duke of Argyll, in discussing some aspects of aeronautical propulsion, noted that the steam engine was the greatest motive force then available, but it was heavy and cumbersome, requiring large supplies of water and fuel, precluding any hope that it could ever power an airplane: "Still, the absence of the lighter motive power required ought not to stop us from investigating the principle upon which it is to be applied." That call for continuing investigations of the principles of flight, even though a feasible engine did not yet exist, was an early recognition of the value of engineering research (i.e., to provide fundamental information on physical principles, even though applications were not yet at hand).

Sometimes, total misconceptions led to useful conclusions. At a meeting of the society on June 25, 1868, during the exhibition at the Crystal Palace, illustrations of the rotary motions of a bird's wings were presented by a speaker identified only as Mr. Young. He argued that air never pressed on the back of the wing and that the rush of air acted as a sustaining force beneath the curvature of the wing. He concluded that the best means of flight would involve a curved wing. That physical picture of the action of the air on the wing was totally erroneous, but his conclusion that an airfoil should be curved was correct. Of course, that conclusion had been reached by George Cayley 60 years earlier on the basis of a more accurate conception.

Another interesting statement, quite prophetic, appeared in the third annual report in 1869: "A large machine is more likely to succeed than a model." It was based on data showing that the "effect produced on one area will not be produced on another." That presaged what today are called *scale effects* in aerodynamics. Scale effects are associated with the Reynolds number. If we define a Reynolds number based on the chord length of the wing, c, as

$$\frac{\rho V c}{\mu}$$

we know from Chapter 1 that the drag *coefficient* due to skin friction, defined as

$$c_f = \frac{\tau}{\frac{1}{2}\rho V^2}$$

becomes smaller as the Reynolds number becomes larger. Thus, for a large airplane such as the Boeing 747, for which the Reynolds number is very large, the skin-friction coefficient will be comparatively smaller, giving a higher lift-to-drag ratio for the airplane – a favorable improvement in aerodynamic efficiency.

Also appearing in the third annual report were statements by both Wenham and Stringfellow that the airscrew would be the "best method of propelling through the air," despite the fact that no real experience had proved that. The publications of the society were peppered with debates concerning the best propulsive mechanism – a propeller or a beating wing. In regard to an engine to power the airscrew (propeller), the statement was made that "steam was, undoubtedly, the most economical, but in their present state, *gas* would answer better." That did not mean gasoline, but rather an engine run by some type of hot gas, such as might be formed from carbonic acid. Cayley had pursued a similar line of thinking.

The technical limitations faced by members of the society were summed up in the third annual report, which certainly reflected the state of the art in 1868:

> With respect to the abstruse question of mechanical flight, ... we are still ignorant of the rudimentary principles which should form the basis and rules for construction. No one has yet ventured to give a correct experimental definition of the primary laws and amount of power consumed in the flight of birds; neither, on the other hand, has any tangible evidence been brought forward to show that mechanical flight is an impossibility for man.... We are equally ignorant of the force of the wind exerted on surfaces of various sizes, forms, and degrees of inclinations: these are generally *assumed* on the mathematical laws of the resolutions of forces, ... which convey but a very distant idea, relative to the conditions of the present inquiry, where the elastic and yielding nature of the air is the cause of such unforeseen results.

The mysteries of aerodynamics were ascendant, and frustration was rampant. But optimism was not totally lacking. In the fourth annual report in 1869, from a paper delivered in Paris by M. De Lucy, translated and printed by the society, we find the following: "Science is ripe, industry is ready, everybody is in expectation; the hour of aerial locomotion will soon arrive."

More aerodynamic misunderstanding was revealed in the fifth annual report in 1870: "Although the relation existing between the velocity and pressure of water, was, for practical purposes, well known; still we were in total ignorance of the connection between velocity and pressure in an elastic medium like air." That was a clear reference to the Bernoulli effect as applied to water – probably the first statement concerning the Bernoulli effect to be found in the *aerodynamic* literature. The last part of that statement reflects a concern that was then common throughout the aerodynamic literature: concern over the "elastic" properties of air, the assumption being that any findings obtained in working with inelastic water would be inapplicable to air. Such concern was unfounded. We know today that the behavior of an airflow at Mach numbers less than 0.3 is essentially like that of a constant-density fluid, like water. There are no untoward effects from the "elastic" properties of air; such effects become important only at high speeds, near and above the speed of sound.

Muddled thinking was epitomized in a paper published in the eleventh annual report, 1876, by D. S. Brown. After a discussion about air density being constant for flows that are "unresisted" (i.e., with no aerodynamic body in the flow), Brown stated that "this uniformity of density, however, is at once broken when a wing-plane is presented to the current. Against the windward face the air compresses itself by the force of its momentum, and thereby forms a cushion of resistance. In the time of the compression, the weight of air compressed has expended its motive force, and the force of the impact, whether derived from air thus in motion against the wing, or from the wing beating against the still air, determines the value of the support that the wing receives." Again we see a total lack of understanding of the mechanism for generation of lift. There was no realization that lift comes from low pressure on the top of the wing and high pressure on the bottom. Also, we see the prevailing idea that the air was "compressed" by the wing, a product of the unnecessary preoccupation about the "elasticity" of air.

In summary, the field of applied aerodynamics during the last half of the nineteenth century developed along a course quite separate from that of basic fluid dynamics. The basic equations of fluid dynamics were known by that time – the Navier-Stokes equations. The work of Helmholtz was known, and the research efforts of Kirchhoff and Rayleigh were in progress. Fluid-dynamics experiments, such as those of Reynolds, were shedding light on the most fundamental aspects of fluid flow. *But there was no technology transfer to the field of applied aerodynamics.* Indeed, it appears that the first use of Bernoulli's equation in the aerodynamic literature was in a paper by Albert Zahm in the *Aeronautical Journal* in October 1904. The work of the Aeronautical Society of Great Britain had elevated aeronautics to a more prestigious plane, but had done little to promote technology transfer.

Francis Wenham and the Development of the Wind Tunnel

The October 1908 issue of *The Aeronautical Journal* contained a memorial to Wenham: "Members of the Society will hear with great regret of the death of Mr. F. H. Wenham, . . . one of the leading members of the Society at its foundation [who] may almost be called the 'father' of aeronautics in his country. His paper on 'Aerial Navigation (1866)' is one of the most important ever published on aeronautical science." Of particular interest is the phrase "may almost be called the 'father' of aeronautics in his country." Was that a valid statement? What was meant by the qualifier "almost"? Let us look closer at Wenham.

Figure 4.20   Francis Wenham.

Francis Wenham (1824–1908) (Figure 4.20) typified one group of nineteenth-century technologists – not necessarily college-educated, but self-taught in the basic principles of mechanics, with interests spanning a wide spectrum of engineering applications (George Cayley would fit in that category). Wenham was born in 1824 in Kensington, which then was an agricultural district outside London. The son of an army surgeon, during childhood he showed an interest in mechanical systems and a questioning, analytical mind. His professional career began at the age of 17, when he was apprenticed to a marine engineering firm in Bristol. For the rest of his life he would consider himself a marine engineer, specializing in screw propulsion for steamships. His first major engineering design was a high-pressure tubular steam boiler for a small streamer, which he also designed. After tests in England, Wenham's steamer was shipped to Alexandria, Egypt, where he proceeded to navigate the Nile River. Wenham's first patent had nothing to do with steamships. It was for a device designed to prevent a sporting gun from firing accidentally.

Wenham's diversity of interests is illustrated by his work on microscopes: "Wenham's place in aeronautical history is a high and world-wide one, but his place in the history of the microscope is at least as high."[44] At the age of 26 he was elected as a fellow of the Microscopical Society. He developed metal parabolic reflectors, one of which is still on display in the society's cabinet of ancient instruments. He advanced the application of binocular vision to the microscope and published papers on the theory of illumination under the microscope. He was especially active in aperture measurements of object glasses. Moreover, Wenham directed his attention to the objects being studied under the microscope, publishing a number of papers on the structure of plant samples, including seaweed.

Other areas of activity for Wenham were photography, gas lamps, and musical instruments. He received a number of patents for methods to improve the intensity of the light from gas lamps, and his ideas were incorporated in designs widely used for both outdoor and indoor lighting. He obtained a patent for an "improved mechanism for mechanically playing pianofortes and other musical instruments where notes are sounded by percussion." He published ideas on how to improve the tone of a violin. In short, Wenham had widely diverse interests, a characteristic common to many men of technology in the nineteenth century. That is in stark contrast to the situation in science and technology today, where the state of high technology is so complex that one must become highly specialized in a narrow subject area if one is to make a contribution to the state of the art.

Wenham's personality made him controversial. Sometimes combative in his interactions with peers, he tended to lash out at more established scientists who dared to criticize his work. For example, during his measurements of apertures, he developed a theory about such measurements that was soon found to be incorrect. Nevertheless, commenting on measurements by other investigators, Wenham wrote in a rather arrogant tone that "I hold that the list of numerical apertures since published is entirely fallacious." Wenham's paper to substantiate that statement was turned down by the Royal Microscopical Society; he promptly resigned from the society, 44 years after having been made a fellow.

Wenham was not a good mathematician, a continual detriment in much of his work. A colleague, Edward Nelson, wrote in the *Transactions of the Royal Microscopical Society* in 1908 that "this lack of familiarity with elementary mathematics was a cloud which obscured his vision on many important points. If only it could have been lifted, what an inventor he would have been!" That, in part, explains why Wenham's 1866 paper to the Aeronautical Society of Great Britain contained no mathematical equations. Wenham emphasized experimental work, as in a paper in 1854 dealing with the microscope: "As I have no predilection for a theory that does not bear upon a practical result, in that which I bring forward I shall endeavor as much as possible to support it by experiment." That was an indication of the philosophy of the man who later would build the first wind tunnel.

Wenham's contributions to aerodynamics, and aeronautics in general, began with the Aeronautical Society of Great Britain, of which he was a charter member and a member of the original seven-number council of the society. Wenham's interest in aeronautics is not surprising, considering his lifelong career as a marine engineer, with special interest in screw propulsion for ships – a close cousin to the technology then perceived as necessary for flying machines. Wenham and three other members of the original council were either fellows or officers of the Microscopical Society. Another council member had been president of the Institution of Mechanical Engineers, and yet another helped to make the laying of the Atlantic telegraph cable a success. The founders of the Aeronautical Society of Great Britain were no upstarts. In terms of intellectual power and vigorous activity, Wenham soon dominated the council and most of the society's early meetings. Wenham's 1866 paper was the first to point out the aerodynamic advantage of high-aspect-ratio wings. In 1903, in the American Journal *Aeronautical World,* Wenham was still emphasizing that point: "Flight is impossible with flat supporting surfaces, as nearly all the lift is confined to a narrow portion of the front edge." That was consistent with the idea of making the wings long and narrow – narrow because most of the lifting action takes place near the leading edge, and hence a long chord length (relative to the span) is simply extra baggage insofar as lift is concerned. The aerodynamic advantage of a high-aspect-ratio wing seems to have been lost on subsequent inventors in the nineteenth century (with the exception of Samuel Langley at the Smithsonian Institution in Washington). Even the Wright brothers failed to appreciate that in their early 1900 and 1901 glider designs, which had a rather low aspect ratio of 3. Only after their wind-tunnel tests in 1901–2 did they increase their aspect ratio to 6, and that was one of the primary reasons for the aerodynamic success of their 1902 glider.

Wenham followed his 1866 paper with other lectures, as well as frequent comments and criticisms of the work of others in the society. On April 17, 1867, in a speech to the society, Wenham assessed the state of the art of aerodynamics: "Our knowledge of Aeronautics, as far as regards the navigation of the air by mechanical means, amounts to but very little, and the information recorded is of a contradictory character . . . without a definite law of the acting and counteracting forces of the elastic air, we have not even entered the threshold of aeronautical discovery, and attempts at obtaining mechanical flight cannot be foreseen in their results" (p. 580).[44]

Wenham clearly was venting a sense of frustration about that state of affairs. However, he had in mind a course of action:

> A series of experiments is much needed, in order to furnish the data for construction. Should these establish the law of the capability of an atmospheric stratum for supporting heavy weights by means of very slightly inclined surfaces travelling at high speeds, we shall then have a certain fact to start from, and let it be borne in mind that the air as a means of transit has in one respect an unequalled advantage, that of a ready made highway, without hills, turnings, or irregularities to damage or break machinery; consequently, the only limit on speed is not in safety, but the amount of propelling force that can be applied [p. 580].[44]

Wenham was following his natural tendency to resolve technical questions by experiment: "I propose shortly to try a series of experiments by the aid of an artificial current of air of known strength, and to place the Society in possession of the results." That statement foretold the development of the first wind tunnel. Its development was further encouraged at the May 1870 meeting of the council, where the absence of data on "reactions and lifting forces" was noted, as well as the opinion that experiments to obtain such data might not be very difficult. By June, the Aeronautical Society formed a committee to acquire such experimental data, consisting of four reputable engineers, including Wenham, and funds were raised to carry out that program. Wenham designed the experimental device: a rectangular duct 10 ft long, with a square cross section 18 in. on a side. The air was driven through the duct by a fan powered by a steam engine. The device was fabricated by John Browning, an optician (not surprisingly) and a member of the society. The wind tunnel was located at Penn's Marine Engineering Works at Greenwich. The first wind-tunnel experiments took place in the shadow of the famous Greenwich Observatory – a fitting venue for an experimental device that, in more modern versions, later would have an important role in the aeronautical progress of the twentieth century.

By modern standards, Wenham's wind tunnel was primitive. The maximum velocity was only 40 mph, and the airstream was unsteady, making accurate, replicable measurements virtually impossible. Even the mean direction of the airstream was in question, because there were no vanes for guiding the air. It was a straight, constant-area, rectangular duct, with no convergent section resembling a nozzle. A crude balance was designed, consisting of a vertical steel rod with an open eye near the top, a horizontal rod inserted through that eye supported by a cross-pin axle allowing up-and-down and back-and-forth movements of the rod, the test model (a planar lifting surface at some angle of attack to the airstream) mounted at one end of the horizontal rod, and a sliding counterweight on the other end of the rod to balance the weight of the model. Lift and drag were measured by vertical and horizontal springs, respectively, the former attached to the horizonal rod, and the latter to a lever from the vertical rod. Lift and drag were measured simultaneously, requiring two people to take the readings from the spring scales. The "test section" of the tunnel was actually the region in the open air 2 ft downstream of the duct exit; today we call this type of wind tunnel an open-jet facility. Only the test-model lifting surface was in the airstream; the rest of the balance was shielded by a wooden covering. The lifting surfaces were flat plates, the largest of which spanned 18 in. (the full width of the wind tunnel). The angle of attack for the flat lifting surfaces ranged from 15° to 60°. Wenham needed to obtain data at lower angles of attack than 15° but the aerodynamic force at low angles of attack was too small to be measured accurately by the rather crude balance.

Despite those difficulties, the wind-tunnel data, being the first of their kind, were welcomed by the members of the society. They showed that meaningful lift was created at low angles of attack – lift that was considerably larger than the accompanying drag. Wenham proved that lift-to-drag ratios considerably greater than 1 could be achieved. That was big news for the aeronautical community. (Today, a modern airfoil section will have a maximum lift-to-drag ratio of about 100, and a complete airplane will have a ratio of 16–20.) He also found that the center of pressure was near the leading edge, and consistent with that finding he was able to demonstrate the advantage of a high-aspect-ratio wing. Later, on April 18, 1872, Wenham set up the wind tunnel in London for its first public demonstration, explaining in detail the operation of the tunnel and the balance. He was candid in underscoring the potential inaccuracies associated with the rather crude measurements.

The most insightful comments of that day came from James Glaisher, after Wenham's presentation. Glaisher had participated in some of the experimental measurements at Greenwich and had the beginning of a true understanding of the aerodynamic force on a body immersed in a flowing fluid:

> Anyone who has not considered with care the nature of the pressure produced by the flow or rush of a fluid, elastic or incompressible, against a plane surface placed in its course, might imagine that the system of parallel forces was merely equivalent to a single resultant force acting at the center of pressure, and capable of resolution according to the ordinary parallelogram law. But this of course is not the case, for the particles of the fluid which come in contact with the plane, have somehow or other to get out of the way, by gliding along the surface of the plane (as they cannot get through it) and this produces a complication in the neighborhood of the surface of such a kind as cannot be theoretically predicted. One thing, however, is quite clear, and that is that the directions of all the small forces acting on the surface certainly are not parallel, and we must therefore have recourse to experiment [1872 annual report, Aeronautical Society of Great Britain].

Thus we see the state of fundamental aerodynamic understanding in 1872 – it was partly right and partly wrong. Recall from Chapter 1 that nature exerts an aerodynamic force through only two mechanisms: (1) the local pressures exerted over the surface, where each local pressure is perpendicular to the surface at that point, and (2) the local shear stresses due to the frictional action of the airflow over the surface, where each local shear stress is tangent to the surface at that point (Figures 1.3 and 1.4). The net aerodynamic force is the integrated sum of the local pressure and shear-stress distributions acting over the surface. Interpreting Glaisher's comments in that light, we make the following observations:

(1) He was correct in visualizing the net aerodynamic force on the plane as being due to the sum of local phenomena acting on the small segments of the surface.

(2) If Glaisher was thinking only about the effect of pressure, then his statement that "the directions of all the small forces acting on the surface certainly are not parallel" was wrong for the specific case of a flat plate (which was the only geometric shape tested in the wind tunnel). Because pressure acts perpendicular to the surface, and the surface was a flat plate, the directions of the local pressure forces on the small segments of the plate were parallel to each other and could easily be added to

obtain the net force due to pressure. Moreover, that net force would, by definition of the center of pressure, act through the center of pressure.

(3) On the other hand, when Glaisher was talking about the particles of fluid "gliding along the surface of the plane (as they cannot get through it) and this produces a complication in the neighborhood of the surface of such a kind as cannot be theoretically predicted," he was, whether or not he realized it, beginning to address the action of frictional effects near the surface. Friction produces shear stress at the surface that acts parallel to the surface. If the forces due to pressure and shear stress are added vectorially at a given local segment of the surface, then the resultant will be at some angle to the surface, neither perpendicular nor tangent to the surface. Moreover, because both pressure and shear stress will vary in magnitude from one point to another along the surface, the directions of the resulting local forces on local segments of the surface indeed will not be parallel. That was what Glaisher was saying at the end of his comments. However, that is true only because frictional shear stress acts along with pressure at each point on the surface. Neither Glaisher nor Wenham explicitly mentioned any consideration about the effects of friction. (Throughout the nineteenth century, most technical investigators in aerodynamics tended to consider the effect of friction to be so small as to be negligible.) We must strongly suspect that Glaisher was considering only the physical mechanism of pressure, and given that interpretation, his comment at the end was wrong.

Once again we observe the confused conceptions that prevailed among those who were most fervently trying to understand aerodynamic phenomena. In that regard, Wenham's wind-tunnel experiments were not of much help.

In other respects, however, Wenham's experiments in the wind tunnel were quite important – revealing the large lift-to-drag ratios at small angles of attack, and showing the advantage of high-aspect-ratio wings. In regard to the movement of the center of pressure as the angle of attack is decreased, Wenham stated that "we were not able to ascertain very accurately . . . . We found as the angle became more acute, the center of pressure came nearer the front edge." (That was indeed true. As the angle of attack for a flat plate is decreased, the center of pressure moves from near the center of the plate to a location nearer the leading edge, a location about 25% of the chord length from the leading edge – the $\frac{1}{4}$-chord point.)

There is no record of subsequent use of Wenham's wind tunnel to collect additional data, neither by Wenham nor by anyone else.

Wenham continued to be active in the society for another 10 years, effectively critiquing the work of others, and generally adding a major degree of technical competence to the meetings of the society. Then, in July 1882, he suddenly resigned – the result of a long-standing feud with Fred Brearey, the honorary secretary since the society's beginning in 1866. Brearey's lack of technical knowledge and his arrogant attitude had caused problems in the society. After Wenham's resignation, the society went downhill. Brearey died in 1896, and a year later the society was totally rejuvenated by Captain B. F. S. Baden-Powell. The first volume of *The Aeronautical Journal* appeared in July 1897, and it continues today as the main publication of the Royal Aeronautical Society. In September 1899, Baden-Powell wrote to Wenham, stating that the council of the society was offering him honorary membership. Wenham, by then retired from business, but not retired intellectually, accepted.

Wenham died August 11, 1908. He was active and interested in flying machines to the end. His last paper, dealing with various construction techniques for parts of airplanes, was finished only a few days before he died. It was published in the October 1908 issue of *The Aeronautical Journal,* the same issue that contained the notice of his death, as quoted at the beginning of this section.

We have come full circle, back to the questions posed at the beginning of this section. Is it reasonable to say that Wenham "may almost be called the 'father' of aeronautics in his country"? And what was meant by the qualifier "almost"? To answer those questions, we need to interpret the work of Wenham in light of the accomplishments of Cayley half a century earlier.

Wenham: Father of Aeronautics?

In light of Wenham's accomplishments, it is easy to understand why the Aeronautical Society of Great Britain would properly hold him in great esteem. However, memorials to the recently deceased tend to be excessive. There is no doubt that Wenham was the next important figure in technical aeronautics following George Cayley. But in comparison with Cayley's pioneering contributions, the work of Wenham was not quite up to the mark. (Perhaps recognition of that discrepancy was why the society used the qualifier "almost.")

We have discussed how calculations based on the Newtonian sine-squared law result in gross underpredictions of lift for surfaces inclined at small angles of attack to the flow. If that law were applicable, then to produce the lift necessary to balance the weight of a flying machine, either or both of the following steps would be necessary: (1) A much higher angle of attack would have to be used. (2) The surface area of the wings would have to be increased. In both cases, the consequence would be much higher drag. Hence the lift-to-drag ratio would be low. Calculations in the eighteenth and nineteenth centuries had indicated that the ratio would be so low that the technical feasibility of powered flight would be in doubt. Thus, when Wenham's wind-tunnel data indicated the generation of a high lift-to-drag ratio at small angles of attack, there was considerable excitement among the members of the Aeronautical Society. However, Cayley and others had already shown the sine-squared law to be invalid for small angles of attack. Cayley's whirling-arm measurements of the aerodynamic force on an inclined flat surface had been reasonably accurate, considering the experimental techniques used at that time. Cayley's data should have swept away, once and for all, any doubts generated by the spurious results obtained using the sine-squared law. His data clearly demonstrated that reasonable lift could be obtained from a lifting surface at small angles of attack. Although Cayley did not emphasize the lift-to-drag ratio in discussing his results, that message clearly was implicit in his data. Cayley's findings, published in 1809, had preceded those of Wenham by about 60 years. It is difficult to conceive that Wenham and the other members of the society would have been unfamiliar with Cayley's data. However, we know that direct familiarity with the findings in Cayley's triple paper (1809–10) appears to have declined rapidly, as if by exponential decay. It appears that many inventors and technologists in the last half of the nineteenth century had little or no knowledge of the details of Cayley's experiments. In Wenham's 1866 paper – the paper cited in the society's memorial to Wenham as being "one of the most important ever published on aeronautical science" – there was not a single reference to Cayley's experimental data, nor any mention of Cayley by name, nor any allusion to any contributions that could be traced to Cayley. Also, Wenham's advocacy of a high-aspect-ratio wing implied his recognition of the superiority of the fixed-wing concept for flying machines – the concept championed by Cayley, and mentioned by da Vinci three centuries earlier. So, in perspective, Wenham's contributions were natural extensions of the work of Cayley, whether or not Wenham was aware of it. Clearly, Wenham was not the father of aeronautics in England. In implying that, the society was a bit overenthusiastic about one of its illustrious members.

This is not meant to detract from Wenham's role in the advancement of aeronautics in general, and aerodynamics in particular. The design, development, and use of the first wind tunnel were sufficient to permanently establish Wenham's name in the history of aeronautical engineering. Moreover, his advocacy of the high-aspect-ratio wing – though he hadn't a clue to the aerodynamic reason that such wings are advantageous (the reduction of induced drag) – was an ingenious insight, well before its time. And finally, his enduring enthusiasm for the technical aspects of aeronautics and his almost infectious transfer of that enthusiasm to the Aeronautical Society of Great Britain during its early days were contributions to the overall atmosphere of aeronautical progress that defy any quantitative measure. Wenham may not have been the father of aeronautics, but certainly he was an influential member of the family.

Horatio Phillips: Cambered Airfoils and the Second Wind Tunnel

Among those in attendance during Wenham's report to the society on his wind-tunnel experiments was a young man by the name of Horatio Phillips, who at the impressionable age of 27 was not impressed. He was critical of the quality of the flow in the wind tunnel and of Wenham's use of only flat lifting surfaces. Six years later, the council of the Aeronautical Society commissioned two members, Thomas Moy and R. C. Jay, to expand on Wenham's data using a large whirling arm with a radius of 14 ft (the commission was endowed with the sum of £15). The data from those whirling-arm experiments verified Wenham's finding that the center of pressure for a flat surface would move closer to the leading edge as the angle of attack was decreased, but they contributed little else to the state of the art. Once again, Phillips was not impressed; he decided to try his hand at such matters.

Horatio F. Phillips (1845–1912) was the son of a gunsmith. Exactly how he became interested in flying machines is not known, but Pritchard reported that "from his earliest years he was experimenting and carrying out research, at much cost to himself, in aeronautics."[20]

In the early 1880s, dissatisfied with the quality of the experimental data from Wenham's wind tunnel and Moy and Jay's whirling arm, Phillips designed and operated a second wind tunnel (Figure 4.21). Phillips was much concerned with the flow imperfections in Wenham's tunnel; in an effort to avoid flow fluctuations, Phillips chose a steam injector as the means

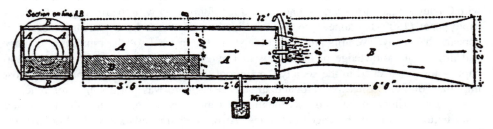

Figure 4.21   Phillips's sketch of his wind tunnel showing, from left to right, cross-sectional views from the front and side.

to suck air in through the entrance of the wind tunnel. The injector was located at the exact center of the wind tunnel, and the flow direction was from left to right in Figure 4.21. To the left of the injector was a rectangular box 6 ft long, with a square cross section 17 in. on each side. Mounted inside the rectangular section was a large block of wood (*D* in Figure 4.21) that reduced the flow area; hence the region above block *D* was a "throat region" where the flow velocity was at a maximum value – up to 60 ft/s (about 41 mph). The aerodynamic model to be tested was mounted in the throat region. The steam injector was a ring of iron pipe perforated with a number of holes facing in the downstream direction (toward the right in Figure 4.21). Steam from a large Lancashire boiler (32 ft long and 7 ft in diameter) at a pressure of 70 lb/in.^{2} would expand at high speed through the holes in the injector ring, entraining the surrounding airflow and creating a local region of low pressure in the center of the tunnel, which in turn would suck air into the entrance of the tunnel at the left. Both the air and expanded steam would then be exhausted at the right through a circular duct (*B* in Figure 4.21). Although not perfect, the steam injector produced an airflow through the test section of better quality than had been achieved with Wenham's tunnel.

Unsatisfied with Wenham's use of flat-plate lifting surfaces, Phillips experimented with cambered (curved) airfoils. Taking his cue from the shapes of birds' wings, Phillips designed a series of cambered airfoils with greater curvature on the top than on the bottom – so-called double-surface airfoils (Figure 4.22). Phillips measured the aerodynamic performances of those airfoils in his wind tunnel and compared them with that for a flat plate tested in the same tunnel. The measurements were made by a lift-and-drag balance (Figure 4.23). The airfoil model (*A* in Figure 4.23) was attached by wires to the pivot *B*. A weight *W* was attached at the $\frac{1}{3}$-chord location on the airfoil by means of a wire and suspended below the model. Phillips assumed that the center of pressure for the airfoils would be at that $\frac{1}{3}$-chord location; hence, with the wind tunnel turned on, he could measure the flow velocity at which

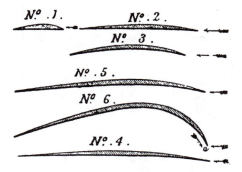

Figure 4.22 Phillips's sketches of his cambered-airfoil shapes patented in 1884.

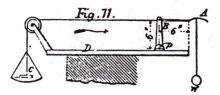

Figure 4.23 Phillips's sketch of his lift balance.

Table 4.2. *An analysis of Phillips's wind-tunnel data*

Airfoil shape	Original data from Phillips (1885)				Data added by Chanute (1893)	Data added by Anderson (this volume) (1993)			
	Airspeed (ft/s)	Dimensions (in.)	Lift (ounces)	Thrust (drag) (ounces)	Foot-pounds per pound	L/D	C_L	C_D	Reynolds number
Flat plate	39	16×5	9	2	8.67	4.5	0.56	0.124	1.03×10^5
Cambered airfoil #1	60	16×1.25	9	0.87	5.80	10.3	0.947	0.092	0.4×10^5
Cambered airfoil #2	48	16×3	9	0.87	4.64	10.3	0.616	0.060	0.76×10^5
Cambered airfoil #3	44	16×3	9	0.87	4.25	10.3	0.733	0.071	0.70×10^5
Cambered airfoil #4	44	16×5	9	0.87	4.25	10.3	0.440	0.043	1.17×10^5
Cambered airfoil #5	39	16×5	9	0.87	3.77	10.3	0.560	0.054	1.03×10^5
Cambered airfoil #6	27	16×5	9	2.25	6.75	4	1.169	0.292	0.72×10^5
Rook's wing	39	0.5 ft^2	8	1.00	4.87	8	0.553	0.070	—

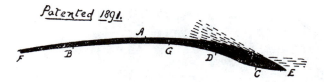

Figure 4.24 Phillips's sketch of his cambered-airfoil shape patented in 1891.

the airfoil would "fly" (i.e., the velocity at which the lift would equal the weight W). In that sense, Phillips's device was not truly a lift balance; he could not change the conditions in the wind tunnel (say the velocity or the airfoil angle of attack) and simultaneously measure the corresponding change in lift. He could only suspend a fixed weight W below the model and measure the flow velocity at which the airfoil would sustain that particular weight. The drag was measured using a scale pan C attached to a wire that passed over a pulley and was attached to the leading edge of the airfoil. Whenever the airfoil "flew" (when the lift equaled the weight W), the corresponding drag could be measured by adding or taking away various measuring weights in the scale pan. The complete balance sat inside the wind tunnel on top of block D; the scale pan hung just outside the entrance to the tunnel, and the weight W hung in a slot cut in block D, outside the main airstream.

The design and operation of the balance determined the format in which Phillips presented his findings (Table 4.2), which appeared in the August 14, 1885, issue of the journal *Engineering*, published in London. Examining the data in Table 4.2, we see that the choice of data points and Phillips's rationale for his experimental plan were greatly flawed, as will be discussed in the next section. Despite the flaws, the basic message from Phillips's data was that cambered airfoils produced considerably more lift than did a flat plate. That was the first *quantitative* demonstration of that fact, although George Cayley had earlier recorded a slight inference in that direction. Phillips clearly recognized the significance of his work, because in 1884 he obtained a patent for the airfoil shapes shown in Figure 4.22 – a year before he published his findings for all to see in the journal *Engineering*. Seven years later, he patented another cambered airfoil (Figure 4.24). His designs were the first truly modern airfoils, and his rationale for the basic design principle was sound. He recognized that when the flow moved over the curved upper surface of the airfoil, the pressure would decrease; hence the lifting action of the airfoil was due to a combination of the lower pressure exerted on the upper surface and the higher pressure exerted on the lower surface. Although Cayley had alluded to those facts, the prevailing intuition throughout most of the nineteenth century was that the lifting action of an inclined plane moving through the air was due to the "impact" of air on the lower surface – a misconception that was erroneously reinforced by the Newtonian flow model. Phillips, by designing double-surface airfoils, with more curvature for the top surface than for the bottom surface, was qualitatively trying to achieve lower pressure over the top surface to obtain a more efficient airfoil. Phillips's findings were widely disseminated, and thereafter all serious flying-machine developers used cambered airfoils.

Phillips is best known for developing the second wind tunnel and for designing and testing cambered airfoils, but later he began to build his own flying machines. In 1893 he constructed a large device consisting of 50 wings, each with a span of 19 ft and a chord length of 1.5 in. – giving an amazingly high aspect ratio of 152 (Phillips clearly had adopted Wenham's philosophy of wing design). The wings were arrayed vertically above

Figure 4.25   Phillips's flying machine (1893).

a long, cigar-shaped fuselage; the machine (Figure 4.25) resembled a huge venetian blind. Powered by a 6-hp steam engine connected to a single pusher propeller, that unmanned flying machine lifted a total weight of 385 lb at a speed of 40 mph, tethered to move on a circular track with a 628-ft circumference. On the basis of that tethered flight experiment, Phillips felt that the viability of cambered airfoils had been established, and he halted his testing until the turn of the century. His last effort at flight was in 1907, when he became airborne for about 500 ft in a larger derivative of his 1893 venetian-blind aircraft. That machine had four venetian blinds in tandem, powered by a 22-hp engine turning a tractor propeller. It was anticlimactic to the success of the Wright brothers four years earlier. After that, Phillips disappeared from the aeronautical scene. He died in 1924, having witnessed the explosive growth of aviation during World War I.

Gibbs-Smith called Horatio Phillips "one of the great men of flying history."[25] In view of the impoverished status of applied aerodynamics in the nineteenth century, Phillips's compelling demonstration of the superiority of cambered airfoils over flat plates was a monumental development in wing design. Of course, his findings were totally empirical; theoretical calculations of the performances of those airfoils were beyond the capacity of anyone at that time. But it had become clear that man-made flying machines should have cambered airfoils, like those that most of nature's flying machines had always had – the curved shape of birds' wings.

Phillips's Wind-Tunnel Data: Flaws in the Interpretation

In light of our modern understanding of airfoils, Phillips's interpretation of his data was flawed in substance, although the final general conclusion was correct. That flawed interpretation of his data has been carried through to the present day, transmitted via various historical papers on his work. We shall reexamine Phillips's data and discuss where the flaws in its interpretation occurred.

Table 4.2 shows Phillips's data for eight different shapes: a flat plate, six cambered airfoils (corresponding to the numbered shapes in Figure 4.22), and the wing of a rook, a crowlike Old World bird. To those wind-tunnel airfoil models was attached a weight W (Figure 4.23) of 9 ounces (8 ounces for the rook wing). When an airfoil was successfully "flying" in the wind tunnel, the lift produced by the wing equaled the weight, namely,

9 ounces, plus the weight of the airfoil. However, Phillips consistently ignored the weight of the airfoil in his tabulations. Hence the column in Table 4.2 labeled "Lift" has the consistent entry of 9 ounces for the first seven shapes. Phillips could vary the flow velocity in his tunnel by adjusting the mass flow of steam through the injector. Above a certain threshold velocity, an airfoil would "fly" (i.e., the lift would sustain the 9-ounce weight and the weight of the airfoil). As the flow velocity was increased above that threshold value, Phillips thought that the mechanical design of the balance (Figure 4.23) would always guarantee that the lift would equal the weight suspended below the airfoil (ignoring the weight of the airfoil itself). If the weight W was 9 ounces, then the lift would be 9 ounces, no matter what the flow velocity, so long as the airfoil was "flying." That would have been the case if the wire attaching the weight to the airfoil had been at the center of pressure of the airfoil. Phillips made the assumption that the center of pressure was at the $\frac{1}{3}$-chord location and attached the suspending wire accordingly. But the center of pressure will vary with the angle of attack of the airfoil and also will depend on the airfoil shape. Therefore, in Phillips's experiments, the weight W was not suspended at the exact center of pressure of the airfoil. That compromised the accuracy of his lift values. For example, in Table 4.2, in the column labeled "Lift," the recorded value of 9 ounces was not exactly equal to the actual lift. In reality, for all of the test cases reported, the actual lift values would have been different, none being exactly 9 ounces. That inaccuracy in Phillips's data was recognized by Octave Chanute as early as 1894. In *Progress in Flying Machines* (p. 167), Chanute stated that "all the results obtained were probably somewhat vitiated by assuming that the center of pressure was uniformly one-third of the distance back from the front edge, and therefore applying the load at that point. We have already seen that this center of pressure varies with the angle of incidence . . . , and the load should have been attached accordingly."[45]

But that experimental inaccuracy was not as serious as a fundamental flaw in Phillips's interpretation of his data – a flaw that would affect the choice of the optimum airfoil shape from among those he tested. To understand that flaw, let us look at some aspects of our modern understanding of airfoil aerodynamics. It is well understood that meaningful comparisons between various data sets can be made only in terms of proper *dimensionless* quantities, usually identified by a technique called *dimensional analysis*.[1,17] In that technique, the lift and drag *coefficients*, C_L and C_D, respectively, are much more fundamental than the lift and drag *forces* per se, where

$$C_L = \frac{L}{\frac{1}{2}\rho_\infty V_\infty^2 S}$$

and

$$C_D = \frac{D}{\frac{1}{2}\rho_\infty V_\infty^2 S}$$

These definitions were discussed in Chapter 1. Moreover, for a low-speed subsonic flow, these coefficients, *for a body of given shape,* are functions of only the angle of attack α and the Reynolds number (Re):

$$C_L = f_1(\alpha, \text{Re}) \quad \text{and} \quad C_D = f_2(\alpha, \text{Re})$$

In Phillips's wind-tunnel experiments, the values for Re at the various data points did not vary greatly. As seen in the last column in Table 4.2, Re $\approx 10^5$ for most of the cases. Hence, for the conditions of Phillips's experiments, we can comfortably state that

$$C_L = f_1(\alpha) \quad \text{and} \quad C_D = f_2(\alpha)$$

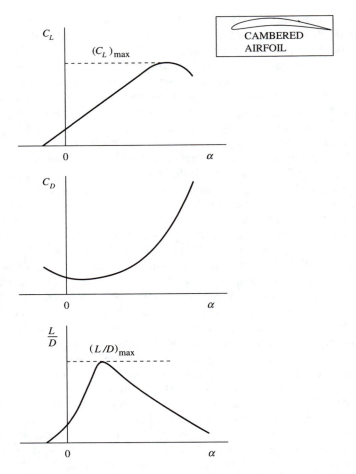

Figure 4.26 Sketches of lift, drag, and lift-to-drag curves versus angle of attack for a cambered airfoil.

Typical qualitative variations of C_L and C_D with α are sketched in Figure 4.26, which also shows the resulting variation of the lift-to-drag ratio (L/D) with the angle of attack. These sketches are for a *cambered* airfoil, for which there is positive lift at a zero angle of attack. Figure 4.26 shows the two most important quantities that determine the effectiveness of an airfoil:

(1) The maximum value of the lift coefficient is denoted by $(C_L)_{max}$. At an angle of attack slightly higher than that for $(C_L)_{max}$, the lift drops sharply (i.e., the airfoil *stalls*). The value of $(C_L)_{max}$ determines the lowest speed at which an aircraft can fly (i.e., the stalling velocity). The higher the value of $(C_L)_{max}$, the lower the stalling velocity.

(2) The maximum value of the lift-to-drag ratio is denoted by $(L/D)_{max}$. $(L/D)_{max}$ is a direct measure of the aerodynamic efficiency of an airfoil. The higher the value of $(L/D)_{max}$, the longer the *range* of the airplane.

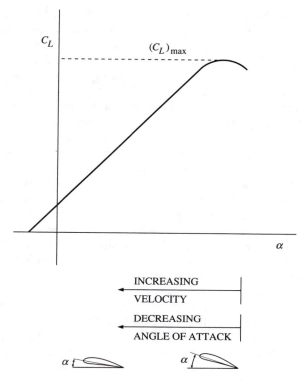

Figure 4.27   Lift curve showing that as the velocity increases, the angle of attack must decrease in order to maintain lift equal to the weight in steady, level flight.

Another important aspect of the generation of lift is illustrated in Figure 4.27, which shows the variation of the lift coefficient with the angle of attack, as well as the directions of increasing flow velocity and decreasing angle of attack for a lifting body of fixed weight. The equation

$$L = W = \tfrac{1}{2}\rho_\infty V_\infty^2 S C_L$$

relates the lift to the lift coefficient. If the lift (hence weight) is held constant, clearly the lift coefficient (and hence the angle of attack) will decrease as the flow velocity increases.

With the foregoing modern perspective in mind, we are ready to examine Phillips's airfoil data. The advantage of expressing lift and drag in terms of the lift and drag coefficients was not appreciated by Phillips and his contemporaries. Because he could not do so, Phillips produced a series of data points that, for the most part, are not related in a logical fashion, which led him to a flawed interpretation. Phillips was not aware of his problem. He carried out a series of measurements that, at first glance, seemed intuitively logical to him, but that in reality led to an improper comparison between different airfoils.

Phillips's experimental approach is revealed by his data in Table 4.2. First he examined the performance of a flat plate. He suspended a 9-ounce weight below the flat plate and brought the wind-tunnel flow velocity up to a value sufficient for the flat-plate model to "fly" (i.e., for the aerodynamic lift produced by the plate to equal the 9-ounce weight). Note that an entry of 9 ounces appears in the "Lift" column in Table 4.2 for the flat plate. Referring to Figure 4.27, the lowest velocity at which the lift produced will equal the weight corresponds to flight at the maximum lift coefficient, $(C_L)_{max}$. At that velocity, the flat plate is at an angle of attack corresponding to the incipient stall of the airfoil. Moreover, the drag of the airfoil will be that corresponding to the stalling angle of attack; referring to Figure 4.26, that drag will be high. Phillips next increased the wind-tunnel velocity. Because of the design of his force balance, the same 9-ounce weight was used for a range of flow velocities; thus the lift remained the same, namely, 9 ounces. Therefore, as the flow velocity was increased, the lift coefficient and angle of attack decreased, according to the variation shown in Figure 4.27. Indeed, the wind-tunnel model, which was allowed to rotate freely about the pivot point on the balance shown in Figure 4.23, naturally sought that lower angle of attack that would keep the lift equal to 9 ounces as the flow velocity was increased. In turn, as the velocity increased, the drag also changed. Placing a weight on the scale pan C shown in Figure 4.23, Phillips adjusted the flow velocity until the drag on the airfoil equaled 2 ounces. He then noted that particular value of the flow velocity to be 39 ft/s. Table 4.2 shows Phillips's entry of 39 ft/s, at which the flat plate (which had a span of 16 in. and a chord of 5 in.) produced a lift of 9 ounces and a drag of 2 ounces. Phillips almost parenthetically noted that the flat-plate angle of attack for that condition was about 15°.

Phillips then proceeded to test a series of cambered airfoils (numbers 1–6, as labeled in Figure 4.22 and Table 4.2) using the same experimental approach. That is, he kept the lift constant, equal to 9 ounces, and then adjusted the wind-tunnel flow velocity until the drag equaled 0.87 ounce. That flow velocity was recorded in the "Airspeed" column in Table 4.2. However, the six different airfoil models did not have the same surface area S. Instead, their chord lengths were 1.25, 3, or 5 in. Right away, those different surface areas cause a problem when using the airspeed values in Table 4.2 as the figure of merit to ascertain the "best" airfoil shape (as was done by Phillips and others after him). For example, the lower airspeeds listed for airfoils 4, 5, and 6, as compared with airfoils 1, 2, and 3, were in part due to the larger surface areas of the models; they were not necessarily due to any intrinsic values of the shapes of those airfoils.

For airfoils 1–5, the experimental parameter that Phillips held constant was the lift-to-drag ratio, namely, $L/D = 10.3$. In addition, for airfoils 2 and 3 the surface areas were the same; for airfoils 4–6, the surface areas were the same, but larger than for 2 and 3. What was the point in comparing the different airspeeds associated with different airfoils when the model surface area and the lift-to-drag ratio were kept constant? How could one choose a "best" airfoil from Phillips's data? The answer is provided by the following hypothetical: Consider two airfoils with the same thickness but different cambers. The lift-coefficient variations and L/D curves for these airfoils are sketched in Figure 4.28. Airfoil A has more camber than airfoil B; hence the lift-coefficient curve for A is shifted to the left relative to B. Furthermore, because the airfoil thickness more or less determines $(C_L)_{max}$, the $(C_L)_{max}$ values will be essentially the same for airfoils A and B. Let us assume that the values of $(L/D)_{max}$ for A and B are essentially the same. Thus neither airfoil is superior to the other, because they have the same values for $(C_L)_{max}$ and $(L/D)_{max}$, which are the most important criteria in determining the "goodness" of an airfoil shape. Airfoils A and B are equally good. What would Phillips have concluded from testing airfoils A and B using his experimental approach? They would have had the same L/D (points a and b in Figure 4.28). However, airfoil A would have had a higher lift coefficient (point c) than airfoil B (point d). For the same weight (hence lift), Phillips would have recorded a lower

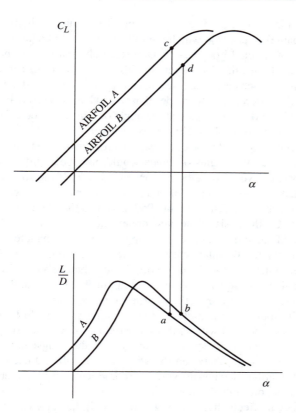

Figure 4.28   Aerodynamic comparison of cambered and symmetric airfoils.

airspeed for airfoil A, because of its higher lift coefficient. Thus he would have concluded that airfoil A was better than airfoil B, but we have already shown that they are equally "good" based on modern knowledge. We have to conclude that Phillips's designation of the airfoil in Table 4.2 with the lowest airspeed (but with the same value of $L/D = 10.3$) as the best airfoil was definitely a flawed interpretation. This assessment is based on the nature of his experimental procedure and is a totally separate consideration from the earlier question of the experimental inaccuracy caused by the weight W being suspended from the $\frac{1}{3}$-chord location.

If we were to use the lowest airspeed in Phillips's tabulation as the criterion to determine the best cambered-airfoil shape, as did Phillips, then airfoil 5, with an airspeed of 39 ft/s, clearly would be the best. (Airfoil 6 was listed at a lower airspeed of 27 ft/s, but for that data point the drag was much higher than 0.87 ounce. Obviously Phillips found no velocity at which the drag would be as low as 0.87 ounce for airfoil 6; instead, he recorded a data point for which the drag was 2.25 ounces, giving a low lift-to-drag ratio of 4. Thus it was clear from the start that airfoil 6 was not as good as the other cambered shapes.) Because we have already shown that the way in which Phillips interpreted his data was flawed, we cannot say with any degree of certainty that airfoil 5 was indeed the best shape among those tested.

The confusion surrounding the interpretation of Phillips's data has carried through to the present. Pritchard stated that "the table may give the wrong impression that lift and drag (thrust) were the same for planes 1 to 5. The 9-ounce and 0.87-ounce weights were the same for the tests, to compare the speeds at which the ratio L/D would be highest for the *same* angle of incidence throughout, 15 degrees."[20] Unfortunately, Phillips did not report the angles of attack associated with the data points in Table 4.2. But most certainly the angle of attack was different for each case, because each data point corresponded to a fixed value of $L/D = 10.3$, and for different airfoil shapes that would have occurred at different angles of attack (Figure 4.28 clearly illustrates that the angle of attack will be different for different airfoils at the same value of L/D – witness points a and b).

Phillips's failure to report the angles of attack for his cambered airfoils was quickly noted by Chanute.[45] Chanute knew that the angles of attack would have been different for the different cases: "the comparison would have been more satisfactory if the soaring angles of incidence had been stated." The only angle of attack reported by Phillips was 15° for the flat plate (that may have been why Pritchard, in 1957, assumed that the angle of attack had been 15° for all cases). On the basis of the Newtonian rule that the drag-to-lift ratio, D/L, for the flat plate should equal the tangent of the angle of attack, Chanute calculated an angle of attack of 12.5° for the flat-plate case in Table 4.2 and stated that Phillips's measurement of 15° "agrees fairly well with calculation." But Chanute's calculation was flawed, because the Newtonian sine-squared law was not applicable for those flow conditions, as had been recognized by Cayley and others earlier in the nineteenth century. Even in modern aerodynamics it would be a complicated task (although it could be done) to calculate the angle of attack associated with Phillips's flat-plate case. The complication arises because the flat plate did not totally span the wind tunnel; the wingspan was 16 in., in a 17-in. test section. The $\frac{1}{2}$-in. gap between the wind-tunnel wall and each wing tip would have allowed the formation of wing-tip vortices (complicated by interaction with the walls), which would have contributed a nontrivial amount of induced drag and other induced flow effects – physical effects that were not known until the early twentieth century. Some modern low-Reynolds-number data for flat plates are shown in Figure 3.24, in conjunction with an earlier discussion. Taking the lift coefficient of 0.56 (as added here to Phillips's original table), we see from Figure 3.24 that the corresponding flat-plate angle of attack should be about 6.5° – less than half of what Phillips reported. There could be several factors involved in the discrepancy. The weight of the airfoil model was ignored in the calculation of the lift coefficient. With that weight included, the lift coefficient (hence angle of attack) would be larger. Also, the induced flow effects due to interactions among the wing-tip vortices, the rest of the wing, and the wind-tunnel walls would have led to a larger geometric angle of attack for Phillips's flat plate than that obtained from Figure 3.24, but larger by a factor of 2 is too much to swallow. What seems more likely is that Phillips's wind tunnel had a substantial degree of "flow angularity" in the test section (i.e., the local flow most likely was not aligned with the axis of the wind tunnel). If Phillips did not measure that flow angularity and take it into account in his determination of the angle of attack, that would have contributed further inaccuracy to his measurement of a 15° angle of attack. For Phillips to have taken that phenomenon into account in his data would have required a higher degree of technical sophistication than appears to have existed at that time.

It is to Phillips's credit that he was interested in some of the more detailed flow-field characteristics associated with the cambered airfoils. For example, he used something like modern-day tufts (very thin ribbons attached to the wing surface) to observe the local flow directions near the surface. With that technique, he observed a region of reversed flow just underneath the leading edge of airfoil 6, as shown by the arrow labeled a in

Figure 4.22. That airfoil had such extreme curvature that it is no surprise that the flow would separate just underneath the sharp leading edge. That appears to have been the first localized measurement of flow separation on an airfoil, although for more conventional airfoil shapes the flow separation usually occurs over the top surface.

From a modern perspective, we have to view the series of data points obtained by Phillips (Table 4.2) as simply a random collection, from which we cannot select the "best" airfoil shape. Although Chanute[45] was suspicious of the accuracy of Phillips's data, he basically accepted the theory that led to the interpretation that airfoil 5 had the best shape. Indeed, from Phillips's data, Chanute calculated the amount of work that would be required to lift a unit weight for each of the airfoils (the column labeled "Foot-pounds per pound"). Chanute stated that "the most efficient shape is, of course, that which requires the least expenditure of power, or the smallest number of foot-pounds per pound of weight to keep it afloat, and this is seen to be shape No. 5, which soared with 3.77 foot-pounds per pound, or at the rate of 146 lbs. sustained per horse power, while the flat plane absorbed more than twice as much power." Chanute, along with Phillips, did not realize that had each of the cambered airfoils been tested over a range of lift-to-drag ratios, rather than at a fixed $L/D = 10.3$ (Phillips could easily have done that by varying the flow velocity, and changing the scale weights in the pan C accordingly, to measure the corresponding drag at the fixed lift of 9 ounces), then one of the other airfoil shapes might have exhibited a higher *maximum* value for L/D, and therefore would have been judged as the best airfoil shape. The approach by both Phillips and Chanute to use the minimum airspeed in Table 4.2 as the criterion to judge the "best" airfoil simply was not correct. However, it is understandable that flying-machine enthusiasts at that time would have been inclined to think that way. Their main concern was simply to have enough engine power to get a flying machine off the ground. Because the power required to fly a given airplane varies as the cube of the velocity, the early inventors would have embraced any design feature that would have resulted in lift-off at a lower velocity. The point here is that had Phillips relaxed his constraint of a fixed $L/D = 10.3$, he might have found that one of the other airfoils, if allowed to operate at higher L/D values, would have given a lower velocity than those in Table 4.2.

We have been examining Phillips's airfoil data, and their interpretation by Phillips, from the modern point of view – as Monday-morning quarterbacks. We must not forget that Phillips did the best he could, given the very limited understanding of aerodynamics that existed in 1884. Perhaps the most meaningful comment that can be made here is that our reexamination simply endorses Chanute's statement in 1894 that "the shapes patented by Mr. Phillips are not absolutely the most efficient forms." Phillips's major contribution was to alert the technical community in his day to the superiority of cambered airfoils over a flat plate. The quantitative data proving that fact were presented in his table, and that is perhaps the only real significance of Table 4.2. There are two cases in Table 4.2 that featured the same velocity and wing area, and thus the same lift coefficient, namely, the flat-plate case and that for airfoil 5. They had the same lift coefficient (not just the same lift) of 0.56. That was a truly meaningful comparison, indicating that the flat plate was definitely an inferior shape, because its drag coefficient for the same lift coefficient was much higher (0.124 for the flat plate, compared with 0.054 for the cambered airfoil). Another way of saying the same thing is that the flat-plate L/D ratio was much lower than that of airfoil 5 (4.5 compared with 10.3) at the same lift coefficient. That is not to say that the data points for the flat plate and airfoil 5 were determined at the optimum conditions for either shape, but at least, because they were compared at the same lift coefficient, we can be more confident about that comparison. By presenting quantitative data on cambered airfoils versus a flat plate and by demonstrating the superiority of the cambered airfoil (a correct conclusion, despite inaccurate data and a flawed interpretation of the data), Horatio Phillips made a substantial contribution to the state of the art of applied aerodynamics, and Gibbs-Smith's statement that Phillips was "one of the great men of flying history" was certainly well founded.

Interim Summary: Aerodynamics in Its Infant Stage

In the first 75 years of the nineteenth century, aerodynamics struggled through its infant stage. George Cayley's emphasis on fixed-wing lifting surfaces for flying machines was an important development in applied aerodynamics. However, applied aerodynamics and its parent discipline fluid dynamics took separate directions. We have seen how advances in theoretical and experimental fluid dynamics led directly to the fundamentals of modern fluid dynamics; two classic examples of that increasing maturity were the theoretical equations derived independently by Navier and Stokes for the flow of a viscous fluid, and the pioneering experiments by Reynolds showing transition from laminar flow to turbulent flow. Those who were mainly responsible for the great advances in fluid dynamics were university-educated scientists of some stature and repute.

Developments in applied aerodynamics followed a quite different path. The approach to applied aerodynamics was almost completely empirical, and because there was a basic lack of understanding of the fundamentals, much of the empirical work yielded only random, isolated bits of information, none of them constituting any kind of lasting data base. Wenham and Phillips pioneered the use of wind tunnels in aerodynamics, certainly a vital contribution, but the usefulness of their data for a *quantitative* understanding of aerodynamics was virtually nil. The demonstration by Phillips that a cambered airfoil made a more effective lifting surface than a flat plate was of major importance, but that finding seems to have been lost to many investigators at the end of the nineteenth century, for the superiority of a cambered shape was "rediscovered" by Lilienthal near the turn of the century. Moreover, those who were working on applied projects in aerodynamics usually were from the ranks of the technicians – self-educated, intelligent, and mechanically adept practitioners of trades – quite different from the class of university-educated scientists working in fluid dynamics. It was that rapidly growing class of technicians who nurtured the beginnings of engineering as an identifiable profession. Their banding together in the Aeronautical Society of Great Britain was an indication of that growing professionalism.

The last 25 years of the nineteenth century brought much more important advances in applied aerodynamics, particularly with respect to the quality of the experimental data and the direct impact of such data in the design of flying machines. The men responsible for those important developments will be discussed next: Otto Lilienthal, a college-educated German mechanical engineer; Samuel Langley, one of the most distinguished American scientists of the time; and Octave Chanute, the most respected civil engineer in America. Their work brought the infancy of aerodynamics to a close.

A Leap Forward in Applied Aerodynamics: The Work of Otto Lilienthal

The two dominant figures most responsible for the seminal developments in applied aerodynamics in the nineteenth century were George Cayley, at the beginning of the century, and Otto Lilienthal, at the end. Lilienthal ranks as high as Cayley in terms of the importance of his contributions.

In 1866, Otto Lilienthal, with the help of his younger brother Gustav, began a lengthy series of aerodynamic measurements of the lift and drag on a variety of lifting surfaces of different shapes, continuing, with some long interruptions, until 1889. Those measurements fell into two categories: those obtained with a whirling-arm device, and, later, those obtained outside in the wind. The importance of Lilienthal's aerodynamic data derives from two factors:

(1) It was the first body of data in applied aerodynamics to be obtained and plotted in a systematic fashion that respected the basic rules of fundamental fluid dynamics – rules for which Lilienthal obviously had some appreciation.
(2) The data base proved to have "lasting power." It was later used by Langley and Chanute, and it was particularly useful to the Wright brothers in the wing designs for their 1900 and 1901 gliders.

The first period of Lilienthal's work concerned measurements obtained with a whirling arm, as illustrated in Figure 4.29, from *Der Vogelflug als Grundlage der Fliegekunst (Birdflight as the Basis of Aviation).*[46] Identical lifting surfaces were placed at the two ends of a horizontal arm, which was rotated by two falling weights whose cords passed over a series of pulleys. Several whirling-arm devices of different sizes were constructed, with air paths ranging from 2 to 7 m in diameter. The velocities of the lifting surfaces ranged from 1 to 12 m/s. The drag on the lifting surfaces was determined "by reducing the propelling weight for the centers of air pressure of the surfaces." It is unclear what Lilienthal meant by "reducing" the propelling weight. However, we can suggest that if one took moments about the vertical rod connecting to the center of the horizontal arm while the whirling arm was moving at some steady or equilibrium rotational velocity, the torque due to the falling weights would exactly balance the torque due to aerodynamic drag, and hence knowledge of the magnitude of the weights would allow direct calculation of the drag. In the language of experimentalists, the drag data would be "reduced" by means of a calculation using the measured weight data. It is difficult to imagine that Lilienthal did otherwise. Also, Lilienthal took into account in his data reduction the "air resistance due to the arms only, and the frictional loss" associated with the machine – a mark of a good experimentalist. The lift was measured by means of a horizontal balance lever (Figure 4.29). That lever was balanced on the left end by a spherical counterweight and on the right end by the weight of the horizontal arm, pulley, and vertical spindle assembly, along with a scale. When the whirling arm was in motion, upward lift reduced the counterbalancing force on the right end of the lever, and hence weights had to be placed on the scale to maintain the horizontal balance of the lever. The lift was then directly equal to the magnitude of the weights placed on the scale.

From the measured lift and drag, Lilienthal calculated the magnitude and direction of the resultant aerodynamic force. He then plotted that resultant force from two different perspectives. First, he looked at the resultant force relative to the horizontal or free-stream direction. That perspective is shown in Figure 4.30a, where the free-stream direction is indicated by the horizontal arrow, and the lifting surface *ab* (in this case, a flat plate) is inclined at the angle of attack α. The resultant aerodynamic force is given by $0g$, obtained as the resultant of the lift $0f$ and the drag $0l$. The perpendicular to the plate is given by $0N$. (Lilienthal quite correctly noted that the resultant aerodynamic force was not perpendicular

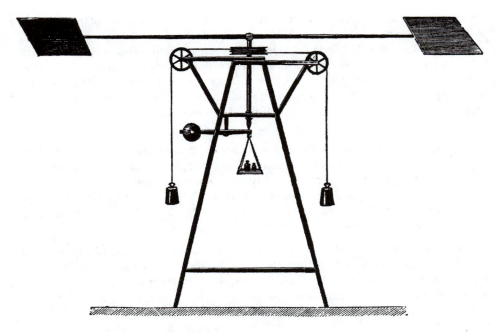

Figure 4.29   Lilienthal's whirling-arm device.

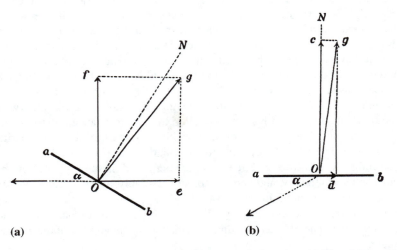

(a)   **(b)**

Figure 4.30   Resolution of the resultant aerodynamic force into (a) lift and drag and (b) normal and axial forces.

to the plate.) He measured the resultant force as a function of the angle of attack from zero to 90° and plotted the results, as shown in Figure 4.31, where the free-stream direction is along the horizontal axis, and the magnitude and direction of the resultant force are indicated by an arrow, each arrow corresponding to a different angle of attack, printed at the end of each arrow. The outer scale is simply a protractor for giving the exact angular direction of the resultant force; the direction of each resultant force is extended via a dashed line to the outer

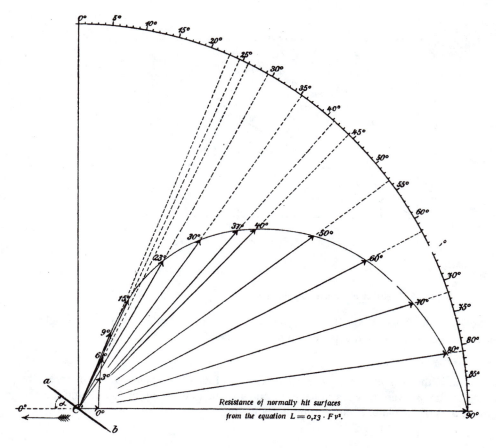

Air pressure on plane, inclined surfaces.

Figure 4.31 Lilienthal's drag-polar diagram for a flat plate.

scale for easy reading. Note that the angular direction of the resultant force (relative to the vertical) is not equal to the angle of attack, again a graphic illustration that the resultant aerodynamic force is not perpendicular to the plate.

The *format* of the diagram in Figure 4.31 has been of tremendous importance in the history of aerodynamics in particular, and aeronautical engineering in general. Note that Lilienthal has connected the tops of the arrows for all the measured resultant forces with a solid curve. Today, we call such a curve a *drag polar* (i.e., a plot of lift coefficient versus drag coefficient), as sketched in Figure 4.32. In modern aeronautical engineering, drag polars illustrate aerodynamic information essential for analysis of the performance of an airplane. Theoretical calculation and/or experimental measurement of the drag polar for an airplane are at the very heart of aerodynamic design and airplane performance. Otto Lilienthal was the first person to measure and construct a drag polar – a pioneering development in aeronautical engineering.

Note that in Figure 4.31 the aerodynamic force measured at an angle of attack of 90° is given by the horizontal vector labeled 90°; that is the drag force on a flat plate oriented

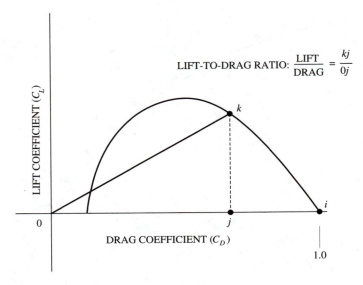

Figure 4.32   Elements of a drag-polar diagram.

vertically in the flow, as sketched in Figure 3.18. In our earlier discussion we argued that the aerodynamic force on that plate is approximately

$$F = (p_0 - p_\infty)S$$

where p_0 is the stagnation pressure on the forward face of the plate, p_∞ is the pressure acting on the rearward face (assumed to be equal to the free-stream pressure), and S is the surface area of the plate. From Bernoulli's equation,

$$P_0 - P_\infty = \tfrac{1}{2}\rho V_\infty^2$$

Combining those two equations, the force on the perpendicular plate is

$$F = \tfrac{1}{2}\rho V_\infty^2 S$$

Because the plate is oriented perpendicular to the flow, that force is the drag D. Hence we write

$$D = \tfrac{1}{2}\rho V_\infty^2 S$$

Comparing that with the conventional way of writing drag in terms of the drag coefficient C_D, that is, comparing it with

$$D = \tfrac{1}{2}\rho V_\infty^2 S C_D$$

we see that the drag coefficient for a flat plate oriented perpendicular to the flow is essentially

$$C_D = 1^\dagger$$

> † Modern experimental measurements for a flat plate oriented perpendicular to the flow show that C_D is a function of Reynolds number and plate aspect ratio. According to Hoerner,[38] for Reynolds numbers above 1,000 and aspect ratios below 30, a reasonable value is $C_D = 1.17$. However, for the elliptical-like planform with pointed wing tips used by Lilienthal, the three-dimensional relieving effect is stronger than for rectangular planforms, and the resulting drag coefficient should be smaller, closer to 1.0. For this reason, in light of the lack of modern aerodynamic data for Lilienthal's wing configuration, we can feel comfortable with the value $C_D = 1.0$.

Returning to Figure 4.31, we note that Lilienthal drew all his resultant-force arrows *relative* to the magnitude given for the force at 90°. If we assume that the length of the arrow at 90° is a *unit* length, equivalent to $C_D = 1$, then the vertical and horizontal components of all the other arrows can directly give the values for the lift and drag coefficients for all the different angles of attack. That construction is further illustrated in Figure 4.32. Let the length $0i$ be equal to 1, equivalent to the length of Lilienthal's horizontal vector in Figure 4.31. Hence, in Figure 4.32, the point i corresponds to a drag coefficient $C_D = 1$. Now consider the case for another angle of attack less than 90°, such as point k in Figure 4.32. For this case, the value of C_L will be equal to the length kj, and C_D will be equal to the length $0j$. Therefore, if the length of Lilienthal's vector at 90° is scaled to be unity, then the lift and drag coefficients for any other angle of attack can be measured directly from the vertical and horizontal coordinates of that particular point on the curve. Clearly, Lilienthal's curve (Figure 4.31) was, in the purest sense, a drag polar.

The foregoing discussion sets the stage for clarification of the role played by Smeaton's coefficient in the reduction of Lilienthal's data. In Chapter 3 we discussed the origin of Smeaton's coefficient. Dating from 1759, it was simply a correlation factor obtained from experimental data for the force on a plate oriented perpendicular to an airflow:

$$F = kSV^2$$

where F was the force in pounds, S was the surface area in square feet, and V was the wind velocity in miles per hour. The correlated value for k was 0.005. The constant k became known as Smeaton's coefficient, and the value $k = 0.005$ propagated through the next century. However, the accuracy of that determination was soon questioned. George Cayley had doubts about the published value of Smeaton's coefficient (on the basis of Cayley's data, we have calculated $k = 0.0037$). Nevertheless, the value $k = 0.005$ was used by countless investigators during the nineteenth century. Lilienthal was no exception. The metric expression of Smeaton's coefficient, where F is in kilograms force, S is in square meters, and V is in meters per second, is $k = 0.13$. The value $k = 0.13$ appears throughout Lilienthal's book; indeed, Lilienthal printed the equation for force on a perpendicular flat plate directly on his drag polar (Figure 4.31) using the accepted value of 0.13 for Smeaton's coefficient. (Do not be confused by the equation shown in Figure 4.31; in Lilienthal's notation, L is the symbol for drag force, and F is the area.) Lilienthal's data on aerodynamic coefficients, such as those in Figure 4.31, were later questioned by the Wright brothers, at first being rejected outright, but later partially confirmed in their wind-tunnel tests. It has long been the conjecture of aeronautical historians that the perceived problem with Lilienthal's data was due in part to the error in Smeaton's coefficient.[24] However, I would disagree with that assessment. Once again we must look at Lilienthal's mode of presentation of his data in Figure 4.31. Clearly he has scaled all his data to the force data obtained at an angle of attack of 90°. By doing so, he has divided out any influence of Smeaton's coefficient. If we scale the drag at 90° to be equivalent to $C_D = 1$, then the values for C_L and C_D obtained from Figure 4.31 for any other angle of attack are not affected by any error in Smeaton's coefficient. Lilienthal's data, as presented in Figure 4.31, were subject to a host of experimental uncertainties because of the use of a whirling arm and his apparatus for measuring lift and drag. However, the error in Smeaton's coefficient was not one of those factors. Smeaton's coefficient had no effect on the accuracy of Lilienthal's data, given the mode in which those data were presented in Figure 4.31.

The aerodynamic data for a flat plate shown in Figure 4.31 were just the beginning of Lilienthal's work. Those data were obtained purely for reference purposes. Lilienthal's

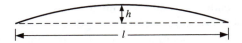

Figure 4.33   Circular-arc airfoil.

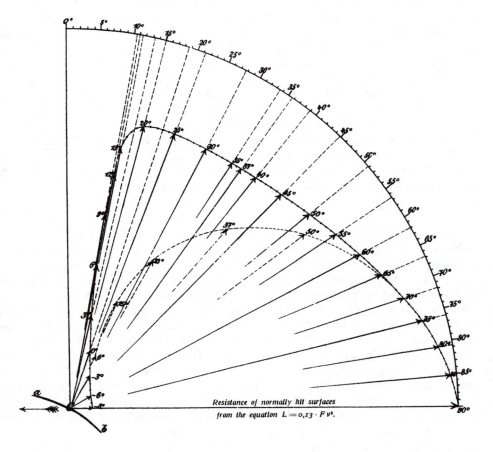

Figure 4.34   Lilienthal's drag-polar diagram for a $\frac{1}{12}$-cambered airfoil.

major objective was to examine the aerodynamic properties of cambered (curved) shapes. In his book he gave the drag polars for airfoils shaped in the form of a circular arc, with the maximum height at the center of the airfoil (Figure 4.33). He also showed plots of whirling-arm data for circular-arc airfoils with cambers (h/l) of $\frac{1}{40}$, $\frac{1}{25}$, and $\frac{1}{12}$, where h and l are as defined in Figure 4.33. The drag polar for the $\frac{1}{12}$-cambered airfoil is shown in Figure 4.34, where the solid curve is the drag polar from Lilienthal's data. The dashed curve is simply a repetition of the flat-plate data from Figure 4.31 for purposes of comparison. Clearly the cambered shape produces more lift for a given drag. The graphic representation shows that the lift-to-drag ratios for the cambered airfoil in Figure 4.34 were much higher

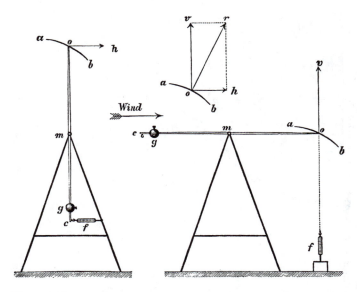

Figure 4.35   Lilienthal's devices for measuring lift and drag in the natural wind.

than those for the flat plate in Figure 4.31. Commenting on those data, Lilienthal stated that "the advantages of a curved surface over a plane, for flight purposes, are clearly shown." Of course, Lilienthal was not the first person to indicate the superiority of a cambered airfoil over a flat plate; Cayley had alluded to that, and Phillips had proved it in his wind-tunnel experiments. However, Lilienthal's data provided the most extensive and coherent proof of the superiority of the cambered airfoil to date; from that time on, there was no question that future flying machines would have cambered wings.

In his book, Lilienthal noted that "all the experiments confirmed the law of the increase of the air resistance with the square of the velocity." Any question about that fact should have been put to rest by the experimental measurements of Mariotte and Huygens in the seventeenth century, as repeatedly confirmed by Robins, Cayley, Phillips, and others. But Lilienthal felt compelled to mention it again, perhaps in an effort to further validate his findings by linking them to existing knowledge, rather than simply presenting an additional proof of the velocity-squared variation. However, his use of the word "confirmed" seems to imply the latter.

The second phase of Lilienthal's experimental work in aerodynamics began in 1874 and involved measurements of the aerodynamic force on a stationary airfoil placed outdoors in the natural wind. The location was a flat plain devoid of trees behind the Charlottenburg Palace in Berlin. The apparatus used for his measurements is shown in Figure 4.35. The technique for measuring drag is shown at the left. The airfoil ab exposed to the wind experiences a drag $0h$. The magnitude of that drag is measured by the horizontal spring scale f. The measurement of lift is shown at the right, where the lift on the airfoil ab exposed to the wind is measured by the vertical spring scale f. Because the wind velocity and direction varied with time, sometimes changing many times over the interval of a minute, Lilienthal needed instantaneous and simultaneous measurements of wind velocity and aerodynamic force. He measured the wind velocity with the anemometer device shown in Figure 4.36, where a flat plate F is mounted on a rod ik and is connected to point i by means of a spiral spring. The wind exerts a force on plate F, which slides along the rod until

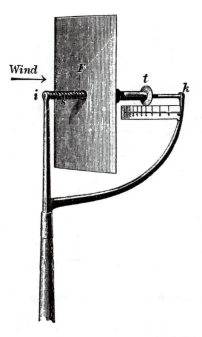

Figure 4.36   Anemometer device built by Lilienthal for measuring airflow velocity.

an equilibrium position is reached (where the drag force is balanced by the spring force).
The index t passes over a scale that is calibrated directly in terms of velocity. In Lilienthal's
words, "from the area of the plate F, it is easy to calculate the wind pressure which exists at
various wind velocities, and since the elasticity of the spiral spring is known, it is possible
to calibrate the scale with accuracy." Lilienthal did not realize it, but his calibration was by
no means accurate. The calculation he made was based on the familiar law for the drag on
a vertical flat plate discussed earlier, namely,

$$D = kSV^2$$

where k is Smeaton's coefficient. From that equation, the velocity is obtained as

$$V = \sqrt{D/kS}$$

However, Lilienthal used the value $k = 0.13$ (in metric units) for Smeaton's coefficient. We
have already seen that that value for k is too high by almost a factor of 2. Thus Lilienthal's
velocity measurements were too low by a factor of $\sqrt{2}$.

However, that error in measuring the velocity did not, per se, have any effect on his
measured lift and drag coefficients, because the aerodynamic data obtained in the natural
wind were plotted in the same form as his earlier whirling-arm data (i.e., the data were
ratioed to the aerodynamic force at a 90° angle of attack). Figure 4.37 shows a drag polar
for a circular-arc airfoil with a camber of $\frac{1}{12}$, where the solid curve is for the measured
data obtained in the natural wind. Because the data for all angles of attack are shown in
relation to those for a 90° angle of attack, the effects of both the velocity and Smeaton's
coefficient have been divided out. For comparison, the dashed curve in Figure 4.37 repeats
the whirling-arm data from Figure 4.34.

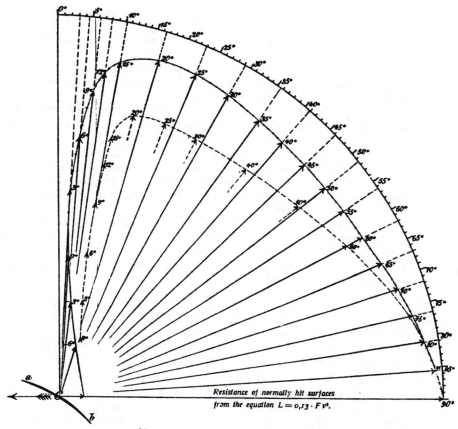

Camber $^1/_{12}$ width.

*Air pressure on curved surfaces as determined
by measurements made in a wind, but disregarding
the additional lifting effect of wind.*

Figure 4.37   Lilienthal's drag-polar diagram for a $\frac{1}{12}$-cambered airfoil; data taken in the natural wind.

The obvious discrepancy between the natural-wind data and the whirling-arm data in Figure 4.37 deserves some comment. Clearly the data obtained in the natural wind give a higher lift coefficient and a higher lift-to-drag ratio over the whole range of angles of attack. Why? The answer does not lie in the error in Lilienthal's wind-velocity measurements, because the effect of the velocity was divided out in the presentation of the data. The answer most certainly lies in the glaring experimental uncertainties inherent in both the whirling-arm and natural-wind devices – devices that later found no place in twentieth-century aerodynamic laboratories. Lilienthal himself understood the experimental inadequacies of those devices:

> For the determination of air pressures on curved wings under different inclinations, we must have recourse to experiment; only actual measurements of forces can give us useful figures for the explanation of birdflight and for aviation.

We may either move the experimental wing in still air, or we may cause air to impinge on the wing.

In the former case we are limited to a circular movement of the wing, and have to employ an apparatus similar to [Figure 4.29].

Rectilinear movement of the surfaces would necessitate a mechanism with greater secondary resistances, and consequently would introduce greater sources of error. On the other hand, the whirling machine does not allow us to investigate rectilinear motion; after half a rotation, the wind under test enters a region of disturbed air, thus introducing errors. Both disadvantages decrease as the diameter of the circular path increases, so that such machines ought to be made as large as possible.

The second method – that of exposing the test surface to the wind – though giving the effect of rectilinear motion, suffers from this defect: the strength of the wind varies almost every second, and it is extremely difficult to seize those moments during which the anemometer indicates the proper velocity. It is, therefore, necessary to determine by means of very numerous experiments good average values.

We have repeatedly employed both methods of measurement because we were impressed with the importance of obtaining the most accurate knowledge of the air pressures on curved surfaces, and because we desired to check the various methods one with the other, as we had no knowledge of similar experiments made by others which would have permitted us to check our own results [p. 63].[46]

Clearly, Lilienthal had no illusions about the accuracy of his data. However, he believed that the data obtained in the natural wind were closest to the true values. Specifically in regard to the comparison between the natural-wind data (solid curve) and the whirling-arm data (dashed curve) in Figure 4.37, he stated that "we can now appreciate the difference between the two modes of experimenting. The discrepancies are due to the errors introduced by the whirling arm machine." Lilienthal's partiality for the natural-wind data resulted in those data being propagated to other investigators and in being seen as the main findings associated with Lilienthal. Lilienthal put his findings from Figure 4.37 in tabular form, and they were published in 1895 by Hermann Moedebeck,[47] a fellow member of the German Society for Advancement of Airship Travel. Subsequently, Octave Chanute made that table available to American investigators by printing it in his paper "Sailing Flight," which appeared in *The Aeronautical Annual* in 1897, published in Boston by James Means. Figure 4.38 is a photocopy of Lilienthal's table as it appeared in Chanute's article. It was that table that was used by the Wright brothers to design their 1900 and 1901 gliders. In the Wright brothers' voluminous correspondence, whenever they referred to the "Lilienthal Table," they meant the table shown in Figure 4.38, which gives Lilienthal's data obtained in the natural wind – not whirling-arm data. Furthermore, those data pertained to a circular-arc airfoil of camber $\frac{1}{12}$. The planform shape (top view) of the wing is shown in Figure 4.39. Unfortunately, Chanute failed to note the details of the shape of Lilienthal's wings and airfoils in his 1897 paper, simply identifying them as "surfaces arched upward [in section] about $\frac{1}{12}$ of their width." As a result, subsequent investigators, including the Wright brothers, often applied the data from Lilienthal's table to configurations for which they were never intended. Later, Wilbur Wright gently took Chanute to task for that oversight (Wilbur's letter to Chanute, November 22, 1904): "In the Lilienthal table I would suggest that it be stated in connection with the table that it is for a surface 0.4 meters by 1.8 meters, of the shape [here Wilbur sketched a shape like that shown in Figure 4.39] and a curvature of 1 in twelve, and measured in the natural wind. Any table is liable to great misconstruction if the surface to which it is applicable is not clearly specified. No table is of universal application."

TABLE OF NORMAL AND TANGENTIAL PRESSURES

Deduced by Lilienthal from the diagrams on Plate VI., in his book "Bird-flight as the Basis of the Flying Art."

α Angle.	η Normal.	ϑ Tangential.	α Angle.	η Normal.	ϑ Tangential.
$-9°$	0.000	$+0.070$	$16°$	0.909	-0.075
$-8°$	0.040	$+0.067$	$17°$	0.915	-0.073
$-7°$	0.080	$+0.064$	$18°$	0.919	-0.070
$-6°$	0.120	$+0.060$	$19°$	0.921	-0.065
$-5°$	0.160	$+0.055$	$20°$	0.922	-0.059
$-4°$	0.200	$+0.049$	$21°$	0.923	-0.053
$-3°$	0.242	$+0.043$	$22°$	0.924	-0.047
$-2°$	0.286	$+0.037$	$23°$	0.924	-0.041
$-1°$	0.332	$+0.031$	$24°$	0.923	-0.036
$0°$	0.381	$+0.024$	$25°$	0.922	-0.031
$+1°$	0.434	$+0.016$	$26°$	0.920	-0.026
$+2°$	0.489	$+0.008$	$27°$	0.918	-0.021
$+3°$	0.546	0.000	$28°$	0.915	-0.016
$+4°$	0.600	-0.007	$29°$	0.912	-0.012
$+5°$	0.650	-0.014	$30°$	0.910	-0.008
$+6°$	0.696	-0.021	$32°$	0.906	0.000
$+7°$	0.737	-0.028	$35°$	0.896	$+0.010$
$+8°$	0.771	-0.035	$40°$	0.890	$+0.016$
$+9°$	0.800	-0.042	$45°$	0.888	$+0.020$
$10°$	0.825	-0.050	$50°$	0.888	$+0.023$
$11°$	0.846	-0.058	$55°$	0.890	$+0.026$
$12°$	0.864	-0.064	$60°$	0.900	$+0.028$
$13°$	0.879	-0.070	$70°$	0.930	$+0.030$
$14°$	0.891	-0.074	$80°$	0.960	$+0.015$
$15°$	0.901	-0.076	$90°$	1.000	0.000

Figure 4.38 The Lilienthal table of aerodynamic coefficients.

Lilienthal's table was a historic document in applied aerodynamic and aeronautical engineering for another reason: It introduced the concept of aerodynamic-force *coefficients*. It tabulated values for the coefficients η (labeled "Normal") and θ (labeled "Tangential") versus the angle of attack, the values for η and θ being less than or equal to unity. Lilienthal defined those coefficients via the following relationships:

$$N = \eta \times 0.13 \times F \times V^2$$
$$T = \theta \times 0.13 \times F \times V^2$$

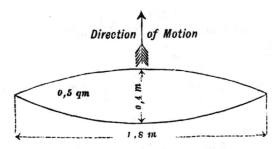

Figure 4.39 Planform size and shape of Lilienthal's model used during his aerodynamic measurements.

The nomenclature is Lilienthal's: N is the "normal pressure," defined as the component of the resultant aerodynamic force normal to the chord; T is the "tangential pressure," defined as the component of the resultant aerodynamic force tangential to the chord; F is the planform area of the wing; V is the free-stream velocity; 0.13 is Smeaton's coefficient in metric units. Lilienthal did not give names to the coefficients η and θ; later, Chanute, in his article in the 1897 *Aeronautical Annual,* referred to them simply as "Lilienthal's coefficients." In modern aeronautical engineering, the normal and axial (tangential) forces are written as

$$N = C_N \left(\tfrac{1}{2}\rho\right) F V^2$$

and

$$T = C_A \left(\tfrac{1}{2}\rho\right) F V^2$$

using Lilienthal's notation for the forces N and T, and F for the area. C_N and C_A are the normal-force and axial-force coefficients, respectively, commonly used in modern aeronautical engineering. Comparing these equations with Lilienthal's equations, and interpreting Smeaton's coefficient as essentially $\rho/2$, clearly Lilienthal's coefficients η and θ are the normal-force and axial-force coefficients, respectively. That was the first time that such force coefficients were defined and used in aerodynamics. The lift and drag coefficients are related to the normal-force and axial-force coefficients through simple trigonometric relations[1] (using Lilienthal's nomenclature):

$$C_L = \eta \cos \alpha - \theta \sin \alpha$$
$$C_D = \eta \sin \alpha + \theta \cos \alpha$$

Lilienthal's table shows that for an angle of attack of 90°, $\eta = 1$ and $\theta = 0$. From the foregoing equation, that gives $C_D = 1$ for a 90° angle of attack, consistent with our earlier discussion.

The definition and first use of aerodynamic coefficients are credited to Lilienthal. He tabulated them in the form of normal-force and axial-force coefficients. However, Lilienthal fully understood the resolution of aerodynamic force into lift and drag components as well; that was the essence of his drag-polar diagrams, although he did not put the lift and drag coefficients in tabular form. The fact that Lilienthal freely moved back and forth between resolution of aerodynamic force into normal and axial components, on one hand, and lift and drag components, on the other hand, is clearly seen in his illustrations. Figures 4.34 and 4.37 reproduce only the right sides of his original figures, namely, the drag polars dealing with resolution of the force perpendicular and parallel to the free-stream direction – the lift and drag, respectively, as illustrated in Figure 4.30a. There the free-stream is fixed in the horizontal direction, and the airfoil rotates through various angles of attack, with the direction of the aerodynamic force changing appropriately. Lilienthal also plotted graphs of the aerodynamic force relative to the airfoil itself; Figure 4.30b shows the normal and axial forces. In this perspective, the airfoil is fixed with its chord in the horizontal direction, and the change in angle of attack is drawn as a change in the direction of the relative wind. Lilienthal plotted his data from that perspective on the left side in his illustrations – to the left of his drag polars. Figure 4.40 shows Lilienthal's juxtaposition of data for a flat plate. On the right is the familiar drag polar, with the horizontal being the direction of the relative wind. On the left is the resultant aerodynamic force plotted relative to the plate, with the plate always oriented in the horizontal direction. The resultant aerodynamic forces are shown by the arrows, each of which is labeled at the tip with a number in degrees; that number gives the angle of attack for which that arrow applies. Note that the arrow labeled 90° is the aerodynamic force on the plate when the plate is at a 90° angle of attack. As expected, at an angle of attack of 90°, the aerodynamic force is perpendicular to the plate. However, for all other angles of attack, Lilienthal's data show that the resultant force is *not* perpendicular to the plate (Figure 4.40, left). Since the advent of Newtonian impact theory in the late seventeenth century, which had predicted the force to be perpendicular to the plate, the question of the direction of the resultant aerodynamic force relative to the airfoil had been a matter of debate. Was it perpendicular to the airfoil chord, or not? Lilienthal's definitive experiments finally put that question to rest. The left side of Figure 4.40 shows that for any angle of attack, the vertical component of the arrow is the normal force, and the horizontal component is the tangential or axial force. Furthermore, the lengths of all the arrows are scaled relative to that for 90°. If the length of the arrow at 90° is taken as unity, corresponding to $C_D = C_N = \eta = 1$, then the vertical and horizontal components of any other arrow at any other angle of attack will directly give the values for η and θ (or C_N and C_A in modern nomenclature), respectively.

In light of this interpretation, let us examine the data for Lilienthal's $\frac{1}{12}$-cambered circular-arc airfoil (Figures 4.41 and 4.37). There are two important points to be made in regard to Figure 4.41. First, if the length of the arrow at 90° is taken unity, then the vertical and horizontal components of all the arrows will directly give the values for η and θ. The entries in Lilienthal's table (Figure 4.38) were obtained directly from Figure 4.41 in that fashion. Second, unlike the case for the flat plate (Figure 4.40), where the resultant aerodynamic force was always inclined *behind* the perpendicular to the plate (i.e., with an axial force always oriented toward the right), Lilienthal's data for the cambered airfoil show that at some angles of attack the resultant aerodynamic force will be inclined *ahead* of the perpendicular to the chord (i.e., with an axial force pointing to the left). Lilienthal called that leftward force a "pushing component" and cited its existence as further evidence of the superiority of cambered airfoils (today, a forward-facing axial force is a well-known property of a body with camber). It is a further confirmation of the viability of Lilienthal's data that he observed such a case. Lilienthal was the first person to observe that phenomenon, another example of his contributions to applied aerodynamics.

Lilienthal's data proved, beyond doubt, that a cambered airfoil was superior to a flat plate. But did Lilienthal fundamentally understand why? Rather than simply accepting the trends shown by his data, Lilienthal sought an intellectual understanding of the basic nature of the flow fields around a flat plate and a curved shape. He wanted to know what

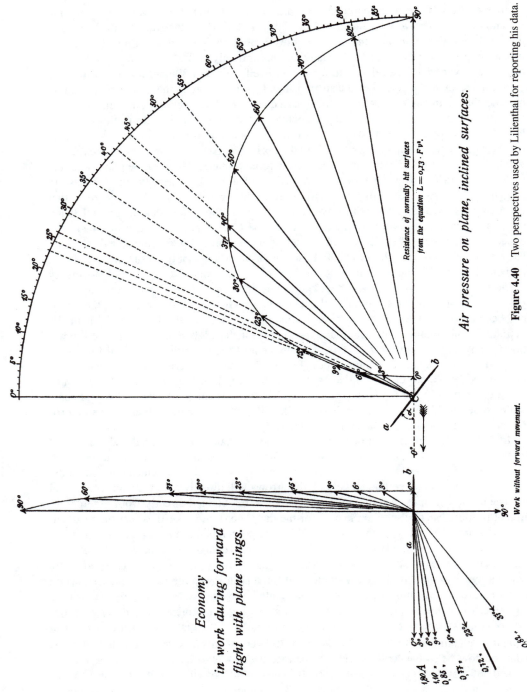

Economy
in work during forward
flight with plane wings.

Work without forward movement.

Resistance of normally hit surfaces
from the equation $L = 0.3 \cdot F v^2$.

Air pressure on plane, inclined surfaces.

Figure 4.40 Two perspectives used by Lilienthal for reporting his data.

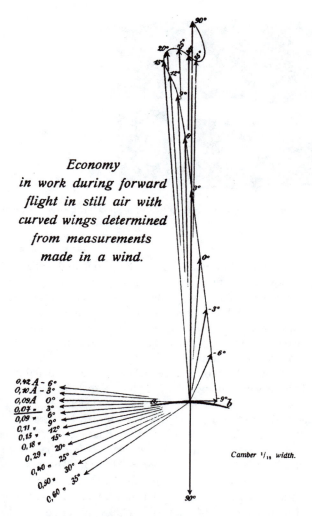

Figure 4.41 Lilienthal's diagram showing that for cambered airfoils, the aerodynamic-force vector can sometimes lean forward, giving a positive axial force.

was different about those flow fields that would cause the difference in aerodynamic performance. He devoted a whole section of his book to that matter, prefacing his discussion with the following statement: "Experiment enables us to determine the actual degree of superiority of curved over plane wings, but in view of the importance of this matter we must obtain a perfect conception of the nature of this phenomenon."[46] Lilienthal had to theorize about the qualitative nature of the flow fields over flat and curved surfaces; he had no experimental data on flow-field visualization. He sketched the streamline patterns around the two shapes, based on what he thought was reasonable (Figure 4.42). In his theorized flat-plate flow field (Figure 4.42a), Lilienthal was on the right track, although he was wrong about the details. He believed that a deviation in the flow due to the presence of the flat plate would take place suddenly at the leading edge. The flow that was uniform and parallel in the free stream ahead

Figure 4.42 Lilienthal's sketches of his conceptions of the flow fields over a flat plate and a cambered shape.

of the plate would suddenly be turned downward at the leading edge of the plate. Lilienthal surmised that "the deviation downward of the air stream takes place generally at the leading edge of the plane surface, and rather suddenly, giving rise to shocks and the formation of eddies." Lilienthal's use of the word "shocks" suggests an almost discontinuous change in flow properties at the leading edge. Figure 4.42a shows his idea of a uniform, parallel free stream ahead of the leading edge, and then a sudden change in direction at the leading edge. In Lilienthal's words, the flow would be "torn, crushed, or broken by plane surfaces." In reality, that model of the flow was grossly incorrect; the real flow pattern over a flat plate at an inclination in a low-speed flow is shown in Figure 3.23, which reveals that the free stream is forewarned about the presence of the plate far before it reaches the leading edge, and it begins to move upward well ahead of the leading edge. That is characteristic for subsonic flow over any lifting airfoil. In a subsonic flow, the flow prepares itself for the leading edge long before it actually encounters the leading edge, but that is twentieth-century information that Lilienthal had no way of knowing. On the other hand, his concept of "eddies" being formed at the leading edge was analogous to the real situation, in which the flow readily separates from the top of the leading edge, bathing the top surface of the plate in a region of recirculating eddies, such as those sketched by Lilienthal (Figure 4.42a). Lilienthal's indication of eddies on the top surface was qualitatively correct, but his supposition that such eddies and separated flow would also occur on the bottom surface of the plate was totally erroneous. Basically, Lilienthal's postulated flow field in Figure 4.42a had a grain of truth to it (separated flow does occur over a flat plate at an inclination), but most of the flow details shown there were not correct. In contrast, for a cambered airfoil at an inclination (Figure 4.42b), because the leading edge is closer to being tangential to the incoming free stream, and because the flow is turned downward much more gently, Lilienthal assumed that no "shocks" or "eddies" would occur. Hence the flow over the curved airfoil would not suffer a "diminution of the desired effect" as was the case for the flat plate. Therefore, a curved surface was a more efficient lifting surface.

In retrospect, the comparison shown by Lilienthal's sketches (Figure 4.42) was not totally off the mark. We understand today that the reason a flat plate is an inefficient lifting surface is that the flow readily separates over the top surface, thus decreasing the lift and increasing the drag – as Prandtl would say much later, the separated flow over a flat plate is an "unhealthy" flow. Lilienthal's sketch of the flat plate in Figure 4.42a certainly shows an "unhealthy" flow, in comparison with the smooth, attached flow over the curved surface in Figure 4.42b. Lilienthal was correct in the broadest sense, but not in the details. He knew enough about the flow field to make the following statement: "In the avoidance of eddies lies the principle which may some day enable us to actually fly." Even a modern aeronautical engineer would agree with that if the word "efficiently" were added at the end.

In Lilienthal's discussion of eddies and their adverse impact on aerodynamic properties, he noted that the flow over a flat plate was noisier than that over a cambered airfoil and quite correctly connected that noise with the presence of eddies: "Even the ear tells us whether we deal with pure wave motion or with wasteful eddies. We therefore prefer those surfaces which, even at considerable velocities in air, do not produce noise to those which produce a rushing noise under the same conditions." Another instance of prescience. Today, one technique used by aerodynamicists to determine the point of transition from smooth laminar flow to rough turbulent flow is to use a stethoscope to listen for the location of a noise increase. Lilienthal seems to have been the first person to have formally recognized that eddies create noise and to have used that fact in a diagnostic sense.

A Leap into the Air: Otto Lilienthal, the Glider Man

In addition to his important contributions to applied aerodynamics, Lilienthal played another role: inventor, builder, and pilot of the first practical gliders. By 1896, he had made more than 2,000 glider flights. His work coincided with important developments in photography and printing. In 1871 the dry-plate negative was invented, and by 1890 one could photograph moving objects without blurring. Also, the halftone method of printing had been developed. Lilienthal was the first person to be photographed in flight (Figure 4.1), and photos of his flights were widely distributed. Otto Lilienthal (1848–96) (Figure 4.43) was multifaceted, with all the qualities necessary for success in the flying-machine business: a scientist interested in unraveling the technical secrets of flight; an educated engineer who knew how to apply scientific findings in designing a machine; a dedicated enthusiast of flight with the dexterity to fly those machines. There is a standard biography of Lilienthal by Schwipps.[48]

Figure 4.43   Otto Lilienthal.

Otto Lilienthal was born in Anklam, Prussia, May 23, 1848, to middle-class parents. His mother had been educated in Berlin and Dresden and was knowledgeable about artistic and cultural matters. She had turned down opportunities in the musical theater to become a voice teacher. His father had attended high school and then become a cloth merchant. After their marriage in 1847, economic problems arose, and the business failed. Otto's father turned to alcohol and gambling, dying in 1861, when Otto was only 13 years old. The Lilienthals had eight children, only three of whom survived infancy: Otto, Gustav, and Marie. Frau Lilienthal was a strong motivating force in the lives of her children, inculcating a strong set of intellectual and artistic values. Later, when Otto and Gustav were beginning experimental work on flying machines, she cheered them on. She died of pneumonia in 1872, but not before she had seen her children safely educated.

Otto attended high school in Anklam, earning below-average marks. But in the fall of 1864 he entered the provincial vocational school in Potsdam and excelled, achieving the best examination scores in the history of the school. After working the following year as an apprentice in a machine factory, he entered the Berlin Trade Academy (now the Technical University of Berlin), graduating in 1870 with a degree in mechanical engineering. The Franco-Prussian War had broken out 10 days earlier, and he enlisted in a Prussian regiment. After his one-year enlistment, he turned down an offer of a position at the Trade Academy in order to work for the C. Hoppe machine factory in Berlin. By the time of his mother's death in 1872, Otto was already poised on the brink of a successful career.

However, Otto was simultaneously developing a second, parallel career: work on the principles of heavier-than-air flight. His passion for flight had been shared by his younger brother Gustav as early as 1861, when both were teenagers. They studied birds in flight, and they constructed wings, which they strapped to their arms in attempts to fly (the materials were readily supplied by their mother). Their "flight experiments," all unsuccessful, were carried out at night and in private, "in order to escape the jibes of our schoolmates." After the hiatus of their formal education and the Franco-Prussian War, their aeronautical work began anew, on a much more mature level. During a stay in London, Gustav was introduced to the Aeronautical Society of Great Britain. Both he and Otto became members, quickly learning all about the aeronautical progress in Britain. By 1873 the Lilienthals' experiments were well under way with the construction of their first whirling-arm device (Figure 4.29). By 1874 the Lilienthals had collected the first definitive data showing the aerodynamic superiority of cambered airfoils over flat surfaces, predating the work of Phillips in England by 10 years, but their findings were not published until 1889, four years after Phillips published his findings. Otto Lilienthal did not know about the work of Phillips until 1889, when he explored the possibility of a patent on cambered airfoils and found that Phillips already held such a patent. The question of why the Lilienthals did not publish their results until 1889 was answered by Otto in an 1893 article in the German *Journal for Airship Travel*:

> The long delay in publishing our aviation discoveries was nothing more than the natural result of the accompanying circumstances. While we were dedicating every hour of our free time to the flight question and were already on the trail of the laws which would free the solution from its guarded summit, most people in Germany considered anyone who would waste his time on such a profitless art to be a fool.... At that time it had just been confirmed once and for all by a particularly learned government-appointed commission that man could not fly, which did not particularly lift the spirits of those working on the problem of flight.... In addition, as young people, completely without means, we had to

> save our money penny by penny, by skipping breakfast, in order to be able to carry out our
> experiments.... We were absolutely not in any position to produce a good publication of
> our achievements.

That government-appointed commission had been headed by Hermann von Helmholtz, the most respected scientist in Germany. The major conclusion of that commission was that *human-powered* flight was highly improbable, but that conclusion was interpreted by the wider public to include any form of human flight in a heavier-than-air machine, and it reinforced the prevailing attitude that would-be inventors of flying machines were misguided and were wasting their time. Otto Lilienthal, a trained mechanical engineer, wanted to be totally confident of his findings before exposing them to the public.

Otto Lilienthal married Agnes Fischer in 1878. Music brought them together. Agnes was trained in piano and voice, and Otto played the French horn and had a good tenor voice. They sang together at various civic events and at local inns. After their marriage, they took up residence in Berlin. Otto continued as an engineer and sales representative for the Hoppe machine factory, but he had begun looking for an independent income. He had obtained a patent for a compact, efficient, low-cost spiral-tube boiler, and in 1881 he opened a factory to manufacture the boiler – his major source of income for the rest of his life. The factory grew to employ 60 workers, producing not only boilers but also steam engines, steam heaters, and forged pulleys. Lilienthal had many other personal interests; he was a social activist and progressive thinker. In 1890 he introduced a profit-sharing plan for his employees, one of the first in Germany. His income from that factory was never great, rising and falling with the economic tides that swept Germany during the latter part of the nineteenth century. However, the family was able to move to the pleasant Berlin suburb of Gross-Lichterfelde in 1886. Their house on Booth Street, designed by his brother Gustav (who by then had become an artisan and architect), and built in part by Otto's own hands, was one of the more modest in the neighborhood. By 1887, Otto's immediate family was complete, with two sons and two daughters.

After a substantial hiatus, Otto resumed his aeronautical experiments in 1888. Two years prior to that, both Otto and Gustav had joined the German Society for Advancement of Airship Travel, in which Otto would be an active participant; it became a public forum for much of his aeronautical work. The society had been founded in 1881 by a few people interested in steerable balloons, and it published a journal entitled *Journal of Aviation*. In 1889 Otto was elected a member of the technical commission of the society, and after 1892 he served on the editorial committee for the *Journal of Aviation*. By 1889 the society had about 100 members.

Otto and Gustav resumed their aerodynamic experiments in 1888 to verify their 1873–4 measurements, using larger and better instruments. The experiments were carried out in the workshop and garden of the house on Booth Street, generally in the early morning and late evening, when there was essentially no wind. A new whirling arm, 7 m in diameter, was constructed, and thousands of tests were conducted, using a variety of airfoil shapes and angles of attack. The wing surfaces were rather large, about 2 m in span and 0.5 m at maximum chord. The arm rotated at a height of 4.5 m (about 15 ft) above the ground. The new findings were consistent with their 1873-4 whirling-arm data. In addition, the Lilienthals repeated their natural-wind experiments in an open plain, again confirming their earlier measurements. Their testing continued into 1889, until finally they felt comfortable about announcing their findings to the public. That presentation took two forms. First, Otto gave three lectures before the society in 1888–9 attracting some attention beyond the membership of the society, including newspaper coverage. Second, the material from those lectures evolved into Otto's book.[46] It was the second most important aeronautical publication of the nineteenth century, after Cayley's triple paper of 1809–10. The book was published by R. Gaertner's company in Berlin, but Lilienthal had to pay the printing costs; 1,000 copies were printed, but by the time Otto died seven years later, fewer than 300 copies had been sold. As late as 1909, copies of the first printing could be found in bookstores for 10 marks;[48] today, a first printing sells for several thousand marks.

With his book finished, Lilienthal was ready to begin aeronautical activities in the real world of flying-machine design. At first he took halting, careful steps. In the summer of 1889 he built a large wing, 11 m long, with a 1.4-m maximum chord; the planform was patterned after that of a bird's wing (including pointed wing tips). The wing sections were cambered, reflecting the basic finding of Lilienthal's aerodynamic experiments. An opening in the middle of the wing was large enough to hold a pilot. Lilienthal never left the ground in that wing; he used it to conduct experiments to determine the strength of its lifting force (noted by Lilienthal to be considerable) and to estimate how the wing could be balanced in the natural wind. A year later, he was still conducting such tests with a slightly modified wing. Finally, in a lecture to the society in March 1891, Lilienthal outlined a plan for flying. He suggested trying only short, downhill hops. By the late spring, he put his words into practice. Using a glider that had both a wing and a tail, Lilienthal was jumping into the air for short hops, and by the end of the summer he had jumped into the air more than 1,000 times. In his annual report to the society in 1891 concerning those tests, Lilienthal stated that "in this way I acquired the ability to glide down the gentle slopes of the hill in moderate winds and land at the foot of the hill with no accident of any kind." In 1898 the French aviation pioneer Ferdinand Ferber wrote that "I conceive of the day in 1891 when Lilienthal first sliced fifteen meters through the air as the moment in which humanity learned to fly."

For the years 1892 and 1893, Lilienthal moved his glider experiments to a new practice field at Rauh Hill in Steglitz, within walking distance of his house on Booth Street. His gliders had progressed through five design evolutions, featuring differences in wingspan, area, and planform, and some structural changes. Because Lilienthal's gliders were hang gliders, for which the only mode of control was to shift the position of his body and hence shift the center of gravity, the gliders could not be made so large that weight shifts would not be effective. In 1893 his latest glider had a wingspan of 7 m, with a wing area of $14\,\text{m}^2$, yielding an aspect ratio of 3.5. It weighed only 20 kg (44 lb). The design was sufficiently advanced that it was used as the basis for Lilienthal's first aircraft patent in 1893 (Figure 4.44). An English patent was granted in 1894, and an American patent in 1895.

In the summer of 1893, Lilienthal began to fly from elevated sites in the Rhinow hills, about 100 km northwest of Berlin, terrain he considered ideally suited for gliding flights. His machines and flying expertise had matured to the extent that it was worth the hour-long train ride and wagon trip to get to the more challenging flying field. In 1894 he paid for the construction of a conical-shaped hill at Lichterfelde; its advantage was that he could always fly from the top of the hill no matter what the wind direction. Lilienthal's hill was 15 m high (about 50 ft) and was built from an existing rubble heap; its cost has been estimated at around three thousand marks. He constructed a windowless shed at the top of the hill where his gliders were stored. That became the main location for his flying experiments after 1894. In 1932 it was made an official monument to Lilienthal. Today it is in the middle of a park (Figure 4.45), providing a panoramic view of the Berlin suburbs. At the top of the hill is a small pavilion to shield a stone globe on a square basalt base, and at the

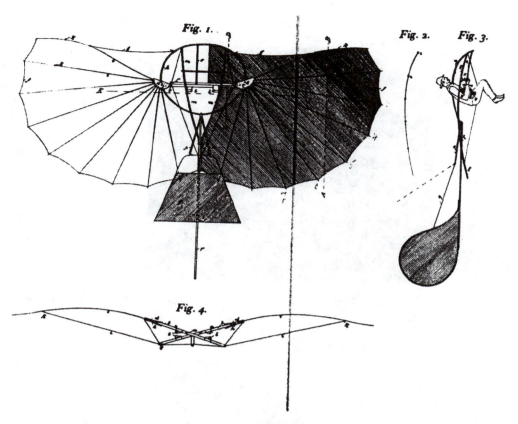

Figure 4.44   Patent drawing for Lilienthal's glider (1893).

foot of the hill, stone plaques commemorate some of Lilienthal's supporters and helpers. Dedicated in 1932, it stands as Germany's tribute to its first aeronautical pioneer.

Lilienthal continued to hone his flying skills and improve his glider designs until his death in 1896. His gliders were predominantly single-wing monoplanes (Figures 4.44 and 4.46), but in 1895 he designed and built several biplane gliders. He continued to fly both types of gliders. Lilienthal accumulated a great deal of flight time, completing more than 2,000 flights in both monoplane and biplane machines.

Lilienthal's ultimate objective was an engine-powered, manned flying machine. However, his approach toward that objective was technically unsound – in stark contrast to his innovative and well-conceived programs for aerodynamic experiments and glider development. Lilienthal was convinced that the route to powered flight was to directly emulate bird flight, and all of his work toward a powered machine focused on ornithopters. An ornithopter design was included as part of his 1893 patent. The idea was to have an engine (a one-cylinder engine in his patent) drive the up-and-down beating motions of the outer portions of the wings. The beating sections of the wings would have a slat-like design that would close during the downstroke (to maximize the "lifting" action) and would open like a fan during the upstroke to minimize resistance. Lilienthal built such a machine and began to test it as a glider at his flying hill in 1894. With its engine, the machine weighed

Figure 4.45 Lilienthal's flying hill at Lichterfelde, built in 1894. In 1932 it was designated a monument to Lilienthal.

Figure 4.46 Lilienthal flying his normal glider (1895).

more than 90 lb, twice the weight of his unpowered gliders. The result was steeper glide paths and faster landing speeds, making it difficult to fly as a glider. When he attempted to operate the engine, which was to be powered by compressed carbon dioxide, it froze after a few strokes. In spite of that failure, Lilienthal began building a second ornithopter in 1896, using a new engine, to be mounted on a larger airframe, with a wing area of more than $20\,\text{m}^2$. Lilienthal was killed before he could finish the second ornithopter.

It seems incomprehensible and totally out of character that Lilienthal, a sensible, accomplished, practical mechanical engineer, would so blindly pursue the concept of a powered ornithopter. But he did, despite knowing of other work carried out in the nineteenth century on propeller-driven fixed-wing designs, emanating from Cayley's seminal concept of separating the mechanisms for producing lift and propulsion. Indeed, there had been many reports in the German *Journal of Aviation* about work on propeller-driven, fixed-wing machines. However, Lilienthal continued to preach the virtues of ornithopters to his colleagues; he was concerned that the slipstream following the addition of a propeller would affect the flying qualities of a fixed-wing glider. Because no propeller-driven fixed-wing aircraft had succeeded in flying at that time, Lilienthal felt that his views were confirmed.

Some reflection on Otto Lilienthal's personality may be of help in seeing his technical aeronautical accomplishments in perspective. Otto lost his father at the age of 13, as well as five siblings in early childhood. The surviving siblings remained close for the rest of their lives. In particular, Otto and Gustav, in spite of following different career paths, maintained exceptionally close ties, though the brothers were quite different in stature and outlook. Otto "was the picture of an ever-happy, optimistic and successful man. He had a sanguine temperament. He could be relaxed and cheerful even when economic concerns were pressing on him."[48] That picture of Otto as outward-looking, constantly optimistic, may help to explain his unflagging pursuit of success with his flying machines, in spite of economic and family constraints, and against massive technological odds. Gustav, in contrast, "was by his own estimation a rather sober and serious soul. It was difficult for him to make the transition from serious life to merriness . . . . By his own admittance, he was proud, even argumentative, and could be rude and fervent. He would express his opinion straight out, without considering whether he was hurting anyone's feelings."[48] Their quite different personalities perhaps counterbalanced in a synergistic fashion, for they were an inseparable team in generating their aerodynamic data. Their data usually were presented and published under Otto's name. That may have reflected the fact that Otto was a degreed mechanical engineer, whereas Gustav was an artist and an architect. In any event, their aerodynamic experiments clearly were joint efforts. On the other hand, when Otto began building gliders in 1891, Gustav remained in the background, generally with both feet firmly planted, except for a few flights in the Rhinow Hills.

Otto had another passion: the theater. He believed that theater should be more accessible to the general public, at ticket prices that would be affordable by the masses. In 1892 he became a co-owner of the East End Theater in Berlin and converted it to a "people's theater" with a new name: the National Theater. Otto's circle of friends included many actors. He also wrote a play, full of social criticism about unscrupulous business practices in Berlin and their adverse effects on innocent people; it ran for nine performances. From time to time Otto would even take small acting parts in plays.

Otto and Gustav were supporters of social reform in Germany. They became followers of the reformist and ethicist Mortiz von Egidy, who campaigned for the reconciliation of humanity and the harmonization of life and morality. Otto carried his ethical concerns into the world of flight, hoping that human flight could be a great equalizer, promoting peace among peoples, bringing "the possibility that continued investigation and experience will bring us ever nearer to that solemn moment, when the first man will rise from earth by means of wings, if only for a few seconds, [marking] that historical moment which heralds the inauguration of a new era in our civilization."[46] In a January 1894 letter to Egidy, Otto elaborated on his view of a new era brought about by conquest of the air: "The borders between countries would lose their significance, because they could not be closed off from each other; linguistic differences would disappear as human mobility increased. National defense would cease to devour the best resources of nations . . . and the necessity of resolving disagreements among nations in some other way than by bloody battles . . . would secure for us eternal peace." Two world wars in the twentieth century would make Otto's beliefs seem naive. However, that outlook was in keeping with his optimistic personality, and we cannot fault him for having the best of hopes. Besides, jet travel has indeed shrunk the world, with millions of travelers visiting the corners of the world, and, one hopes, bringing the peoples of different nations closer together. In that sense, Otto's hopes were not completely futile.

On a more mundane level, Otto hoped that he could make money in the flying-machine business. His boiler factory had not been doing well financially, and he hoped to make flying a sport in which other people could participate. He began to receive orders for his gliders, which he priced at 300 marks. By 1895 the price was increased to 500 marks for an improved design: his "normal glider," the monoplane design used in the majority of his flights. He was also interested in selling his airplane patents in different countries. After he was granted his U.S. patent, he attempted to sell it for $5,000, enlisting the support of Octave Chanute, who made several inquiries on behalf of Lilienthal. All of that came to very little. He never sold his U.S. patent, and he sold only eight of his normal gliders. One went to Nikolai Joukowski in Moscow, who would become the most famous Russian aerodynamicist, primarily because of his contributions to the circulation theory of lift. The one Lilienthal glider sold in the United States went to newspaper publisher William Randolph Hearst. After passing through several hands, and being restored in 1967, it now hangs in the Early Flight Gallery at the National Air and Space Museum of the Smithsonian Institution in Washington, D.C.

Lilienthal's numerous glider flights and his book attracted attention among certain elements of the German public and gained him the curiosity, and later the respect, of the technical aeronautical community around the world. The list of technical people who corresponded with Lilienthal, and even beat a path to his door, was impressive. The growing interest in flying machines was promoted by Octave Chanute in the United States, and Chanute and Lilienthal corresponded frequently. Lilienthal corresponded with James Means in Boston, editor of the *Aeronautical Annual,* a compilation of papers on various aspects of flight published during 1895–7. That correspondence dealt mainly with several papers by Lilienthal published in the *Aeronautical Annual.*

Of particular note was Lilienthal's interaction with Samuel P. Langley, then secretary to the Smithsonian Institution. Langley had begun a series of aerodynamic experiments in 1886, as reported in *Experiments in Aerodynamics*[49] in 1891. Lilienthal owned a copy of that book. Langley traveled in Europe once each year, and one of his assistants, George Curtiss, was delegated to follow aeronautical progress in Europe and to keep Langley informed. So Langley was well aware of Lilienthal's work; indeed, a paper by Lilienthal had been published in the 1893 annual report of the Smithsonian Institution. In 1895, Langley visited Lilienthal's factory, where flying machines were constructed and the powered ornithopter was under construction. Communication was difficult because they spoke no language in common. Langley observed several of Lilienthal's flights, using both the monoplane and biplane gliders. Langley was not greatly impressed with what he saw; he found the flight demonstrations interesting, but could not see that much could be learned from them. Langley told an assistant that he had been more taken by the construction of the flying hill than by the glider flights themselves. In that interaction with Lilienthal, Langley revealed an attitude that foreshadowed the failure of his own attempts at powered flight. Langley did not appreciate the value of learning to fly *before* attempting powered flight; he was concerned primarily with power and lift, like so many failed inventors of flying machines during the nineteenth century. Lilienthal was convinced that it was necessary to experience flying through the air well before attempting powered flight – the true airman's philosophy. That was also the approach of the Wright brothers, who ultimately succeeded.

Lilienthal was also visited by Nikolai Joukowski, who was developing an aerodynamic laboratory in Moscow. Joukowski observed several of Lilienthal's flights and was most impressed. After returning to Moscow, he spoke before the Society of Friends of the Natural Sciences: "The most important invention of recent years in the area of aviation is the flying machine of the German engineer Otto Lilienthal."

Clearly, Otto Lilienthal had captured the attention of the international aeronautics community, but he was not viewed by the German public as anyone of special importance, only one among many who were attempting what was believed to be impossible – to fly under power through the air. After the Wright brothers had shown that flight was possible, and Wilbur's spectacularly successful demonstration flights in Europe in 1908, Lilienthal began to be accorded posthumous acclaim in Germany. With the clouds of war building on Europe's horizons, Germany was looking for an instant national pioneer of flight, and it found Otto Lilienthal. On June 17, 1914, a large stone memorial with a winged figure at the top was dedicated in Berlin, commemorating Lilienthal's pioneering contributions to flight – and creating a national hero out of whole cloth only two months before the Great War broke out. The memorial was funded primarily by public subscription. (Orville and Wilbur Wright were solicited for a contribution. Holding no value in memorials, the Wrights instead gave $1,000 to Otto's wife, who by that time was virtually impoverished.) Other memorials followed: The crash site on Gollenberg Hill where Lilienthal was killed was marked by a circle of stones by the town's inhabitants in the 1930s; in 1954, a memorial stone was added. In 1932 the flying hill in Lichterfelde became a major monument to Lilienthal. In 1940, under the Nazi regime, Lilienthal's grave site was remodeled according to a design funded by the Lilienthal Association for Aeronautical Research. Another memorial, of a different sort, was commissioned by the Nazis: a third edition of *Birdflight as the Basis of Aviation*, prefaced by Ludwig Prandtl's glowing appreciation of Lilienthal's work in aeronautics.

On Sunday morning, August 9, 1896, Otto Lilienthal left Berlin for the Rhinow hills. At noon, he took his first flight of the day, a long glide from a takeoff point high on Gollenberg Hill. It took half an hour to lug the machine back to the takeoff point, and then he took off for a second time. A thermal eddy took him by surprise. The glider completely stalled, and then nosed down, crashing into the ground from a height of 50 ft. Lilienthal was lifted from the glider, which was only slightly damaged, but Lilienthal's spine was broken. Gustav was notified by telegram and was at Otto's side by Monday morning. Otto recognized his brother, but soon lost consciousness. He was taken by train to Berlin, where he died, August 10, 1896, without regaining consciousness.

If Lilienthal had not crashed, would he have been the first to develop manned, powered flight? In an article in the *Aeronautical Annual* in 1897, Chanute stated that "had he lived, success would probably not have been denied him." In flying gliders, as well as in basic aerodynamic knowledge, Lilienthal had a considerable head start. But his idea for a powered machine focused on the ornithopter principle; he was going down the wrong track with a full head of steam. Had he lived longer, he most likely would have spent much effort and time trying to perfect his flapping-wing machine, for he was a strongly optimistic person and would not have given up easily. However, eventually he would have been forced to redirect his efforts toward a fixed-wing design with a propeller, and by that time he most likely would have lost his head start on the Wright brothers.

There is no doubt that Otto Lilienthal was an important figure in nineteenth-century applied aerodynamics and that he did more to advance manned, heavier-than-air flight than anyone except Cayley. Wilbur Wright's last article before his untimely death from typhoid fever in 1912, in the bulletin of the Aero Club of America (September 1912, published posthumously), cited Lilienthal:

> Of all the men who attacked the flying problem in the 19th century, Otto Lilienthal was easily the most important. His greatness appeared in every phase of the problem. No one equaled him in power to draw new recruits to the cause; no one equaled him in fullness and dearness of understanding of the principle of flight; no one did so much to convince the world of the advantages of curved wing surfaces; and no one did so much to transfer the problem of human flight to the open air where it belonged. As a scientific investigator none of his contemporaries was his equal.

Samuel Langley's Experiments in Aerodynamics

In 1886 the most important applied aerodynamic research was being conducted in western Europe. The use of wind tunnels had been pioneered by Wenham and Phillips. The aerodynamic superiority of cambered airfoils over flat surfaces had been demonstrated by Phillips and verified by Lilienthal, whose seminal book was only three years from publication. Western Europe had hosted most of the activity in aerodynamics up to that time. But in that same year, things were about to change; across the Atlantic the seeds for a new center of aerodynamic activity were being sown. In August 1886 the American Association of the Advancement of Science (AAAS) met in Buffalo, New York. Through the efforts of Octave Chanute, a vice-president of the AAAS, the topic of aeronautics had been placed on the schedule. An amateur experimentalist, Israel Lancaster, had been invited to present his work on "soaring effigies" of birds – models that he launched in the air. Although Lancaster's presentation was not as spectacular as expected, one member of the audience was intrigued by what he had heard and began to think seriously about the idea of manned flight: Samuel Pierpont Langley, director of the Allegheny Observatory in Pittsburgh.

After his return from Buffalo, Langley obtained permission from the observatory's board of trustees to construct a whirling-arm device for aerodynamic experiments. Although the observatory's function was astrophysical observation, and Langley's reputation was built on his contributions in astronomy, especially his studies of the sun and sunspots, Langley was allowed to construct and operate a major facility for the sole purpose of obtaining aerodynamic data. Funding for the whirling arm and the initial experiments came from a wealthy friend, William Thaw. At its completion in September 1887, Langley's whirling arm was the largest yet built. Its arms swept out a circle 60 ft in diameter, revolving 8 ft above the ground (Figure 4.47). By comparison, Lilienthal's largest whirling arm had a diameter of 7 m (23 ft). Both men recognized the importance of having a large diameter, so as to minimize the effects of centrifugal force on the airflow over the lifting surface mounted at the end of the arm, and, more important, to minimize various flow nonuniformities created by the circular motion of the arm. In 1887 Langley began a series of carefully designed and executed aerodynamic experiments that continued for more than four years and resulted in a book that elevated Langley to world-class status among aerodynamic researchers in the late nineteenth century. Langley's *Experiments in Aerodynamics*[49] was the first substantive American contribution to aerodynamics. That book and Langley's subsequent work on flying machines after he became secretary of the Smithsonian Institution in 1887 began the challenge to western Europe's virtual monopoly in aerodynamic experimentation.

There is no doubt about the goal of Langley's experiments: He intended to explore the basic physical laws of aerodynamics and demonstrate scientifically the practicability of powered, heavier-than-air flight:

> To prevent misapprehension, let me state at the outset that I do not undertake to explain any art of mechanical flight, but to demonstrate experimentally certain propositions in aerodynamics which prove that such flight under proper direction is practicable. This being understood, I may state that these researches have led to the result that mechanical sustentation of heavy bodies in the air, combined with very great speeds, is not only possible, but within the reach of mechanical means we actually possess, and that while these researches are, as I have said, not meant to demonstrate the art of guiding such heavy bodies in flight, they do show that we now have the power to sustain and propel them [p. 3].[49]

Those comments reflected Langley the scientist. Later, Langley the engineer would design a series of flying machines in order to confirm his conclusions from his whirling-arm data.

When Langley began his whirling-arm experiments in 1887, he had read the works of Wenham and Phillips, but was not aware of the experiments of Lilienthal, who had not yet published. In any event, Langley believed that he was breaking new ground, entering a sphere that was essentially a technological void. Much later, in 1897, he wrote of the situation and attitudes that had prevailed when he had begun his earlier experiments: "The whole subject of mechanical flight . . . was generally considered to be a field fitted rather for the pursuits of the charlatan than for those of the man of science. Consequently, he who was bold enough to enter it, found almost none of those experimental data which are ready to hand in every recognized and reputable field of scientific labor" (p. 2).[50]

Langley's published aerodynamic data obtained with the whirling arm were all for flat plates, although he mentioned, in various places, some unpublished work on cambered surfaces. His attention to the flat plate was due in part to his desire to examine the accuracy of the Newtonian sine-squared law, used since the eighteenth century to calculate the normal force on a flat plate.

Langley took virtually nothing for granted. Even the fact that the aerodynamic force varied as the square of the free-stream velocity he treated as a *theoretical* finding from Newtonian theory, to be used only "in the absence of any wholly satisfactory assumption" – an example of the dysfunctional status of aerodynamics at that time. The use of the Newtonian sine-squared law for calculating the variation of aerodynamic force with the angle of attack had been discredited by numerous previous investigators, including Cayley, but as late as 1887, Langley felt that he had to add his voice to that chorus. The variation of aerodynamic force with the square of the velocity had been proved experimentally by Mariotte and Huygens two centuries earlier and verified by many investigators, including Cayley, but Langley viewed that as only a theoretical result from Newtonian theory. That Langley would take nothing for granted perhaps reflected his view of aerodynamics as a technical void, as well as his nature as a thorough and exacting experimentalist.

Our examination of Langley's aerodynamic experiments can be divided into six phases: (1) preliminary considerations, (2) direct aerodynamic-force measurements, (3) the

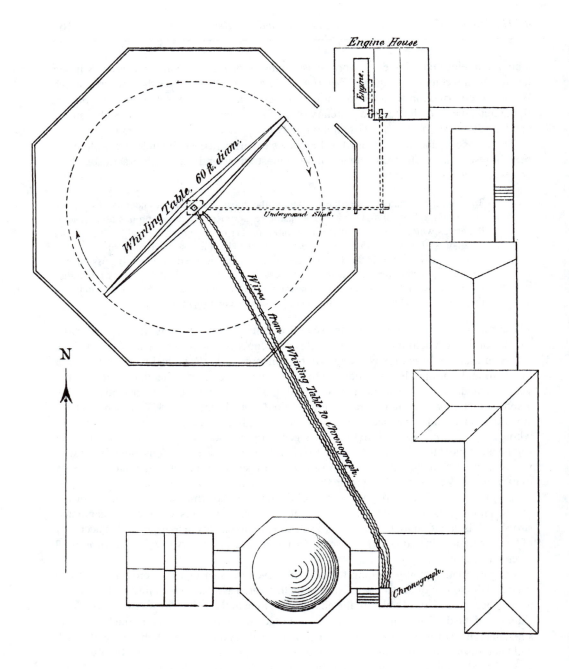

Engine House

Engine

Whirling Table. 60 ft. diam.

Underground Shaft.

Wires from Whirling Table to Chronograph.

N

Chronograph.

ALLEGHENY OBSERVATORY

Plan of Grounds.

SCALE: 1 INCH — 20 FEET

Figure 4.47 Langley's whirling-arm device at the Allegheny Observatory.

"plane-dropping" experiments, (4) "soaring experiments," (5) propeller experiments, and (6) center-of-pressure measurements.

Preliminary Considerations

Langley was concerned about experimental inaccuracies inherent in the whirling-arm setup. He recognized that as the flat-plate lifting surface at the end of the arm whirled around in a circular path, the outer edge would face a greater free-stream velocity than the inner edge. In Figure 4.48, the arm rotates at an angular velocity ω. The inner and outer edges of the lifting surface are at radii R_1 and R_2, respectively, from the center of rotation, and the outer edge of the lifting surface moves through the air at a higher velocity (V_2) than the inner edge (V_1). For a lifting surface of given span (ΔR), the larger the value of R, the smaller the relative difference in velocity between the inner and outer edges, and hence the smaller the flow nonuniformity across the span of the surface (see Appendix C

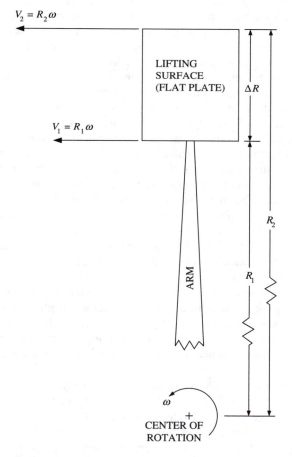

Figure 4.48 Sketch showing the different velocities at the outside and inside edges of a flat plate mounted at the end of a whirling arm.

for mathematical proof). That is one of the advantages to having a large-radius arm on a whirling-arm device. Langley reported a calculation for a flat plate with a span of 30 in. mounted at the end of a whirling arm with a radius of 30 ft. He first calculated the pressure distribution over the span of the plate, assuming a local application of Newtonian theory, and then integrated that distribution to obtain the net aerodynamic force on the plate. He compared that with a calculation of the force assuming a constant pressure over the span of the plate equal to the pressure at the center of the plate. The difference in the forces calculated by the two methods was less than 0.2%. Langley stated that "such disturbing effects of air-pressure arising from circular motion are for our purposes negligible." Another disturbing effect addressed by Langley was that if the device was housed indoors, the "rotating arm itself sets all the air of the room into slow movement, besides creating eddies which do not promptly dissipate." He believed that "the erection of a large building specifically designed for [the experiments] was too expensive to be practicable." Therefore Langley carried out his whirling-arm experiments in the open air, conducting tests only when the outside air was calm. However, Langley lamented that "these calm days almost never came, and the presence of wind currents continued from the beginning to the end of the experiments, to be a source of delay beyond all anticipation, as well as of frequent failure." Such problems were inherent in the operation of a whirling arm; even in modern aerodynamics it would be difficult to reduce them to negligible levels or account for their effects without great effort. It is no wonder that whirling arms quickly fell out of favor for aerodynamic research at the beginning of the twentieth century. In spite of those difficulties and sources of error, nineteenth-century investigators who used whirling arms (e.g., Cayley, Lilienthal, and Langley) somehow obtained data that were meaningful for that time.

In interpreting his data, Langley made an assumption that was plainly wrong: He neglected the influence of friction on his measurements of aerodynamic force. At the end of the nineteenth century, the methods used to calculate skin-friction drag were unreliable. Even the basic physical mechanism was a mystery; there was continuing debate about the applicability of the no-slip condition at the surface (i.e., the assumption of zero relative velocity between a surface and the air immediately over that surface). Langley expressed an opinion about that – the correct opinion – in a footnote in *Experiments in Aerodynamics*: "There is now, I believe, substantial agreement in the view that ordinarily there is no slipping of a fluid past the surface of a solid, but that a film of air adheres to the surface, and that the friction experienced is largely the internal friction of the fluid – i.e., the viscosity." Langley went on to calculate skin-friction drag using a friction formula stated by Clerk Maxwell. He compared the friction drag on a plate at a zero angle of attack with the pressure drag on the same plate at a 90° angle of attack and concluded that the former was negligible compared with the latter. Of course, that was like comparing apples and oranges, and we stand amazed by Langley's uncharacteristic faulty logic. As we shall see, Langley's intentional neglect of skin friction greatly compromised his data interpretations for plates at small angles of attack.

Langley took many measurements of the aerodynamic force on a flat plate over a large range of angles of attack, including 90° (i.e., with the plate oriented perpendicular to the flow). From those 90° data he calculated values for Smeaton's coefficient – "values" because the numbers he obtained varied moderately from one test to another. Using the standard formula for the force on a flat plate oriented perpendicular to the flow,

$$F = kV^2 S$$

the average pressure on the plate (the measured force divided by the area of the plate) can be expressed as

$$p = kV^2$$

where k is Smeaton's coefficient (Langley identified k as the "coefficient of normal pressure"). Taking an average of his measurements, Langley declared that the "final value" for k was

$$k = 0.08$$

when p is in kilograms force per square meter, and V is in meters per second. The original, conventional value was $k = 0.13$, based on the table in Smeaton's paper, and, curiously, was accepted by Lilienthal (although Lilienthal had doubts about the accuracy of that value). If we convert Langley's value of $k = 0.08$ to English engineering units, with p in pounds force per square foot, and V in miles per hour, we obtain

$$k = 0.003$$

That value is very close to the modern value of $k = 0.0029$ established by the Royal Aeronautical Society. It is a far cry from the earlier accepted value of 0.005 obtained from Smeaton's tables – a value that had been shown by several investigators over two centuries to be too high. Thus Langley's measurement was quite accurate – a testimonial to his experimental technique.

Direct Measurements of Aerodynamic Force

Langley was a master instrument designer. In contrast to the simple weight, pulley, and spring mechanisms developed by Lilienthal for his measurements of aerodynamic force, Langley designed rather sophisticated electromechanical instruments for measuring various types of forces, such as his resultant-pressure recorder, which measured both the direction and magnitude of the resultant aerodynamic force on the flat plate; both the recorder and the flat plate were mounted at the end of the whirling arm, and they moved in unison. Detailed descriptions of all his measuring devices, with elaborate mechanical drawings, were included in *Experiments in Aerodynamics*.[49] Langley reported his force measurements in both tabular and graphic form. In the same spirit as Lilienthal, Langley referenced his force measurements to the measured force on a flat plate at a 90° angle of attack; hence, his recorded ratios are simply the *resultant-force coefficient* C_R, defined as

$$C_R = \frac{R}{\frac{1}{2}\rho V^2 S}$$

Langley quoted values for the ratio P_α / P_{90}, where P_α is the force at an angle of attack α, and P_{90} is the force at $\alpha = 90°$. Using the same argument as in our discussion of Lilienthal's data, we have

$$\frac{P_\alpha}{P_{90}} \equiv C_R$$

Earlier we credited Lilienthal as the first to use aerodynamic-force coefficients (1889), but Langley was not far behind (1891). Together they established the use of aerodynamic-force coefficients as standard practice in applied aerodynamics.

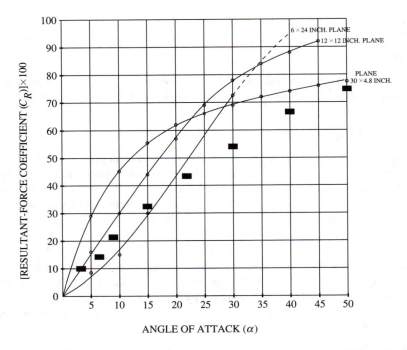

Figure 4.49 Resultant aerodynamic force on a flat plate as a function of the angle of attack; comparison of Langley's data (circles) and Lilienthal's data (squares).

Langley's first major measurements were taken using a 1-ft^2 flat plate, wherein the magnitude and direction of the resultant aerodynamic force were measured with the resultant-pressure recorder over a range of angles of attack from 5° to 90°. The linear velocity of the center of the flat plate ranged from 4.5 to 11.1 m/s for different tests. The findings, presented in force-coefficient form, were independent of velocity, as clearly demonstrated by the entries in Langley's tables. Those square-plate data, obtained over the period from August to October of 1888, are shown by the curve labeled "12×12 inch plane" in Figure 4.49, which is essentially Fig. 10 from *Experiments in Aerodynamics,* with some of Lilienthal's data points added, as discussed later. From Langley's point of view, the main value of those measurements was that they showed that the Newtonian sine-squared law was totally wrong for all cases that he considered. He pointed out, for example, that at a 5° angle of attack, the experimental data gave a resultant force 20 times that predicted from the Newtonian sine-squared law. Although Langley was not the first to point out the limited applicability of that law, making that point was especially important to him because the practicability of sustained powered flight would depend in part on that finding.

In Figure 4.49, the curve that Langley faired through his square-flat-plate data predicts the value $C_R = 0.1$ at a 3° angle of attack. At small angles of attack, C_R is essentially the same as the lift coefficient C_L. Early in the nineteenth century, Cayley had measured the force on a 1-ft^2 flat plate, and his data translate to a lift coefficient of $C_L = 0.11$ at a 3° angle of attack – remarkable consistency for two measurements so far apart in time and laboratory technique.

Also shown in Figure 4.49 are Langley's measurements for two other flat plates with different aspect ratios. The three curves taken together illustrate the strong effect of the aspect ratio on the resultant aerodynamic force. In the angle-of-attack range below 20° (the range for practical flight), Langley's data show that the highest-aspect-ratio plate (30 × 4.8 in., aspect ratio 6.25) gives the highest values for C_R, and that the lowest-aspect-ratio plate (6 × 24 in., aspect ratio 0.25) gives the lowest values for C_R. The data for the 1-ft^2 plate (aspect ratio 1) lie between the other two values. Viewed from our modern understanding of aerodynamics, the variations in Langley's data due to aspect ratio were qualitatively correct.

In Figure 4.49, for the sake of comparison, Lilienthal's whirling-arm data for a square flat plate have been added to Langley's data. At low angles of attack, on the order of 5° or less, Lilienthal's measurements were in agreement with an extrapolation from Langley's measurements. That is important, because cruising flight normally takes place at such low angles of attack. However, at higher angles of attack there was some discrepancy between the two sets of data, with Lilienthal's data falling some 20–25% below those of Langley, most likely because of deficiencies in Lilienthal's particular whirling-arm setup – deficiencies that he suspected, prompting his alternative setup to measure aerodynamic force using a stationary device in the natural wind. Lilienthal's tests of cambered airfoils in the natural wind (Figure 4.37) consistently gave higher force coefficients than did his whirling arm – on the order of 20%, which is essentially the difference shown in Figure 4.49. Lilienthal did not report any flat-plate data from his natural-wind experiments; had he done so, it seems probable that they would have been higher than those obtained with his whirling arm, and hence would have been in closer agreement with Langley's data than are the data shown in Figure 4.49. The comparison shown in Figure 4.49 illustrates the unreliable nature of whirling-arm experiments.

Plane-dropping Experiments

Another novel device designed by Langley was his plane-dropping apparatus: Attached vertically at the end of his whirling arm was an iron frame on which was mounted an aluminum falling piece that ran up and down on rollers. He attached his flat-plate lifting surface to that falling piece, with the lifting surface parallel to the ground. With the lifting surface locked into its highest position, the whirling arm was started, and when the desired airspeed over the plate was reached, the plate was released. It would then fall a maximum distance of 4 ft (as allowed by the height of the iron frame). The time it took the plate to fall that distance was recorded by Langley. When the plate was locked in position at the top of the frame and the whirling arm was in motion, the relative airflow over the plate was parallel to the plate (i.e., the angle of attack was zero), and no lift was generated on the plate (Figure 4.50a). When the plate was released and fell vertically, there was a component of air velocity in the vertical direction that when added to the forward motion resulted in a relative wind inclined slightly upward relative to the plate (i.e., the flat plate was then at some angle of attack to the resultant airflow), and thus some lift was generated (Figure 4.50b). When lift was produced, it would act counter to the weight of the falling plate, and thus the plate would take a longer time to fall the distance of 4 ft. The more lift, the longer the time. The time required to fall the distance of 4 ft was an index of the lifting capacity of the plate.

The most important finding from Langley's plane-dropping tests was that a wing with a high aspect ratio would produce more lift than a wing with a low aspect ratio (Figure 4.51). The time required to fall 4 ft was plotted versus the horizontal velocity for three plates of

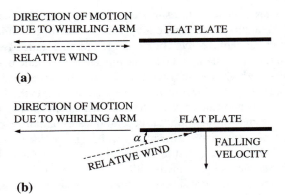

Figure 4.50   Relative wind directions for a flat plate in (a) horizontal motion (no lift is produced) and (b) combined horizontal motion and vertical dropping motion (lift is produced).

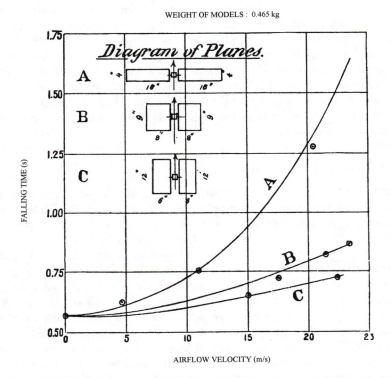

Figure 4.51   Langley's data for flat plates of different aspect ratios; plane-dropping tests.

the same weight and surface area but different aspect ratios. Clearly, at any given velocity, the higher the aspect ratio, the longer the falling time. Wenham had preceded Langley in appreciating the aerodynamic efficiency of high-aspect-ratio wings, but Langley was the first to produce an organized set of experimental data clearly showing the superiority of such wings. In Figure 4.51, the model with the highest aspect ratio consisted of two planes 18 × 4 in., with the aspect ratio for each plane being 4.5 – a fairly high value, given the state of the art at that time. Influenced by those findings, Langley later designed his successful aerodrome no. 5 with a relatively high aspect ratio of 5.

Langley used the word "aspect" to describe the orientation of a plate (in regard to which edge of the plate was perpendicular to the flow). For example, consider a plate 12×6 in. If the plate was oriented such that the 12-in. side was the advancing (leading) edge (i.e., the 12-in. side was perpendicular to the flow), then in Langley's nomenclature that was a different "aspect" compared with when the plate was oriented such that the 6-in. side was the advancing edge. Although Langley did not define *aspect ratio* as we know it today, could Langley's use of the word "aspect" in that context have been the source for the term "aspect ratio"? The word "aspect" had not been used in the aeronautical literature prior to Langley's book.

In his explanation of the plane-dropping tests, Langley revealed an almost complete lack of understanding of the nature of the flow over the lifting surface, as well as the nature of the generation of lift. He thought of the air below the falling plane as simply providing a cushioning effect due to the inertia of the air. As for the reason that the lift increased when the velocity increased, Langley stated that "in the case of any heavy body which will fall rapidly in the air if it fall from rest, the velocity of fall will be more and more slow if the body be given successively increasing velocities of lateral translation and caused to run (so to speak) upon fresh masses of air, resting but a moment upon each." Nowhere in his explanation of his experiments did Langley show recognition of the geometric relationship sketched in Figure 4.50, namely, that by falling at the same time it is moving horizontally through the air, the plate assumes an angle of attack relative to the resultant airflow, and hence lift is produced. However, Langley was a premier experimentalist, not a theoretician. In those days, so little was known about the flow field over a lifting surface that he cannot be seriously faulted for misunderstanding the basic nature of such a flow. Moreover, Langley was interested primarily in *empirical findings* about the force on a lifting surface; a detailed understanding of the flow-field physics was not important to his objectives. An irrefutable demonstration that high-aspect-ratio wings were superior to low-aspect-ratio wings, and by approximately how much, was good enough for Langley, and that determination was one of his major contributions to applied aerodynamics.

Soaring Experiments and the Component-Pressure Recorder

Langley's experiments featured yet another experimental technique and another measuring instrument. The technique involved the "soaring" of his flat-plate models, and the measuring instrument was his specially designed component-pressure recorder – essentially a balance arm that was supported in the middle on a knife-edge bearing. In addition to being able to move up and down vertically on the knife-edge, the arm could oscillate horizontally about a vertical axis. A flat-plate lifting surface was mounted at one end of the balance arm, with the plate mechanically locked to a specified angle of attack. The whole apparatus was moved through the air at the end of Langley's whirling arm. The speed of the flat plate through the air was adjusted so that for the given angle of attack, the lift generated by the plate would exactly equal the weight of the plate; in that situation, the plate was "soaring," to use Langley's term. Also in that situation, the measuring arm of the component-pressure recorder was exactly balanced in the horizontal position (i.e., the arm was parallel to the ground). The drag force on the plate would tend to rotate the balance arm about the vertical axis of the recorder. The drag ("horizontal pressure" in Langley's terms) was measured by the extension of a spring that resisted the horizontal oscillation. The component-pressure recorder was designed to record a measurement only when the lift on the plate balanced the weight; when that situation existed, an electrical contact was joined, and the horizontal force (drag) was recorded. In short, the component-pressure recorder was an ingeniously designed device that would measure the drag on the plate when the flight condition was such that the lift exactly equaled the weight. Given the fixed angle of attack of the plate, that flight condition could be achieved only at one particular translational velocity of the plate through the air. Hence, for the given angle of attack, the component-pressure recorder measured both the lift and drag coefficients via the equations

$$C_L = \frac{L}{\frac{1}{2}\rho V^2 S} = \frac{W}{\frac{1}{2}\rho V^2 S}$$

$$C_D = \frac{D}{\frac{1}{2}\rho V^2 S}$$

where W was the known weight of the plate, D was the measured drag force from the recorder, and V was the translational velocity at which the plate would "soar" (i.e., the velocity at which the lift equaled the weight). That velocity was found by simply varying the rotational speed of the whirling arm. By repeating a series of such tests, each for a different angle of attack α, he could plot the variations in C_L and C_D with α. Finally, from the geometric relationship involving lift, drag, and resultant force (Figure 1.5), Langley extracted from his lift and drag data a value for the resultant aerodynamic force R. Thus he was able to plot C_R as a function of the angle of attack α.

All the data plotted in Figure 4.49 were obtained with the component-pressure recorder using the "soaring" test technique. Langley compared his data from the resultant-pressure recorder experiments and the component-pressure recorder experiments for the square plate and found them to agree within 2%, at worst. Considering the state of experimental aerodynamics at the end of the nineteenth century, that is amazing.

The consistency of Langley's measurements of C_R in two totally different sets of experiments tends to lend validity to his findings. It speaks highly of Langley as an instrument designer and a careful organizer of experiments. However, that consistency lends no validity to Langley's whirling-arm data in general, because the two sets of data were obtained with the same whirling arm and thus were subject to the same experimental uncertainties characteristic of a whirling-arm device. It has already been suggested that such uncertainties were responsible for the discrepancies between the data sets of Langley and Lilienthal (Figure 4.49).

Langley was well aware that the power required to drive a flying machine through the air would be equal to the product of drag and velocity: DV. In order to estimate the size of the engine that would be needed to power a flying machine, he needed reliable estimates for aerodynamic drag, and therefore he placed particular emphasis on the measurements of drag in his soaring experiments. Figure 4.52 is a plot from Langley's *Experiments in Aerodynamics* showing the measured variation of drag with the angle of attack for a flat plate (30 × 4.8 in.) at soaring speeds. Langley understood the difficulty of accurately measuring drag at low angles of attack α. The absolute magnitude of the drag would be small at low α, and the relative experimental error in measuring that small force would be large:

> The horizontal pressures on the inclined planes diminish with decreasing angles of elevation, and for angles of 5° and under are less than 100 grammes. Now, for a pressure less than 100 grammes, or even (except in very favorable circumstances) under 200 grammes, the

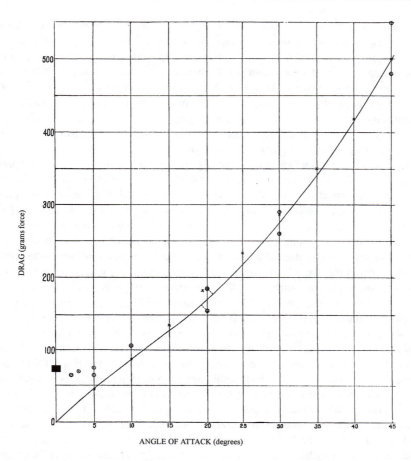

Figure 4.52 Langley's data for the drag on a flat plate as a function of the angle of attack. The black square shows the calculation from Appendix D.

various errors to which the observations are subject become large in comparison with the pressure that is being measured, and the resulting values exhibit wide ranges. In such cases, therefore, the measured pressures are regarded as trustworthy only when many times repeated [p. 63].[49]

His understanding of that problem was another example of his seemingly natural talent for carrying out and interpreting experimental observations. However, in one respect Langley faltered: "On the 30 × 4.8 inch plane, weight 500 grammes, fifteen observations of horizontal pressure have been obtained at soaring speeds. These values have been plotted [Figure 4.52], and a smooth curve has been drawn to represent them as a whole. For angles below 10° the curve, however, instead of following the measured pressure, is directed to the origin, so that the results will show a zero horizontal pressure for a zero angle of inclination. This, of course, must be the case for a plane of no thickness, . . ." There Langley committed a sin that had bedeviled scientific investigators since the dawn of science: fairing a curve to show what the investigator thinks the right answer should be, even though the data show otherwise. In Figure 4.52, for angles of attack less than 10°, Langley ignored his data and faired his curve to go through the origin (zero drag at zero α), even though the experimental data clearly were converging to a *finite* drag at a zero angle of attack. In fact, almost incredibly, his measured data at low α were converging to the correct value in spite of the experimental uncertainty discussed earlier. A modern calculation of the drag at zero α for Langley's conditions is shown in Appendix D. That calculation predicts a total drag force of 76 g at a zero angle of attack, shown as the black square in Figure 4.52. Note that Langley's experimental measurements were nicely converging toward that computed value. There were two physical phenomena that were contributing to the finite drag at a zero angle of attack: (1) The flat plate in the experiment had a finite thickness of $\frac{1}{8}$ in., and when it was at a zero angle of attack, the blunt face of the front edge perpendicular to the flow caused pressure drag. (2) The viscous shear stresses exerted over the top and bottom surfaces parallel to the flow caused skin-friction drag. To Langley's credit, he was well aware of the pressure drag, as shown by the completion of the last sentence in the preceding quotation: "This, of course, must be the case for a plane of no thickness, and cannot be true for any planes of finite thickness with square edges, though it may be and is sensibly so with those whose edges are rounded to a so-called 'fair' form" (p. 65).[49] Langley went on to state that his calculations showed that the pressure drag due to plate thickness was responsible for most of the drag at low angles of attack, and when that calculated pressure drag was subtracted from his experimental data, there was good agreement with the faired curve. Langley was partially correct in that assessment, because the calculation in Appendix D for a zero angle of attack shows a pressure drag of 61 g and a friction drag of 15 g. Clearly the pressure drag in his case was a large percentage of the total drag. But it is equally clear that the friction drag was not trivial and should not have been ignored. It was in that respect that Langley was wrong. Throughout all of his aerodynamic work, Langley intentionally ignored friction drag. He stated his reason near the beginning of *Experiments in Aerodynamics*: "Most of the various experiments which I have executed involve measurements of the pressure of air on moving planes, and the quantitative pressures obtaining in all of the experiments are of such magnitude that the friction of the air is inappreciable in comparison." Once again, we should not blame Langley too much for that wrong impression; in 1891 there was no reliable formula for accurate calculation of friction drag. On the other hand, Langley's outright neglect of friction reinforced the beliefs of others that friction played no significant role in the net aerodynamic force, and thus it was a disservice to the next generation of aerodynamicists.

Langley's belief that the effect of friction was negligible had a second impact on his conclusions: For a flat plate at some angle of attack, if pressure was assumed to be the only factor imposing a force on the plate (and, locally, pressure always acts perpendicular to the surface), then the resultant aerodynamic force on an infinitely thin plate would be perpendicular to the plate. Newtonian theory also predicted that the resultant aerodynamic force would be perpendicular to the plate. Langley was interested in measuring the direction of the net aerodynamic force as a means of verifying those arguments. Indeed, the preponderance of evidence from all his experiments gave him reason to suspect that the resultant force was perpendicular to the plate. If that was true, then the geometry shown in Figure 4.53a would lead to the relation

$$D = L \tan \alpha$$

For Langley's soaring experiments, where the lift equaled the weight, that relation would become

$$D = W \tan \alpha$$

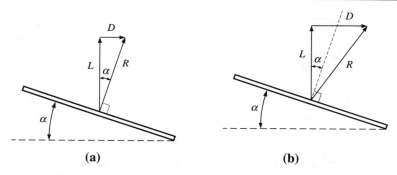

(a)   **(b)**

Figure 4.53 Resultant force on a flat plate: (a) The case with no friction drag, where the resultant force is perpendicular to the plate. (b) The case with friction drag, where the resultant force is inclined behind the local perpendicular.

Langley used that simple relation to calculate D as a function of α for the known weight of 500 g for the flat-plate model. He plotted those data as the points denoted by "×" in Figure 4.52. Those "calculated" points fell close to the curve that he faired through the experimental data:

> For the purpose of comparing the points given by this equation [$D = W \tan \alpha$] with the curve deduced from the observed pressures, the former are shown by crosses on the diagram with the curve. The agreement between the two is remarkably close, and, according to the standpoint from which the subject is viewed, we may say that the formula is actually identifiable, as it appears to be, with a simple case of the resolution of forces, or that the accuracy of the harmonized experiments is established by their accordance with an unquestioned law of mechanics [p. 65].[49]

Essentially, Langley was saying that as the formula had been shown to be accurate because its data compared favorably with his experimental data, and the formula assumed the resultant aerodynamic force to be perpendicular to the plate, then such must be the case in reality. For the cases at large angles of attack, where the pressure force is much larger than the friction force, the resultant aerodynamic force will indeed be almost perpendicular to the plate, and Langley's "verification" will be reasonable. However, at very low angles of attack, friction plays a stronger role, and the resultant aerodynamic force will be inclined behind the local perpendicular (Figure 4.53b). (Recall that Lilienthal's flat-plate data clearly showed that the resultant aerodynamic force was inclined behind the local perpendicular; Langley made no mention of Lilienthal's findings.) For that case, the actual drag will be larger than that predicted by the formula, as seen in Figure 4.52. For angles of attack of 10° or lower, Langley's experimental measurements of drag were consistently higher than the predictions given by the formula, as denoted by the crosses. (In that region, Langley obviously used the crosses to fair his curve.) With our modern aerodynamic knowledge, we can immediately recognize the influence of friction in Langley's low-angle-of-attack data. Langley had convinced himself otherwise.

From Langley's point of view, his aerodynamic experiments accomplished their objectives. He reaffirmed that the Newtonian sine-squared law was not valid for predictions of lift and drag on aerodynamic surfaces at the small angles of attack pertaining to actual flight. He clearly verified that aerodynamic force varies as the square of the velocity. His data on flat plates showed that the lifting action of an inclined surface was much stronger than predicted by the Newtonian law, and therefore he was confident that mechanical flight was possible, needing only the proper engineering. However, he knew that to convince the public that powered flight was possible, the proof would have to be in the pudding – he would have to build and fly such a machine.

Propeller Experiments

In order to propel a given flying machine at a velocity sufficient to produce enough lift to sustain the vehicle in the air, a certain amount of power would be required. Langley was determined to calculate that power. Indeed, by Langley's time, the "power required" had become a pivotal concern of those seeking to achieve powered flight. That was one of the practical uses of his flat-plate drag measurements; Langley knew that the power required to overcome drag was given by the product of drag and velocity (DV), and his book *Experiments in Aerodynamics* presented many power calculations based on his drag measurements. He believed that a propeller driven by an engine would be the most realistic mechanical means for providing that power. He recognized that a propeller was not going to be 100% efficient; some of the power delivered by the engine to the propeller was going to be lost because of friction and other effects. Therefore Langley proceeded to carry out a series of propeller tests using his whirling arm. In his usual thorough fashion, he designed a special instrument, his dynamometer-chronograph, to measure the power output from a propeller driven by an independent motor. The whole apparatus was translated through the air at the end of his whirling arm. Uncharacteristic of Langley, he reported few details about the shapes of the various propellers tested. It is clear that he was guided by the design of steamship propellers: "These considerations very intimately connect themselves with the theory of the marine screw-propeller, and the related questions of slip and rate of advance." He was confused about one essential aspect of propeller performance: "Instead of aiming to set in motion the greatest amount of air, as in the case of the fan-blower, the most efficient propeller is that which sets in motion the least." Quite the opposite: A propeller is efficient because it derives its thrust by taking a *large amount of air* and increasing its velocity through the propeller disk by only a small amount. Perhaps such misunderstandings account in part for the fact that Langley's propeller tests yielded an efficiency of only 52%, compared with more than 70% efficiency obtained a decade later by the Wright brothers. Propeller design was not Langley's forte, nor did he contribute anything of substance to the understanding of propeller aerodynamics.

Center-of-Pressure Measurements

The center of pressure for an aerodynamic surface is that point on the surface through which the resultant aerodynamic force acts. Knowledge of the center of pressure (or equivalent information) is essential in understanding the stability, control, and balance of a flying machine. Langley was acutely aware of that, and in the design of his later aerodromes he went to great lengths to ensure inherent static stability. A chapter in *Experiments in Aerodynamics* was devoted to measurements of the center of pressure on a flat plate as a function of the angle of attack. For that purpose, Langley designed his " counterpoised eccentric plane," which was a flat-plate lifting surface mounted in a square frame that was attached to the whirling arm via a pivot. The plate was free to rotate about the pivot, and the location of the axis of rotation relative to the leading edge of the plate could be changed from one test to another. As the flat plate was moved through the air, the plate would naturally assume that equilibrium angle of attack that would place the center of pressure at the axis of rotation. A pencil fixed on the lower part of the plane recorded the angle of attack on a tracing board perpendicular to it. By running a series of tests with the pivot axis fixed at different locations relative to the plate's leading edge, Langley was able to tabulate the variation of the location of the center of pressure with the angle of attack. With the plate perpendicular to the flow (i.e., a 90° angle of attack), he found the center of pressure to be at the center of the plate, as expected. As the angle of attack was decreased, he observed that the center of pressure moved forward: "The extreme forward position of the center of pressure in these experiments was for the smallest angle, one fourth the length of the plane from the front edge." Langley was measuring the proper behavior (on a flat plate, the center of pressure at very high angles of attack is near the $\frac{1}{2}$-chord location, and it moves forward as the angle of attack is decreased). Findings from thin-airfoil theory reported by Ludwig Prandtl and Max Munk in the period 1915–22 would prove that at small angles of attack, the center of pressure for a flat plate is at the $\frac{1}{4}$-chord point, as found experimentally by Langley. (The fact that the center of pressure for a flat plate generally lies ahead of the $\frac{1}{2}$-chord location had been observed in 1808 by Cayley in his experiments with large kites, but Cayley never produced a coherent set of experimental data in this regard.) No practical use was made of Langley's flat-plate center-of-pressure measurements; the variation for cambered airfoils is quite different, and all flying machines (including Langley's aerodromes) would use cambered lifting surfaces. Later, in the *Langley Memoir on Mechanical Flight,*[50] Langley made it clear that he was aware of the different center-of-pressure travel for cambered airfoils. Langley's contribution regarding center-of-pressure location was an absolute appreciation that it moves as the angle of attack is changed, and that movement is different for cambered airfoils than for flat plates.

The Langley Law

Perhaps Langley's most interesting and most controversial conclusion from his experimental data was the Langley law, which stated that the power required for a vehicle to fly through the air decreased as the velocity increased. He considered that to be one of his most important contributions, stating it on the first page of *Experiments in Aerodynamics:*

> These new experiments (and theory also when reviewed in their light) show that if in such aerial motion, there be given a plane of fixed size and weight, inclined at such an angle, and moved forward at such a speed, that it shall be sustained in horizontal flight, then the more rapid the motion is, the *less* will be the power required to support and advance it. This statement may, I am aware, present an appearance so paradoxical that the reader may ask himself if he has rightly understood it.

That was repeated three more times in his book, twice in italics. That conclusion did indeed fly in the face of intuition. It was considered to be misleading, at best, by some contemporaries, and outright wrong by others. Crouch[51] stated that Lilienthal and the Wright brothers rejected that conclusion outright. In a meeting of the British Association for the Advancement of Science at Oxford in 1894, Langley presented a short paper summarizing his work and conclusions, only to be criticized and taken to task by both Lord Kelvin and Lord Rayleigh. Indeed, even today mention of Langley's power law can evoke derision.

Langley's conclusion was based on his experimental data, and those data consistently supported it. In Appendix B of *Experiments in Aerodynamics* he presented a theoretical

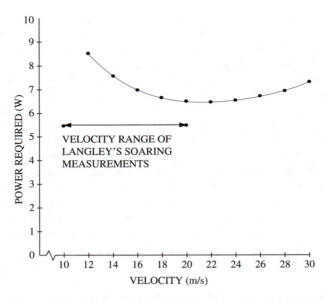

Figure 4.54 Calculated values (using modern aerodynamics) of the power required by Langley's flat-plate model, proving that all of Langley's data points were on the back side of the power curve.

"proof" of that law. As a means to assess the validity of Langley's conclusion, Appendix E herein shows a calculation of the power-required curve for Langley's flat plate in soaring flight. *It clearly shows that all of Langley's experimental data were obtained on what today is called the back side of the power curve – the region where the power required for steady, level flight indeed decreases with an increase in velocity.* Figure 4.54 shows the calculated power-required curve for a flat plate of aspect ratio 6.25, planform area 1 ft^2, and weight 500 g. All conventional flight vehicles have curves of the same general shape. It has a local minimum point, for the minimum power required. In Figure 4.54, that local minimum occurs at a velocity of about 22 m/s. At velocities below and above that point, the power required increases. The higher-velocity side of the curve, that part to the right of the minimum point, is dominated by parasite drag, which increases essentially as the square of the velocity. The lower-velocity side, that part to the left of the minimum point, is dominated by the "drag due to lift" (i.e., the pressure drag that is associated with the pressure difference that creates the lift). The drag due to lift actually increases as the velocity decreases; one can associate that trend with the rapidly increasing angle of attack as the velocity decreases – the increase in α is necessary to maintain the lift equal to the weight as the velocity decreases. All of Langley's data were taken at velocities of 20 m/s or less, as shown in Figure 4.54. *Clearly, all of Langley's data were obtained on the back side of the power curve,* and thus his conclusions that led to the Langley power law were *correct* for his range of test velocities. Had his whirling arm allowed testing at velocities greater than 22 m/s, he would have seen a reversal in his data trend, and most likely the Langley power law would never have been proposed.

Langley's work was the first meaningful aerodynamic research in America, and because of his reputation among scientists, even abroad, he was responsible for shifting the epicenter of aerodynamic investigation slightly west of its former late-nineteenth-century location in

Europe. Langley's experiments contributed little of practical value to the design of flying machines. In *Experiments in Aerodynamics,* all his data pertained to flat plates and were, for the most part, of only academic interest. Especially counterproductive was his emphasis in the Langley power law. Although we have shown that the Langley law was an appropriate conclusion from his data, it pertains only to the back side of the power curve, which is avoided as much as possible in real flight. In any event, the Langley law was against intuition, and criticism of Langley on that account tended to diminish the credibility of the rest of his findings in the eyes of many.

Crouch said of Langley's aerodynamic experiments that "the work did serve a very useful purpose. The fact that a man of Langley's stature believed in the possibility of the flying machine was enough to convince most laymen that aeronautics was no longer the pastime of fools." Langley was a careful organizer of well-thought-out experiments, and his data were all internally consistent from one type of experiment to another. Although his unique measuring instruments were never used by anyone else, and the whirling arm quickly disappeared as an aerodynamic testing device, Langley's experiments reported in *Experiments in Aerodynamics* provided inspiration for experimental aerodynamicists.

Langley's Aerodromes

Samuel Pierpont Langley (1834–1906) (Figure 4.55) was born at Roxbury, Massachusetts, August 22, 1834, to a family of some wealth and influence. His father was in the produce business. After attending the prestigious Boston Latin School and graduating from Boston High School, and declining college, Langley moved to the Midwest, where he worked as a civil engineer and architect for a dozen years. For the rest of his life, Langley continued to learn from self-study. During the American Civil War, Langley returned to Boston and directed his attention to astronomy. As part of his self-education process, he toured Europe, visiting a number of European astronomical observatories. Back in Boston

Figure 4.55　Samuel Pierpont Langley.

in 1865, Langley accepted an invitation from the director of the Harvard Observatory to be an assistant. A year later, with the help of that same director, Langley became an assistant professor of mathematics at the U.S. Naval Academy. Biddle[52] commented that the academy must have been hard-pressed for faculty to have offered a mathematics professorship to a man with only a high-school education and no experience in mathematics or teaching. Within a year, Langley took a position as professor of physics and director of the Allegheny Observatory at the Western University of Pennsylvania (now the University of Pittsburgh).

The Allegheny Observatory was only a few years old, but when Langley arrived it was in a state of disrepair. The observatory had been started by a group of private citizens in Pittsburgh who had bought an expensive German telescope, but none of the supporting equipment to properly point it. They soon found that they were in over their heads, and they deeded it to the local university, along with a meager endowment for a professorship. Langley quickly sought out a wealthy railroad executive, William Thaw, who provided funding to properly equip the observatory. Thaw also presented the university a grant of $100,000, with the stipulation that Langley be relieved of any teaching duties, leaving him free to pursue his observatory research. Thaw continued to be a lifelong benefactor to Langley's work, and Langley paid tribute in the preface to *Experiments in Aerodynamics:* "If there prove to be anything of permanent value in these investigations, I desire that they may be remembered in connection with the name of the late William Thaw, whose generosity provided the principal means for them."

Over the next 20 years, Langley earned a distinguished reputation as director of the observatory and as an observational astronomer. Langley's lack of a mathematical background pointed him in the direction of experimental rather than theoretical astronomy, and that same experimental emphasis would later dominate his aeronautical work. On the practical side, Langley set up a source of income for the observatory by providing the exact time of day to the railroads, making astral observations and then telegraphing the results to his customers twice a day. On the scientific side, Langley specialized in studies of the sun, especially the sunspots and the energy produced by the sun. In the late 1870s he developed the bolometer, an instrument to measure the spectral variation of the sun's energy incident on the earth. In 1884 Langley organized an expedition (funded by Thaw) to Mount Whitney in the Sierra Nevada of eastern California to measure the heat-absorbing characteristics of the atmosphere. Langley's data also enabled him to determine a value for the solar constant (a measure of the amount of energy reaching the earth's atmosphere). That work brought Langley lavish praise from his scientific colleagues and solidified an international reputation. (In 1914 the scientific community found that Langley's value for the solar constant had been too large by 50%.) Langley's scientific reputation peaked in 1886, when he was awarded the Rumford Medal by both the Royal Society in London and the American Academy of Arts and Sciences, and the Henry Draper Medal by the National Academy of Sciences.

It was at that high point that Langley accepted the job of secretary of the Smithsonian Institution. It was also at that time that Langley, motivated by Lancaster's paper at the 1886 meeting of the American Association for the Advancement of Science, began a totally new scientific career: the study of mechanical flight. It is testimony to Langley's prestige as a scientist, and Thaw's influence on the board of trustees, that Langley received permission to construct an aerodynamic laboratory at the observatory in Pittsburgh: "The board of trustees of the university and observatory could hardly be expected to fund a study so far removed from astronomy. In fact, to spare the institution and himself any possible embarrassment in the event of total failure, Langley consistently referred to his 'work in pneumatics' in communications to the trustees as well as in annual reports" (p. 47).[51] After Langley became secretary of the Smithsonian in 1887, he moved to Washington, but the aerodynamic work using the whirling arm continued at the Allegheny Observatory under his direction. Langley maintained close supervision by mail and telegraph. By the time the whirling-arm flat-plate tests were completed in 1890, Langley had amassed a body of aerodynamic data unparalleled in the United States and, in terms of bulk and variety, rivaling anything in Europe. That body of data convinced Langley that mechanical flight would be possible.

Langley's aeronautical work can be divided into four stages: (1) experiments in aero-dynamics using the whirling arm (which have already been discussed) and experiments with (2) small, rubber-powered models, (3) steam-powered subscale aerodromes, and (4) a full-scale aerodrome. Some of those stages overlapped chronologically.

Langley's original goal in his whirling-arm experiments was to understand and demon-strate the physical phenomena involved in flight to an extent that would show that powered mechanical flight would be possible. At the end of *Experiments in Aerodynamics* he summa-rized the impact of his data: "The most important general influence from these experiments, as a whole, is that, so far as the mere power to sustain heavy bodies in the air by mechanical flight goes, *such mechanical flight is possible with engines we now possess*" (Langley's italics). At the 1894 meeting of the British Association for the Advancement of Science at Oxford, Lord Rayleigh commented on the validity of Langley's aerodynamic data: "if he . . . succeeded in doing it [flying] he would be [proved] right." Rayleigh's comment provided some support for activities that Langley already had under way, activities aimed at the design, construction, and use of a heavier-than-air machine.

Toward that end, Langley could already draw on experience with small, rubber-powered models of aircraft. That work had begun in 1887 at the Allegheny Observatory and continued for four years, mostly in Washington. The early models were made of pine; later, light metal tubes were tested, proving too heavy; still later, shellacked paper tubes proved to have the best strength-to-weight ratio. The wings were made of paper stretched over a supporting frame. During the tests of those flying models, Langley experimented with nearly a hundred different configurations, some of which are shown in Figure 4.56. Langley's efforts with those model aircraft were not greatly productive. The difficulty was "that it was almost impossible to build the model light enough to enable it to fly, and at the same time strong enough to withstand the strains which flight imposed upon it."[50] The models were easily broken by their falls, and the conditions of observation could not be exactly repeated. Finally Langley gave up on the rubber-powered models: "The final results . . . were not such as to give information proportioned to their trouble and cost, and it was decided to commence experiments with a steam-driven aerodrome on a large scale."[51]

The next stage of Langley's aeronautical work, namely, the development of steam-powered, subscale flying machines, was to prove much more successful. Langley had called his rubber-powered models "aerodromes," and he carried that name over to all his subsequent flying machines. Langley chose that name after consulting a scholar of the classics, adopting the Greek work *aerodromoi* ("air runner"). Strictly translated, the Greek word *aerodrome* would mean a place from which such a machine would fly, rather than the machine itself.

Langley believed that steam was the power source of choice, and over the next four years he built seven steam-driven aerodromes, numbering them consecutively from 0 to 6. Numbers 0–3 were quickly abandoned because they were too heavy and underpowered,

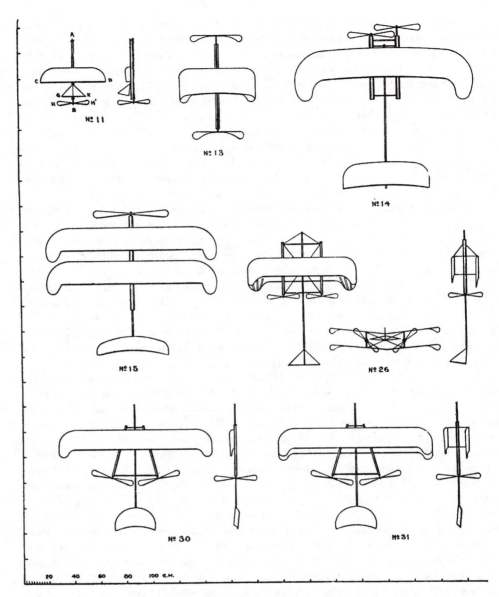

Figure 4.56   Some of Langley's rubber-powered model aircraft for testing different wing configurations.

but the lessons learned led to more successful designs. Aerodrome no. 5 was the most successful.

Aerodrome no. 5 was, in general aerodynamic design, representative of all of Langley's aerodromes (Figure 4.57). It had a tandem-wing design, a feature derived from some of the rubber-powered models. The tandem wings were of the same shape and size. The planform was rectangular, with a relatively high aspect ratio of 5 (Langley was applying one of the

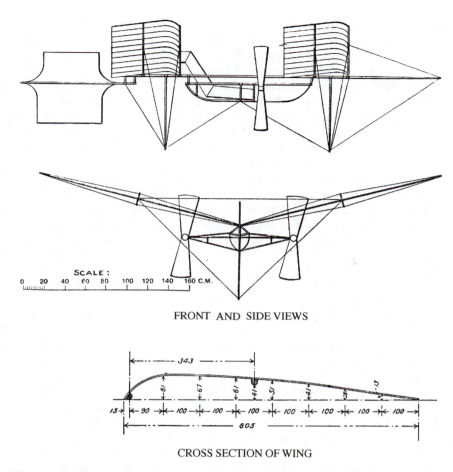

FRONT AND SIDE VIEWS

CROSS SECTION OF WING

Figure 4.57 The Langley steam-powered aerodrome: front and side views and a cross section of the wing (airfoil section).

important conclusions from his whirling-arm experiments). Both wings had wingspans of 4 m (13.1 ft), and the total sustaining wing area was 6.4 m^2 (68.9 ft^2). The total flying weight of the aerodrome was 26 lb, yielding a wing loading (weight divided by wing area) of 0.38 lb/ft^2. The airfoil shape was highly cambered (a ratio of 1 : 12), with the maximum camber at the 23.8% chord location (quite close to the $\frac{1}{4}$-chord point).

Why did Langley choose a cambered wing for his steam-powered aerodromes when all of his detailed aerodynamic data from the whirling arm had been obtained for flat plates? A curious matter indeed. Although his data were for flat plates, in the *Langley Memoir*[50] he referred to the existence of technical data for curved surfaces. In the *Langley Memoir* he promised a sequel: "It is expected later to publish a third part of the present memoir, to consist largely of the extensive technical data of other tests of the working of various types of curved surfaces, propellers, and other apparatus." Part III never appeared. However, Langley had gained some experience with cambered wings in testing his rubber-powered models. For example, in a test on November 20, 1891, Langley noted that the center-of-pressure behavior for a cambered wing was quite different from that for a flat plate. We note that Langley's first mention of tests with a cambered wing came two years after Lilienthal had published his dramatic findings clearly showing the aerodynamic superiority of a cambered airfoil over a flat plate.[46] Langley was aware of Lilienthal's work; Langley's handwritten notes contain direct comparisons of some of Lilienthal's cambered-airfoil data with Langley's flat-plate data. On page 234 of Langley's "wastebook" no. 11 (in the Ramsey rare-book room of the National Air and Space Museum) he noted receipt of the October 1893 issue of the German journal *Zeitschrift für Luftschiffahrt*: "Oct. 1893 has just come in. This contains an elaborate treatise on the resistance of curved surfaces by G. Wellner. Original data obtained both in wind and on trains are fully supplied. It is in the main corroborative of Lilienthal's work." However, in spite of Lilienthal's data and their confirmation by others, including Wellner, for some reason Langley remained skeptical about cambered airfoils: "I have made numerous experiments with curves of various forms upon the whirling-table, and constructed many such supporting surfaces, some of which have been tested in actual flight . . . . I do not question that curves are in some degree more efficient, but the extreme increase of efficiency in curves over planes understood to be asserted by Lilienthal and by Wellner, appears to have been associated either with some imperfect enunciation of conditions which gave little more than an apparent advantage, or with conditions nearly impossible for us to obtain in actual flight."

Incredibly, Langley was attempting to dismiss the work of Lilienthal. Furthermore, in calculating the performances of his aerodromes, Langley continued to use his flat-plate data, claiming that their wings were "very nearly plane." In fact, their wings were far from plane; a $\frac{1}{12}$-cambered airfoil is highly cambered by today's standards, and its aerodynamic performance is quite different from that of a flat plate. Langley continued to equivocate. He was reluctant to admit the clear advantages of cambered airfoils, but he readily used them in his steam-powered aerodromes (and later in his full-scale manned aerodrome of 1903). Moreover, after his feeble effort to discredit Lilienthal's data, Langley chose a $\frac{1}{12}$ camber for his aerodromes, precisely the camber value identified by Lilienthal as optimum. Langley placed the maximum camber near the $\frac{1}{4}$-chord location, whereas Lilienthal's airfoils had maximum camber at the $\frac{1}{2}$-chord point. Uncharacteristic of Langley as a scientist, he published not a shred of technical evidence for that choice. We can only wonder what was going on in Langley's mind on that issue. Perhaps he was not satisfied with the validity of his own data for cambered surfaces, and therefore was unwilling to publish.

Langley's equivocation on the camber issue leads us to the issue of his personality. He was said to be self-confident, self-centered, arrogant, and pompous. He was a prim man, regularly inspecting Smithsonian facilities wearing a morning coat and striped pants. He was a strict taskmaster, sometimes putting unreasonable pressure on those who worked for him. Crouch summed up the harsher side of Langley's nature: "What a friend would characterize as 'an eagerness to push on in specific pursuits which amounted at times to impatience', the secretary's subordinates viewed less kindly. At best, Langley was a difficult man to work for. He was an impatient, demanding perfectionist who insisted on absolute obedience" (p. 147).[51] Langley ruled the Smithsonian autocratically. Worse, he was sometimes accused of taking credit for work done by others. He demanded that his subordinates always walk behind him. However, there were some who revered Langley. His closest assistant and collaborator after 1898, Charles Manly, praised Langley posthumously: "He had given his time and his best labors to the world without hope of remuneration."[50] In the preface to the *Memoir,* five years after Langley's death, Manly wrote that "he began his investigations at a time when not only the general public but even the most progressive men of science thought of mechanical flight only as a subject for ridicule, [and] he helped to transform into a field of scientific inquiry what had before been almost entirely in the possession of visionaries."[50]

Langley moved in the best intellectual and scientific circles in Washington. One of his closest friends and supporters was Alexander Graham Bell, who would contribute to aeronautics by forming the Aerial Experiment Association in 1907. Another close friend was Albert Zahm, head of the Department of Physics and Mechanics at Catholic University, a few miles north of the Smithsonian. Zahm was responsible for building the first aerodynamics laboratory in an American university. Langley's status can be no better illustrated than by noting that he was a member of and a resident at the Cosmos Club, still one of the most prestigious addresses in Washington. Langley remained a bachelor all his life.

All those traits, good and bad, led Langley to glorious success in 1896. The engineering development of his steam-powered aerodromes has been described in great detail.[50,51] To launch his aerodromes, Langley used a catapult mounted on top of a houseboat in the middle of the Potomac River: "In the present stage of experiment, it was desirable that the aerodrome should – if it must fall – fall into water where it would suffer little injury and be readily recovered, rather than anywhere on land, where it would almost certainly be badly damaged." The four factors that posed problems for Langley were weight, structural strength, power, and vehicle stability. The matters of strength and weight combined into the strength-to-weight ratio, and Langley was able to deal with that problem for the subscale aerodromes, but not without great effort. For example, in some abortive tests of aerodromes nos. 4 and 5 in 1894, "the wings did not remain in their original form, [and] at the moment of launching there was a sudden flexure and distortion due to the upward pressure of the air." Langley fixed that problem for the subscale aerodromes, but it would arise again, bringing disaster for Langley's full-scale aerodrome in 1903. As for power, Langley was able to design and build a small steam engine that could produce a maximum of 1 hp (he had calculated the power required for the theoretical soaring velocity of 24 ft/s to be 0.35 hp). Langley was most concerned for the inherent stability of the aerodromes in flight. He understood the basic principles of static stability, and to achieve lateral stability he designed the wings with a substantial dihedral angle of 15°. It was much more difficult to achieve longitudinal stability, because Langley could not predict with any certainty the location of the center of pressure for the wings, let alone that for the complete aerodrome. His flat-plate data were of no help in that regard; he knew that the center-of-pressure variation for a cambered airfoil was quite different. Moreover, Langley was concerned about the interactions of the tandem wings and the effect of the propeller slipstream on the center-of-pressure location.

When Langley was finally prepared to fly his aerodromes, he often adjusted the location of the center of gravity by moving various components (the tail, the hull, etc.), and he could adjust the location of the center of pressure by changing the tail inclination angle, as well as the angle that the wing made with the fuselage. Through his system of structural guy wires, he could set the inclination angle of the wing tip to one value, and the inclination angle at the wing root to another. One typical combination of root and tip angles was 8° and 20° respectively. By a trial-and-error process with such adjustments, Langley would find the arrangement that would provide longitudinal stability as well as balance for horizontal flight.

On May 6, 1896, after three years of frustrating failures in attempting to fly his aero-dromes, Langley finally had a success. At a location about 30 miles below Washington, the Potomac River afforded a wide expanse of water over which he could test his aerodromes. The launch crew, with both Langley and Alexander Graham Bell in attendance, attempted to fly aerodrome no. 6 in the early afternoon, but it became fouled in the catapult mechanism, breaking the left wing and plunging into the water. After recovering that aerodrome, the crew mounted aerodrome no. 5 on the catapult. At 3:05 p.m., from a height of 20 ft above the water, no. 5 was launched into a gentle breeze. After slowly descending 3–4 ft, it began to rise and circle to the right, executing a spiral path at heights of 70–100 ft above the river. After about 80 s, the engine began to run out of steam, the propeller slowed down, and the aerodrome began to descend. By the time the machine settled gently to the water, it had been airborne for a minute and a half and had covered a distance of 3,300 ft. The aerodrome was immediately recovered from the water and was launched for a second time at 5:10 p.m. The ensuing flight was very much like the first; no. 5 remained aloft for 90 s and covered a distance of 2,300 ft. Langley and Bell were justifiably elated. What happened that afternoon was the most important advance and the most dramatic event in powered flight to that time. Fifteen years later, in Part I of the *Langley Memoir,* Manly described that event: "Just what these flights meant to Mr. Langley can be readily understood. They meant success! For the first time in the history of the world a device produced by man had actually flown through the air, and had preserved its equilibrium without the aid of a guiding human intelligence. Not only had this device flown, but it had been given a second trial and had again flown and had demonstrated that the result obtained in the first test was no mere accident."[50]

Langley lost no time in spreading the news. By May 26 he was at the Academy of Sciences in Paris reporting on his success, emphasizing that it proved that heavier-than-air powered flight was possible and that it could be achieved with existing technology. To further substantiate his report to the academy, Langley appended a letter from Bell describing in detail the flights that Bell had observed. Bell ended his letter with this statement: "It seemed to me that no one could have witnessed these experiments without being convinced that the possibility of mechanical flight had been demonstrated." That had been Langley's goal all along. And who more prestigious than Bell to have confirmed the realization of that goal?

The Huffaker-Langley Cambered-Airfoil Experiments

Langley was strikingly ambivalent about the relative advantages of cambered air-foils over flat plates. On the same page of the *Memoir* where he was denigrating the cambered-airfoil findings of Lilienthal, Langley admitted that "other things being equal, somewhat more efficiency can be obtained with suitable curved surfaces than with planes."[50] Langley never published the data on which that statement was based, and I have not been able to find such data in Langley's personal wastebooks for the period 1890–6. However, a footnote in the *Memoir* stated that "more recent experiments conducted under my direction by Mr. Huffaker give similar results, but confirm my earlier and cruder observations that the curve, used alone, for small angles, is much more unstable than the plane."[50] Edward Chalmers Huffaker was an assistant hired by Langley in December 1894 at the recommen-dation of Octave Chanute. Huffaker had a bachelor's degree in mathematics from Emory and Henry College and a master's degree in physics from the University of Virginia. Af-ter beginning to work for the Smithsonian, Huffaker was assigned, in part, to study the characteristics of cambered airfoils. Much of the data gathered can be found in Huffaker's wastebooks from June 23 to October 10, 1897, a full year after Langley's successful tests of the small-scale aerodromes. Huffaker did more than simply take aerodynamic measurements; he was an innovative thinker who may have been more interested in understanding the physical nature of the flow field over an airfoil than Langley himself.

Although one would think that Langley's *Experiments in Aerodynamics* would have been the definitive treatise on the aerodynamics of flat plates, in 1897 Huffaker was still examining such plates, along with cambered airfoils. Huffaker's work fell into three categories. First, he presented an expression for the lift coefficient for a flat plate that was amazingly accurate. On the basis of his own flat-plate data, on page 186 of his wastebook Huffaker advanced the following expression for the lift of a square flat plate as a function of the angle of attack:

$$P_\alpha = P_{90} \sin 2\alpha$$

where he stated that P_α was the weight of the plate (which, because the plate was soaring, was also equal to the lift), and P_{90} was the force at a 90° angle of attack, which we have seen to be approximately equal to $q_\infty S$, where q_∞ is the dynamic pressure, and S is the plate area. Hence the foregoing equation can be written in modern terms as

$$L = q_\infty S \sin 2\alpha$$

where L is the lift. Because the lift coefficient C_L is defined from

$$L = q_\infty S C_L$$

by direct comparison with Huffaker's formula we have

$$C_L = \sin 2\alpha$$

At small angles of attack, where such a formula will find its most practical use, we can employ the "small-angle approximation" for the sine function, namely,

$$C_L = \sin 2\alpha \approx 2\alpha$$

with α in radians (not degrees). So this is basically Huffaker's suggestion for a lift coefficient. How accurate is it? We can address that question by using a result from the more modern Prandtl lifting-line theory,[1] which for a finite flat plate gives

$$C_L = \frac{2\pi\alpha}{1 + 2/\text{AR}}$$

where AR is the aspect ratio. (This equation is less accurate for low aspect ratios, but for the present case at low α it is not too bad.) For a square flat plate, AR $= 1$, and hence

$$C_L = (2/3)\pi\alpha$$

Because $\pi = 3.14$, that expression gives

$$C_L \approx 2\alpha$$

which essentially confirms Huffaker's original lift equation. Unfortunately, that result was useless for designing a wing with a different aspect ratio, and of course there was no theory at that time for aspect-ratio corrections. However, it is testimony to the quality of Huffaker's work that he was able to obtain a relatively accurate formula for the lift coefficient, no matter how limited in scope.

The second category of Huffaker's work involved some conceptual thinking about the nature of the flow over an airfoil at some angle of attack. Indeed, he reached a conclusion that was far advanced for that time, concerning what today is called the "hysteresis effect" for airfoils. Most airfoils exhibit the following hysteresis effect: As the angle of attack is increased, the lift increases until stall occurs, beyond which the lift decreases (sometimes precipitously). As the angle of attack is decreased from a high value to a low value, the lift in the region of intermediate angles of attack remains low, more or less equal to its value at stall, until a much lower angle of attack is reached. For example, when the airfoil is pitched upward through a range of increasing angles of attack, its lift coefficient may be 1.4 as it passes through a 15° angle of attack. Then, as the same airfoil is brought back down to lower angles of attack, its lift coefficient may be 1.0 as it passes through a 15° angle of attack. That is the hysteresis effect, caused by the stability characteristics of the separated flow that occurs at stall, and it causes nonunique flow patterns and hence nonunique lift coefficients at angles of attack near stall. Huffaker never observed the hysteresis phenomenon, but he *theorized* that it would occur: "If a rectangle plane be made to rotate about an axis while exposed to a current of air, the pressure at any instant and for any assumed angle of attack will be greater for an increasing than for a decreasing cross section of the intercepted current" (wastebook, May 5, 1896, p. 220). Of course, Huffaker was saying that the effect operated for *any* angle of attack, which is not the case. Huffaker elaborated further on the same page:

> In front of the plane is an area of compression, in the rear of rarefaction. Now the pressure upon the plane depends upon the extent of this area of disturbance. And my argument is that as the angle is increased, or as the crossection of the intercepted current is increased, the disturbed area is pushed out into the undisturbed current at the same time, deflecting still further the streamlines already existing, with the formation of new ones; while as the angle (α) decreases the area of disturbance is contracted, no new streamlines are formed, those previously existing are but little altered, and the compressed air in front finds access to the rear decreasing the rarefaction and lessening the pressure upon the plane. So that on the whole the pressure is greater at any given instant and given value for α for increasing than for decreasing values of α.

That piece of intellectual property was signed "E. C. Huffaker" at the bottom – one of the few times that Huffaker signed his name to a particular idea in the wastebook. He went on to develop some analytical formulas to justify his hypothesis (the development is not convincing). Nevertheless, Huffaker's thinking was certainly headed in the direction of the hysteresis effect. Moreover, Huffaker's idea derived solely from intellectual reasoning and was not due to any experimental evidence on the effect. To have measured that effect would have required instrumentation far beyond anything available at that time.

The third category of Huffaker's work concerned his measurements of the aerodynamic characteristics of cambered airfoils. Most of the work focused on a $\frac{1}{12}$-cambered airfoil similar to that in Figure 4.57, and the experiments were carried out with the same whirling arm used for Langley's flat-plate measurements seven years earlier. Huffaker's experimental data for cambered airfoils are scattered in his wastebook, dating from June 23 to October 10, 1897 (pp. 262–356). The following data, from July 6, 1897, are representative. They pertain to a wing 24 × 6 in., weighing 250 g. The first two columns show Huffaker's data for soaring speed versus angle of attack. The third column shows the corresponding lift coefficients (calculated herein from Huffaker's data). The fourth column shows corresponding

lift coefficients calculated from Langley's earlier flat-plate findings[49] for the same planform
(24 × 6 in.).

α (degrees)	Soaring Velocity (ft/s)	Lift coefficient (cambered airfoil)	Lift coefficient (flat plate)
5	25	0.74	0.3
8	22	0.96	0.44
10	21	1.05	0.48
15	21	1.05	0.58
20	20	1.16	0.66
30	18	1.43	0.75

From the third and fourth columns, the cambered airfoil is clearly seen to have much higher
lift coefficients than the flat plate – by a factor of 2. Huffaker's comment on his cambered-
airfoil data was a gross understatement (July 6, 1897): "The curves used appear to possess
slightly greater lifting forces than the plane." Let us compare Huffaker's data with those
of Lilienthal (Figure 4.38). Although both sets of data were for $\frac{1}{12}$-cambered airfoils,
Lilienthal's airfoil was a circular arc, with maximum camber at the $\frac{1}{2}$-chord point, whereas
Huffaker's airfoil was more of a parabolic shape, with maximum camber at approximately
the $\frac{1}{4}$-chord point. Lilienthal's lift coefficient for a 5° angle of attack can be obtained
by multiplying the normal-force coefficient (Figure 4.38) by the cosine of the angle of
attack: For $\alpha = 5°$, $C_L = 0.65 \cos 5° = 0.648$. Returning to the foregoing tabulation of
Huffaker's data, we see that Lilienthal's value for the lift coefficient at $\alpha = 5°$ was smaller
than the value of 0.74 obtained from Huffaker's data. That trend holds for all the angles
of attack tabulated. This comparison does not at all substantiate Langley's earlier claim
that Lilienthal's findings were too optimistic; rather, the Huffaker-Langley data were more
optimistic.

On page 282 of his wastebook, July 10, 1897, Huffaker reported a comparison between
soaring speeds for cambered and plane wings, as well as their "drift" (drag) values. From
that he calculated the "power expended" for both wings and found that the cambered wing
required about 30% less power to achieve soaring speed.

Not all of Huffaker's experiments contributed to the science of flight. One set of experiments was "to determine the soaring speed of a curved surface when lined upon the under
surface with feathers from a buzzard's wing." That idea was "suggested by O. Chanute, to
ascertain whether or not Wenham is correct in his contention that the ribbed under surface
of the feathers add to their lifting power." Huffaker measured equal soaring speeds for the
wings with and without feathers, indicating that lift was unaffected. However, the feathers
increased the measured drag by almost 30%.

In summary, Langley's ambivalence about cambered airfoils was inexplicable, not only
because of Lilienthal's convincing data but also because of the experiments carried out by
Huffaker. The only thing about a cambered airfoil that might have been construed as a
disadvantage was the more complex behavior of its center of pressure; Langley may have
been overconcerned with how that complexity would affect the stability of his aerodromes.

The State of the Art as Reflected in Lilienthal's Gliders and Langley's Aerodromes

The representative flying machines for this chapter are Lilienthal's gliders (Figure 4.1) and Langley's steam-powered aerodromes (Figure 4.2). We have assessed the state of the art of aerodynamics during the nineteenth century. We are now in a position to address the question of the extent to which the existing aerodynamic knowledge was used in the design of those machines.

Virtually nothing from theoretical aerodynamics was used in those designs. Although the governing equations of motion for a viscous fluid (the Navier-Stokes equations) were in place by the middle of the nineteenth century, there had been no applications of those equations in the design of flying machines. There were two reasons for that situation: First, the university-educated scientists and academicians who developed and studied those equations were, for the most part, totally uninterested in powered flight (indeed, some firmly believed that such flight was not possible). Second, the equations were complex, and techniques for their solution (usually some simplification to a more manageable form) were just beginning to be developed at the end of the nineteenth century. Thus, what was known about theoretical aerodynamics was virtually useless to Lilienthal and Langley in the design of their flying machines.

The situation in applied aerodynamics was quite different. By the end of the nineteenth century there had been useful applications of applied aerodynamics in the design of flying machines. The reason is obvious: The men defining the state of the art were the same men designing the flying machines. Lilienthal's definitive data on cambered airfoils led to their use in his gliders and in Langley's aerodromes. Langley's experiments in aerodynamics clearly showed the superior aerodynamic characteristics of high-aspect-ratio wings, and his aerodromes incorporated aspect ratios as high as were allowed by structural limitations. Langley's measurements of center-of-pressure movements allowed him to make good decisions about where to place the center of gravity for his aerodromes and how to shift it when necessary. Lilienthal and Langley had access to each other's published data within a reasonable time period. For the first time in the history of aerodynamics, the state of the art was beginning to be reflected in the design of flying machines.

Developments in Aviation between 1804 and 1896: Chanute's Progress in Flying Machines

Against the backdrop of the developing science of aerodynamics in the nineteenth century there was a parade of various characters who attempted to fly in heavier-than-air machines. Most noteworthy were the powered "hops" by Felix Du Temple, in 1874 in France, and Alexander Mozhaiski, in 1884 in Russia. Du Temple was a French naval officer and engineer who in 1857–8 experimented with small model airplanes powered by clockwork mechanisms. In 1874, a larger machine designed by Du Temple made the world's first powered takeoff with a pilot on board. That airplane had swept-forward wings and was powered by a hot-air engine (Figure 4.58). Piloted by a sailor, it was launched down an inclined plane and left the ground for a moment, but did not come close to sustained flight. It represented the first "powered hop" in aviation history. Ten years later, near St. Petersburg, Russia, a large, steam-powered flying machine designed by Alexander Mozhaiski was launched down an inclined plane (Figure 4.59). Piloted by I. N. Golubev,

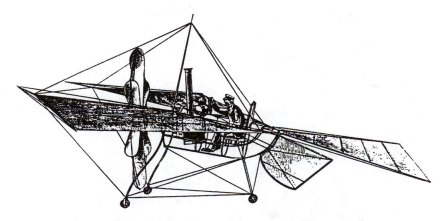

Figure 4.58 Felix Du Temple's flying machine, which made the first powered hop (1874).

Figure 4.59 Alexander Mozhaiski's flying machine that made the second powered hop (1884).

the airplane momentarily left the ground – it was the second "powered hop." At various times it has been claimed that Mozhaiski launched the first powered flight, but it was not a sustained, controlled flight, and the claims that it was the first true flight are not accepted by professional historians of aviation. However, Du Temple and Moshaiski can be credited with the first and second *assisted* powered takeoffs – "assisted" because they were launched down inclined planes, building up momentum to assist the powered takeoffs.

Prior to those powered hops, some progress in flying machines had been made with small models. Among the most important and influential were the rubber-band-powered models of Alphonse Pénaud, a Frenchman devoted to aeronautics (Figure 4.60). Pénaud was the first designer to achieve inherent stability, both longitudinally and laterally. Longitudinal stability (pitching stability) was provided by setting the horizontal tail to a negative angle of 8° relative to the chord line of the wing, and lateral stability (roll stability) was provided by bending the wing tips up in a dihedral angle. Pénaud demonstrated that model in the Tuileries gardens in Paris, August 18, 1871; it flew 131 ft in 11 s. Pénaud is credited with solving the basic problems of longitudinal and lateral stability for flying machines. In 1880,

Figure 4.60   Alphonse Pénaud's model (1871).

Figure 4.61   John Stringfellow's model triplane (1868).

at the age of 30, in poor health and spirits, he committed suicide, cutting short a promising aeronautical career.

Another experimenter with models was John Stringfellow, an English engineer who in 1843, along with William Samuel Henson, formed the Aerial Transit Company to design, build, and operate a large flying machine for worldwide commercial purposes. Their "aerial steam carriage" was to have a 150-ft wingspan and a 30-hp steam engine. It was never built, but Stringfellow and Henson got as far as building a 20-ft-wingspan model that was tested in 1845–7. It was a failure, unable to sustain itself in the air. After a hiatus, Stringfellow built a triplane model that he displayed at the Crystal Palace exhibition in 1868 sponsored by the fledgling Aeronautical Society of Great Britain (Figure 4.61). The model moved on a cable; successful lifting flight seemed to elude it. On the other hand, the widely distributed photographs of that model popularized the idea of superposed wings; it was the forerunner for biplanes and triplanes.

The nineteenth century was awash with flying machines, both ornithopters (flapping wings) and fixed-wing aircraft, but they were all eclipsed by the success of Lilienthal's gliders during 1891–6 and Langley's aerodromes in 1896. One of the best contemporary compilations describing those various flying machines was Octave Chanute's *Progress in Flying Machines*.[45]

Octave Chanute (1832–1910) (Figure 4.62) was born in Paris, February 18, 1832. The family moved to the United States in 1838 when his father, a distinguished scholar, became vice-president of Jefferson College in Louisiana (now Loyola University). Rather than going to college, Chanute chose to become an apprentice in civil engineering. He worked in railroading, and by 1853 he was a division engineer in charge of laying track for the Chicago and Mississippi Railroad. He went on to become one of the most successful American engineers of the nineteenth century. He designed and built the first bridge over the Missouri River and the stockyards in Chicago and Kansas City. In 1873 he became chief engineer of the Erie Railroad. At about that time, Chanute became interested in the quest for powered flight, and he began to apply his professional knowledge to that problem.

Figure 4.62 Octave Chanute.

He conducted a long and exhaustive survey of flying-machine development, published in a series of articles for the *Railroad and Engineering Journal* and later compiled into *Progress in Flying Machines*.[45] Although Chanute began to design and experiment with gliders as early as 1894 (frequently corresponding with Lilienthal), his major contribution to the advancement of aeronautics was as ambassador, spreading the word about new developments in aeronautics. He corresponded with and was respected by virtually all of the principal workers in aeronautics at the time. He served as a catalyst, inspiring and encouraging others in their efforts toward powered, manned flight. His book was read by the Wright brothers, and later he became a friend and advocate of their cause. When Chanute died in 1910, he had seen his expectations for powered flight become reality. When his book was published in 1894, it was the definitive publication to date on the history and current status of flying machines: "Eighty years after the original publication, *Progress in Flying Machines* remains one of the most comprehensive and reliable histories of pre-Wright aeronautics available."[51] In terms of the history of aerodynamics, his book was noteworthy in several respects:

First, the state of the art of applied aerodynamics was laid out in the initial 10 pages of the book – there was not much to relate. Chanute tabulated lift and drag coefficients for a flat plate of aspect ratio 1 at various angles of attack, calculated by Chanute from a formula attributed to Duchemin:

$$P = P' \frac{2 \sin \alpha}{1 + \sin^2 \alpha}$$

where P is the aerodynamic force on the inclined surface, and P' is the force when the plate is perpendicular to the flow. As explained earlier, the ratio P/P' is essentially the resultant-force coefficient C_R on the plate, which from Duchemin's formula is

$$C_R = \frac{2 \sin \alpha}{1 + \sin^2 \alpha}$$

From that, the lift and drag coefficients were obtained as $C_L = C_R \cos \alpha$ and $C_D = C_R \sin \alpha$, as tabulated in Chanute's book (pp. 4–5). In the same table, Chanute compared those calculations with Langley's measurements and found the correlation quite good. Because of that, Chanute offered his table as a reliable source of estimates for lift and drag coefficients for the wings of various flying machines, and he continually drew from that table for design calculations throughout the rest of his book. In so doing, he was intentionally ignoring the effects of aspect ratio and camber – a significant neglect on both counts. (Recall that Chanute published Lilienthal's table of normal-force and axial-force coefficients for a cambered airfoil just three years later in the *Aeronautical Annual*.)

Second, Chanute formalized some terminology that would become standard for future aeronautical literature – some terms had more lasting power than others. For example, in discussing the aerodynamic force on a wing, he clearly defined the formal term "lift," which is the component of aerodynamic force perpendicular to the free stream (perpendicular to the flight path of a moving vehicle), and that term has prevailed to the present. In contrast, the component of aerodynamic force parallel to the free stream he denoted as "drift," whereas Langley called that simply the "horizontal pressure" or sometimes the "horizontal component of pressure" or the "resistance to advance." The term "drift" prevailed for a short time, and the Wright brothers used it in their writings, but none of those terms is used today, having been universally replaced by the term "drag." Chanute reserved the term "drift" for the drag of only the wing. For the overall "resistance" of a flying machine, Chanute listed three elements: (1) the hull resistance (drag due to the "solid body or hull to contain the machinery and the cargo"), (2) the drift (the drag of the wing), and (3) the skin friction (which, following Langley, was dismissed as negligible). Clearly, by ignoring friction, Chanute was saying that hull resistance and drift were (in today's terms) simply types of "pressure drag." In Chanute's nomenclature, an "aeroplane" was a "thin fixed surface, slightly inclined to the line of motion, and deriving its support from the upward reaction of the air pressure due to the speed, the latter being obtained by some separate propelling device"[45] (i.e., an "aeroplane" is simply a wing). The term "aeroplane" gradually came to signify the whole flying machine before World War I. As another example, Chanute introduced the term "aeronautical engineer," a label that lasted (generally replaced by the term "aerospace engineer" for the past three decades). Like any book that becomes the standard reference in a fledgling field, Chanute's definitive *Progress in Flying Machines* had a head start in coining the formal nomenclature – an essential process for the growth and propagation of any technology.

Third, Chanute's book had a philosophical cast – assessments of the present and hopes and forecasts for the future of flying machines.

Chanute was, in part, a disciple of Langley, and the book served to disseminate and celebrate Langley's views concerning aerodynamics:

> The important thing for us to know is to ascertain what pressure exists under a wing (of a bird) or, to simplify the question, under a plane surface, when it meets the air at a certain velocity and with a certain angle of incidence.
>
> This has been, until the recent publication of Professor Langley's most important labors, a subject of uncertainty, which uncertainty he has done much to remove. We had had glimpses of the law; but notwithstanding very many experiments by physicists, its numerical values were a subject of doubt and controversy among the few who gave any attention to the subject. It was the missing link, which rendered nearly unavailable the little that was known in other directions [p. 3].[45]

However, Chanute was also conservative in his thinking. He recognized that Langley's findings were limited to flat plates. Referring to a table of coefficients in Langley's *Experiments in Aerodynamics* (p. 58), Chanute emphasized that "it should be borne in mind that the table only purports to apply to thin planes one foot square, and hence is given as containing only approximate percentages of normal pressures. For other shaped planes, for curved surfaces, and for solids the percentages may be different, because a great many anomalies have been found in experimenting upon air resistances, and we yet know painfully little about them," Chanute went on to list nine anomalies, one of which was the Langley power law.

Chanute's book cited the advantages of testing flying machines over water. Referring to the fourteenth-century Italian mathematician J. B. Dante, who was said to have sailed over a lake near Perugia on artificial wings, Chanute commented that "the selection of a sheet of water to experiment over was very happy, as it would furnish a yielding bed to fall into if anything went wrong, as is pretty certain to happen upon the first trials." Chanute went on to state emphatically that the selection of water for a testing area "cannot be too strongly urged upon any future inventor who desires to make similar experiments." Could that be where Langley got the idea of flying his aerodromes over the Potomac River? Chanute's book does not mention Langley's work with aerodromes; it must be assumed that in 1894 Chanute was unaware of Langley's work along those lines. Thus it seems improbable that Langley had discussed the idea with Chanute. Langley most likely got the idea from Chanute's writings.

Chanute's lengthy final section should be read by all serious students of the history of aeronautics. Chanute suggested that in 1894 the technology for powered flight was almost in hand: "It will be seen that the mechanical difficulties are very great; but it will be discerned also that none of them can now be said to be insuperable, and that material progress has recently been achieved toward their solution." Chanute was optimistic regarding the problems of understanding the basic nature of aerodynamics and the considerable lack of such understanding by all investigators then in the field: "Science has been awaiting the great physicist, who, like Galileo or Newton, should bring order out of chaos in aerodynamics, and reduce its many anomalies to the rule of harmonious law." That great physicist was already waiting in the wings. In the first half of the twentieth century, Ludwig Prandtl would singlehandedly revolutionize the way that aerodynamics was understood and practiced.

PART III

Aerodynamics Comes of Age

CHAPTER 5

Applied Aerodynamics Comes of Age
The Wright Brothers

Isn't it astonishing that all these secrets have been preserved for so many years just so that we could discover them!

Orville Wright (June 7, 1903)

The year 1896 brought a turning point in the history of aeronautics. Otto Lilienthal was killed in 1896, ending the most promising aeronautical activity up to that time. Samuel Langley's small, steam-powered aerodromes flew in 1896, demonstrating beyond any doubt the technical feasibility of heavier-than-air powered flight. In 1896, prompted in part by Lilienthal's death, Wilbur and Orville Wright decided to join the quest for manned flight: "In 1896 we read in the daily papers, or in some of the magazines, of the experiments of Otto Lilienthal, who was making some gliding flights from the top of a small hill in Germany. His death a few months later while making a glide off the hill increased our interest in the subject, and we began looking for books pertaining to flight" (p. 3).[55] From that beginning, the Wrights came onto the scene representing a new interest, a new group of challengers entering the list at a time when the development of manned, powered flight was viewed by some as just around the corner, and by others as a physical impossibility.

This chapter will attempt to make the case that applied aerodynamics came of age with the work of the Wright brothers. The Wrights' contributions to applied aerodynamics composed only one part of their overall inventive efforts that ultimately led to the first manned, powered, heavier-than-air flight, December 17, 1903. The unusual nature of the Wrights' inventive methods and the reasons they led to success have been thoroughly discussed by Peter Jakab in *Visions of a Flying Machine*.[24] Jakab attempted to discover what had been different about the Wrights and their approach to aeronautics: There were "specific personality traits, innate skills, and particular research techniques present in the Wrights' approach that came together in a unique way and largely explain why these two men invented the airplane. In short, the Wrights had a definable method that in very direct terms led them to the secrets of flight" (p. xv).[24]

The obvious choice for a representative flying machine for this chapter is the 1903 Wright Flyer (Figure 5.1), and again we seek to determine the state of the art of aerodynamics at that time and to discover to what extent that state of the art was reflected in the Wright Flyer.

The Wright Brothers to 1896

The story of the two men who built the first practical flying machine has filled many books, among them the definitive biography of the Wright brothers by Tom Crouch: *The Bishop's Boys*.[53] Wilbur and Orville Wright were the sons of a bishop of the Church of the United Brethren in Christ, and Crouch tells a most interesting story of the family and household environment in which Wilbur and Orville grew up, absorbing a work ethic that would drive them to success. A confidant of the Wrights' heirs, Crouch had access to family correspondence and diaries overlooked by, or simply not available to, previous

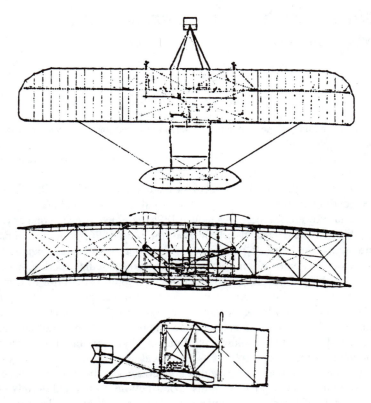

Figure 5.1   Three-view of the 1903 Wright Flyer.

writers. Because there are so many good sources of biographical material on the Wright
brothers, here we shall consider only those aspects of their lives that can contribute to an
understanding of the way they worked and thought.

Wilbur and Orville were products of the post–Civil War era. Wilbur was born near
Millville, Indiana, April 16, 1867, and Orville in Dayton, Ohio, August 19, 1871. There
were seven children in the family: two older brothers, Reuchlin and Lorin; a set of twins
who died in infancy; a younger sister, Katharine; and Wilbur and Orville. Their father,
Milton Wright, was an active minister, church administrator, and eventually bishop in the
United Brethren church and was frequently away from home on business.

> Milton was not adept at the skills required to win friends and influence people. As an
> administrator he had "personally offended" a number of presiding elders. His limitations
> as a politician were apparent. Reconciliation, negotiation, and compromise, the tools of
> the effective vote-getter, were foreign to him. Moreover, he would never trust men who
> possessed those skills. His written descriptions of various Dayton political contests over
> the years are studded with words like "scheming," "malicious," and "treacherous." There
> were no moral gray areas in his world. Right was right. Wrong was wrong [pp. 60–2].[53]

Their mother, Susan Koerner Wright, was born in Virginia. She married Milton in 1859,
at the age of 28. She was shy and scholarly, and before marrying Milton Wright, she had
attended Hartsville College, coming within three months of graduation. She was gifted
with considerable mechanical ability:

"She designed and built simple household appliances for herself and made toys, including a much-treasured sled, for her children. When the boys wanted mechanical advice or assistance, they came to their mother. Milton was one of those men who had difficulty driving a nail straight" (p. 33).[53] Her enjoyment in working with things mechanical would appear to have been transmitted to Wilbur and Orville. Susan died of tuberculosis in 1889. Katharine, who had been looking after the family during her mother's illness, inherited the formal responsibility of caring for the family after Susan's death. Reuchlin and Lorin were already out on their own, so there were four people left in the house at 7 Hawthorn Street in Dayton: Milton, Wilbur, Orville, and Katharine. They became a self-support group of extreme closeness, sharing all problems and collectively working out solutions: whenever one of them was away, letters were posted almost daily. In spite of her family responsibilities, Katharine graduated from nearby Oberlin College in 1898, after which she took a job teaching the classics at Steele High School in Dayton. (The closeness of the Wright family would reach an unpleasant extreme in 1926, when Katharine married Henry Haskell, editor and part owner of the Kansas City *Star*. Orville considered that a rejection of him – Wilbur had been dead for 14 years – and cut off all contact with Katharine. In his mind, Katharine had broken the sacred pact forged after their mother's death. Even when Katharine was dying of pneumonia in 1929, Orville refused to visit her until the very end, when he was at her bedside, but only at the insistence of his older brother Lorin. The unusually close family ties of the Wrights' proved a double-edged sword.)

Wilbur and Orville never officially graduated from high school. Wilbur attended high school in Richmond, Indiana, where the family was living at the time. He was an accomplished scholar, taking courses in Greek, Latin, geometry, natural philosophy, geology, and composition and earning grades well above 90% in them all. In addition, he was an athlete, excelling in gymnastics. He was doing so well that Milton and Susan were considering sending him to Yale. Unfortunately, just before Wilbur was to graduate in June 1884, Milton abruptly moved the family to Dayton; there were compelling political reasons having to do with church business for that quick move. Wilbur was not able to complete the courses required for graduation. Consequently, he never received a high-school diploma and never attended Yale. However, he continued his own education, ultimately becoming better-read than most college graduates. Orville attended Central High School in Dayton. Unlike Wilbur, Orville was not an outstanding student, but he managed to earn grades in the 70–90% range. Just before he was to return for his senior year, Orville's mother died. However, Orville had already decided not to return, mainly because he had taken several advanced courses in his junior year that were not part of the required curriculum, and as a result he was going to be several credits shy of the number necessary for graduation. Because he would not be able to graduate with his classmates, Orville decided not to return. He was interested in becoming a printer, and he began to pursue that activity, including building his own printing press. However, his commercial ventures in the printing business were not successful. In short, the inventors of the first practical flying machine never earned high-school diplomas.

By 1892, both Wilbur and Orville were bicycle enthusiasts, competing in local races. In December of 1892, they opened a bicycle shop on West Third Street in Dayton. At first they sold and repaired bicycles, but by 1895 they were manufacturing and selling their own line of bicycles. It was a profitable business. The bicycle had become a major mode of transportation, and at the height of its popularity in the 1890s there were more than 300 companies in the United States manufacturing over a million bicycles each year. Their talent and experience in designing and building bicycles helped prepare them for their later flying-machine work:

> While the major bicycle manufacturers were employing mass-production techniques adopted from the firearm and sewing-machine industries, the Wrights remained small scale and continued to produce handmade originals. At a time when manufacturing was becoming increasingly mechanized and rapidly rushing toward the twentieth century, the Wrights stayed firmly within the classic artisan tradition of handcrafted, carefully finished individual pieces. This kind of attention to detail and craftsmanship would be a hallmark of their flying machines. Every component of their aircraft was designed and built with great care and served a specific and essential function. It is a bit ironic that an invention that has been so influential in the twentieth century was the product of men whose approach was so firmly anchored in the nineteenth century [p. 9].[24]

The Wrights also typified another nineteenth-century phenomenon: a disparate group of self-educated mechanics and experimenters who, without benefit of university education, devised a wide range of gadgets and useful inventions and made significant contributions to the progress of engineering. The Aeronautical Society of Great Britain had originally been organized by such men. The term "aeronautical engineer" was beginning to appear more widely, and both Lilienthal and Chanute used it in their writings. The Wright brothers were the first to deserve that designation in its fullest sense.

In the year 1896 the brothers Wright were securely entrenched in the bicycle business in modest surroundings at 1127 West Third Street in Dayton (Figure 5.2) and were living comfortably in the house at 7 Hawthorn Street (Figure 5.3). (Both the bicycle shop and the house are now located in Greenfield Village, Dearborn, Michigan, having been purchased and moved there by Henry Ford in 1936–7 as part of the museum that Ford created to honor American ingenuity and industrial achievement.) Wilbur and Orville Wright (Figure 5.4) were intelligent and widely read, possessed of a Puritanical moral attitude, considerable mechanical aptitude, a dedication to quality craftsmanship, and a tireless work ethic – fertile ground indeed for nurturing the seeds of the idea of flying and meeting the engineering challenge of powered flight.

Figure 5.2   The Wrights' bicycle shop (ca. 1900).

Figure 5.3   The Wright family home at 7 Hawthorn Street (ca. 1900).

Figure 5.4   Wilbur (left) and Orville Wright on the back steps of their home in Dayton.

The Existing State of the Art (1896–1901): The Wrights Make Some Errors

When the Wrights began their activities in aeronautics, they were not faced with a vacuum – indeed, in the field of aerodynamics, they inherited the considerable legacy described in Chapter 4 that pointed them in the general direction of ultimate success:

> The phase of aeronautical development just prior to the Wrights' entry into the field had yielded much productive research and had laid a foundation of critical inquiry that would be invaluable to the next generation of experimenters. The status of the invention of the airplane was still much like searching for a needle in a haystack, but now at least it was known in which haystack to look.... The experience of their predecessors did more to

reveal fundamental questions than to provide answers, but wading through the failures and the misunderstandings of others aided the brothers in focusing quickly on the basic problems that needed to be addressed. Much of what the Wrights accomplished was highly original, but the findings of the late nineteenth century definitely gave them several useful pieces to the puzzle, ... If the brothers had been a generation older, it is not at all certain that they would have avoided the stumbling blocks of those who were working in the second half of the nineteenth century. The Wrights were especially talented to be sure, but there is no reason to believe their genius operated in a vacuum, and that they would have invented the airplane no matter when they took up the problem [pp. 16–17].[24]

My assessment agrees with that of Jakab. In the discipline of applied aerodynamics, the stage was set for almost immediate clarification and achievement. The Wright brothers were the right people at the right time, taking advantage of the unusual opportunity: "The Wrights' genius lay as much in their insightful analysis and adaptation of what had come before them as it did in their own innate creativity" (p. 38).[24]

Insight into the Wrights' thinking processes, their aspirations and feelings about failures and successes, can be gained by reading their correspondence and diary entries.[55] The Wrights expressed their feelings and ideas well, both in their personal correspondence and in the few instances in which they prepared technical papers for a technical audience. McFarland's collection[55] of the Wrights' papers is an essential source for any serious student of the history of aeronautics. Although those papers deal with many subjects, the Wrights' understanding of and contributions to applied aerodynamics are clearly there for all to see. Much of what follows is gleaned from those papers.

Although the brothers began reading and thinking about flying machines in 1896, serious study of the important sources of information did not begin until 1899. On May 30, 1899, Wilbur wrote to the Smithsonian Institution requesting a list of the available reference sources dealing with flight. In that letter, he mentioned that he had access to several books and magazine articles and some encyclopedia information on flight. It is quite telling that he did not mention any of the truly important sources that we discussed in Chapter 4, such as Cayley's papers, Lilienthal's and Langley's books, and Chanute's *Progress in Flying Machines*. It appears that none of those had been read by the Wrights at that time; indeed, the indications are that they were unaware of the existence of most of those works. Of course, Lilienthal's *Birdflight as the Basis of Aviation* was available only in German at that time, but the other basic sources were in English. It is difficult to imagine that none of the articles on Lilienthal in the popular press would have mentioned *Birdflight*. Perhaps that was testimony to how obscure any writings on aeronautics were at that time; recall that in Lilienthal's lifetime, fewer than 300 copies of his book were sold. In any event, Wilbur's letter of May 30 indicated no knowledge of such material, but it was strong on purpose and modesty:

> I am about to begin a systematic study of the subject in preparation for practical work to which I expect to devote what time I can spare from my regular business. I wish to obtain such papers as the Smithsonian Institution has published on this subject, and if possible a list of other works in point in the English language. I am an enthusiast, but not a crank in the sense that I have some pet theories as to the proper construction of a flying machine. I wish to avail myself of all that is already known and then if possible add my mite to help on the future worker who will attain final success [p. 5].[55]

Note that Wilbur used the singular "I"; there was no hint that Orville was working with him. That carried through much of Wilbur's early correspondence.

At that time, any flying machine was a hot topic at the Smithsonian. Langley had experienced considerable success with his aerodromes only three years earlier, and he was feverishly working on a large, manned aerodrome for the War Department. The design of flying machines was a major activity along the Mall in Washington. Within three days, Richard Rathbun, assistant secretary of the Smithsonian, replied to Wilbur's letter with a list of sources, including Chanute's *Progress in Flying Machines,* Langley's *Experiments in Aerodynamics,* and James Means's *Aeronautical Annual* (for 1895–7). Rathbun also sent four Smithsonian pamphlets: *Empire of the Air,* by Louis-Pierre Mouillard; *The Problem of Flying and Practical Experiments in Soaring,* by Lilienthal; *Story of Experiments in Mechanical Flight,* by Langley; and *On Soaring Flight,* by Huffaker. In turn, Wilbur lost no time; on June 14, 1899, he wrote back:

> I have to thank you for your letter of June 2, in which you kindly enclosed a list of selected books treating on aerial navigation. I also wish to thank you for pamphlets #903, 938, 1134 and 1135 from the Smithsonian Reports.
>
> I enclose one dollar currency for which you may send me *Experiments in Aerodynamics* by Langley [p. 5].[55]

So it was not until June 1899 that the Wrights began to focus on the important nineteenth-century works in aeronautics, obtaining a copy of *Experiments in Aerodynamics.*

Some historians have suggested that the most important aeronautial contribution to emanate from the Smithsonian, all of Langley's work notwithstanding, was that simple transfer of knowledge to the Wright brothers in 1899. There is no doubt that it focused the efforts of the Wrights. Indeed, much later, in February 1920, Orville wrote that "after reading the pamphlets sent to us by the Smithsonian we became highly enthusiastic with the idea of gliding as a sport." By the summer of 1899 the Wright brothers were on a track that would take them to success beyond their imagination.

Once they began, progress was rapid: By August 1899 the Wrights had constructed a small glider that incorporated a wing-warping technique for lateral control conceived by Wilbur. (Their attention to lateral control, and the use of wing warping to achieve it, was one of the most important features of the Wrights' flying machines.) That glider, with a 5-ft wingspan, was flown only as a kite, but the kite flights vindicated the use of the wing-warping concept, and the Wrights committed themselves to the design and construction of a full-size manned glider.

It was natural that the Wrights would rely on the existing state of the art of aerodynamics in the design of their 1900 and 1901 gliders. That is still conventional engineering practice today: Utilize the tried and proven, and perhaps improve on it. When the Wrights' 1900 glider was first flown in early October at Kitty Hawk, it embodied the best of the existing knowledge in applied aerodynamics, at least insofar as the Wrights perceived and understood it.

It is clear that the Wrights did their homework with the reading list provided by the Smithsonian. Some terms from Chanute's *Progress in Flying Machines* began to be used in the Wrights' correspondence (e.g., in their notebook A from the September–October period, frequent discussions center around lift and drift and the drift-to-lift ratio). In addition, Wilbur had initiated a correspondence with Octave Chanute, beginning with a long letter of introduction dated May 13, 1900. That correspondence would continue, off and on, until Chanute's death in 1910. Thus the Wrights became well versed in Chanute's views on technical aeronautics. Another resource, arriving indirectly by way of Chanute, was the

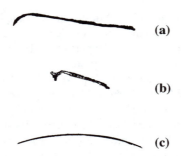

Figure 5.5 Sketches of airfoil shapes: (a) Wilbur's sketch of their 1900 glider airfoil; (b) Wilbur's sketch of their 1901 glider airfoil; (c) circular-arc airfoil.

Lilienthal table of normal-force and axial-force coefficients, first published in Germany in 1895 in Moedebeck's handbook, and then reproduced by Octave Chanute in 1897 in the *Aeronautical Annual*. By 1900 the Wrights had a copy of the Lilienthal table, but they did not have access to a copy of Lilienthal's book until after they had designed and flown their 1900 glider. We know that they had bought a copy of Langley's *Experiments in Aerodynamics* in 1899 and presumably had read it before designing their 1900 glider. Also, they had read the summary of Langley's findings given at the beginning of Chanute's *Progress in Flying Machines*. By the time the Wrights began construction of their 1900 glider in August, they had available on their bookshelf all of the relevant aerodynamic data of that day.

It is no surprise that the Wrights chose to use a cambered airfoil rather than a flat surface for their wings. Given Lilienthal's data and Langley's success with his aerodromes, the superiority of cambered airfoils must have been quite apparent to the Wrights. The shape of their airfoil was sketched by Wilbur in a letter to Chanute on November 16, 1900 (Figure 5.5a). The 1900 glider had a camber of $\frac{1}{23}$, and the location of maximum camber was close to the leading edge. The glider was a biplane, with a total wing planform area of 165 ft^2, and the span of each wing being 17 ft. The corresponding aspect ratio was 3.5. The shape of the wings was essentially rectangular.

The Wrights began flying their 1900 glider in October at Kitty Hawk, North Carolina. Its aerodynamic performance fell far short of what had been predicted. The lift was only one-third to one-half of that calculated on the basis of the Lilienthal table. That unexpected poor performance severely curtailed their progress. The glider was flown for 2–4 h each day as a kite; in that mode they were able to measure the lift and "drift" simply by measuring the net pulling force along the string, using a grocer's-type spring scale, and measuring the angle between the string and the horizontal. From those two measurements, and knowledge of the glider's weight, they could calculate the corresponding values of lift and "drift" (see Appendix F for the calculational details). Only when the wind was very strong was Wilbur able to get off the ground for a few seconds of real flight in the glider. (During 1900 and 1901, Wilbur made all the manned glider tests; Orville did not glide until 1902.) Some of the manned glider flights were made with the glider tethered as a kite, and others were free flights down the slopes of sand dunes.

In spite of the glider's poor performance, the Wrights were generally satisfied with their efforts at Kitty Hawk, and they left for home on October 23. Later, in a paper delivered to the Western Society of Engineers in 1901, Wilbur discussed the 1900 glider trials:

> Although the hours and hours of practice we had hoped to obtain finally dwindled down to about two minutes we were very much pleased with the general results of the trip, for

> setting out as we did, with almost revolutionary theories on many points, and an entirely
> untried form of machine, we considered it quite a point to be able to return without having
> our pet theories completely knocked in the head by the hard logic of experience, and our
> own brains dashed out in the bargain.[56]

One of the "revolutionary theories" was the concept of wing warping for lateral control; the
Wrights believed that the method had been proved a success by their 1900 glider tests.

What do you do when your glider is not producing enough lift? One obvious suggestion
would be to make the glider larger, and that was what the Wrights did. They returned in
July of the next year with a completely new machine. The 1901 glider had a wing area of
290 ft^2, almost 75% more than that of the previous year. (The Wrights must have believed
that such a large increase in wing area would go a long way toward compensating for the
factor-of-2 shortfall in lift of the 1900 glider.) The aerodynamic design of the wing reflected
the same data used for the 1900 glider. The Wrights were still depending on the Lilienthal
table, and the aspect ratio was 3.3 – even smaller than that for the 1900 glider. In an effort
to more closely replicate Lilienthal's findings, the Wrights increased the airfoil camber to
$\frac{1}{12}$; however, unlike Lilienthal, who dealt exclusively with circular-arc airfoils (maximum
camber at the $\frac{1}{2}$-chord location), the Wrights continued to place their maximum camber
near the leading edge. A crude sketch of their airfoil shape appears in Wilbur's diary for
July 27, 1901 (Figure 5.5 b). In the course of their 1901 flight trials they found that the
excessive center-of-pressure movement over the $\frac{1}{12}$-cambered airfoil caused major problems
in controlling the pitch of the glider. For any cambered airfoil, as the angle of attack is
progressively decreased, the center of pressure will first move forward on the airfoil, and
then reverse itself. For a high camber such as $\frac{1}{12}$, the angle of attack at which the reversal
will occur will be relatively high. The Wrights found that to be unacceptable, and they
changed the camber to a smaller value, on the order of $\frac{1}{19}$. They had constructed the wings
such that there was a convenient way to change the camber, so that they could do so at will.
With that smaller camber, the glider handled better and responded faster.

However, compared with their calculations based on the Lilienthal table, the lift generated
by the 1901 glider was too small; the discrepancy was essentially the same as in the case of
the 1900 glider. Wilbur's diary, July 29, contains the following entry: "Afternoon spent in
kite tests. Found lift of machine much less than Lilienthal tables would indicate, reaching
only about 1/3 as much" (p. 76).[55] The entry for the next day was more emphatic:

> The lift is not much over 1/3 that indicated by the Lilienthal tables. As we had expected
> to devote a major portion of our time to experimenting in an 18-mile wind without much
> motion of the machine, we find that our hopes of obtaining actual practice in the air are
> decreased to about one fifth of what we hoped, as now it is necessary to glide in order to
> get a sustaining speed. Five minutes' practice in free flight is a good day's record. We have
> not yet reached so good an average as this even [p. 77].[55]

The Wrights were encountering major discrepancies in drag as well. On one hand,
the measured total drag for the 1900 glider was about one-half of what the Wrights had
calculated, based in part on formulas provided by Chanute for the calculation of the "head
resistance of the framing." On the other hand, the drag for the 1901 glider was higher
than expected; the drag for the framing alone was about double the calculated value. The
Wrights had reason to be discouraged.

The Wrights were particularly concerned about the discrepancies between their calculations of lift and the measured values. Given our modern understanding of aerodynamics,

the reasons the Wrights' calculations gave lift values larger than the measured values by more than a factor of 2 are quite clear. There were three such reasons:

First, when converting the aerodynamic coefficients from the Lilienthal table to actual values for the lift and drag of their gliders, the Wrights used the wrong value for Smeaton's coefficient. The classic value for Smeaton's coefficient had been handed down over 150 years as $0.005 \, (\text{1b} \cdot \text{ft}^{-2})(\text{h} \cdot \text{mile}^{-1})^2$ or, in metric terms, $0.13 \, \text{kgf} \cdot \text{s}^2 \cdot \text{m}^{-4}$. We have discussed how that value was in question even at the time it was published. Much later, Langley published a more accurate value of $0.003 \, (\text{1b} \cdot \text{ft}^{-2})(\text{h} \cdot \text{mile}^{-1})^2$ or, in metric terms, $0.08 \, \text{kgf} \cdot \text{s}^2 \cdot \text{m}^{-4}$. Langley's value was 40% smaller than the classic value and was very close to the modern value. The Royal Aeronautical Society gives the modern value as $0.00289 \, (\text{1b} \cdot \text{ft}^{-2})(\text{h} \cdot \text{mile}^{-1})^2$. Langley highlighted his many repetitive determinations of Smeaton's coefficient in *Experiments in Aerodynamics,* and the Wrights must have seen it. However, they were more strongly influenced by Lilienthal and Chanute, who had continually invoked the classic value in their books. Lilienthal frequently cited its value as $0.13 \, \text{kgf} \cdot \text{s}^2 \cdot \text{m}^{-4}$ in *Birdflight,* but as discussed earlier, Lilienthal presented his data in such a form that Smeaton's coefficient was divided out – the wrong value for Smeaton's coefficient did not compromise Lilienthal's table of normal-force and axial-force coefficients. However, it is clear that the Wrights did not understand that. They clearly believed that the value of $0.13 \, \text{kgf} \cdot \text{s}^2 \cdot \text{m}^{-4}$ had been used by Lilienthal to obtain his force coefficients and that for consistency they should use the same value when employing the Lilienthal table. That opinion was reinforced by reading Chanute, who used the classic value of $0.005 \, (\text{1b} \cdot \text{ft}^{-2})(\text{h} \cdot \text{mile}^{-1})^2$ throughout his calculations in *Progress in Flying Machines.* In their early stages of designing gliders, the Wrights clearly felt more comfortable in following the example of Lilienthal and Chanute in using the classic value for Smeaton's coefficient. The Wrights used the Lilienthal aerodynamic coefficients and simply multiplied them by 0.005; they should have used 0.003. That error by itself caused their calculations of lift to be too high by a factor of 1.67. This is not new information. Jakab[24] and Culick and Jex[57] pointed out that the Wrights' use of the wrong Smeaton coefficient explained a major part of the discrepancy between their calculations of lift and their measurements of lift in 1900 and 1901. But there were mistakes concerning two other aerodynamic phenomena that contributed to the error in the Wrights' calculations. Those two additional factors, so far as I can determine, have not been discussed to any extent in relation to the Wrights' work.

The second source of error was that the aspect ratios for the wings of the Wrights' gliders (AR = 3.5 for the 1900 glider, and AR = 3.3 for the 1901 glider) were considerably smaller than the aspect ratio of 6.48 for the wing used by Lilienthal to collect his data, as subsequently entered in the Lilienthal table. The aspect ratio affects the value of the lift coefficient: The lower the aspect ratio (everything else begin equal), the lower the lift coefficient. Hence, the coefficients in the Lilienthal table pertain *only* to a wing of aspect ratio 6.48, and by using those coefficients without correcting for the aspect ratio, the Wrights were using coefficients for their gliders that were too large. The Wrights apparently did not realize that, and even if they had they would not have known how to make the correction. The proper correction formula for aspect ratios was first derived by Prandtl 15 years later.

Let us examine the magnitude of that source of error. The lift curves for two wings with different aspect ratios are sketched in Figure 5.6. The upper line is for AR = 6.48, which pertains to the Lilienthal wing; the lower line is for AR = 3.5, which pertains to the Wright wing. In their preliminary calculations for their 1900 glider, the Wrights focused on an angle of attack of 3°. In Wilbur's presentation to the Western Society of Engineers

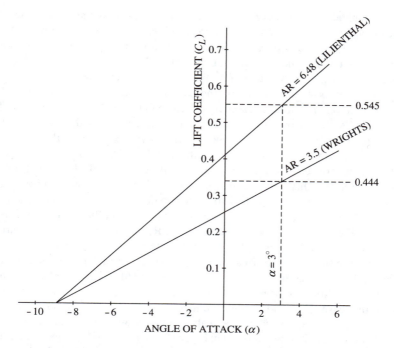

Figure 5.6 Isolated effect of an aspect-ratio difference on the lift coefficient obtained from the Lilienthal table and applied to the Wrights' 1900 glider.

later in 1901, he stated, concerning their 1900 glider, "we were compelled to make it only 165 square feet in area, which according to the Lilienthal tables, would be supported at an angle of three degrees in a wind of about 21 miles per hour." From the Lilienthal table, at an angle of attack of 3°, the lift coefficient C_L is 0.545; of course, that value pertains to the Lilienthal wing, with an aspect ratio of 6.48, as shown in Figure 5.6. Using the aspect-ratio correction formula obtained from Prandtl's lifting-line theory, we can readily adjust that value to pertain to the 3.5 aspect ratio of the Wrights. The calculation is carried out in Appendix G, and it shows that the lift coefficient obtained from the Lilienthal table should be reduced by a factor of 0.814 to pertain to the Wrights' 1900 glider. Hence, at a 3° angle of attack, the aspect-ratio effect reduces the lift coefficient to $(0.814)(0.545) = 0.444$, as shown in Figure 5.6. That is, to account for the aspect-ratio effect, the Wrights should have reduced the values obtained from the Lilienthal table by about 19%.

Although the Wrights did not know how to adjust Lilienthal's coefficients for different aspect ratios, they should have been aware of the effect. We have emphasized that one of Langley's contributions from his aerodynamic measurements was the demonstration that a lower aspect ratio would reduce lift. That effect was emphasized by Langley in *Experiments in Aerodynamics,* which the Wrights had available to them. Why did the Wrights not pay attention to that finding by Langley? There is no mention in the Wrights' correspondence of Langley's aspect-ratio experiments. I suspect that the answer lies in the relative technical immaturity of the Wrights at that time in regard to aerodynamics. Neither Lilienthal nor Chanute mentioned aspect-ratio effects in their writings, and it is clear that the Wrights were most strongly influenced by those two "authorities." Also, at that time the Wrights were using the Lilienthal table as if it were a listing of some kind of universal values; they clearly were not aware that it did not apply to all situations. It is easy to understand how the Wrights could have paid no heed to aspect-ratio effects at that early stage in their understanding of aerodynamics, although had they done their homework a little more thoroughly, they might have noticed the message that was there for all to see in Langley's data.

The third source of error had to do with the fact that the airfoils for the 1900 and 1901 gliders had the location of maximum camber very close to the leading edge (Figure 5.5a-b), whereas Lilienthal's airfoil, on which the Lilienthal table was based, was a circular-arc shape, with maximum camber at the $\frac{1}{2}$-chord location (Figure 5.5c). Because of their radically different maximum-camber locations, the Wrights' airfoils and Lilienthal's airfoil were aerodynamically different – another reason why the data from the Lilienthal table were not applicable to the Wrights' gliders. At the time, the Wrights did not realize that. The aerodynamic differences between such airfoils have to do with the angle of attack that will produce zero lift, denoted by $\alpha_{L=0}$. The variation of the lift coefficient with the angle of attack is shown in Figure 5.7 for three different airfoils. Recall that the lift varies linearly with the angle of attack until stall is approached. Also note that the angle of attack at which there is zero lift is, by definition, the zero-lift angle of attack $\alpha_{L=0}$. For a flat plate, $\alpha_{L=0}$ is zero, as shown in Figure 5.7. For a cambered airfoil, positive lift is still generated at a zero angle of attack, and such an airfoil must be pitched to some negative angle of attack if it is to produce no lift, and in that case the zero-lift angle of attack $\alpha_{L=0}$ is negative. The value of $\alpha_{L=0}$ depends on the magnitude of the camber and also on the location of the maximum camber. Consider the Lilienthal and Wright airfoils in Figure 5.7, each with $\frac{1}{12}$ camber (for purposes of discussion). The lift curves for both cambered airfoils are shifted to the left relative to that for the flat plate. However, the airfoil that has its maximum camber farthest away from the leading edge will be shifted the most. Hence the Lilienthal airfoil is shifted farther than the

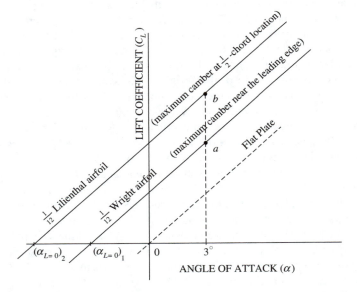

Figure 5.7 Schematic showing how the location of the maximum camber affects the lift coefficient for an airfoil.

Wright airfoil. That is, $(\alpha_{L=0})_2$ for the Lilienthal circular-arc airfoil is a larger-magnitude negative number than is $(\alpha_{L=0})_1$ for the Wright airfoil. The implication of that is as follows: Imagine both airfoils at the same angle of attack, say 3°. The lift coefficient for the Wright airfoil is given by point *a* in Figure 5.7, and the lift coefficient for the Lilienthal airfoil is higher, given by point *b*. Therefore, when the Wrights used Lilienthal's table for any given angle of attack, they read lift coefficients that were too high for their airfoil shape.

In order to estimate just how high their readouts were, we shall consider some data on the shift of $\alpha_{L=0}$ due to a shift in the location of the maximum camber for several families of modern airfoils.[37] On the average, those data show that when the location of the maximum camber is moved forward from $0.5c$ to $0.2c$ (*c* is the chord length), the value of $\alpha_{L=0}$ decreases by about 0.2° for each 1% of camber. A camber of $\frac{1}{22}$ is 8.3% camber; Hence the corresponding shift in $\alpha_{L=0}$ is $(8.3)(0.2) = 1.7°$. Figure 5.8, an uncluttered version of Figure 5.7, shows the shift $\Delta\alpha_{L=0} = 1.7°$. Because we are simply isolating the effect of the location of the maximum camber, we assume that the two lift curves have the same aspect ratio and thus have the same slope. In Figure 5.8, we assume the value of the lift slope is that which pertains to the Wrights' wing, obtained from equation (G.1) in Appendix G as 0.065 per degree. From the geometry shown in Figure 5.8, for a shift in the zero-lift angle of attack of $\Delta\alpha_{L=0} = 1.7°$, the corresponding change in the lift coefficient at a given angle of attack is $\Delta C_L = (0.065)(1.7) = 0.11$. Hence, the Wrights should have reduced the lift coefficient obtained from the Lilienthal table by an amount of 0.11 just to take into account the different location of the maximum camber.

Of course, the Wrights hadn't a clue about what effect the location of the maximum camber would have on the lift coefficient. I suspect that they assumed that it would have no effect. Indeed, the reason the Wrights placed the maximum camber close to the leading

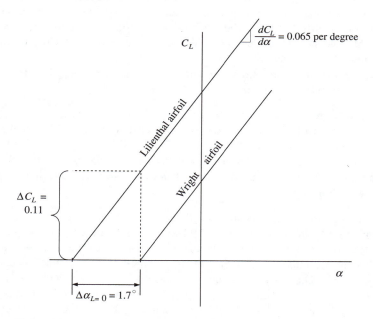

Figure 5.8   Isolated effect of the location of the maximum camber on the lift coefficient obtained from the Lilienthal table and applied to the Wrights' 1900 glider.

edge had to do with the movement of the center of pressure; there is no evidence that they were concerned about the effect on lift. In his 1901 paper, Wilbur clearly stated the rationale for the airfoil design used with their gliders:

> In deeply curved surfaces the center of pressure at 90 degrees (angle of attack) is near the center of the surface, but moves forward as the angle becomes less, till a certain point is reached, varying with the depth of curvature. After this point is passed, the center of pressure, instead of continuing to move forward, with the decreasing angle, turns and moves rapidly toward the rear. The phenomena are due to the fact that at small angles the wind strikes the forward part of the surface on the *upper* side instead of the lower, and thus this part altogether ceases to lift, instead of being the most effective part of all, as in the case of the plane. Lilienthal had called attention to the danger of using surfaces with a curvature as great as one in eight, on account of this action on the upper side; but he seems never to have investigated the curvature and angle at which the phenomena entirely cease. My brother and I had never made any original investigation of the matter, but assumed that a curvature of one in twelve would be safe, as this was the curvature on which Lilienthal based his tables. However, to be on the safe side, instead of using the arc of a circle, we had made the curve of our machine very abrupt at the front, so as to expose the least possible area to this downward pressure.[56]

Clearly, the Wrights believed that the choice for the location of the maximum-camber point should be determined by center-of-pressure considerations, and they must have assumed that any effect on lift was secondary. On that point, the Wrights made a serious error of omission, but that is understandable, for at that time nobody knew about that effect on lift. That was another illustration of their early acceptance of the Lilienthal table as a set of universal values for cambered airfoils. Such an impression could easily have been gained from Chanute's 1897 article in the *Aeronautical Annual,* wherein he reproduced Lilienthal's table with only the qualification that it applied to a camber of $\frac{1}{12}$; he did not mention wing shape, aspect ratio, or airfoil shape. The Wrights would soon realize the limitations of the Lilienthal table; we saw in Chapter 4 that Wilbur gently reproached Chanute for that situation in a letter.

In summary, the Wrights made three significant errors in calculating the lift for their 1900 and 1901 gliders using the Lilienthal table:

(1)   They used the wrong value for Smeaton's coefficient.
(2)   They did not correct for the differences in aspect ratio between Lilienthal's wing and the wings of their gliders.
(3)   They did not account for differences in the location of the maximum camber between Lilienthal's circular-arc airfoil and their own airfoils with maximum camber near the leading edge.

The net effect of those three errors is shown in Figure 5.9. For an angle of attack of $3°$, the Lilienthal table gives the lift coefficient as 0.545. The aspect-ratio effect reduces that value to 0.44. The effect of the location of the maximum camber further reduces the value to 0.33. The Wrights used 0.545 to calculate the lift for their 1900 and 1901 gliders; they should have used 0.33. On that basis alone, their calculations of lift were in error by a factor of $0.33/0.545 = 0.60$. When the error in Smeaton's coefficient is factored into this calculation (the Wrights used 0.5, but should have used 0.3), the Wrights' lift calculations were in error by a factor of $(0.3/0.5)(0.60) = 0.36$. That is, the actual lift was lower than their calculated value by a factor of 0.36, or one-third, which was precisely what the

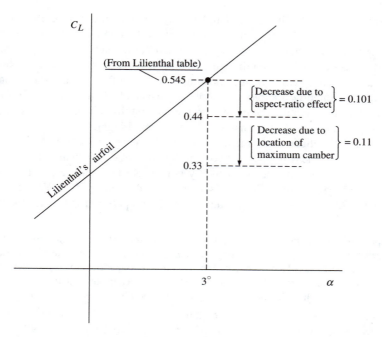

Figure 5.9   Cumulative corrections to the Lilienthal data for application to the Wrights' glider.

Wrights observed, as recorded in Wilbur's diary July 29, 1901: "Afternoon spent in kite tests. Found lift of machine much less than Lilienthal tables would indicate, reaching only about 1/3 as much." The discrepancy is explained by their errors in using the Lilienthal table, not inaccuracies in the table itself.

There is another aerodynamic consideration lurking in the background that we have not yet considered: the Reynolds-number effect. Ordinarily, at the high Reynolds numbers (Re) associated with airplane flight (Re values in the millions to hundreds of millions), the lift slope for an airfoil is essentially not sensitive to changes in Re. (The value of the *maximum* lift coefficient, which depends on the phenomenon of flow separation at stall, is dependent on Re, but that is not of concern here, because we are not dealing with the maximum lift coefficient in this discussion.) On the other hand, for exceptionally low values, say Re on the order of 100,000, the slope of the lift curve is noticeably less than at very high Re. We addressed that trend in Chapter 3 in our discussion of Cayley's aerodynamic measurements, where we noted the modern experimental observation that for Re on the order of 100,000, airfoil lift coefficients are about 70% of those measured at much higher Re. For Lilienthal's whirling-arm data, Re was 330,000 (based on a velocity of 12 m/s and a wing chord length of 0.4 m). For the Wrights' 1901 glider, Re was 1,100,000 (based on a velocity of 25 ft/s and a wing chord length of 7 ft). Clearly, the Wrights were in a higher Re regime than that for which the Lilienthal table applied. That should have given them an increase in lift over that calculated earlier. But by how much? The answer is not easy; simple analysis will not suffice. All that we can say with any conviction is that the difference will be smaller than that discussed in Chapter 3. I suspect that we are talking about an effect of approximately 10%, maybe less; and that will not change any of our earlier conclusions.

We note that the Wrights, like Lilienthal, Langley, and Chanute before them, made no mention of Reynolds-number effects in regard to any of their aerodynamic measurements or calculations. In 1901, no one working in applied aerodynamics had any familiarity with Reynolds-number effects, even though it had been 18 years since Osborne Reynolds had carried out his classic experiments showing the effect of Re on the transition from laminar flow to turbulent flow – another example of the lack of technology transfer from the academically educated scientific community to the community of self-educated engineers who were groping their way toward answers about powered flight.

At the end of their glider experiments of 1901, the Wrights were quite despondent, particularly because of the discrepancy between the amount of lift their calculations had promised and the meager amount they actually got. From the vantage point of modern aerodynamics, we have accounted for that discrepancy. But to what did the Wrights attribute the discrepancy? Some answers can be found in Wilbur's paper in 1901 discussing their 1900 glider:

> It appeared sadly deficient in lifting power as compared with the calculated lift of curved surfaces of its size. This deficiency we supposed might be due to one or more of the following causes: (1) That the depth of the curvature of our surfaces was insufficient, being only about 1 in 22, instead of 1 in 12. (2) That the cloth used in our wings was not sufficiently airtight. (3) That the Lilienthal tables might themselves be somewhat in error.[56]

Later in that same paper, Wilbur also pointed a finger at Smeaton's coefficient, stating that "the well-known Smeaton coefficient of $0.005 V^2$ for the wind pressure at 90 degrees is probably too great by at least 20 percent." In regard to Smeaton's coefficient, Wilbur was on the right track. Also, it was natural for the Wrights to begin to suspect the Lilienthal table; it was a key factor in their calculations, and obviously something was wrong with their calculated values. However, blaming the Lilienthal table was not the answer. Indeed, later the Wrights would vacillate back and forth on the question of the validity of the Lilienthal table, finally reaching the conclusion that the data in the table were reasonable.

Wind-Tunnel Experiments (1901–2): The Wrights Discover "the Right Aerodynamics"

Wilbur and Orville were in a state of despair when they left Kitty Hawk on August 20, 1901, to return to Dayton. The discrepancy in lift values was not their only problem. Their predictions of drag were also unreliable, as well as inconsistent. During the 1900 glider trials the measured drag on their machine had been lower than they had calculated. For example, in Wilbur's letter to Chanute on November 16, 1900, he noted that the measured drag had been 8 lb in a 20-mph wind. The Wrights followed a procedure suggested by A. M. Herring in an article entitled "Recent Advances toward a Solution of the Problem of the Century," published in the *Aeronautical Annual* in 1897, wherein the total drag on the machine was expressed as the sum of the drag on the wings plus the resistance due to the framing. The Wrights' measurement of the 8-lb total drag was less than the estimate of the framing drag alone. Also, Wilbur commented on the drag on the wings alone: "We found the drift of the surfaces [wings] under full load was greater than the Lilienthal tables would indicate, but it may be that this is due to the fact that our curve was only one in twenty-three instead of one in twelve" (p. 42).[55] (It is not clear how the Wrights singled out the wing drag from their total drag measurements.) A year later, the Wrights measured the drag on their 1901 glider and, in contrast to the previous year, found it to be higher than calculated. Wilbur's diary for July 27, 1901, noted that the "resistance of framing apparently much in excess of that estimated beforehand." On the same page, he noted that "at speed of 18 miles, resistance (from all causes) of machine alone about 15 lbs. We estimate a drift of surfaces [the wings] at 1/12 of 98 lbs. = 8 lbs. Leaving a head resistance of 7 lbs. for framing. This would indicate that the unfavorable results in total resistance observed in gliding were due rather to drift of surfaces than to resistance of framing, as first supposed, thus leading to doubts of the correctness of [the] Lilienthal table of ratio of lift to drift" (p. 71).[55] Clearly, the Wrights were groping blindly in regard to the problem of drag prediction, and that simply reinforced their sense of despair.

Even in modern aerodynamics, accurate prediction of drag is an imprecise science; it is no surprise that the Wrights had difficulty. They labored under a primitive and incomplete understanding of the basic physical mechanisms of drag production. For example, they had read Langley and Chanute claiming (incorrectly) that friction was negligible. They had no real conception of flow separation and its effect on drag, and the phenomenon of induced drag associated with vortices from the tips of a finite wing had not been observed at that time. (The low aspect ratios of the Wrights' 1900 and 1901 gliders led to fairly high induced drag.) It is no wonder that the Wrights were not able to make accurate calculations of drag.

On the encouraging side, the Wrights had come to an understanding of how the center of pressure would move with a change in the angle of attack for a cambered airfoil. In 1901 they designed their glider so that the airfoil camber could be readily changed in the field. They found that an airfoil with a camber of $\frac{1}{12}$ had undesirable center-of-pressure behavior, but they mitigated that effect by going to smaller camber. Also, they may have received some help from Edward Huffaker, who joined the Wrights at their Kill Devil Hills camp for a short period during the 1901 glider experiments. Huffaker had earlier worked for Langley and had carried out a series of aerodynamic measurements on cambered airfoils. Although Langley had never published those data, Huffaker must have brought the Wrights some corporate knowledge of that work, including some appreciation for center-of-pressure movement and reversal. (There is, however, no mention of any such technology transfer in the Wrights' papers. The Wrights, like Langley before them, soon found Huffaker to be a rather unmannered and unlikable fellow, who later became unwelcome at the Wrights' camp.)

Also during that time, Wilbur published his first technical paper, and in a most prestigious journal: *The Aeronautical Journal* (July 1901) of the Aeronautical Society of Great Britain (now the Royal Aeronautical Society). Entitled "Angle of Incidence," it addressed the ambiguity surrounding the term "angle of incidence" in the existing aeronautical literature. Some investigators defined "angle of incidence" as the angle between the chord line of a wing and the horizontal, using that as the angle of attack for the wing. Wilbur recognized, quite rightly, that the aerodynamically significant angle was the angle between the chord line and the relative wind direction, and he made a case for unambiguously defining that angle as the "angle of incidence." That definition prevails today, except that the modern terminology is "angle of attack" rather than "angle of incidence." Wilbur ended his paper as follows: "*Rule:* The angle of incidence is fixed by area, weight, and speed alone. It varies directly as the weight, and inversely as the area and speed, though not in exact ratio." Wilbur was correctly expressing a basic fact of airplane aerodynamics. Figure 5.10 illustrates an airplane in steady, climbing flight. The flight path is inclined to the horizontal at the angle θ, called the *climb angle*. The relative wind direction felt by the airplane is the same as

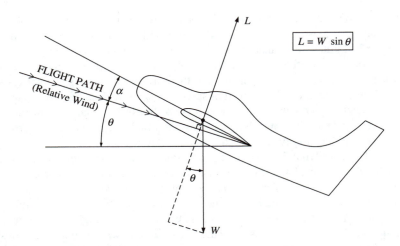

Figure 5.10   Illustration of some of the forces on an airplane in climbing flight.

the flight-path direction. The angle between the chord line of the wing and the relative wind is the angle of attack α. In some of the aeronautical literature contemporary with the Wrights, the combined angle $\theta + \alpha$ would have been called the "angle of incidence." Wilbur was arguing that the only angle of significance for the aerodynamic properties of the airplane was α, and it should be defined as the "angle of incidence." In Figure 5.10, for steady, climbing flight at the climb angle θ, the lift is balanced by the component of weight in the direction opposite to the lift (the sum of forces perpendicular to the flight path must be zero). That is,

$$L = W \sin \theta$$

The lift can be expressed in terms of the lift coefficient C_L:

$$L = \tfrac{1}{2} \rho_\infty V_\infty^2 S C_L$$

Combining those two equations,

$$\tfrac{1}{2} \rho_\infty V_\infty^2 S C_L = W \sin \theta$$

Solving for C_L,

$$C_L = \frac{2W \sin \theta}{\rho_\infty V_\infty^2 S}$$

The lift coefficient is a function of the angle of attack α: $C_L = f(\alpha)$. For a fixed C_L, there is a fixed angle of attack α:

$$f(\alpha) = \frac{2W \sin \theta}{\rho_\infty V_\infty^2 S}$$

That equation provides direct mathematical proof of Wilbur's "rule." It says that "the angle of incidence [α] is fixed by area [S], weight [W], and speed [V_∞] alone. It varies directly as the weight, and inversely as the area and speed, though not in exact ratio." It appears that Wilbur Wright was the first to formulate and publish that relationship, which is pivotal to an analysis of airplane performance.

Clearly, the Wrights' experience with the applied science of aerodynamics through the end of their glider tests in 1901 was a checkered affair – some enlightenment, but also a lot of confusion. That confusion was evidenced by a statement in Wilbur's diary on July 30, 1901, regarding their measurements of drag: "Indications are that the total resistances decrease with increased velocity of wind." That was not necessarily a confirmation of Langley's power law (see Chapter 4), because power is equal to drag times velocity; it remains to be seen whether or not, if Wilbur had made that calculation, the power required would also have decreased with velocity. However, the Wrights knew about Langley's power law, and in that light the measured drag variation quoted by Wilbur must not have been too surprising to him. At the relatively low airspeeds they were dealing with in 1900 and 1901, some of their data points could very well have been on the back side of the power curve. Indeed, considering only the drag itself for steady, level flight, where lift equals weight, there is, for every flying machine, a portion of the drag-versus-velocity curve where the drag actually decreases as the velocity increases – a variation analogous to the back side of the power curve. That portion of the drag curve is found only at low airspeeds, and that situation is avoided by pilots – it is a somewhat unstable regime of flight. The Wrights' glider tests may have partially spanned that regime, which would account for Wilbur's observation that drag decreased with increased velocity. However, just as for Langley, that must have been counterintuitive to the Wrights and most likely was a source of additional confusion.

Things were so bad that on the train back to Dayton in August 1901, Wilbur told Orville that men would not fly for another fifty years. Later, in the 1940s, Orville embellished that for Fred Kelly, one of his biographers: "not within a thousand years would man ever fly." The Wrights were clearly disappointed and discouraged.

But their discontent would soon fade. Wilbur and Orville made a bold decision that would turn things around. Problem: They had been using the best aerodynamic data available at that time (at least to the extent that they understood the data), and yet their gliders had been performing considerably below par (in comparison with their calculations). Conclusion: Something must be wrong with the existing data. Solution: Start over, and compile their own aerodynamic data. That decision – essentially to do everything themselves, starting with the very basics – was a primary reason for the success of the Wright Flyer in 1903.

Also at that time, they received invaluable moral support from Octave Chanute – tangible moral support. In a letter to Wilbur dated August 29, 1901, Chanute invited Wilbur to give a presentation to the Western Society of Engineers – a show of interest and confidence in the value of the Wrights' work:

> I have been talking with some members of the Western Society of Engineers. The conclusion is that the members would be very glad to have an address, or a lecture from you, on your gliding experiments. We have a meeting on the 18th of September, and can set that for your talk. If you conclude to come I hope you will do me the favor of stopping at my house. We should have the photos you want to use about a week before the lecture in order to get lantern slides. The more the better. Please advise me [p. 91].[55]

Wilbur accepted Chanute's invitation. Only 13 days after boarding the train to return home from Kitty Hawk, shrouded in a cloud of gloom, the Wrights had a new focus.

Wilbur's paper, "Some Aeronautical Experiments,"[56] is reprinted in McFarland's collection of the Wright papers.[55] It is well worth reading, showing Wilbur to have been an accomplished technical writer and a practitioner of Midwestern humor. Only a small part of the paper dealt with aerodynamics, but Wilbur did not shy from disclosure of the discrepancies between their calculations and their experiments for both lift and drag, nor from offering possible explanations for those discrepancies.

Whether or not the Lilienthal table was accurate, the important point at that time was that the Wrights *thought* that it might not be accurate. The first written indication of their suspicions appeared in Wilbur's diary, July 27, 1901, where he referred to the discrepancies between the drag calculations and the measurements: "... thus leading to doubts of the correctness of [the] Lilienthal table of ratio of lift to drift." Later, in his paper to the Western Society of Engineers, he suggested that "the Lilienthal tables might themselves be somewhat in error." Their concern about the accuracy of the Lilienthal table was one of their strongest motivations for compiling their own aerodynamic data. Another factor was their uncertainty about the value of Smeaton's coefficient. Wilbur questioned its accuracy in his paper to the Western Society. Also, in a letter to Chanute on September 26, 1901, Wilbur noted that "Prof. Langley and also the Weather Bureau officials found that the correct coefficient of pressure was only about 0.0032 instead of Smeaton's 0.005." Those doubts, whether reasonable or unfounded, forced the Wright brothers to conduct their own experiments in aerodynamics. During those experiments they found what constituted "the right aerodynamics." Their wind-tunnel test program in the fall of 1901 and winter of 1902 was the decisive factor that led to a new, remarkable successful glider in 1902, and in the next year to the powered Wright Flyer and even greater success. Without those wind-tunnel tests, human flight might have been considerably delayed, and it might not have involved the Wright brothers. In that light, it was perhaps a good thing that the Wrights did not use the Lilienthal table correctly and that they used the wrong value for Smeaton's coefficient, for without the discrepancies between their calculations and their measurements on their 1900 and 1901 gliders, they might not have been driven to discover "the right aerodynamics."

Their initial concern was to determine whether or not Lilienthal's data were accurate. Toward that end, they first attempted to take some comparison measurements by mounting a small model wing with a cambered airfoil on the rim of a bicycle wheel, with a flat plate at a location on the rim 90° from the wing, the plate being oriented perpendicular to the wind direction (Figure 5.11). When the lift on the wing equaled the drag on the plate, the two moments about the center of the wheel would be equal and opposite; in that case the wheel would be balanced (i.e., it would not rotate). Wilbur's objective was stated in a letter to Chanute, September 26, 1901:

> I am arranging to make a positive test of the correctness of the Lilienthal coefficients at from 4°–7° in the following manner. I will mount a Lilienthal curve of 1 sq. ft. and a flat plane of 0.66 sq. ft. on a bicycle wheel in the position shown [Figure 5.11]. The view is from above. The distance from the centers of pressure to center of wheel will be the same for both curve and plane. According to Lilienthal's tables the 1 sq. ft. curve at 5° will just about balance the 0.66 sq. ft. plane at 90°. If I find that it really does so no question will remain in my mind that these tables are correct. If the curve fails to balance the plane I will cut down the size of the plane till they do balance [p. 121].[55]

The Wrights first tested that apparatus in the natural wind, with no conclusive findings. They then mounted the wheel rim horizontally on the front of a bicycle and pedaled the apparatus through the air. The semiquantitative findings from those tests were described in detail in a letter from Wilbur to Chanute, October 6, 1901. The major purpose of the tests was to compare the lifting action of a curved airfoil with that of a flat plate and to determine the accuracy of the Lilienthal table. They calculated, from the Lilienthal table,

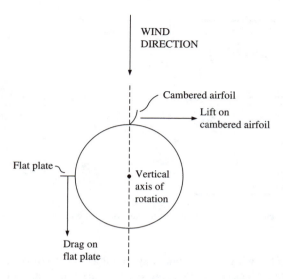

WIND
DIRECTION

Cambered airfoil

Lift on
cambered airfoil

Flat plate

Vertical
axis of
rotation

Drag on
flat plate

Figure 5.11   Schematic of the Wrights' bicycle-wheel balance device.

that to balance the drag on the flat surface, the curved surface would have to be at a 5°
angle of attack, but their measurements showed that an 18° angle of attack was necessary –
indicating that the lift on the curved surface was less than half of that predicted by using
the Lilienthal table. That was consistent with their findings from the glider tests, for the
reasons we have already discussed. When the Wrights replaced the curved surface with a
flat plate, they found that the plate had to be pitched to an even higher angle of attack (24°)
in order to balance the wheel rim. Like Phillips, Lilienthal, and Langley before them, with
that test the Wrights simply demonstrated once again the superiority of a cambered airfoil
over a flat plate.

Wilbur's October 6 letter to Chanute was important; it recorded some watershed decisions
that turned the Wrights in the direction of success. For example, after his description of the
wheel-rim tests, Wilbur stated that the findings had prompted them to search carefully for
possible sources of error in their glider tests at Kitty Hawk. They decided that the source
of the error was "in the use of the Smeaton coefficient of 0.005, or 0.13 metric system."
Wilbur then decided to adopt the value measured by Langley: "I see no good reason for
using a greater coefficient than 0.0033." From that time on, the Wrights used that reasonably
accurate value for Smeaton's coefficient.

Another important topic discussed in the October 6 letter was their venture into wind-
tunnel testing: "Although the curve was found to be far less effective than Lilienthal's table
would indicate, it was so much in excess of the plate that we considered it important to
obtain tests of greater exactness at smaller angles" (p. 124).[55] Wilbur went on to describe
a small, makeshift wind tunnel that Orville constructed from an old starch box and a small
fan turning at 4,000 revolutions per minute; the whole apparatus was only 18 in. long.
That small tunnel was transitional, used for only one day, but it was an important stepping-
stone to their next generation of aerodynamic testing equipment. The Wrights had devised
an innovative technique for comparing two different wing surfaces in the small tunnel
(Figure 5.12): Two wings were mounted on a bracket in a vertical fashion, one above the

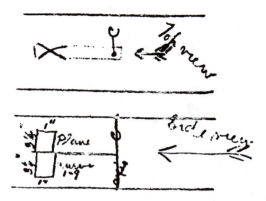

Figure 5.12 Wilbur's sketch of a simple balance mechanism for use in their first wind tunnel. That wind tunnel was used for only one day.

other (Figure 5.12, bottom). One wing was mounted at an angle to the left of the center plane, and the other to the right at the same angle (Figure 5.12, top). The bracket on which the wings were mounted pivoted about a vertical rod, labeled C, much like a weathervane. When the whole apparatus was placed in the airflow in the tunnel, the vane would rotate to an angular position about the rod C such that the amounts of lift generated by the two wings would be equal, but in opposite directions. With two different wings, one mounted above the other, the vane would rotate to a position where the less efficient lifting surface would always be at a higher angle of attack. Thus the Wrights were able to carry out comparison tests between different airfoils. With that apparatus, in one day the Wrights were able to confirm "the correctness of Lilienthal's claim that curved surfaces lift at negative angles." However, again the Wrights' quantitative predictions based on the Lilienthal table did not agree with their experimental findings. With those findings from the makeshift wind tunnel in hand, Wilbur's October 6 letter revealed a new confidence: "I am now absolutely certain that Lilienthal's table is very seriously in error, but that the error is not so great as I had previously estimated" (p. 127).[55] Perhaps the most important point in Wilbur's October 6 letter was his announcement of their decision to build a more sophisticated wind tunnel:

> The results obtained, with the rough apparatus used, were so interesting in their nature, and gave evidence of such possibility of exactness in measuring the value of ($P_{(\text{tang} \cdot \alpha)} / P_{90}$) [essentially the lift coefficient], that we decided to construct an apparatus for making tables giving the value of ($P_{(\text{tang} \cdot \alpha)} / P_{90}$) at all angles up to 30° and for surfaces of different curvatures and different relative lengths and breadths [pp. 126–7].[55]

That was a dramatic step by the Wright brothers. They were saying that any comparative assessments of Lilienthal's table were behind them; they were discarding Lilienthal's table. Instead, they were going to compile their own detailed tables of aerodynamic coefficients – a whole series of tables for different airfoil shapes and wing planform shapes, thus going well beyond the Lilienthal table.

Wilbur's letter of October 6 to Chanute was indeed a watershed document. There Wilbur revealed a new maturity of approach and technical self-confidence well beyond anything seen in their previous correspondence. By then, they were using a reasonably correct value for Smeaton's coefficient, and using it with conviction. And they were convinced that the wind tunnel could yield accurate, detailed aerodynamic data. They were ready

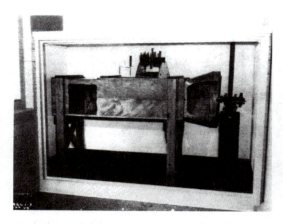

Figure 5.13   The Wrights' second wind tunnel, in which virtually all of their wing and airfoil data were obtained.

to produce new tables of aerodynamic coefficients, to make an original contribution to applied aerodynamics. Most important, those data would be directly applicable to their future glider designs. The Wrights were beginning to exhibit the philosophical approach so clearly delineated in the study by Jakab,[24] namely, that of the engineer who focuses on obtaining only those data necessary to accomplish a specific aim: "The goal was less to understand *why* in principle these forces behaved as they did than to learn *how* in actual practice they acted with respect to one another, and in turn to use this information to construct a successful flying machine. This was engineering in its most basic form and the supporting foundation for all other aspects of the Wrights' inventive method" (pp. 1–2).[24]

By October 6, 1901, the Wrights were excited about the prospects for their wind-tunnel tests. The new wind tunnel was built and operating by mid-October (Figure 5.13). The flow duct was 6 ft long, with a square cross section 16 in. on a side. There was a glass window on top for observing the tests. The airflow was driven by a fan powered by the central power plant of the Wrights' bicycle shop, a 1-hp gasoline engine connected to the fan via shafts and belt drives. The maximum velocity attainable in the wind tunnel was about 30 mph. The tunnel was housed on the second floor of the bicycle shop, where all the testing took place. With their new tunnel the Wrights were following in the tradition of Wenham, Phillips, and a dozen other developers who had built and operated wind tunnels, but the Wrights' tunnel featured some improvements:

> The Wright wind tunnel of 1901 . . . marked a significant departure from prior instruments. Besides being well-designed and producing accurate results, it was the first used to systematically collect specific data on a wide range of prospective wing shapes to be used in conjunction with the established lift and drag equations. Moreover, this was the first time anyone had used such an instrument to obtain aerodynamic data in a form that could be incorporated directly into the design of an actual aircraft. Aside from the technical superiority of the tunnel, it was the manner in which the brothers used their device that made it so much more effective than anything that had preceded it [p. 127].[24]

The Wrights' tunnel and the instrumentation used for their aerodynamic measurements have been described in detail.[24] The ingenuity shown by the Wrights in the design of their force balances deserves some comment here. What looked like a rather crude and lashed-up

Figure 5.14   The Wrights' clever device for measuring the lift coefficient directly.

device was, in reality, a precise instrument for measuring lift coefficients (Figure 5.14). The wing model where lift coefficient was to be measured was mounted vertically, and the entire device was mounted in the tunnel such that the flow direction was parallel to the left and right sides of the wood base (or perpendicular to the ruler shown resting on the base). The four "fingers" hanging below the wing were flat surfaces oriented perpendicular to the flow. The mechanical details of the balance have been described by Jakab.[24] Suffice it here to say that the lift force on the wing and the drag on the flat surfaces created torques about the vertical pivot rods on the instrument. With the airflow turned on, there would be one angular position of the pivot rods where those torques would be balanced. An indicator on the bottom of the device would register that angle. A mathematical calculation based on the torque balance will show that the lift coefficient is equal to the sine of the indicated angle (see Jakab[24] for the details). Note that the device registered the *lift coefficient* directly, not the lift force. When the drag on the flat surfaces oriented perpendicular to the flow was properly combined with the lift generated on the wing, the effects of velocity, surface area, and (most important of all) Smeaton's coefficient were completely divided out of the measurement. Thus the Wrights had designed a force balance that directly measured the lift coefficient in such a manner that the measurement was independent of Smeaton's coefficient and the airflow velocity. They did mechanically what Lilienthal had done mathematically in the presentation of his data.

The Wrights built a second balance that would directly measure the drag-to-lift ratio D/L (Figure 5.15). The wing model was mounted vertically on one of the arms of the balance. An angular measurement of the orientation of the balance was the basic determination, and from that the D/L ratio could be obtained directly. Again, the measurement was independent of velocity and Smeaton's coefficient. Knowing the lift coefficient (from the first balance) and the D/L ratio (from the second balance), the drag coefficient C_D fell out directly from the relation

$$C_D = (D/L)C_L$$

From mid-October to December 1901 the Wrights tested more than 200 different wing models with different planform and airfoil shapes (Figure 5.16). They tested camber ratios from $\frac{1}{6}$ to $\frac{1}{20}$; the locations of maximum camber ranged from near the leading edge to the $\frac{1}{2}$-chord position. The planform shapes included squares, rectangles, ellipses, surfaces

Figure 5.15   The Wrights' instrument for direct measurement of the D/L ratio.

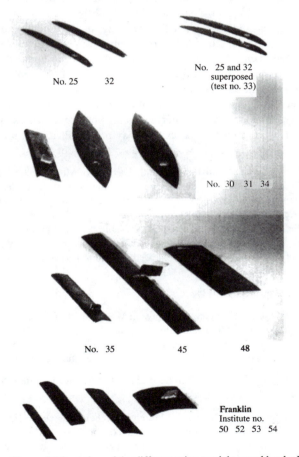

No. 25   32

No. 25 and 32
superposed
(test no. 33)

No. 30   31   34

No. 35   45   **48**

Franklin
Institute no.
50 52 53 54

Figure 5.16   A few of the different wing models tested by the Wrights in the wind tunnel.

with raked tips, and circular-arc segments for leading and trailing edges meeting to form sharp points at the tip. They also examined tandem wing configurations (as in Langley's aerodromes), biplanes, and triplanes. Finally they had to end those experiments because of the press of business. A letter from Katharine to their father, December 7, 1901, announced that "the boys have finished their tables of the action of the wind on various surfaces, or rather they have finished their experiments. As soon as the results are put into tables, they will begin work for next season's bicycles" (p. 171).[55] Those experiments, conducted over a period of less than two months, produced the most accurate and practical aerodynamic data on wings and airfoils thus far. They epitomized applied aerodynamics, in that the data were obtained specifically to provide aerodynamic information from which to design a flying machine.

The Wrights tabulated the data for 48 of their wing models. One table showed lift coefficients versus angles of attack α. Another gave D/L ratios as functions of α. The Wrights' tables supplanted the Lilienthal table in all respects. At that time, they held the most valuable technical data in the history of applied aerodynamics.

Unfortunately, during the critical period in the birth of the airplane, when those tables could have found widespread use, they were known only to the Wrights and Chanute. That does not appear to have been the original intention of Wilbur and Orville. They had been feeding Chanute bits and pieces of their lift and drag measurements during the course of their wind-tunnel experiments. In a letter to Wilbur dated November 27, 1901, Chanute mentioned that he had been asked to write a new chapter for an updated version of Moedebeck's handbook for aviators – where the Lilienthal table had first been published in 1895. Chanute gently hinted that some of the Wrights' data might be included in his contribution to the handbook. Chanute went on to make Wilbur an offer: "Now, I will either prepare the whole of the notes, including your experiments, or prepare the notes up to the latter point, and let you describe your own work, as you may prefer. Please let me know" (pp. 165, 168).[55] Wilbur's rather modest reply was penned on December 1:

> I think very well of your plan to republish the Lilienthal section of Moedebeck's handbook substantially in its present form and add your own notes as a supplementary article . . . . You will be better able than we to preserve a proper perspective in describing our experiments, so you had better keep the matter in your own hands. It is a question whether any table additional to his should be inserted, but if deemed advisable it should preferably be of a surface of a markedly different character so that instead of contradicting Lilienthal it should emphasize the necessity of considering shape, relative dimensions, and profile in calculating the expected performance of a machine [p. 172].[55]

Chanute agreed with Wilbur, responding on December 11:

> I quite agree with you that it will be preferable to give in Moedebeck's handbook the coefficients for a surface differing markedly from Lilienthal's, and to emphasize the necessity of considering shape. Please furnish me the necessary data and comments when you consider that you have arrived at such definite results as to warrant publication [p. 172].[55]

Wilbur did not answer until January 5, 1902, when he included in a letter to Chanute tables of aerodynamic data for 17 different models:

> In a recent letter you inquired what of our tables I thought ought to be given to Moedebeck. My failure to answer sooner was for the simple reason that I did not know what to say. On the one hand, the value of our tables lies chiefly in the opportunity they afford of comparisons of the effect of aspect, curvature, thickening, and chord, upon the lift, the tangential, and

the angle at which the maximum occurs. But on the other, there is the objection that to include all would make more than would be advisable in so brief a work as a handbook. And then there is the further and greater objection that to insert in an authoritative work like a handbook a set of tables which are not claimed to be perfect, in advance of their general public acceptance, would entail a personal responsibility on your part which ought not to be assumed lightly. Although I have great confidence myself in their substantial accuracy, yet there comes the haunting thought that all previous experimenters in this line have made mistakes and that though we have avoided or corrected ninety-nine sources of error there may be one that has escaped attention. We could assume the responsibility of issuing them with a clear conscience, but the case would be somewhat different with you even though you should disclaim personal responsibility. However, when you have figured and studied all our measurements, you will be better able to determine the best course, whether to specifically publish a few typical tables, or whether to make a general statement of the tendency of the results, or whether to say nothing at all. We will send on the data of the measurements of the other surfaces shortly [pp. 195–7].[55]

Wilbur's letter reveals various degrees of modesty, caution, conservatism, and deference to Chanute. He seemed to be leaving to Chanute the decision whether or not to publish the Wrights' data, but there were hints that he was going to drag his feet on the matter. In his letter to Wilbur, February 6, Chanute appeared to detect veiled reluctance, and he emphasized that Moedebeck was rushing him for completion of the new chapter: "I had come to quite the same conclusion as yourself, i.e., that it would be unwise to give the public full information as to the properties of curved surfaces at present, and I wrote him [Moedebeck] that the article would not contain your data" (p. 209).[55] After that, Wilbur obviously considered the publishing of the Wrights' tables in the Moedebeck handbook to be a dead matter, commenting to Chanute on February 7 that "in considering the matter of publishing our tables of pressures and tangentials, . . . I think I shall prepare to make them public some time during this summer" (p. 213).[55] But he never did. After all that correspondence and discussion, the complete aerodynamic tables compiled by the Wrights were never published in their lifetimes. Finally, from the Wrights' notebooks, currently in the collection of the Franklin Institute in Philadelphia, McFarland included the tables for 48 different wing models as an appendix to *The Papers of Wilbur and Orville Wright*,[55] 51 years after the data were taken.

That leads to a recurring question: Did those tables represent a contribution to the state of the art? In Chapter 2 we asked the same question about da Vinci's notes on aerodynamics, which were not published until centuries after his death, and thus da Vinci made no contribution to the state of the art of aerodynamics during the period in which he lived. In the case of the Wrights' aerodynamic data, there is the fundamental difference that da Vinci never parlayed his thinking into a successful flying machine, but the Wrights did. So one could easily argue that the contributions of the Wrights to advance the state of the art of aerodynamics were embodied in their successful airplanes, and hence were imparted to the public in a dramatic and compelling fashion – much more dramatic than the publication of their tables would have been.

As for the accuracy of the Wrights' aerodynamic tables, they are now available for any-one to see.[55] A definitive assessment of their accuracy could be made only by repeating the experiments in a modern wind tunnel, or by calculating the detailed flow field and hence the aerodynamic force for their wing models by means of modern computational fluid dynamics (CFD). Neither has been done. (Modern wind-tunnel tests and CFD calculations have been carried out for the complete Wright Flyer,[57] but not for the various wing models the Wrights

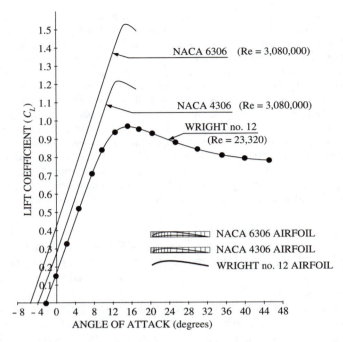

Figure 5.17 Comparison of lift curves for the Wright no. 12 airfoil and two standard NACA airfoils, with an aspect ratio of 6 for all three cases.

examined in their wind tunnel.) However, we can perhaps obtain a partial assessment from the following comparison: The variation of the lift coefficient versus the angle of attack for the Wrights' model no. 12 is plotted in Figure 5.17 with data directly from the Wrights' tables.[55] Also shown is the shape of their no. 12 airfoil section; that airfoil had a $\frac{1}{20}$ camber (i.e., a 5% camber), with the location of maximum camber near the leading edge. For comparison, Figure 5.17 shows lift curves for two standard National Advisory Committee for Aeronautics (NACA) airfoils that are about as close to the Wrights' airfoil shape as can be found in the standard NACA series. The NACA 6306 has 6% camber located at the 30% chord point. The 6% and 4% cambers of the NACA airfoils straddle the 5% camber value for the Wrights' airfoil. Both of the NACA airfoils are thin airfoils, with only a 6% thickness ratio, but they are certainly thicker than the Wright airfoil, which was made from 20-gauge steel (0.0319 in. thick, which, with the 1-in. chord, gives a thickness ratio of only 3.2%). Each of the three model wings used in Figure 5.17 had a rectangular planform and an aspect ratio of 6. The NACA data are from NACA Technical Report 460,[58] obtained in 1933 using model wings in the NACA variable-density wind tunnel (VDT) at the NACA Langley Memorial Laboratory. The Wright airfoil yields essentially the same lift-curve slope as the NACA airfoils, which should be the case for equal-aspect-ratio wings. However, the NACA curves are shifted to the left (their zero-lift angles of attack are larger), and they show larger maximum lift coefficients. That is explainable on the basis of Reynolds-number effects. Note that there is a huge Re difference between the Wright airfoil (a very low Re of 23,320 for their 1-in.-chord model) and the NACA airfoils (a high Reynolds number of 3,080,000). (The purpose of the NACA VDT was to run at high pressures in order to achieve the high Re values pertinent to actual flight conditions.) It is well known that the effect of increasing the Reynolds number is to increase the magnitude of both the zero-lift angle of attack and the maximum lift coefficient; both trends are seen in Figure 5.17. Also, the shape and behavior of the plot of the Wrights' data follow the classic trends for very thin airfoils, as explained in Chapter 4 of *Fundamentals of Aerodynamics*.[1] Very thin airfoils have low values for the maximum lift coefficient and rather mild decreases in the lift coefficient beyond the point of stall; that is precisely the trend shown by the Wright airfoil. In summary, although the comparison in Figure 5.17 is somewhat like an apples-and-oranges situation, there is enough substance to indicate that the Wrights' data were reasonable.

By December 1901 the Wrights had found what was for them "the right aerodynamics." In particular, they had proved, to their own satisfaction, the aerodynamic advantage of a high aspect ratio, using wind-tunnel models with aspect ratios from 1 to 10. Although Langley's data in *Experiments in Aerodynamics* had clearly shown that higher-aspect-ratio wings produced more lift, the Wrights either overlooked Langley's finding or mistrusted it. There was no mention of Langley's aspect-ratio data in the correspondence between Wilbur and Chanute. In any event, the advantage of a high aspect ratio was clear to the Wrights from their own wind-tunnel tests. Of all the models tested, Wilbur noted that model no. 12 (Figure 5.17) had "the highest dynamic efficiency of all the surfaces shown," with a 5% camber ($\frac{1}{20}$) and an aspect ratio of 6. The data from model no. 12 had a strong influence on the wing design for their 1902 glider.

There is no doubt that by 1902 the Wrights possessed far more aerodynamic data and understanding than anyone before them. In the field of applied aerodynamics, the Wrights were unquestioned leaders. Orville was justified in making the following statement years later, in a deposition dated February 2, 1921:

> Cambered surfaces were used prior to our experiments. However, the earlier experimenters had so little accurate knowledge concerning the properties of cambered surfaces that they used cambered surfaces of great inefficiency, and the tables of air pressures which they possessed concerning cambered surfaces were so erroneous as to entirely mislead them. They did not even know that the center of pressure traveled backward on cambered surfaces at small angles of incidence, but assumed that it traveled forward. I believe we possessed in 1902 more data on cambered surfaces, a hundred times over, than all of our predecessors put together [p. 551].[55]

Orville was short-changing Langley; the reversal of the center-of-pressure movement had previously been observed and measured by Langley and Huffaker, but their data were never published. Nor was Orville showing sufficient respect for the reasonable accuracy of the Lilienthal table for air pressures (a matter we shall discuss in the next section). However, in the main, Orville was justified in emphasizing that he and Wilbur had been the sole owners of the most advanced, most precise, and most useful data base in applied aerodynamics at that time – one of the reasons that I consider the Wright brothers the first true aeronautical engineers.[17]

Lilienthal's Table and the Wrights' View of It

Much has been said, both here and in other books, about the impact of the Lilienthal table on the work of the Wright brothers, and over the past century the table has been praised by some and condemned by others – engineers and historians alike. This section offers a slightly different perspective on the Lilienthal table, especially in regard to how it was viewed by the Wrights.

As mentioned earlier, a definitive assessment of the accuracy of the Lilienthal table would require modern wind-tunnel tests and/or accurate CFD calculations using the Lilienthal model. In lieu of such definitive tests, we can at least make an educated appraisal. The whirling-arm technique usually used by Lilienthal introduced some degree of uncertainty into his data, but he also conducted extensive tests using the natural wind, and it was those data that were presented in his table, not whirling-arm data. Moreover, his tabulated entries were *averages* from numerous repeated tests, and that tended to diminish any random errors. For whatever it is worth, Lilienthal was a trained mechanical engineer, steeped in the German tradition of precision and thorough preparation in carrying out experiments. We have to assume that every effort was made to achieve the highest degree of accuracy possible under the circumstances. The variation of the lift coefficient versus the angle of attack for the Lilienthal wing is plotted in Figure 5.18. Recall that the Lilienthal table gives normal-force and axial-force coefficients, C_N and C_A, respectively, and that the lift coefficient must be obtained from

$$C_L = C_N \cos \alpha - C_A \sin \alpha$$

where α is the angle of attack. Also recall that the Lilienthal table pertains only to a circular-arc airfoil (maximum camber at the $\frac{1}{2}$-chord location) with a camber ratio of $\frac{1}{12}$. The planform shape of the wing model is sketched at the top of Figure 5.18. For comparison,

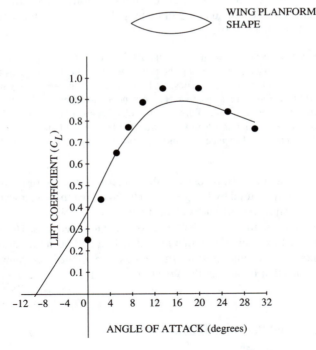

Figure 5.18 Comparison between Lilienthal's lift-coefficient data (solid curve) for a circular-arc airfoil and the Wrights' data (dots) for a parabolic-arc airfoil (camber ratio $\frac{1}{12}$ for both airfoils).

the data for the Wright no. 31 wing model are shown as the black circles. Model no. 31 was identified by McFarland[55] as the "Lilienthal model airfoil, i.e., an adaptation of the model aerofoil on which Lilienthal's tables were based." However, the only similarities between the Wright no. 31 and the Lilienthal wing were the same planform shape and the same camber ratio of $\frac{1}{12}$. The airfoil shapes were different; the Wright no. 31 had a parabolic airfoil, with maximum camber near the leading edge. Thus Figure 5.18 shows an apples-and-oranges comparison. However, it is the closest we can come to a direct comparison between the Lilienthal data and the Wright data. The Wrights apparently made an exact copy of the Lilienthal wing, using a circular-arc airfoil; that is listed as model no. 34 by McFarland.[55] However, data on lift and drag coefficients for the no. 34 model were not given in the Wrights' tables. McFarland noted that in the Wrights' 1901 notebook those entries had been crossed out, as if the entries had been made incorrectly. Did the Wrights delete those data out of respect for Lilienthal? Recall that Wilbur had suggested to Chanute that in the new edition of Moedebeck's handbook, any of the Wrights' data that might be presented should pertain to a model different from Lilienthal's, plainly stating that that choice should be made "instead of contradicting Lilienthal." Did the Wrights wind-tunnel data actually contradict Lilienthal's findings? And did the Wrights suppress some of those data by crossing out those entries in their notebook? And how does all that relate to Wilbur's statement in his October 16, 1901, letter to Chanute that "it would appear that Lilienthal is very much nearer the truth than we have heretofore been disposed to think"? At present, we have no information that could satisfy our curiosity about those seeming inconsistencies.

Returning to Figure 5.18, the apples-and-oranges comparison is not as useless as it might seem. The difference between the two airfoils concerned the location of the maximum camber. We have already discussed the impact of such differences: As the location of the maximum camber is shifted farther away from the leading edge, the zero-lift angle of attack becomes larger (i.e., $\alpha_{L=0}$ shifts to the left). That is indeed the situation shown in Figure 5.18: The zero-lift angle for the Lilienthal airfoil was larger than that for the Wright airfoil. (The Wrights did not record the zero-lift angle for their no. 31 airfoil, whereas they did for most of their models. However, we can easily extrapolate the Wright data to a zero-lift angle between $-5°$ and $-6°$.) Lilienthal measured his zero-lift angle to be $-9°$. That seems rather large; for today's conventional airfoils, $\alpha_{L=0}$ is typically less than half that measured by Lilienthal. But Lilienthal's circular-arc airfoil was not conventional. Hoerner and Borst[37] reported a theoretical expression for the zero-lift angle of attack for circular-arc airfoils as a function of camber ratio, f/c, as

$$\alpha_{L=0} = -1.15(f/c)$$

where $\alpha_{L=0}$ is in degrees, and f/c is expressed in percent. For the $\frac{1}{12}$ camber of Lilienthal's airfoil, that yields $\alpha_{L=0} = -9.6°$. Lilienthal's experimental measurements gave $\alpha_{L=0} = -9°$. That was essentially the same value as predicted by modern airfoil theory – strong evidence that Lilienthal's measurements were accurate.

Further evidence can be found in other comparisons of the Wrights' data and Lilienthal's data. If we ignore the large difference in their zero-lift angles of attack, which we have just shown to have been real and expected on the basis of the different locations of the maximum cambers for the two airfoils, and if we therefore ignore the comparison at small angles of attack (which is dominated by the difference in zero-lift angles of attack), then Lilienthal's data and the Wrights' data were in reasonably close agreement – they did better than just fall on the same piece of graph paper. The values for their maximum lift coefficients agreed within 5%, and the angles of attack at which maximum lift would occur were essentially the same, around 17°. Both airfoils featured the rather soft stall behavior (i.e., the gradual decrease of lift beyond stall) that is characteristic of very thin airfoils. In my judgment, the agreement between the two sets of data is reasonable, and it tends to confirm the accuracy of both.

In short, I believe that the Wrights' data and the Lilienthal table are not contradictory. Their differences are explainable on sound technical bases. In fact, I find it amazing, from the vantage point of late-twentieth-century aerodynamics, that two sets of data collected in Berlin and Dayton almost a century ago, using completely different techniques, would agree as well as they do (Figure 5.18). This comparison attests to the accuracy of both sets of data.

The Wrights' enthusiasm for the Lilienthal data ran hot and cold. In a letter to Wilbur, December 2, 1900, Chanute stated that "I tested the Lilienthal coefficients with models, and found them to agree closely." The Wrights accepted that opinion and used the Lilienthal table for their 1900 and 1901 gliders. But they used Lilienthal's coefficients incorrectly, which led to a factor-of-2–3 discrepancy between their calculated lift and measured lift. We have seen how that discrepancy can be explained on a sound physical basis. However, the Wrights had no way of knowing about their mistaken interpretation; all they saw was a major discrepancy. As a result, in 1901 the Wrights began to have serious doubts about the validity of Lilienthal's data. On the basis of consistent disagreement with Lilienthal's data, Wilbur was emboldened to say, during his presentation to the Western Society of Engineers, September 18, 1901, that "the Lilienthal tables might themselves be somewhat in error."[56] That impression was reinforced by their wheel-rim tests, after which Wilbur wrote in his October 6 letter to Chanute that "I am now absolutely certain that Lilienthal's table is very seriously in error, but that the error is not so great as I had previously estimated" (p. 127).[55]

Of course, the problem all along was that the Wrights were uninformed about some important aspects of the Lilienthal data. They, like Chanute, seem to have interpreted the Lilienthal table as universal in its application to cambered airfoils, ignoring any possible effects of airfoil shape and planform shape. However, at that stage they began to receive some clarification. On October 10, 1901, Chanute told Wilbur of his discovery that "Lilienthal's coefficients were obtained in natural wind, which he claims to give much greater values than for the same surfaces driven in still air" (p. 128).[55] Until then, the Wrights had thought that the Lilienthal table was based on whirling-arm experiments. After the Wrights began their wind-tunnel tests, Wilbur began to reconsider, as in his October 16 letter to Chanute: "It would appear that Lilienthal is very much nearer the truth than we have heretofore been disposed to think" (p. 135).[55] It is reasonable to believe that by that time the Wrights had made the type of comparison shown in Figure 5.18, and they probably judged the agreement to be good.

Finally, after a very long time, on October 24, 1901, the Wrights got their hands on a copy of Lilienthal's book – Chanute's copy, which he lent to them. Chanute enclosed an English translation of certain sections; the translation had been commissioned by Langley. The Wrights were well into their wind-tunnel tests and were just beginning to read Lilienthal's book in detail, insofar as the partial English translation allowed. On November 2, Wilbur wrote to Chanute:

> I have read the Lilienthal translation and examined the illustrations and plates many times. It is certainly a wonderful book. The more it is studied the more wonderful it becomes.

> Although, as I see it, errors are not entirely absent, yet considering that it was a pioneer work, developing an entirely new field, it is remarkably sound and accurate.... We have found that the apparent discrepancies between our calculations based on his tables and our actual experiences may be brought under the following heads: (1) errors of Lilienthal's formula and tables; (2) errors in the use of them; (3) errors in anemometers and other estimates of velocities; (4) errors of our own in overlooking or improperly applying certain things [p. 145].[55]

As discussed earlier, Wilbur was close to the mark with his fourth point: The problem was in part due to their improper application of the data. However, the Wrights still did not know about aspect-ratio effects and the consequences of changing the location of the maximum camber – they did not know the real reasons that they were "improperly applying certain things." That was clear in Wilbur's November 2 letter: "Lilienthal is in error in fixing the negative angle at which lift begins. He makes it $-9°$, while I am certain that with surfaces with dimensions in the ratio of 1×4 the true lift does not begin until $-4.5°$ or $-4°$" (p. 145).[55] We have seen quite the contrary: With the luxury of modern airfoil theory, we have shown that Lilienthal's measurement of $\alpha_{L=0} = -9°$ was quite reasonable.

Wilbur went on to another spurious conclusion:

> Lilienthal's second error, as I find it, is in the very high coefficients he found in the natural wind. This was probably in part due to the anemometer he used. It seems that instead of a standard instrument he constructed his own, and graduated it by calculations based on the Smeaton coefficient 0.005, or 0.13 metric. Consequently his instrument probably under-recorded the true velocity and his coefficients are higher than in his turntable experiment where the velocities were measured by counting the feet traveled per second by the ends of the arms [p. 145].[55]

We discussed that problem in Chapter 4. Lilienthal's anemometer is shown in Figure 4.36. Wilbur was correct about Lilienthal's velocity measurements made with that anemometer; in Chapter 4 we discussed how the use of the classic value for Smeaton's coefficient led to velocity measurements that were too low by a factor of $\sqrt{2}$. But Wilbur was wrong in thinking that that error resulted in an error in Lilienthal's coefficients. That error in measuring the velocity did not, per se, have any effect on Lilienthal's recorded lift and drag coefficients, because the aerodynamic data obtained in the natural wind were plotted in the same form as his earlier whirling-arm data (i.e., the data were *ratioed* to the measured aerodynamic force at a 90° angle of attack). Thus the effects of errors in both velocity and Smeaton's coefficient were divided out.

Wilbur went on to state that Lilienthal's "third error is in the use of 0.13 in his formula. While I have not personally tested this particular point I am firmly convinced that it is too high" (pp. 145–6).[55] Wilbur was stating, quite correctly, that the classic value for Smeaton's coefficient, 0.13, was too high. In Lilienthal's book, that value was mentioned frequently. Indeed, Lilienthal often placed it on his graphs (e.g., Figures 4.31, 4.34, 4.37, 4.40). That could easily have led Wilbur and others to think that the value 0.13 was used by Lilienthal to calculate his data. However, the coefficients in the Lilienthal table were ratios based on the measured force at a 90° angle of attack, and thus the value 0.13, no matter how inaccurate, was simply divided out.

In spite of those criticisms by Wilbur, he was quite taken with the borrowed copy of Lilienthal's book, as shown in his November 2 letter to Chanute: "I must have a copy of Lilienthal's book for myself. The publisher's name is on the title page, but I nowhere find the price stated. Can you give me any information on this? I will return your copy shortly, first making blueprints of his plates, which really are the essence of the book" (pp. 148).[55] Ten days later, Chanute wrote back telling Wilbur to keep the copy of *Birdflight;* he had already ordered another copy. Moreover, Chanute sent Wilbur a set of reports published by the Aeronautical Society of Great Britain – an example of the positive benefit of Chanute's friendship. Chanute was playing his role as the premier disseminator of aeronautical information, and the Wrights were rapidly expanding their collection of aeronautical literature, even as they were feverishly carrying out their extensive series of wind-tunnel tests.

From reading Lilienthal's book as best they could (they had an English translation of only selected sections), the Wrights came to the realization that the Lilienthal table was considerably more limited in application than they had initially thought. Finally, on November 22, 1901, Wilbur gently took Chanute to task for publishing the Lilienthal table without any mention of the wing shape and aspect ratio for which the data had been obtained: "Any table is liable to great misconstruction if the surface to which it is applicable is not clearly specified. No table is of universal application" (p. 162).[55] That announced the Wrights' coming-of-age in regard to technical maturity in applied aerodynamics.

We have seen that both Lilienthal's data and the Wrights' data were reasonably accurate and surprisingly similar, considering their different places and technical methods of collection. The Wrights' problem with the Lilienthal table was misapplication of the data. By November 22, 1901, Wilbur and Orville were aware of their problem, but they did not understand the nature of the corrections that could have made the Lilienthal data applicable to their 1900 and 1901 gliders. However, by that time the whole matter was irrelevant: Because of the discrepancies between their calculations and their glider measurements, they had begun the most comprehensive series of wind-tunnel tests in the history of aeronautical engineering. With those tests the Wrights were discovering "the right aerodynamics" for their future machines. They could discard the Lilienthal table; they had their own data and tables, and they had confidence in the accuracy of their findings. From that time on, there was virtually no reference to the Lilienthal table in their correspondence. It was perhaps fortuitous that the Wrights misused the Lilienthal table and came up with inaccurate estimations of lift and drag. Had the use of that table instead led to more satisfactory findings, they might never have carried out their important series of wind-tunnel tests, and they might not have found the aerodynamic data so essential to the success of the Wright Flyer in 1903.

Langley and the Wright Brothers: Crossed Paths

There is an interesting footnote to the history of applied aerodynamics that concerns Samuel Langley and the Wrights. Octave Chanute corresponded with and was on good terms with most of his contemporary aeronautical colleagues, including Samuel Langley at the Smithsonian. In November 1901, Langley sent Chanute some data on a cambered airfoil, and he asked Chanute not to publish the data. Noting that the Langley airfoil was similar to some of the shapes being tested at that time by the Wrights, Chanute sent Wilbur the letter he had received from Langley, along with the shape of langley's airfoil and the corresponding aerodynamic data that Langley had gathered using his whirling-arm apparatus. Figure 5.19 shows the sketch that Chanute sent to Wilbur. It was identical with that used by Langley for his aerodromes (Figure 4.57), with a camber of $\frac{1}{12}$. Chanute asked Wilbur to examine Langley's data: "I shall be glad to know how your new experiments on surfaces agree with Langley's" (p. 150).[55] The Wrights exceeded their mandate: They constructed a wind-tunnel model identical with that tested by Langley, with an aspect ratio

Figure 5.19   The Langley airfoil that was tested in the Wright brothers' wind tunnel.

of 4. They measured its ratios of "drift to lift" for angles of attack ranging from zero to 50° and compared them with Langley's data. The correlations were, for the most part, within 2% – again, incredibly good agreement for the technology of that day. However, the Wrights found that in regard to aerodynamic efficiency, Langley's airfoil was significantly inferior to some of their own airfoils, as Wilbur reported to Chanute, November 14, 1901: "This surface is by no means an efficient one for flying, as its best angle of gliding [a function of the lift-to-drag ratio] is 8.5°; and with framing and operator about 11°. Among the thirty or forty surfaces we have tested we found nearly all equal or superior to this one in lift and very much better in tangential [the axial force]" (p. 153).[55]

So in late 1901, Wilbur and Orville were actually measuring the aerodynamic properties of an airfoil that later would be used by Langley – their competition – for a flying machine. It is unclear whether or not Chanute ever transmitted that information to Langley. Even if he had, it is my opinion that Langley would have ignored it. He was already in the final phase of designing his large aerodrome and would have been reluctant to change the airfoil shape. Indeed, to what would he have changed it? The Wrights certainly were not inclined to give anyone the data from their airfoil shapes. Besides, Langley was not the kind of person to take anyone's findings on faith. The result was that Langley proceeded to use a relatively poor airfoil for his large aerodrome, and the Wrights knew about it beforehand.

The Wrights' Flying Machines

When their wind-tunnel tests were essentially completed in December 1901, the Wrights were in possession of an aerodynamic data base that exceeded anything previously assembled by orders of magnitude. They made good use of that information.

Wilbur and Orville arrived at Kitty Hawk on August 28, 1902, with the parts for a completely new glider. The new machine was assembled by September 19. Within two days they had made nearly 50 glides with it. Aerodynamically it was performing like a charm, and for the first time they had a glider that was performing precisely as their calculations said it would. In a letter to a friend and colleague, George Spratt, on September 16, Wilbur anticipated success with the almost completed glider (everything in place except the rudder): "The indications are that it will glide on an angle of about 7° to 7$\frac{1}{2}$° instead of 9$\frac{1}{2}$° up 10° as last year. The drift is only about $\frac{1}{8}$ of the weight. In a test for 'soaring' as a kite the chords stood vertical or a little to the front on a hill having a slope of only 7$\frac{1}{2}$°. This is an immense improvement over our last year's machine which would soar only when the slope was 15° to 20°" (p. 253).[55] Figure 5.20 shows Wilbur in the 1902 glider making a right turn during a glide on October 24. The glider even looked beautiful (contributing to the belief common in airplane aerodynamics that if it looks beautiful, it will fly beautifully). The new glider had a total wing area of 305 ft^2 and an aspect ratio of 6.7. Their wind-tunnel tests had clearly shown the benefit of a high aspect ratio, and they wasted no time in applying that new information. The new aspect ratio of 6.7 was about twice that for their previous gliders.

Figure 5.20   Wilbur Wright in the 1902 glider at Kill Devil Hills, October 24, 1902.

Because of that higher aspect ratio, the induced drag was reduced by 50%, and equally important, the lift at a given angle of attack was considerably increased. That increase in aspect ratio was the major reason for the improved aerodynamic performance of their 1902 glider over the 1900 and 1901 gliders.

As in the preceding year, the Wrights had designed the airfoil of the glider so that its camber ratio could be quickly changed in the field. During the 1902 flight trials the Wrights experimented with camber ratios from $\frac{1}{24}$ to the more shallow $\frac{1}{30}$ – camber shapes more shallow than Lilienthal's by a factor of 2 or more. Also, the camber ratios used for the 1902 glider were more shallow than any of the airfoil shapes tested in the wind tunnel – they designed the airfoil *outside* of their data base, which extended only to a camber ratio of $\frac{1}{20}$. Jakab[24] discussed the possible reasons:

(1)   In their tunnel tests, the Wrights had observed an improving trend as camber was decreased: Their no. 12 airfoil, reported as having the "highest dynamic efficiency of all the surfaces shown," had a camber ratio of $\frac{1}{20}$; that was the smallest camber ratio tested in their wind tunnel.

(2)   By 1902 the Wrights had gained enough experience and technical expertise that they could extrapolate outside their data base with some confidence.

Perhaps the most gratifying technical achievement for the Wrights in 1902 was that their measured glider performance agreed with their calculations, which were based totally on their wind-tunnel data. They had completely freed themselves of any baggage, real or perceived, associated with the Lilienthal table.

When Bishop Wright met Wilbur and Orville at the Dayton train station as they returned home on October 31, there was an air of success and optimism, in contrast to the return trip the previous year. They had already decided to build a new machine for 1903, and that new machine would be powered. On December 3, 1902, a number of letters went out under the Wright Cycle Company letterhead to engine manufacturers, inquiring about the availability of a gasoline-fueled engine that could develop eight to nine brake horsepower and would weigh no more than 180 lb. All the replies were negative. On December 11, 1902, Wilbur casually informed Chanute about their plans: "It is our intention next year to build a machine much larger and about twice as heavy as our present machine. With it we will work out problems relating to starting and handling heavy weight machines, and if we find it under satisfactory control in flight, we will proceed to mount a motor" (p. 290).[55] That rather low-key comment belied the Wrights' actual level of activity. They had already begun work on their next two major challenges: the propeller, and the engine to drive it.

The next phase in the Wrights' inventive process has been well described by Jakab.[24] Following the pattern established with the wind-tunnel tests, the Wrights did everything themselves. Orville, with the assistance of Charlie Taylor, a newly hired mechanic for the bicycle shop, designed and built the engine. Its performance was marginal, but sufficient. Wilbur assumed responsibility for the propeller design, although in the process he had many stimulating and sometimes heated discussions with Orville about the aerodynamic complexities of propeller design and performance. Between them, they produced the first viable theory of propeller design. It was essentially a version of what today is called "blade-element theory," and it involved aerodynamic information about airfoils from their earlier wind-tunnel tests. Most important, Wilbur was the first person to recognize that a propeller was nothing more than a twisted wing oriented such that the lift force produced on the twisted wing would be oriented in the general flight direction (i.e., in the *thrust* direction). Orville summed up the results in a long letter to George Spratt, June 7, 1903:

> During the time the engine was building we were engaged in some very heated discussions on the principles of screw propellers. We had been unable to find anything of value in any of the works to which we had access, so that we worked out a theory of our own on the subject, and soon discovered, as we usually do, that all the propellers built heretofore are *all wrong,* and then built a pair of propellers 8 1/8 ft. in diameter, based on our theory, which are *all right!* (till we have a chance to test them down at Kitty Hawk and find out differently). Isn't it astonishing that all these secrets have been preserved for so many years just so that we could discover them!! Well, our propellers are so different from any that have been used before that they will have to either be a good deal better, or a good deal worse [p. 313].[55]

As a rough standard for comparison, the crude propellers employed by would-be flying-machine inventors in Europe in the nineteenth century had efficiencies on the order of 40–50%. (Propeller efficiency is defined as the power output of the propeller divided by the shaft power input from the engine, expressed in terms of percentage.) Samuel Langley had conducted whirling-arm tests with a propeller of his own design and measured a propeller efficiency of 52%; in *Experiments in Aerodynamics* Langley admitted that the "form of the propeller blades" was "not a very good one." On the basis of their propeller theory, the Wrights predicted an efficiency of 66% for their propeller; it was most likely even better. Much later, in an anonymous article published in the November 1909 issue of the magazine *Aeronautics* in New York, entitled "Wrights' Propeller Efficiency," it was reported that a Captain Eberhardt in Berlin had taken detailed measurements of the propeller used by Wilbur in his European flights during 1908 and 1909. From that he was able to ascertain a value of 76% for propeller efficiency.

The Wrights immediately achieved spectacular improvements in the art of propeller design, and it was no fluke: Their blade-element propeller theory was far ahead of any earlier means of designing propellers. The underlying basis of that theory was the Wrights' under-standing of the true aerodynamic function of a propeller – a major contribution to applied aerodynamics in itself. Their propeller theory revealed a degree of technical sophistication in applied aerodynamics never seen before, and their application of the theory in designing a real propeller was masterful. The details are available in the Wrights' notes, as compiled in McFarland's Appendix 3.[55] The lengthy discussion therein is recommended for anyone interested in the details of the Wrights' propeller theory. The role of the Wrights' propeller theory in the history of applied aerodynamics has not always been given its due, and the degree to which their highly efficient propeller design contributed to the success of the 1903 Wright Flyer has not been adequately recognized by the public. However, in my opinion that work was so important that had such an award existed in 1903, the Wrights would have deserved the Collier Trophy for their work on the aerodynamics of propellers.

Little is known about the Wrights' fundamental understanding of the physical details of the flow fields surrounding airfoils and wings. The streamlines in water flows had been observed and recorded by da Vinci as early as the fifteenth century (Figure 2.7), and in 1900 Etienne-Jules Marey had published smoke-flow photographs of streamlines over various shapes (Figure 4.12). By 1903, especially if they studied the papers in the annual reports of the Aeronautical Society of Great Britain sent to them by Chanute, the Wrights would have been aware of the general nature of streamlines in a flow. The only "flow fields" mentioned in the Wrights' notes reflect such an understanding. Figure 5.21 shows three sketches made by Wilbur in a letter to George Spratt, March 28, 1903, comparing the streamlines over cam-bered airfoils with a short chord (Figure 5.21a) and a long chord (Figure 5.21b). Wilbur was arguing, quite correctly, that if the two airfoils were geometrically similar, "the air would be disturbed to a depth proportioned to the chord length" (i.e., the streamlines would be geo-metrically similar no matter the sizes of the airfoils). Such understanding was the beginning of the concept of *dynamic flow similarity,* so important to modern applied aerodynamics. The streamline sketch in Figure 5.21c accompanied Wilbur's discussion of aspect-ratio ef-fects. In the wind-tunnel tests he had observed correctly that a high-aspect-ratio wing was

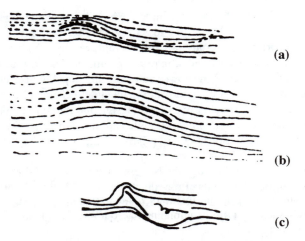

(a)

(b)

(c)

Figure 5.21   Three flow-field sketches made by Wilbur Wright, apparently the only flow-field sketches in the Wrights' correspondence. (From a letter to George Spratt, March 28, 1903.)

aerodynamically more efficient at lower angles of attack (those angles encountered in actual flight), whereas at very high angles of attack, such as 50–60°, a low-aspect-ratio wing was actually better:

> The relative value of spar and chord lengths depends entirely on the angle of incidence. Thus at 0° to 20° a surface 1 × 4 [aspect ratio of 4] is much better than a similar surface 4 × 1 [aspect ratio of 0.25]; but from 30° to 90° the 4 × 1 has the greatest lift. This is because at small angles the wind slips off the ends chiefly [Wilbur was showing some crude knowledge of the flow around the tip of a finite wing, although he seems to have been unaware of the nature of the wing-tip vortex], while at big angles the slip is mostly off the front edge [p. 302].[55]

The flow "slipping off the front edge" at a high angle of attack is illustrated in Wilbur's sketch (Figure 5.21c), where he shows accurate knowledge of the qualitative aspects of flow over a flat plate: the dividing of the streamline flow at some point on the bottom surface (the stagnation point on the bottom surface, not identified by Wilbur in the sketch), the movement of the flow from the bottom to the top (with the flow curling around the leading edge), the separation of that flow at the leading edge, and the dead-air, separated region over the top surface. We have to conclude that along with the Wrights' considerable knowledge of applied aerodynamics, they had the beginnings of an understanding of the qualitative physics of flow fields.

The Wrights' learning process was a remarkable exercise in determination, self-discipline, and genius, leading to the inventive nature described by Jakab.[24] Perhaps some insight on the Wrights' process of learning can be gained from the following letter from Wilbur to Spratt, April 27, 1903:

> It was not my intention to advocate dishonesty in argument nor a bad spirit in a controversy. No truth is without mixture of error, and no error so false but that it possesses some elements of truth. If a man is in too big a hurry to give up an error he is liable to give up some truth with it, and in accepting the arguments of the other man he is sure to get some error with it. Honest argument is merely a process of mutual picking the beams and motes out of each other's eyes so both can see clearly. Men become wise just as they become rich, more by what they *save* than by what they receive. After I get hold of a truth I hate to lose it again, and I like to sift all the truth out before I give up an error [p. 307].[55]

That attitude showed insight into the art of critical thinking and analysis and perhaps helps to explain in part why the Wrights were able to make so much progress in the few years leading up to the Wright Flyer.

The Wrights arrived at their Kill Devil Hills camp, just south of Kitty Hawk, on September 25, 1903, and 14 days later the crated parts of their new machine arrived. On October 9 they began to assemble the flying machine that would herald a new era in technology and bring sweeping changes in human affairs.

Langley and the Wright Brothers in 1903: Failure and Success

About 400 miles to the east of Dayton, during the period of the Wright brothers' enlightenment in aerodynamics, Samuel Langley was busily working on his own flying machine at the Smithsonian. Picking up our thread of history from Chapter 4, there was a short hiatus after Langley's success with his unmanned aerodrome flights in 1896. However, in 1898, prompted by the Spanish-American War, the War Department (with President McKinley's personal backing) offered Langley a $50,000 contract to build a manned, powered flying machine. Langley accepted and immediately hired as an assistant a recent graduate of the Sibley School of Mechanical Engineering at Cornell, Charles Manly. Over the next five years they labored to design and construct a large aerodrome, essentially four times larger than the 1896 aerodromes. During that period, in contrast to the Wrights' intensive learning process, Langley generated no new aerodynamic findings worth noting (except for the work on cambered airfoils by Huffaker that was never published). However, during that period, Langley, who by nature was not an engineer, took on the massive engineering task of actually building a real machine. His mind and heart were fixed on powered human flight, not on advancing the state of the art of aerodynamics.

A detailed account of the technical design and construction of Langley's full-scale aerodrome can be found in the *Langley Memoir*.[50] Langley used the same launching scenario as before: a houseboat on the Potomac River, with a catapult mounted on top of the houseboat, and the aerodrome mounted on top of the catapult. On October 7, 1903, with Charles Manly at the controls, the aerodrome was launched; it simply settled into the river in front of the houseboat. But that failure did not deter Langley and Manly. The aerodrome was fished out of the river and repaired in the Smithsonian laboratory. On December 8, everything was ready, with Manly again at the controls of the repaired aerodrome. Figure 5.22 shows the Langley aerodrome an instant after its launch, going through a 90° angle of attack. The rear wing had totally collapsed. That was Langley's final attempt at powered human flight. His efforts were severely criticized by the press, the public, and even the government. Three years later, Langley died, a broken man.

That part of Langley's career has been discussed in detail by Crouch.[51] After Langley's failure, the War Department issued its final report on the project: "We are still far from the ultimate goal, and it would seem as if years of constant work and study by experts, together with the expenditure of thousands of dollars, would still be necessary before we can hope to produce an apparatus of practical utility on these lines" (p. 293).[51] Only nine days after Langley's second failure, the Wright Flyer would make its first flight.

Figure 5.22   The Langley aerodrome's second failure, December 8, 1903.

Back at Kill Devil Hills, the Wrights finished assembling the Flyer on November 5, 1903. The machine had a total wing area of 510 ft^2 and a wingspan of 40.33 ft, yielding an aspect ratio of 6.4 for each of the two biplane wings. They increased the camber of the wings over that used with their successful 1902 glider, going to a camber ratio of $\frac{1}{20}$ – the camber for their no. 12 airfoil, identified as that with the "highest dynamic efficiency of all the surfaces shown." The engine and propellers were the last components to be attached. The Wrights then began a series of engine tests, but because of failures of propeller shafts and misfiring of the engine, along with extremely bad weather, their efforts to fly were delayed until well into December.

During that time, the Wrights were aware of Langley's work and progress on his aerodrome. In turn, Langley was becoming somewhat aware of the Wrights' activities, mainly via the Chanute conduit for information. The Wrights' first contact with the Smithsonian in 1899, requesting basic information on aeronautics, had been handled by Richard Rathbun, the assistant secretary of the Smithsonian, and it is highly unlikely that Langley was aware of that letter. Perhaps the first time that Langley became aware of the Wright brothers was in an exchange initiated by Wilbur in 1901. In a letter to Chanute on November 14, Wilbur thanked him for the gift of Lilienthal's book: "Do you think Prof. Langley could be induced to reprint this book in the Smithsonian *Reports?* It is a wonderful book and should be given an English publication" (p. 152).[55] Chanute was pessimistic about Langley's possible response, writing to Wilbur on November 16: "I fear Langley will not favor republishing Lilienthal in English." Chanute nevertheless transmitted the idea to Langley, who indeed declined. Chanute's letter of December 19 to Wilbur was somewhat critical of Langley:

> I enclose a letter from Prof. Langley giving his reason for not publishing Lilienthal in English. As it was my understanding that the professor does not read German, I did not think that the supervision of a translator would involve much labor for him. His own book has been practically ready 3 or 4 years [a reference to Langley's writings that eventually were published in the *Langley Memoir*], and I sometimes have felt that he was keeping other students back by not printing it [p. 183].[55]

On December 23 Wilbur replied to Chanute, expressing regret over Langley's answer. Relations between the Wrights and Langley were not off to a good start. It was also during that time that the Wrights tested Langley's airfoil and found it lacking in aerodynamic efficiency. By the fall of 1902, Chanute had informed Langley of the Wrights' successful glider flights. That prompted Langley's first show of interest in the Wrights. Langley wrote to Chanute on October 17: "I should like very much to get some description of the extraordinary results which you told me were recently obtained by the Wright brothers" (p. 282).[55] Langley quickly followed with a second letter to Chanute, October 23: "After seeing you, I almost decided to go, or send someone, to see the remarkable experiments that you told me of by the Wright brothers. I telegraphed and wrote to them at Kitty Hawk, but have no answer, and I suppose their experiments are over" (p. 282).[55] As Wilbur explained to Chanute, November 12: "We received from Mr. Langley, a few days before we finished our experiments at Kitty Hawk, a telegram, and afterwards a letter, inquiring whether there would be time for him to reach us and witness some of our trials before we left. We replied that it would be scarcely possible as we were intending to break camp in a few days. He made no mention of his experiments on the Potomac" (p. 283).[55] Langley became more anxious. Still communicating through Chanute, not directly with the Wrights, Langley wrote to Chanute, December 7: "I should be very glad to hear more of what the Wright brothers have done, and especially of their means of control, which you think better than

Figure 5.23   The first heavier-than-air, powered, manned flight by the Wright Flyer, December 17, 1903.

the Pénaud. I should be very glad to have either of them visit Washington at my expense, to get some of their ideas on this subject, if they are willing to communicate them" (p. 290).[55] Chanute forwarded that invitation to the Wrights, with the comment that it "seems to me cheeky." The Wrights declined, giving the excuse of pressing business.

Langley and the Wrights never met. The main reason appears to have been lack of enthusiasm on the part of the Wrights, possibly because of their growing lack of respect for Langley. Also, Chanute never seemed to encourage such a meeting, although he frequently visited the Wrights during that period. In a letter to Chanute on October 16, Wilbur wasted no ink on pro forma expressions of regret over the failure of Langley's aerodrome on October 7: "I see that Langley has had his fling, and failed. It seems to be our turn to throw now, and I wonder what our luck will be" (p. 364).[55]

On the morning of December 17, 1903, the Wright Flyer was ready for flight. An earlier attempt had been made on December 14 with Wilbur at the controls, but because of unfamiliarity with the flight characteristics of the new machine, Wilbur stalled the aircraft immediately after taking off. On December 17 it was Orville's turn. At 10:35 in the morning, with Wilbur running beside the right wing tip to keep it from dragging in the sand, the Wright Flyer lifted into the air and flew for 12 s, covering 120 ft over the ground. Figure 5.23 shows the Flyer leaving the ground – a truly historic photograph in the annals of aviation history. There were three more flights that day, the third lasting for 59 s and covering 852 ft over the ground. The age of powered, manned, heavier-than-air flight had arrived, opening up a new world of aeronautical engineering.

The State of the Art as Reflected in the Wright Flyer

The Wright Flyer was historic on another count: It was the first flying machine that reflected in its design all of the most recent developments in *applied* aerodynamics – those who built the Flyer had also carried out the experiments that had defined the state of the art of applied aerodynamics. The Wright Flyer had a relatively high aspect ratio of 6.4, which is a typical aspect ratio for modern general-aviation airplanes (the popular Cessna 150 has an aspect ratio of 7). The Flyer's airfoil was a good example of the class of thin, cambered airfoils used in all airplanes until the development of thick airfoils by the Germans toward the end of World War I. The propellers on the Wright Flyer offered an enormous improvement in efficiency derived from the pioneering propeller theory of the Wrights. The Flyer also featured state-of-the-art aircraft controls. The Wrights introduced the concept of exercising control about all three of an airplane's axes – control in pitch, yaw, and roll. Particularly important was their understanding of the importance of roll control and the invention of wing warping to achieve such control.

In regard to the state of the art of *theoretical* aerodynamics as reflected in the Wright Flyer, the story is completely different. No theoretical flow-field solutions stemming from the Euler equations or the Navier-Stokes equations were used in the design of the Wright Flyer – no one had been able to produce relevant solutions for those equations. Indeed, there were no meaningful aerodynamic theories of any kind for prediction of lift or drag at the time of the Wright brothers' aerodynamic research. However, that situation was about to change, and to change dramatically.

CHAPTER 6

Theoretical Aerodynamics Comes of Age
The Circulation Theory of Lift, and Boundary-Layer Theory

The field of hydrodynamic phenomena which can be explored with exact analysis is more and more increasing.

Nikolai Joukowski (1911)

Some New Thinking about Aerodynamic Lift: Frederick Lanchester

At the time that Samuel Langley was attending an aeronautics session at the August 1886 meeting of the American Association for the Advancement of Science and was about to begin two decades of experimental aerodynamics and flying-machine design, a young man was walking through the doors of the Royal College of Science (now the Imperial College of Science and Technology) in South Kensington, London, to begin studies in engineering and mining. That man was Frederick W. Lanchester, and he would later formulate the concepts underlying a scientific breakthrough in our understanding and calculation of aerodynamic lift – the circulation theory of lift.

Lanchester (Figure 6.1) was born October 23, 1868, in Lewisham, England. The son of an architect, Lanchester became interested in engineering at an early age (he was told by his family that his mind was made up at the age of 4). Lanchester spent three years as a student at the Royal College of Science, but never officially graduated. He was a quick study and innovative thinker, and he became a designer at the Forward Gas Engine Company in 1889, specializing in internal-combustion engines. In 1899 he formed the Lanchester Motor Company and sold automobiles of his own design. To this day he is remembered in England for his early automobiles. Lanchester married in 1919, but they had no children. He maintained his interest in automobiles and related mechanical devices until his death on March 8, 1946, at the age of 77.

In the early 1890s, Lanchester became interested in aeronautics. He divided his time between designing and developing high-speed engines and carrying out aerodynamic experiments using model gliders. In particular, during 1891–2 he tested a series of airfoils with curved shapes (i.e., cambered airfoils). He was totally unaware of the earlier work by Phillips and Lilienthal on cambered airfoils.

Lanchester's cambered airfoils came directly from his own theoretical concepts of how lift was generated. Those theoretical models are Lanchester's legacy to the history of aerodynamics, not his testing of cambered airfoils. Lanchester had a fundamentally correct idea of the streamline flow over a lifting surface. In considering the case of a flat plate, Lanchester wrote that fluid particles "will receive an upward acceleration as they approach the aerofoil, and will have an upward velocity as they encounter its leading edge. While passing instead under or over the aerofoil, the field of force is in the opposite direction, viz., downward, and thus the upward motion is converted into a downward motion. Then, after the passage of the aerofoil, the air is again in an upwardly directed field, and the downward velocity imparted by the aerofoil is absorbed" (p. 592).[59] The first publication of those thoughts was in his book *Aerodynamics*,[60] published in 1907, but they reflected his thinking as early as 1890. Lanchester was correctly describing the general features of

Figure 6.1   Frederick W. Lanchester.

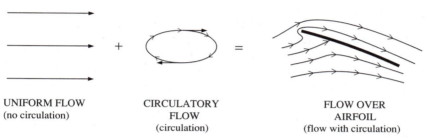

UNIFORM FLOW
(no circulation)

CIRCULATORY
FLOW
(circulation)

FLOW OVER
AIRFOIL
(flow with circulation)

Figure 6.2   Synthesis of an inviscid, incompressible flow over an airfoil by theoretical superposition of a uniform flow and a properly chosen circulatory flow.

the kind of flow field shown in Figure 3.23. Wilbur Wright sketched a similar flowfield (Figure 5.21). Lanchester reasoned that if an airfoil were designed with a curved shape, the surface of the airfoil would be more closely aligned with the natural curved streamlines of the flow and hence would be more efficient than a flat surface. A few years earlier, Lilienthal had expressed similar thoughts in *Birdflight* (Figure 4.42). Lanchester also saw in those upward and downward motions of the flow over a lifting surface a suggestion of some type of circulating motion superimposed on the tangential motion of the free stream.

Let us examine further the idea that the flow over a lifting airfoil can be thought of as a combination of circulatory and translational motions. Sketched at the right in Figure 6.2 is the flow field over a lifting airfoil, which can be artificially dissected into two parts: a uniform flow and a type of "circulatory" flow. Over the top of the airfoil, the circulatory flow and the free stream are in the same general direction, thus increasing the speed of the total flow, whereas over the bottom of the airfoil the circulatory flow is in the direction opposite to the free stream, thus reducing the speed of the total flow. As a result, the new velocity over the top surface is higher than that over the bottom surface; in turn, from the Bernoulli principle, the pressure is lower on the top surface and higher on the bottom surface, thus producing lift.

Lanchester theorized that the circulatory part of the flow (Figure 6.2, center) could be thought of as being produced by the conceptual *vortex filaments* first suggested by Helmholtz in 1858. Those vortex filaments were imagined as running along the span of the wing. Lanchester also theorized that the airflow would curl around the tips of the wings, moving from the higher-pressure region on the bottom of the wing toward the lower-pressure

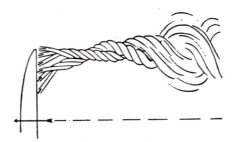

Figure 6.3 Lanchester's concept of a vortex trunk – the creation of a vortex at the tip of a finite wing. (From Lanchester.)[60]

region on the top of the wing. That would result in another circulatory motion in the vicinity of the wing tips, with the vortex filaments trailing downstream from the tips. Lanchester called those trailing vortices "vortex trunks" (Figure 6.3). Hence the flow over a finite wing, according to Lanchester, could be thought of as a synthesis of the uniform flow upstream, the circulatory flow induced by a series of vortex filaments aligned along the span of the wing, and the vortex trunks trailing downstream from the wing tips. Lanchester conceived that model during the years 1891–4, the same period during which Langley was designing his steam-powered aerodromes and Lilienthal was flying his gliders. That model was to provide the basis for a breakthrough in theoretical aerodynamics: the circulation theory of lift.

There were two problems for Lanchester's work at that time. The first was that reputable journals rejected his papers. The first public exposure for Lanchester's model came on June 19, 1894, when he read a paper to the Birmingham Natural History and Philosophical Society; he never published that paper, because at that time he had his eyes on the Royal Society. However, a member of the Royal Society suggested that the work was more appropriate for the Physical Society. When Lanchester submitted his paper to the Physical Society, it was rejected. Lanchester was so disappointed that he never submitted the paper elsewhere. Part, if not all, of the problem with publication was Lanchester's poor writing style. His explanations were not easy to follow. Finally, in 1907 and 1908, he published two books: *Aerodynamics* and *Aerodonics. Aerodynamics* was subsequently translated into German and French, making his ideas on lift available to the general scientific community.

The second problem was that Lanchester's work was essentially not quantitative; he made no substantive aerodynamic calculations of lift and drag. In a review of the state of aerodynamics in 1910, Nikolai Joukowski complimented Lanchester on the merits of his model, "but as regards magnitude and direction of the force of pressure on the body, in the peripteroid motion, Lanchester did not take them into consideration." To Lanchester's credit, he did make some attempt at calculations. Much later, in his Wilbur Wright Memorial Lecture in 1926 to the Royal Aeronautical Society, Lanchester stated that "it was from this makeshift theory that my early experimental aerofoils were designed, and values were tabulated and finally published in [*Aerodynamics*]. In spite of the paucity of experimental data, and the theoretical defects and apparent weakness in the theory, these tabulated results have turned out good forecasts."[59] They may have been useful calculations to Lanchester, but his quantitative analysis, such as it was, had no impact on the advancement of theoretical aerodynamics. Lanchester's airfoil designs, obtained from his calculations, were tested in Prandtl's wind tunnel at Göttingen University in 1912–13 and were found to produce a lift-to-drag ratio of 17; that was a 10% improvement over other models tested in that wind tunnel up to that time. Clearly, Lanchester's calculations were productive.

Because of those problems, recognition for Lanchester's basic model was slow in coming. In 1908 Lanchester visited Ludwig Prandtl and his student Theodore von Kármán in Göttingen. Lanchester spoke no German, and Prandtl spoke no English. Given Lanchester's difficulty in explaining his ideas in English, it would seem likely that there was little understanding between the two parties. However, shortly after that, Prandtl began to develop his own wing theory, which used a vortex model virtually identical with that proposed by Lanchester. Perhaps Lanchester's ideas contributed more than was known, through the minds of others.

Lanchester was bitter about being ignored by his technical contemporaries. Much later, in a letter dated June 6, 1931, to the Daniel Guggenheim Medal Fund, Lanchester wrote that

> so far as aeronautical science is concerned, I cannot say that I experienced anything but discouragement; in the early days my theoretical work (backed by a certain amount of experimental verification), mainly concerning the vortex theory of sustentation and the screw propeller, was refused by the two leading scientific societies in this country, and I was seriously warned that my profession as an engineer would suffer if I dabbled in a subject that was merely a dream of madmen! When I published my two volumes in 1907 and 1908 they were well received on the whole, but this was mainly due to the success of the brothers Wright, and the general interest aroused on the subject.

In summary, the conceptual basis for the circulation theory of lift was laid by Lanchester in the last decade of the nineteenth century, but the quantitative formula relating lift to circulation was developed during the first decade of the twentieth century by two people working independently, and without any knowledge of Lanchester's work. We end this section with an assessment by Giacomelli and Pistolesi: "With regard to Lanchester's contribution to Aerodynamics, there are two great ideas conceived by him: the idea of circulation as the cause of lift, and the idea of tip vortices as the cause of the drag, known today as the induced drag" (p. 344).[10] Clearly, the coming-of-age of theoretical aerodynamics began with Frederick Lanchester.

The Quantitative Development of the Circulation Theory of Lift: Kutta and Joukowski

At the time the Wright brothers were conducting their wind-tunnel tests, which advanced applied aerodynamics toward maturity, Wilhelm Kutta was finishing some work at the University of Munich that would prove an important advance in theoretical aerodynamics. Kutta was born in Pitschen, Germany, in 1867. In 1902, at the age of 35, he received a Ph.D. in mathematics from Munich, with a dissertation on aerodynamic lift. Kutta's interest had been sparked by the glider flights of Otto Lilienthal between 1890 and 1896. Kutta knew that Lilienthal had used a cambered airfoil for his gliders. Moreover, he knew that when the cambered airfoil was put at a zero angle of attack, positive lift was still produced. That was clearly evident from Lilienthal's data in *Birdflight*. Indeed, all of the cambered airfoils tested by Lilienthal had to be pitched to some negative angle of attack in order to reach the point of no lift (the zero-lift angle of attack). The generation of lift by a cambered airfoil at a zero angle of attack was counterintuitive to many mathematicians and scientists at that time, but the experimental data unequivocally indicated it to be a fact. Such a mystery made the theoretical calculation of lift on a cambered airfoil an excellent research topic at the time, and Kutta eagerly took it on. By the time he finished his dissertation in 1902, Kutta had in hand the first mathematical calculations of the lift on cambered airfoils.

At the insistence of his teacher and advisor at Munich, S. Finsterwalder, the essence of Kutta's research was published in a short note entitled "Lifting Forces in Flowing Fluids" in the July 1902 issue of *Illustrierte aeronautische Mitteilungen*. That note was described by Giacomelli and Pistolesi as representing "the first publication on the discovery of lift in the wing of infinite span" (p. 345).[10] By 1902 everyone knew that wings produced lift. However, no previous *theoretical* work had been able to calculate lift values from the equations for fluid motion. Indeed, for a two-dimensional shape (such as an airfoil section of an infinite wing), the governing equations had yielded only the finding of *zero drag* (d'Alembert's paradox, as discussed in Chapter 3). Probably there was always the nagging suspicion in the minds of nineteenth-century theoreticians that the same type of calculations, when properly carried out, might also predict zero lift. Kutta showed that such was not the case; in that sense he was the first to "discover" a finite lift on a two-dimensional body (qualifying as a "discovery" only relative to the world of theory).

In particular, Kutta calculated the lift on a cambered circular-arc airfoil at a zero angle of attack, showing that finite lift was produced even though the chord line was aligned with the free stream. That had already been observed experimentally by Lilienthal, Langley, and the Wrights, but it was a major breakthrough to be able to demonstrate it theoretically. Kutta's equation for lift was published in 1902:

$$L = 4\pi a \rho V^2 \sin^2(\theta/2)$$

where a is the radius of the circular arc, and 2θ is the angle subtended by the arc at the center. The geometry of that construction is shown in Figure H.1 in Appendix H, as is the calculation of the lift coefficient for Lilienthal's $\frac{1}{12}$-camber circular-arc airfoil at a zero angle of attack using Kutta's equation, the result being a lift coefficient of 1.047. By comparison, from the Lilienthal table (Figure 4.38), the experimental value for the lift coefficient at a zero angle of attack is 0.381, considerably lower than the Kutta value. However, the Kutta value is for an infinite aspect ratio; correcting that value for the aspect ratio of 6.4 for Lilienthal's wing, we obtain a theoretical lift coefficient of 0.78, still much higher than the value measured by Lilienthal. However, recall that Lilienthal's data were obtained at low Reynolds numbers. Determination of the precise correction to the theoretical value to take into account the effects of low Re would require modern wind-tunnel data or CFD calculations. In lieu of such extensive efforts, we can only apply, for consistency, the rule of thumb used in some of our earlier discussions and reduce the theoretical lift coefficient further by a factor of 0.68 to take into account Re effects. That brings the theoretical value down to 0.53, still 39% higher than Lilienthal's measurement.

The problem was not any mistake in Kutta's theoretical development. His equation for lift was quite valid for an inviscid, incompressible flow over a circular-arc airfoil, as has been verified using modern thin-airfoil theory. Of course, in 1902 there was no theory that could provide the proper correction to the theoretical findings for a two-dimensional airfoil to take into account the finite aspect ratio of a real wing, and there was no appreciation of the Reynolds-number effect, let alone a proper theoretical correction. As discussed in Chapter 5, the problem was not necessarily with Lilienthal's measurement either; the remaining difference between theory and experiment could very well have been due to low-Re effects. The breakthrough in 1902 with Kutta's equation was that finally a theoretical value for airfoil lift could be obtained, albeit assuming an inviscid, incompressible flow. The fact that the theory gave a lift coefficient of 1.047, whereas experiment yielded a value of 0.381, did not diminish the value of Kutta's contribution. The difference simply underscored the gap between ideal theory and real life – a gap that to some extent still exists today.

Kutta's equation was derived without recourse to the concept of circulation; indeed, at that time, he was not aware of Lanchester's work. Kutta's image of the flow over an airfoil was that described in a more fundamental sense by the governing flow equations (the Euler equations), rather than the more abstract image of a synthesized flow combining uniform flow and circulatory flow. Kutta wrote the stream function for the flow over a circular-arc airfoil (recall that the concept of a stream function was introduced by Lagrange, as discussed in Chapter 3); a constant value of the stream function gives the equation for a streamline, and of course the two streamlines immediately adjacent to the top and bottom surfaces of a curved airfoil must have the same shapes as those surfaces. By differentiating the stream function, Kutta was able to obtain the local flow velocities over the top and bottom surfaces of the airfoil; the pressure distribution, and hence the net lift, followed from Bernoulli's equation. For that calculation, Kutta did not appeal explicitly to the concept of circulation. However, in 1910, again prompted by Finsterwalder, he prepared a second note that revisited certain aspects of his 1902 dissertation: "Ueber eine mit den Grundlagen des Flugsproblems in Beziehung stehende zweidimensionale Strömung," in the proceedings of the Royal Bavarian Academy of Sciences, January 8, 1910. By reinterpreting some of his 1902 theoretical development, he found the classic relation for lift as the product of density, velocity, and circulation, albeit in a form slightly different from that buried in the 1902 dissertation. For that reason, it can be said that Kutta shared in the development of the circulation theory of lift. However, that came to light only in 1910, five years after the relation was clearly and explicitly derived and published independently by Nikolai Joukowski in Moscow.

Kutta was primarily a mathematician whose interest in aerodynamics was sparked by Lilienthal's glider flights. After 1902 he was a professor of mathematics, finally settling at the Technische Hochschule in Stuttgart in 1911, from where he retired in 1935. His death came in 1944 as Germany was rushing headlong into defeat in World War II.

At the time the Wright brothers were carrying out their wind-tunnel tests and Kutta was finishing his dissertation, a 55-year-old professor was directing the construction of the first wind tunnel in Russia. Nikolai Joukowski (Zhukovsky) was professor of mechanics at Moscow University and professor of mathematics at the Moscow Higher Technical School, at which the wind tunnel was being built. A native of Orekhovo, Vladimir province, Russia, Joukowski was born January 17, 1847, the son of a communications engineer. Joukowski earned a bachelor's degree in mathematics from the University of Moscow in 1868 and in 1870 began teaching. In 1882 he completed his Ph.D. at Moscow University with a dissertation on the stability of fluid flows. Four year later he became head of the Department of Mechanics of Moscow University. Joukowski published more than 200 papers in his lifetime, dealing with basic and applied mechanics. By the turn of the century he was one of Russia's most respected scientists, considered the founder of Russian hydrodynamics and aerodynamics. In 1885 he was awarded the N. D. Brashman Prize for major theoretical research in fluid dynamics, and in 1894 he became a member of the St. Petersburg Academy of Sciences. From 1905 until his death in 1921, Joukowski served as president of the Moscow Mathematical Society.

In the late 1880s, contemporary with Otto Lilienthal's aeronautical activities, Joukowski became interested in flying machines. We have already noted that in 1895 he visited Lilienthal in Berlin and purchased one of the eight gliders that Lilienthal sold to the public. This was the first time that a university-educated mathematician and scientist had become closely connected with a real flying machine, actually getting his hands on one. Joukowski was motivated by his interest in flying machines to examine the aerodynamics of flight on a theoretical, mathematical basis.

In particular, he directed his efforts toward the calculation of lift. As early as 1890 he began to conceive a model of the flow over a lifting airfoil as consisting in some way of vortical motions caused by the fluid viscosity. He envisioned bound vortices fixed to the surface of the airfoil, along with the resulting circulation that somehow had to be related to the lifting action of the airfoil. Finally, in 1906 he published two notes, one in Russian and the other in French, in two rather obscure journals: *Transactions of the Physical Section of the Imperial Society of Natural Sciences,* in Moscow, and *Bulletin de l'Institut Aerodynamique de Koutchino,* in St. Petersburg. In those notes he derived and used the following relation for calculating the lift per unit span of an airfoil:

$$L = \rho V \Gamma$$

where Γ is the *circulation,* a technically defined quantity equal to the line integral of the flow velocity taken around any closed curve encompassing the airfoil. That equation was a revolutionary development in theoretical aerodynamics, for the first time allowing calculation of the lift on an airfoil with mathematical precision. Because Kutta was able to show in hindsight that the essence of that relation could be found buried in his 1902 dissertation, that equation has become known as the *Kutta-Joukowski theorem.* It is still taught in university aerodynamics courses and is used to calculate the lift for airfoils in low-speed, incompressible flows.

The apparent simplicity of the Kutta-Joukowski theorem belies the fact that considerable effort usually is required to calculate the value of Γ for a given airfoil at a given angle of attack in a free stream of a given velocity. This is where the model of vortex filaments aligned with the span of the wing comes into the picture. The strengths of the vortex filaments must be calculated precisely so that the resulting flow (the flow induced by the vortices plus the flow due to the free stream) will be tangent to the airfoil along its surface. Once the proper vortex strengths are calculated, they are added together to yield the total circulation Γ associated with the complete airfoil. That is the value of Γ that is inserted into the Kutta-Joukowski theorem to give the lift per unit span.

That is the essence of the circulation theory of lift, composed of elements from the thinking of Lanchester, Kutta, and Joukowski. It provided the foundation for all theoretical aerodynamics for the first 40 years of the twentieth century, after which the advent of high-speed flight required that the compressibility of air be taken into account. The circulation theory of lift is still alive and well today; for example, it is the basis for modern "panel" techniques, carried out with digital computers, for calculating lift values for airfoils in inviscid, incompressible flows. Such panel techniques are continually being revised and improved, and thus the circulation theory of lift is still evolving today, 90 years after its first introduction.

Joukowski went on to become "the father of Russian aviation." He established an aerodynamics laboratory in Moscow during the first decade of the twentieth century and gave a series of lectures on the theoretical basis of aerodynamics, relying heavily on his own theoretical and experimental work – the first systematic course in theoretical aerodynamics. Those lecture notes were recorded by two of his students and were published after Joukowski reviewed them. The first Russian edition appeared in 1912, and the first French edition in 1916; second editions in Russian and in French confirmed the value of those notes. So far as I can determine, there has never been an English translation. Joukowski developed a means of designing airfoils using conformal mapping and the techniques of complex variables. Those Joukowski airfoils were actually used on some aircraft, and today those techniques provide a mathematically rigorous reference solution to which modern approaches to airfoil design can be compared for validation. During World War I, Joukowski's laboratory was used as an instructional school for new military pilots. Shortly before his death, Joukowski founded a new aerodynamics laboratory just outside Moscow called the Central Institute for Aerodynamics. This institute continues to the present time; known as TsAGI, it is Russia's premier aerodynamics facility, the Russian equivalent of the NASA laboratories.

Nikolai Joukowski died in Moscow, March 17, 1921. At the time of his death, he was working in the two areas of high-speed aerodynamics and aircraft stability. Always moving forward, the man who revolutionized the analysis of low-speed airfoils in 1906 would exit the world in 1921 with his attention fixed on analysis of supersonic vehicles and the wave patterns associated with such objects.

Some New Thinking about Drag: Ludwig Prandtl and His Boundary-Layer Theory

The main concern about predicting the lift on a body inclined at some angle to a flow surfaced in the nineteenth century, beginning with George Cayley's concept of generating a sustaining force on a fixed wing. In contrast, concern about drag dates to ancient Greek science. The retarding force on a projectile hurtling through the air has been a major concern for millennia. But the breakthroughs that allowed theoretical predictions of drag and lift occurred at almost the same time, bearing no relationship to how long the two problems had been investigated.

The calculation of aerodynamic drag was a most vexing issue for eighteenth-century scientists. We have discussed d'Alembert's perplexity over the paradoxical findings from his calculations, which indicated that the drag on a two-dimensional body should be zero – totally contrary to real-life observations. We have discussed how some nineteenth-century scientists attempted to resolve that paradox by assuming that surfaces of discontinuity flowed downstream of a body, resulting in a net imbalance in the pressure distribution over the body, yielding a pressure drag on the body. But efforts by Kirchhoff and Rayleigh to calculate the drag using such surfaces of discontinuity did not give accurate results, even though the concept of such surfaces was not far from the truth. In regard to the effect of shear stress on drag, we have seen how such notables as Langley and Chanute believed that skin-friction drag could be neglected. Such was the status of drag calculations as late as the time of the Wright Flyer in 1903. But things were about to change dramatically.

In the fall of 1903, as Wilbur and Orville were laboring over the Wright Flyer at Kill Devil Hills, an introspective young professor at the Technische Hochschule in Hannover, Germany, was considering a revolutionary new idea. Ludwig Prandtl was studying the effect of friction in a fluid flow: How did friction affect a fluid flow? To what extent was the flow field dominated by friction? And how did friction affect the drag on a body moving through a fluid? His ideas were not being developed in a vacuum; some work only a few years earlier by the Italian mathematician T. Levi-Civita had presaged Prandtl's ideas. Whereas Kirchhoff and Rayleigh had postulated that surfaces of discontinuity would be created at sharp edges on a body, Levi-Civita believed that such surfaces of discontinuity would also occur at some point or points on a body with a curved shape and that such surfaces of discontinuity would extend to infinity downstream of the body. Thus, a wake would be formed downstream of the curved body, essentially forming a dead-air region inside the wake, with the motion of the fluid outside the wake being the usual inviscid, irrotational flow considered by d'Alembert and others over the preceding century and a half. In that manner, Levi-Civita was getting close to the idea of flow separation over a curved body, and the attendant pressure drag created by such flow separation. Indeed, Levi-Civita pointed to the smoke-flow photographs of Marey during the 1890s as proof of the existence of points of separation on the backside of a curved body and the ensuing wake. Levi-Civita went further; he began to consider the nature of the surface of discontinuity itself. He thought of the body surface as being covered with a surface of vorticity, with that vorticity layer eventually separating from the body, forming a surface of discontinuity that flowed downstream. In a note published in *Comtes Rendus de l'Académie des Sciences,* June 1901, Levi-Civita discussed that vortex layer:

> New vortex rings must continually emanate from the first part of the surface of discontinuity, in contact with the body, descending along the surface, in substitution for those which separate at the end. To the formation of these vortices, as is apparent, only fluid particles can contribute which were previously in contact with the fore part of the surface of discontinuity and they must contribute to this formation according to a law dependent on the regime of motion of the surface.

Levi-Civita's idea in 1901 that the fluid particles in contact with the fore part of the surface would create a layer of vorticity, which subsequently would separate from the body at some point on the surface, sounded much like Prandtl's boundary-layer concept that would be announced three years later.

Until 1904, the role that friction played in determining the characteristics of the flow over a body was conjectural and somewhat controversial. There was the question of what happened at the surface of the body: Did the fluid immediately adjacent to the surface stick to the surface, giving zero velocity at the surface, or did it slip over the surface with some finite velocity? We have already mentioned that Langley favored the no-slip condition (i.e., the idea that the relative velocity between the surface and the flow at the surface was zero). We have also seen that Langley considered the effect of friction on the aerodynamic drag to be negligible. And there was always the question as to how much of the flow field itself was dominated by the effect of friction.

In 1904, Ludwig Prandtl read a paper before the Third International Mathematical Congress at Heidelberg that was to bring revolutionary changes in aerodynamics: "Ueber Flussigkeitsbewegung bei sehr kleiner Reibung."[61] It was only eight pages long, but it would prove to be one of the most important fluid-dynamics papers ever written. Much later, in 1928, when asked by the fluid dynamicist Sydney Goldstein why the paper was so short, Prandtl replied that he had been given only 10 min for his presentation, and he was under the impression that his paper could contain only what he had time to say.[62]

The important thing about Prandtl's paper was that it gave the first description of the boundary-layer concept. Prandtl theorized that the effect of friction was to cause the fluid

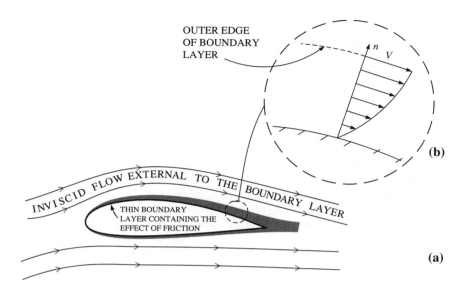

Figure 6.4   Division of a flow into two parts: (a) the thin boundary layer adjacent to the
surface, where the effects of friction are dominant, and an inviscid external flow outside of the
boundary layer; (b) enlarged sketch of the boundary layer showing the variation in velocity
across the boundary layer as a function of the normal distance, perpendicular to the surface.

immediately adjacent to the surface to stick to the surface (i.e., he assumed the no-slip
condition at the surface) and that the effect of that friction was experienced only in the
near vicinity of the surface (i.e., the influence of friction was limited to a thin region called
the boundary layer). Outside the boundary layer, the flow was essentially uninfluenced
by friction (i.e., it was the inviscid, potential flow that had been studied for the past two
centuries). The concept of the boundary layer is sketched in Figure 6.4. In the types of
flows associated with a body in flight, the boundary layer is very thin compared with the
size of the body, much thinner than can be shown in Figure 6.4a. In Figure 6.4b, a portion
of the boundary layer is enlarged to illustrate the variation of the flow velocity through the
boundary layer, going from zero at the surface to the full inviscid-flow value at the outer
edge of the boundary layer. With Figure 6.4 in mind, consider Prandtl's description of the
boundary layer:

> A very satisfactory explanation of the physical process in the boundary layer (*Grenzschicht*)
> between a fluid and a solid body could be obtained by the hypothesis of an adhesion of the
> fluid to the walls, that is, by the hypothesis of a zero relative velocity between fluid and
> wall. If the viscosity was very small and the fluid path along the wall not too long, the
> fluid velocity ought to resume its normal value at a very short distance from the wall. In
> the thin transition layer [*Uebergangsschicht*] however, the sharp changes of velocity, even
> with small coefficient of friction, produce marked results.[61]

One of those "marked results" is that within the boundary layer, there is an enormous change
in velocity over a very short distance, as sketched in Figure 6.4b (i.e., there are very large
velocity gradients in the boundary layer). In turn, as described by Newton's shear-stress
law, which states that the shear stress is proportional to the velocity gradient, the local
shear stress can be very large within the boundary layer, meaning that skin-friction drag

is not negligible (contrary to what Langley and Chanute believed). Indeed, for slender aerodynamic shapes, *most* of the drag is due to skin friction.

Another "marked result" is flow separation:

> In given cases in certain points fully determined by external conditions, the fluid flow ought to separate from the wall. That is, there ought to be a layer of fluid which, having been set in rotation by the friction on the wall, insinuates itself into the free fluid, transforming completely the motion of the latter, and therefore playing there the same part as the Helmholtz surfaces of discontinuity.[61]

Prandtl was referring to the type of flow sketched in Figure 6.5, where the boundary layer separates from the surface. As seen in Figure 6.5a, driven by inviscid-flow conditions of a certain type, the boundary layer can separate and then trail downstream, much like the nineteenth-century concept of a surface of discontinuity. An essentially dead-air region is formed in the wake behind the body. The pressure distribution over the surface of the body is radically changed when the flow separates, such that the altered pressure distribution creates a large, unbalanced force in the drag direction – the pressure drag due to flow separation. When there is massive flow separation (Figure 6.5a), the pressure drag usually is much larger than the skin-friction drag. The type of external inviscid flow that promotes boundary-layer separation is a flow that produces an adverse pressure gradient (i.e., an increasing pressure in the flow direction). Prandtl explained that effect as follows:

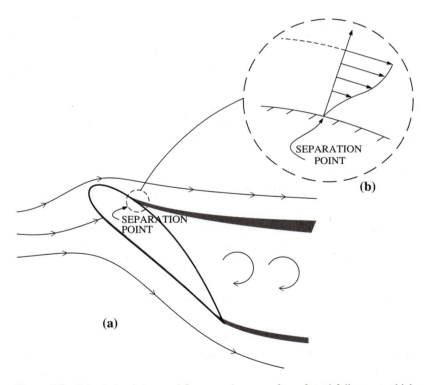

Figure 6.5 Schematic of separated flow over the top surface of an airfoil at a very high angle of attack – beyond stall.

> On an increase of pressure, while the free fluid transforms part of its kinetic energy into potential energy, the transition layers instead, having lost a part of their kinetic energy [due to friction], have no longer a sufficient quantity to enable them to enter a field of higher pressure, and therefore turn aside from it.[61]

That phenomenon is illustrated in Figure 6.5b: At the separation point, the fluid elements deep inside the boundary layer (which have already had substantial portions of their initial kinetic energies dissipated by friction) cannot work their way uphill against a region where the pressure is increasing. Hence, the velocity profile is depleted near the surface. At the separation point, it has an inflection point at the surface, as sketched in Figure 6.5b. Beyond that point, the fluid elements near the surface would actually be pushed backward by the increasing pressure, but nature does not allow that to happen; instead, the boundary layer simply lifts off the surface at that point, as shown in Figure 6.5a.

The overall perspective set forth by Prandtl in his 1904 paper was simple and straight-forward, namely, that an aerodynamic flow over a body can be divided into two regions: a thin boundary layer near the surface, where friction is dominant, and an inviscid flow external to the boundary layer, where friction is negligible. There is a strong effect of the outer inviscid flow on the boundary-layer properties; indeed, the outer flow is what drives the boundary layer. On the other hand, the boundary layer is so thin that it has virtually no effect on the outer inviscid flow. The exception to that is when the flow separates; then the outer inviscid flow is greatly modified by the presence of the separation region. Prandtl's view of those phenomena was as follows:

> While dealing with a flow, the latter divides into two parts interacting on each other; on one side we have the "free fluid," which [is] dealt with as if it were frictionless, according to the Helmholtz vortex theorems, and on the other side the transition layers near the solid walls. The motion of these layers is regulated by the free fluid, but they for their part give to the free motion its characteristic feature by the emission of vortex sheets.[61]

Prandtl used the terms "transition layer" and "boundary layer" interchangeably. Indeed, he used the term "boundary layer" only once in that paper, while frequently referring to the "transition layer." "Boundary layer" is the term that has survived, mainly because of its use in subsequent papers by Prandtl's students.

With the advent of Prandtl's boundary-layer concept it became possible to make rational, and sometimes accurate, calculations of aerodynamic drag. For a boundary layer, the Navier-Stokes equations can be reduced to a simpler form, applicable only to the boundary layer itself. The simpler equations, the *boundary-layer equations,* lend themselves to a multiplicity of solution techniques and thus are much more tractable than the complete Navier-Stokes equations. Such solutions for boundary-layer flows began shortly after Prandtl's presentation in 1904. With those solutions, it became possible to predict with some accuracy the skin-friction drag on a body and, to a certain extent, the locations of flow separation on the surface, and hence the pressure drag due to flow separation (the form drag). In his 1904 paper, short as it was, Prandtl gave the boundary-layer equations for steady two-dimensional flow, suggested some solution approaches for those equations, made a rough calculation of friction drag on a flat plate, and discussed aspects of boundary-layer separation under the influence of an adverse pressure gradient. Those were all pioneering contributions. Goldstein was moved to state that "the paper will certainly prove to be one of the most extraordinary papers of this century, and probably of many centuries."[62]

Early Progress in Boundary-Layer Theory

If Prandtl had presented his paper in our electronic age of almost instant information dissemination, his boundary-layer concept would quickly have spread throughout the aerodynamics community, but at the turn of the century information flowed much more slowly. Also, the International Mathematical Congress was an obscure setting for such an important contribution, and Prandtl's boundary-layer concept went virtually unnoticed for several years. It surfaced again in 1908 when Prandtl's student, H. Blasius, published a paper in the respected journal *Zeitschrift für Mathematik and Physik:* "Boundary Layers in Fluids with Small Friction,"[63] which discussed two-dimensional boundary-layer flows over a flat plate and a circular cylinder. The boundary-layer equations were solved for both cases. For the flat plate, Blasius obtained an even more accurate solution for skin-friction drag – one that is still used. For the circular cylinder, solution of the boundary-layer equations gave the separation points on the back side of the cylinder. The boundary-layer equations, though simpler than the Navier-Stokes equations, are still coupled, nonlinear partial differential equations. Blasius showed that for certain types of external pressure gradients in the flow direction (such as the constant pressure along a flat plate at a zero angle of attack), the boundary-layer equations reduce to a single ordinary differential equation, known today as the Blasius equation.

In spite of the important work by Blasius and the publication of several papers on boundary-layer theory by Prandtl's research group at Göttingen during the next few years, the aerodynamics community paid little attention, especially outside of Germany. That work was not appreciated and was not widely disseminated. Finally, in 1921, Theodore von Kármán, a former student of Prandtl and a professor at the University of Aachen, developed a particular form of the boundary-layer equations called the "integral form," which proved to be directly applicable to a large number of practical engineering problems. With that, the boundary-layer theory finally began to receive more attention and to be more widely accepted in the technical community. The delayed acceptance of the boundary-layer concept is illustrated by the fifth and sixth editions of Sir Horace Lamb's classic text *Hydrodynamics*.[54] The fifth edition, published in 1924, devoted only one paragraph to the boundary-layer concept, describing Prandtl's work as follows: "The calculations are necessarily elaborate, but the results, which are represented graphically, are interesting." In contrast, the sixth edition, published in 1932, had an entire section on boundary-layer theory and the governing equations.

Since the mid-1920s, work aimed at advancing, extending, and applying boundary-layer theory has increased exponentially. It has taken on a life of its own, creating lifetime careers for a large number of fluid dynamicists and aerodynamicists. Dozens of books concerning various aspects of boundary-layer theory have now been written, the classic and best-known being that by Hermann Schlichting: *Boundary-Layer Theory*.[64] Schlichting was a student of Prandtl's during the early 1930s, conducting research on various aspects of flow with friction. He left Göttingen in 1935 to work for the Dornier Company and in 1939 became a professor at the Technische Hochschule in Braunschweig. In 1957 he became the director of the Aerodynamische Versuchsanstalt (AVA), one of Germany's most prestigious aerodynamic laboratories, created in 1919 on the grounds of Göttingen University for studies in aerodynamics and propulsion of airplanes. When graduate students of aerodynamics read Schlichting's book today (and most do), they are exposed to technical material whose roots extend back to Prandtl's 1904 paper, communicated by an author who worked closely with Prandtl – wonderful continuity between the past and the present for an understanding of viscous flows.

Ludwig Prandtl

Prandtl made important contributions to twentieth-century aerodynamics aside from his boundary-layer concept. For example, picking up on the circulation theory of lift, Prandtl developed a theory for calculating the lift and moment coefficients for thin, cambered airfoils; Prandtl's thin-airfoil theory, developed during World War I, allowed the first practical calculations of airfoil properties, and it is still used today. During the same period, Prandtl developed his lifting-line theory for finite wings, the first method for theoretical calculation of aspect-ratio corrections; it also confirmed the existence of induced drag on a finite wing (drag due to the presence of wing-tip vortices) and provided an engineering method for accurate calculation of such drag. Prandtl's lifting-line theory is still in common use. In the area of high-speed aerodynamics, Prandtl and his student, Theodor Meyer, developed the first theory for calculating the properties of oblique shock and expansion waves in a supersonic flow; that was the topic of Meyer's dissertation in 1908, four decades before the first supersonic airplane. In the 1920s, Prandtl developed the first rule for correcting low-speed airfoil lift coefficients to take into account compressibility effects at high subsonic speeds – very useful for the high-speed airplanes of World War II. Also, in 1929 Prandtl and Adolf Busemann first applied the rigorous method of characteristics to design the proper shape for a supersonic nozzle. All designs for supersonic-wind-tunnel nozzles and rocket-engine nozzles use the same basic method today. The contributions cited here were only a few of Prandtl's many contributions to fluid dynamics and to the field of mechanics in general. Several of his other contributions will be discussed at length in subsequent chapters.

Ludwig Prandtl was born February 4, 1874, in Freising, Bavaria. His father, Alexander Prandtl, was a professor of surveying and engineering at the agricultural college at Weihenstephan, near Freising. Although the Prandtls had three children, two died at birth, and Ludwig grew up as an only child. His mother suffered from a protracted illness, and partly as a result of that he became very close to his father. At an early age he became interested in his father's books on physics, machinery, and instruments. Perhaps his remarkable ability to go straight to the heart of a physical problem can be traced to his childhood environment, for his father, a great lover of nature, taught him to observe natural phenomena and to reflect on them.

In 1894 Prandtl began scientific studies at the Technische Hochschule in Munich, where his principal teacher was the well-known mechanics professor August Föppl. Six years later he graduated from the University of Munich with a Ph.D., with Föppl as his advisor. By that time Prandtl was alone; his father had died in 1896, and his mother in 1898.

Prandtl showed no interest in fluid mechanics prior to 1900. Indeed, his Ph.D. work at Munich had been in solid mechanics – unstable elastic equilibrium in which bending and distortion acted together. Prandtl continued his interest and research in solid mechanics through most of his life but that work was overshadowed by his many major contributions to the study of fluid flows. Soon after graduation from Munich, Prandtl had his first major encounter with fluid mechanics. Joining the Nürnberg works of the Maschinenfabrik Augsburg as an engineer, Prandtl worked in an office designing mechanical equipment for the new factory. He was assigned to redesign a suction device to collect lathe shavings. Finding no reliable information in the scientific literature on the fluid mechanics of suction, Prandtl carried out some experiments to answer a few fundamental questions about such flows. The result of that work was his new design for a shavings collector. The apparatus was modified with pipes of improved shapes and sizes, and it operated well at one-third of its original power consumption. Prantl's contributions in fluid mechanics had begun.

Figure 6.6   Kaiser-Wilhelm-Institut für Strömungsforschung (ca. 1937).

A year later, in 1901, Prandtl became a professor of mechanics in the Mathematical Engineering Department at the Technische Hochschule in Hannover (a German "technical high school" is equivalent to a technical university in the United States). At Hannover he developed his boundary-layer theory and began work on supersonic flows through nozzles. After delivering his famous paper on the concept of the boundary layer in 1904, Prandtl's star would rise meteorically. Later that year he moved to the prestigious University of Göttingen to become director of the Institute for Technical Physics, spending the remainder of his life there and building his laboratory into the greatest aerodynamics research center of the 1904–30 period.

In 1925, the Kaiser-Wilhelm-Institut für Strömungsforschung (Kaiser Wilhelm Institute for Flow Investigation) was built on the grounds of Göttingen University, with Prandtl as director, in recognition of his important research achievements in mechanics (Figure 6.6). By the 1930s, Prandtl was recognized worldwide as the elder statesman of fluid dynamics. He continued to do research in various areas, including structural mechanics and meteorology, but his great contributions to fluid dynamics had already been made. He remained at Göttingen throughout World War II, engrossed in his work and seemingly insulated from the politics of Nazi Germany and the privations and destruction of the war. In fact, the German Air Ministry provided new equipment and financial support for Prandtl's laboratory.

Klaus Oswatitsch, one of Prandtl's later students who went on to become famous for his work in high-speed gas dynamics, related an interesting story concerning one of Prandtl's colleagues in the mid-1930s, J. Nikuradse, who was known for some landmark data on turbulent flow through smooth and rough pipes, published in 1932 and 1933, and still used today as a standard for comparison:

> Nikuradse published his test results [1932 and 1933] on turbulent flow through smooth and rough pipes; in order to define a special but reproducible roughness, the so-called sand grain roughness was invented. For many technical applications these two papers proved to be very important and were widely acknowledged. Unfortunately, this increased his self-esteem to such a height that he tried to replace Prandtl as director after Hitler had come to power. It was, indeed, a dangerous attack, for Nikuradse knew at least one man high up in the Nazi regime, whereas neither Prandtl nor Betz [Prandtl's closest assistant] ever became party members in spite of their important positions. Luckily Prandtl was victorious. Nikuradse

Figure 6.7   Ludwig Prandtl.

had to leave the Kaiser-Wilhelm-Institut and – without Prandtl's guidance – he never again wrote a paper worth mentioning.[65]

Prandtl's attitude at the end of the war was reflected in his comments to a U.S. Army interrogation team at Göttingen in 1945: He complained about bomb damage to the roof of his house, and he asked to what extent the Americans planned to support his current and future research. Prandtl was 70 at the time, and still going strong. However, Prandtl's laboratory did not fare well after the war: "World War II swept over all of us. At its end some of the research equipment was dismantled, and most of the research staff was scattered with the winds. Many are now in this country [the United States] and in England, [though] some have returned. The seeds sown by Prandtl have sprouted in many places, and there are now many 'second growth' Göttingers who do not even know that they are."[66]

By all accounts Prandtl (Figure 6.7)[66] was a gracious man, likable and friendly, but studious and totally focused on those things that interested him. He enjoyed music and was an accomplished pianist. One of Prandtl's students, Theodore von Kármán, mentioned in his autobiography that Prandtl bordered on being naive.[67] Perhaps the best illustration is that in 1909 Prandtl decided that he should be married, but he did not quite know how to achieve that. He finally wrote to Mrs. Föppl, the wife of his respected teacher, asking permission to marry one of her two daughters. Prandtl and Föppl's daughters were acquainted, but nothing more than that. Moreover, Prandtl did not stipulate which daughter. The Föppls made a family decision that Prandtl should marry the eldest daughter, Gertrude. They married, lived happily, and had two daughters.

Though perhaps naive, Prandtl was not lacking ego, as shown by his comment on receiving a letter announcing a new honor: "Well, they might have thought of me a bit earlier."[65] Prandtl was considered a tedious lecturer. He could hardly make a statement without qualifying it. Nevertheless, he expected his students to attend his lectures, and he attracted excellent students, many of whom went on to distinguish themselves in fluid mechanics, such as Jakob Ackeret in Zürich, Switzerland, Adolf Busemann in Germany, and Theodore von Kármán in Aachen, Germany, and later at the California Institute of Technology. Writing in 1954, von Kármán commented on Prandtl:

> Prandtl, an engineer by training, was endowed with rare vision for the understanding of physical phenomena and unusual ability in putting them into relatively simple mathematical form. His control of mathematical method and tricks was limited; many of his collaborators and followers surpassed him in solving difficult mathematical problems. But his ability to establish systems of simplified equations which expressed the essential physical relations and dropped the nonessentials was unique, I believe, even when compared with his great predecessors in the field of mechanics – men like Leonhard Euler and d'Alembert [pp. 50–1].[68]

Prandtl led an unpretentious life in Göttingen. He and his family lived in an apartment only 20 minutes walking time from the institute. In addition to his professional interests, Prandtl particularly enjoyed music. He was an accomplished pianist, with a preference for classical music. Oswatitsch[65] tells of Prandtl playing waltzes for dancing when students visited his home.

We are reminded again of Octave Chanute's prophetic statement in *Progress in Flying Machines:* "Science has been awaiting the great physicist, who, like Galileo or Newton, should bring order out of chaos in aerodynamics, and reduce its many anomalies to the rule of harmonious law." Ludwig Prandtl was that great physicist. Chanute died in 1910 without knowing of Prandtl and without knowing that in Göttingen order was being brought out of chaos in aerodynamics.

Prandtl died in 1953. He was clearly the father of modern aerodynamics and a monumental figure in fluid dynamics. The impact of his work will reverberate for centuries to come.

Academic Science Meets the Flying Machine

The beginning of the twentieth century brought numerous major achievements in theoretical aerodynamics that reflected a sociological shift. Wilhelm Kutta, Nikolai Joukowski, and Ludwig Prandtl were academic types, with doctorates in mathematics or the physical and engineering sciences, and all conducted aerodynamic research aimed directly at an understanding of heavier-than-air flight. That was the first time that respected academicians had embraced the idea of flying machines. Indeed, the problems associated with flying machines offered them some of their most interesting research challenges and channeled the directions of much of their research. Kutta, Joukowski, and Prandtl were quite taken with the airplane – what a contrast with the nineteenth century, when respected academicians shunned any association with flying machines, thus leaving a huge deficit in technology transfer from nineteenth-century science to the design of machines for powered flight.

What was it that had begun to change the opinions of academicians? It was the first tenuous achievements of Lilienthal and the Wright brothers. Otto Lilienthal's successful glider flights provided evidence that those fools pursuing the chimera of manned flight might be onto something after all. Both Kutta and Joukowski were intrigued by the image of Lilienthal winging through the air, and when the news of Wilbur and Orville's success with the Wright Flyer in 1903 gradually became known, there was no longer any doubt that flying machines were feasible. Suddenly, research in aeronautics could no longer be left to misguided dreamers and madmen; once aeronautical work became respectable, that opened the floodgates to a whole new world of research problems, to which twentieth-century academicians flocked. After that, the technology-transfer gap, so vast in earlier centuries, began to shrink. In the twentieth century, research in aeronautics, including that in aerodynamics, has been the driving force that has launched a host of technologies to higher levels of achievement, continuing to the present day. That is the subject matter for Part IV of this book.

New Aerodynamic Theories: Impact on Flying Machines

Although the technology-transfer gap between the state of the art of aerodynamics and contemporary flying machines suddenly grew smaller at the beginning of the twentieth century, it would not completely disappear. Prestigious scientists and engineers were beginning to work on the problems of flight, but the extent to which new developments in theoretical aerodynamics and the various fledgling aerodynamic theories were put to use in the design of airplanes was disappointingly small. Reflecting on the design of World War I airplanes, Loftin described the situation:

> The design of successful aircraft, even today, is not an exact science. It involves a combination of proven scientific principles, engineering intuition, detailed market or mission requirements, and perhaps a bit of inventiveness and daring. Aircraft design during World War I was more inventive, intuitive, and daring than anything else. Prototypes were frequently constructed from full-size chalk drawings laid out on the factory floor. The principles of aerodynamics that form so important a part of aircraft design today were relatively little understood by aircraft designers during the War. . . . In an area of engineering in which structural strength, lightweight, and aerodynamic efficiency are so important, it is indeed surprising that a number of relatively good aircraft were produced [pp. 8–9].[69]

When such famous World War I aircraft as the English Sopwith Camel, the French SPAD XIII, and the German Fokker D-VII were designed, the circulation theory of lift and boundary-layer theory had been published, but they were being developed and applied primarily in only two locations: Prandtl's laboratory at Göttingen and Joukowski's laboratory at Moscow. At that time, only a few people understood the essence, let alone the details, of those theories. Such matters were still complete mysteries to airplane designers, especially those outside of Germany.

Let us examine the situation in regard to a few specific airplanes. The British Sopwith Camel (Figure 6.8), a single-seat fighter designed and built in 1917, was a small airplane, with a gross weight of only 1,482 lb. The struts and support wires between the two wings,

Figure 6.8   English Sopwith Camel (1917).

---

characteristic of almost all World War I aircraft, caused considerable drag. They were usually blunt structures, which led to massive flow separations over their back surfaces, and thus the form drag was large in spite of their relatively small diameters. Engineers at the Royal Aircraft Factory at Farnborough, England, knew enough about flow separation to design bracing wires with a streamlined, airfoil-like cross section, in an effort to reduce the form drag (pressure drag due to separation). Such wires were used on the Sopwith Camel. But the mechanisms for drag reduction still were not well known by airplane designers at that time; in the case of the Camel, in spite of the streamlined bracing wires, the overall zero-lift drag coefficient was 0.0378, a fairly high value. By comparison, the drag coefficient for the North American P-51 Mustang in World War II was only 0.0163, smaller by more than a factor of 2. As for induced drag, which is determined mainly by the aspect ratio of the wings, the designers of the Camel had no knowledge of Prandtl's finite-wing theory and therefore had no real appreciation of the effect of aspect ratio. The aspect ratio of the Camel was only 4.11, a relatively small value. That apparent lack of understanding of the importance of using a high-aspect-ratio wing is somewhat surprising. Even though Prandtl's finite-wing theory was essentially bottled up in wartime Germany, the experimental findings of Langley and the Wright brothers were available, clearly showing the aerodynamic benefit of high-aspect-ratio wings. Indeed, the aspect ratio for the Wright Flyer had been 6.4, considerably higher than that for the Camel. So there was still a gap between the state of the art of aerodynamics and the design of aircraft at that time, and the Sopwith Camel suffered because of that gap: Its top speed was only 105 mph.

Such observations are reinforced when we consider the design of the French SPAD XIII (Figure 6.9). SPAD was the acronym for the French aircraft company Société pour Aviation et les Dérivés, headed by the famous aviator Louis Blériot. (In 1909 Blériot became the first person to fly across the English Channel; he flew an airplane of his own design, a monoplane, the Blériot XI.) The SPAD incorporated no new technology; indeed, the bracing wires were simple cables with a circular cross section. The SPAD's drag coefficient was 0.0367, essentially the same high value as that of the Camel. Its aspect ratio was very low (3.69); for comparison, that of the P-51 Mustang was 5.86, not quite twice as large. The SPAD had a relatively high maximum speed of 134 mph because of its powerful engine that simply overcame the relatively poor aerodynamic design features. That and its great structural strength made the SPAD XIII one of the best aircraft of World War I (8,472 were manufactured). Aerodynamics was not always the determining factor in the success or failure of an airplane.

Figure 6.9 French SPAD XIII (1917).

Figure 6.10   German Fokker D-VII (1917).

The German engineers responsible for the Fokker D-VII (Figure 6.10) probably had access to data from Prandtl's laboratory at Göttingen, though it is not clear how much use they made of such data. The zero-lift drag coefficient of the Fokker D-VII was 0.0404, which, surprisingly, was slightly higher than those for the Camel and SPAD. Its aspect ratio was 4.7, which was higher than those for the Camel and SPAD and represented a step in the direction of improved aerodynamics. That probably reflected an appreciation for the effects of aspect ratio on the part of the designers of the D-VII; they most likely had access to Prandtl's lifting-line theory, which clearly spelled out the aerodynamic advantage of high-aspect-ratio wings. The Fokker D-VII was designed in late 1917 and first entered combat in April 1918. Much of Prandtl's lifting-line theory for finite wings was published in a number of confidential technical reports for the German military aviation authorities in 1917. It would be surprising if those reports had not reached the hands of the Fokker engineers in time for the D-VII. One technical innovation used in the D-VII clearly had been derived from wind-tunnel experiments at Prandtl's laboratory: a thick airfoil section. A similar thick airfoil was used on the Fokker Dr-1 triplane. Some typical airfoil sections used in World War I aircraft are shown in Figure 6.11; for most, thin airfoils were used, such as the top three shapes. Those shapes were holdovers from the tradition of very thin airfoils used by Cayley, Phillips, Lilienthal, Langley, and the Wrights. Early wind-tunnel tests, including those of the Wright brothers, comparing thin and thick airfoils had shown that thick airfoils led to higher drag, but those early tests involved only low Reynolds numbers, and the data were misleading. Modern studies of low-Re flows over thick airfoils clearly show high drag coefficients and low lift coefficients because of the creation of a laminar-separation bubble near the leading edge that can burst and cause massive flow separation, even at very low angles of attack. That does not happen at higher Reynolds numbers, especially the higher Reynolds numbers associated with actual flight vehicles. So the early wind-tunnel tests frequently were misleading. That situation was compounded by an undesirable aerodynamic feature associated with very thin airfoils: At the high Reynolds numbers associated with flight, a thin airfoil at even a small angle of attack can experience flow separation right at the leading edge, much like that for a thin flat plate. That effect was not understood in those days. Moreover, it was not picked up in the early wind-tunnel tests because they were at low Reynolds numbers, which masked the effect. In any event, the early wind-tunnel tests were misleading in both respects. During the war, new wind-tunnel tests at Göttingen indicated that thick airfoils produced less drag and more lift than thin

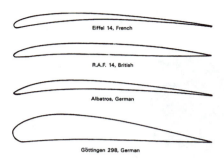

Figure 6.11   Examples of airfoil shapes used in World War I aircraft.

airfoils, especially at an inclination. A family of thick Göttingen airfoils was developed at
Prandtl's laboratory (e.g., the Göttingen 298 in Figure 6.11). The use of this type of airfoil
on the Fokker Dr-1 and the D-VII allowed those aircraft to climb faster and maneuver
better than any of the Allied airplanes, including the Camel and SPAD, because of the
better aerodynamic properties of such thick airfoils at high angles of attack. That rate-of-
climb performance and its excellent handling characteristics made the Fokker D-VII the
most effective of all German World War I fighters. The respect accorded that machine was
clearly indicated in Article IV of the armistice agreement, which listed the war material to
be handed over to the Allies by Germany: The Fokker D-VII was the only airplane listed.

The use of thick airfoils on those airplanes was an unusually rapid transfer of technology
from the laboratory to airplane design, but that transfer was from *applied* aerodynamics.
Direct transfers of technology from *theoretical* aerodynamics to airplane design were still
slow to come, although the gap was getting smaller.

At the beginning of the twentieth century, new aerodynamic concepts derived from the
circulation theory of lift and the boundary-layer theory did not find immediate applications
in the design of airplanes, but soon after World War I those theories achieved worldwide
notice and acceptance. Theoretical aerodynamics had finally caught up with and begun
to complement the earlier-maturing practice of applied aerodynamics, and the stage was
set for the astonishing growth in aerodynamics research and applications over the rest of
the twentieth century. That growth would alternately drive and be driven by the design
of airplanes with performance capabilities never dreamed possible by Wilbur and Orville
Wright in 1903.

Twentieth-Century Aerodynamics

Aerodynamics in the Age of Strut-and-Wire Biplanes

Aeroplane manufacture should be benefitted by laboratory research even more than other branches of industry have been . . . . The results tend to substitute, in aeroplane construction, the judgement of the engineer for the intuition of the mechanic. The latter, for a new type, may chance upon a happy combination, but he is at another time as likely to blunder.

Gustave Eiffel (1910)

At six o'clock on the evening of August 8, 1908, a flying machine took off from the Hunaudières race track near Le Mans, France, and made two seemingly effortless circuits of the field, taking less than 2 min. The machine was the Wright Type A, a larger and more powerful version of earlier Wright aircraft, and the latest in the progression that began with the historic Wright Flyer in 1903. Wilbur Wright was giving the first public performance of a Wright airplane. An excellent pilot, Wilbur made four wide turns, each time banking the aircraft using the coordinated effects of wing warping and rudder control. The early evening was clear and windless, and the small number of people in the grandstand had no trouble seeing the great aeronautical feat. Among the sparse audience were a few French aviators, including Louis Blériot and Ernest Archdeacon, who knew better than most that they were seeing history in the making. Over the next five days Wilbur flew eight more times, the longest flight lasting more than 8 min. By that time the word of the stunning flights had spread throughout France, and the race track was packed with people, the crowds growing larger each day. Orville and Wilbur had first flown in relative obscurity on December 17, 1903, and there had been tenuous flights by others since then (e.g., Santos-Dumont in Paris in 1906, and Glenn Curtiss in 1908 at Hammondsport, New York), but it was not until Wilbur's spectacular performance in France in 1908 that the general public discovered the existence of the airplane.

With that discovery, attitudes about the value of scientific and engineering work in aerodynamics were transformed. Almost overnight it became fashionable, indeed critical, to learn more about the laws of nature that kept these flying machines in the air and to develop engineering solutions and techniques that would yield improved aerodynamic designs. We discussed in Chapter 6 how academic science met the flying machine at the turn of the century. After Wilbur's dramatic 1908 demonstration that the airplane was an established fact, the world of professional engineering suddenly had a new and exciting discipline to be developed.

That technical awakening was announced by the sound of new wind tunnels revving up throughout Europe. In 1908, Prandtl began to operate a wind tunnel at Göttingen with a square test section whose cross section was 2 m on a side, and in early 1909 in Paris, Gustave Eiffel was cranking up a new wind tunnel in the shadow of the great tower he had built 20 years earlier. Eiffel was soon to become France's first great aerodynamicist on the strength of his wind-tunnel experiments. Today the name Eiffel rarely crosses the lips of practitioners and students of aerodynamics, and the general public would not associate Eiffel's name with aerodynamics at all, but his contributions to experimental aerodynamics were as important in the history of technology as were his structural innovations embodied in the design and construction of the Eiffel Tower. Eiffel pioneered some of the experimental techniques in aerodynamics that we still use today, and he made the first quantitative measurement of some of the most basic parameters in airplane aerodynamics.

Gustave Eiffel: Man of Iron and Air

Alexandre-Gustave Eiffel was born December 15, 1832, in Dijon, France. Unable to gain entrance to the Ecole Polytechnique, he graduated with distinction from the Ecole Centrale des Arts et Manufactures at the age of 22, and for the next 50 years his professional reputation was made as a structural engineer, specializing in metal structures. His most famous bridge is the Maria Pia over the Douro River at Oporto, Portugal. Eiffel and the architect L. A. Boileau designed the first cast-iron-and-glass building, a department store in Paris; it had glass walls along three sides, and its open courts were covered by 30,000 ft^2 of skylights. The iron framework for the Statue of Liberty in New York was designed by Eiffel in 1884, but the structure that made Eiffel's name familiar throughout the world was the Eiffel Tower, erected in 1889 in Paris as part of the exhibition celebrating the 100th anniversary of the French Revolution. That imposing landmark, standing 300 m (984 ft) high, was the world's tallest fabricated structure until the Chrysler Building was built in New York in 1930.

During the last 21 years of his long life, Eiffel began a new career as an aerodynamicist. His first venture was a direct spin-off from the Eiffel Tower: measurements of the aerodynamic forces on various objects dropped from the second platform of the tower, 377 ft above ground level.[70] Those drop tests continued for four years and included a series of experiments on flat plates at various angles of attack. That method of aerodynamic testing was not easy, and the kinds of measurements that could be taken were limited. Eiffel measured an object's acceleration during a drop, subtracted its acceleration due to gravity in a vacuum, and from knowledge of the object's mass determined the aerodynamic force via Newton's second law.

Wishing to expand his aerodynamics research, in 1909 Eiffel designed and built a wind tunnel in a small building located on the Champ de Mars, within the shadow of the Eiffel Tower. Electric power to operate the wind tunnel was drawn from the tower's power supply. That wind tunnel was of a new design, the forerunner of a classic style still called the Eiffel-type tunnel (Figure 7.1). Air was sucked in through a convergent nozzle, shown at the right, by means of a blower, shown at the left. The test section was open, but was inside a larger, hermetically sealed enclosure that Eiffel called the "experimental chamber." The open-jet design allowed ready access to the test stream that came from the nozzle. The test stream itself was 1.5 m in diameter, large enough for testing models of reasonable size. A 68-hp motor drove the blower, which allowed a test-stream velocity up to 20 m/s (44 mph), but most of Eiffel's experiments were carried out at an airspeed of 12 m/s, because "for the range of speeds used in this work, it had been demonstrated that the reaction varies as the square of the velocity." Being practical, Eiffel offered a second reason: "In addition, greater precision in reading the manometer levels was secured at this velocity on account of the lack of vibration."

Eiffel was quick to publish his first wind-tunnel data: Only a year after beginning operation, *The Resistance of the Air and Aviation*[71] appeared in Paris – a masterpiece of

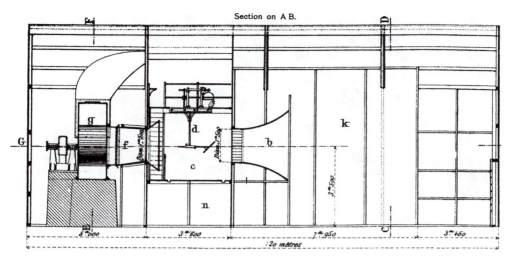

Figure 7.1   Eiffel's drawing of his 1909 wind tunnel.

clear engineering exposition, careful experimentation, and unique creative thoughts about experimental aerodynamics. The importance of Eiffel's work in 1913 is evidenced by the time and effort spent by Jerome Hunsaker in preparing the English translation at that time. Hunsaker was working for the U.S. government as "assistant naval constructor" and was already recognized as a technical expert in aeronautics. He was to become one of the leaders in the development of aeronautics in the United States during the first half of the twentieth century. He certainly did not have time on his hands to translate a work that was not considered to be seminal. *The Resistance of the Air and Aviation* set an important tone for further developments in experimental aerodynamics in the twentieth century, but today the role of Eiffel in aerodynamics is largely unappreciated, and even unknown to some. Eiffel's wind-tunnel experiments at the Champ de Mars laboratory during 1909–10 led to seven substantial contributions:

First, there was the wind tunnel itself. The use of a free jet in a hermetically sealed chamber was unique. Eiffel recognized that the static pressure in the free jet would be the same as in the chamber: "In the case of air passing through the experimental chamber, the streamlines are practically parallel, and the pressure is that of the chamber. The difference in pressure read on a manometer between the air in the hangar [the single-room building in which the wind tunnel was placed] and the air in the chamber represents the kinetic energy imparted to the air" (p. 3).[71] Eiffel recognized that the air in the hangar was the *reservoir* air for the tunnel, at a pressure p_0, and that from Bernoulli's equation the difference between the reservoir pressure and the test-section pressure, $p_0 - p$, was the kinetic energy per unit volume of the airstream:

$$p_0 - p = \tfrac{1}{2}\rho V^2$$

Two years later, Eiffel built a larger aerodynamics laboratory in the fashionable Paris suburb of Auteuil, featuring a larger tunnel with a test section 2 m in diameter and capable of airspeeds of 40 m/s. It had a dramatic design feature: a long, gradually diverging diffuser section downstream of the open jet that would slow the air and increase its pressure before it was dumped into the surrounding room. That made the wind tunnel's operation

Figure 7.2 Eiffel's wind tunnel at Auteuil (1912).

more economical; the free-jet test stream was larger and faster, at the same expenditure of horsepower, than that in his first tunnel. Eiffel's new wind tunnel at Auteuil is shown in Figure 7.2 (the flow was from left to right, and the open test section and gradually divergent diffuser are clearly seen). It was the kind of low-speed, subsonic wind tunnel still found in many aerodynamic laboratories – the Eiffel-style tunnel.

Second, because he had the advantage of the data from his drop tests for purposes of comparison, Eiffel was the first researcher to conclusively put to rest a certain doubt about wind-tunnel testing that had nagged earlier experimenters, a doubt having to do with the basic principle of the wind tunnel, as first stated by da Vinci: "The same force as is made by the thing against air, is made by air against the thing." However, some doubt about that persisted until the twentieth century, even though wind tunnels had been in use since their invention by Francis Wenham in 1871. Perhaps that nagging doubt was one of the reasons that credible investigators such as Lilienthal and Langley chose to use whirling arms rather than wind tunnels. In any event, Eiffel scientifically proved the validity of the wind-tunnel principle: He compared the aerodynamic force on a flat plate measured during his drop tests with the force measured in his wind tunnel and found them to be the same: "The concordance between the results obtained by the two methods shows that a surface moving through still air offers the same resistance as a similar surface held stationary in a wind. This point has been frequently under discussion" (p. 37).[71]

Third, Eiffel made the first detailed measurements of the distribution of pressures over the surface of an aerodynamic body. Earlier, the emphasis in aerodynamic testing had been on direct measurement of the aerodynamic force on an object. It should be noted that Eiffel was not the first to measure pressure distributions of any type in a wind tunnel. That had first been done by Johan Irminger and H. C. Vogt of Copenhagen in 1894. Vogt was a marine engineer who studied the aerodynamics of sails and air propellers, and Irminger was the director of the Copenhagen gas works; they built a crude wind tunnel that was mounted in an opening in the wall of a gas-works smokestack, using the draft of the chimney to draw air through the tunnel. Pressure measurements were made at three points on a flat plate.[32] However, Eiffel's investigations were much more extensive and thorough, and his detailed measurements of variations in surface pressure pioneered a new dimension in experimental aerodynamics, allowing new insight into the nature of aerodynamic flows over wings and bodies and new understanding of how lift and pressure drag are produced. Today, measurements of pressure distributions are among the most important functions of wind-tunnel testing. Eiffel described how he measured pressure distributions:

> In addition to the total resultant pressure [the net aerodynamic force], it is of interest to investigate the distribution of pressure over the back and face of a model. These pressures are measured on a sensitive manometer. The wing is pierced with a number of small holes conveniently distributed, and plugged with small screws. At the place where an observation is to be made, the screw is replaced by a nipple with an internal diameter of 0.5 mm. The nipple opens upon the side under consideration, and the opposite side is connected by a rubber tube with one branch of a manometer. The second branch is open to the still air of the chamber. Since the opening of the screw nipple is very small, the stream lines of air passing it can be considered at any instant parallel to the surface and to each other. Consequently, the stream lines are not affected by the presence of the opening, and the pressure, transmitted laterally, is correctly measured by the manometer [p. 18].[71]

That was the first clear statement of the principle of static-pressure measurement on an aerodynamic surface.

Eiffel carried out a large number of such pressure measurements on a variety of model wings and airfoils. Figure 7.3 shows Eiffel's measurements for a circular-arc airfoil with a camber of $\frac{1}{13.5}$. Figure 7.3a shows the measured chordwise pressure distribution over the top (dashed line) and bottom (solid line) surfaces of the wing at the wing midsection at an angle of attack of $10°$. The pressures are given as gauge pressures (measured above and below atmospheric pressure) in units of millimeters of water and are referenced to a free-stream velocity of 10 m/s. Thus negative pressures are the "vacuum" below atmospheric pressure, and positive pressures are those above atmospheric. Those data show a strong vacuum on the upper surface. Figures 7.3b and 7.3c show the *contours* of static pressure over the top and bottom surfaces, respectively, as plotted by Eiffel.[71] Those findings were dramatic for their time (1910); they were the first published data on detailed pressure distributions over a lifting wing. The contour plots (parts b and c) are most impressive (a contour line is a line of constant property, so those pressure contour lines are lines of constant pressure). Before the development of modern high-speed computers and computer graphic techniques, contour lines had to be plotted by hand using interpolations between the measured data points, a laborious affair. As a result, most representations of experimental data through the first half of the twentieth century did not involve contour plots. Only in the past 30 years, with the advent of computational fluid dynamics (CFD), have contour plots been widely used in aerodynamics, primarily for presentation of CFD data. So the large number of pressure-contour plots in Eiffel's book are quite impressive. From such contour plots, Eiffel correctly concluded that the pressure changed rapidly over the top of the wing near the leading edge (note the close clustering of the contour lines near the leading edge in Figure 7.3b). He also concluded that a wing tip was a region of strong vacuum on the top surface, although he did not make the connection between that measurement and the presence of wing-tip vortices. Lanchester was first to theorize the existence of wing-tip vortices (Figure 6.3), and he published that idea in 1907, but we do not know if Eiffel had read that work before or during his experiments at Champ de Mars.

Eiffel's inspiration for studying such pressure distributions was the earlier experiments by Sir Thomas Stanton, superintendent of the Engineering Department at the National Physical Laboratory (NPL) in England.[72] In 1903 Stanton measured the pressure distributions over the top and bottom of a small flat plate (7.6 × 2.5 cm) at various angles of attack in a

Pressure along centre line

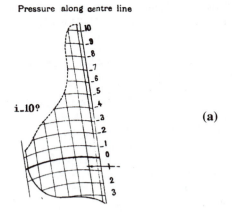

(a)

Curves of equal pressure on back

i.10°

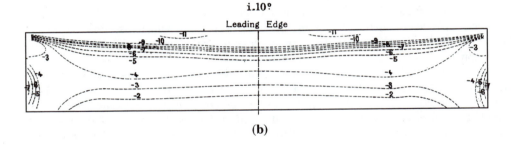

(b)

Curves of equal pressure on face

i.10°

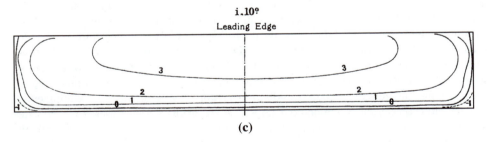

(c)

Figure 7.3   Pressure contours measured by Eiffel on a wing with a circular-arc airfoil.

primitive wind tunnel at the NPL. Stanton's measurements were more in the nature of unique curiosities than of any detailed usefulness. In contrast, Eiffel, because of the sheer mass of the large numbers of pressure measurements he took over a number of model wings more appropriate for airplane application, and because of his insightful interpretation of the findings, must be considered the pioneer of this technique.

Fourth, Eiffel first proved the general principle that the net lift on a body in a flow is the integrated effect of the pressure distributions over its surface. He used his measured pressure distributions and his direct force-balance measurements: "The direct measurement of pressure has given us a result to which we attach great importance; viz., *the summation of the observed pressures was equal in every case to the reaction weighed on the balance*" (p. 20).[71] He did not specify how close the agreement was, but because he was known for precision, we can presume that the agreement was within the range of experimental error, which at that time probably was 5% or less. Again, Eiffel was following Stanton at the NPL in making both pressure measurements and force measurements on a given model and comparing the two, but because of the limited extent of Stanton's measurements, it is Eiffel who deserves credit for pioneering that idea.

Fifth, Eiffel conclusively proved that the majority of the lift on a wing is derived not from the higher pressure exerted on the bottom of the wing but rather from the lower pressure existing over the top of the wing. He measured the aerodynamic force and the pressure distributions on a flat-plate wing as the angle of attack was decreased from 90°: "These figures show that the great increase in pressure [meaning the mean, net aerodynamic force per unit area], in passing from a normal to an inclined position, is due to the vacuum on the back. The vacuum triples in value while the pressure in front is diminished by only one-half." At lower angles of attack, from zero to 20°, Eiffel's data showed that for a flat-plate wing with an aspect ratio of 5.67, one-fifth of the lift came from the higher pressure on the bottom surface, and four-fifths from the vacuum on the top surface. For a wing of aspect ratio 6 and a circular-arc airfoil of camber $\frac{1}{13.5}$, the contributions to lift by the bottom and top surfaces were one-third and two-thirds, respectively: "In conclusion, these experiments have shown that for small angles [0–20°], the reaction of the air on a plane or curved surface is primarily due to the vacuum produced on the back. The variations in pressure and vacuum are most accentuated near the leading edge. The vacuum is greatest near the lateral extremities of the surface [the wing tips]" (p. 73).[71] Eiffel was not the first to observe that a vacuum exists on the top surface of a wing. Horatio Phillips had recognized that the pressure decreased in the airflow over the top surface of a wing, and Irminger and Vogt had shown that a vacuum existed over the top of a flat plate at an inclination, as had Stanton, but Eiffel produced the first conclusive data showing the nature of that vacuum and how it was distributed over the wing and quantifying the magnitudes involved.

Sixth, Eiffel conducted the first wind-tunnel tests that used models of complete airplanes and conclusively showed the correspondence between such tests and the performance of the real airplane in actual flight. In *The Resistance of the Air and Aviation*, Eiffel gave wind-tunnel data for five different airplanes: a Nieuport monoplane, the Balsan monoplane, the Paulham-Tatin Torpedo (Figure 7.4), the Letellier-Bruneau monoplane, and the M. Farman military biplane. For each model, plots of lift, drag, and lift-to-drag ratio were given as functions of the angle of attack. Techniques to measure those parameters in actual flight had not yet been developed, but Eiffel made the following comparisons between the wind-tunnel findings and actual flight: From his wind-tunnel aerodynamic measurements he calculated the power values that would be required for the aircraft to fly at various speeds. He compared those with the maximum horsepower available from the engine and estimated the maximum speed for the airplane. For the Paulhan-Tatin Torpedo, he calculated that the 40-hp engine would power the airplane to a maximum velocity of 34.7 m/s (78 mph). That may sound too high for a 1910 airplane, but the Paulhan-Tatin Torpedo was reasonably streamlined, and of the five aircraft models tested, it had the highest lift-to-drag ratio ($L/D = 6$), still

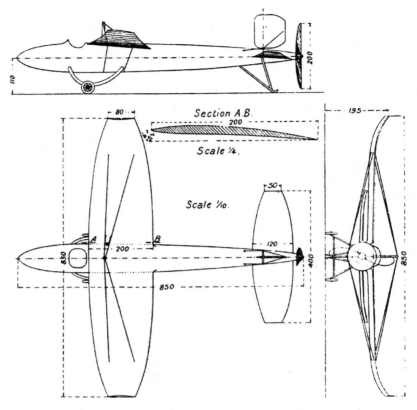

Model of Paulhan-Tatin monoplane.

Velocity	...	...	...	...	...	125 km. or 34·7 m/sec.
Area of Wings	...	..	...	...	12·5 m².
Weight in service	...	...	...	...	420 kg.

Figure 7.4   Wind-tunnel model of the Paulhan-Tatin monoplane used by Eiffel.

considerably below the 15–20 range for modern aircraft – airplane aerodynamics had a long way to go. Wind-tunnel tests of complete airplane models, as pioneered by Eiffel, would become the heart and soul of wind-tunnel applications in the twentieth century. He concluded that "in the preceding examples, the calculations are in each case in complete accord with the actual conditions observed in flight. The conclusion follows at once that, in aeroplane design, careful tests with a model aeroplane, or with model wings, permit the designer to predict the conditions of normal flight" (p. 104).[71]

The only weakness in Eiffel's statement was that he did not understand the Reynolds-number effect on aerodynamic data. He was aware that the size of a body was important in determining its aerodynamic coefficients; he had learned that from his drag measurements for flat plates at a 90° angle of attack, which varied with the size and aspect ratio. Strictly on an empirical basis, Eiffel adopted the following adjustment between his wind-tunnel findings and the performance of the actual airplane: "We have allowed above an augment of 10 percent in the coefficient of sustentation in passing from the model to the full-size wing." Today, we know from dimensional analysis that the lift and drag coefficients are functions of the Reynolds number, $\rho V c / \mu$, where c is a characteristic length, such as the chord of a wing. This is part of the powerful principle of dynamic similarity in aerodynamics. However, in Eiffel's time, the concept of dynamic similarity was still evolving. William Froude, in experiments on the drag of ships' hulls carried out for the British Admiralty from 1868 to 1874, had developed a similarity law for extrapolating model data to full-size shapes, but such similitude ideas were slow to develop. As discussed in Chapter 4, Reynolds had demonstrated the role of the Reynolds number in dictating the transition from laminar flow to turbulent flow, and toward the end of the nineteenth century Rayleigh was exploring the applications of dimensional analysis, but in Eiffel's time such concepts still were not widely known. As Walter Vincenti put it, at that time "dimensional ideas were still catching on among engineers."[36] However, Eiffel did use findings from dimensional analysis to correlate his propeller experiments, as discussed later. So he was not totally ignorant of the idea.

Seventh, Eiffel coined two terms that are used every day in aerodynamics. The first was the term "wind tunnel." Francis Wenham talked about the "artificial current," Horatio Phillips referred to a "delivery tube," and the Wright brothers simply called their wind tunnel "the apparatus." Stanton at the NPL called his device a "channel." The first use of the term "wind tunnel" in print appears to have been in 1910, when Eiffel referred to the flow of air through his duct as *ce qu'on appelle la méthode du tunnel;* in Hunsaker's 1913 translation, the term "wind tunnel" is used. Joseph Black's opinion about Eiffel's reference was that it "would appear to be the first published use of the term, which became in English and U.S. usage *wind tunnel*."[72] It seems appropriate and natural for Eiffel, a civil engineer, to have selected the word "tunnel" in that situation.

The other term coined by Eiffel was "polar diagram" to describe a plot of lift coefficient versus drag coefficient. Otto Lilienthal first published such plots, although he did not give them a name. Today we call such plots "drag polars," and use of the word "polar" in that regard is due to Eiffel. Eiffel made extensive use of polar diagrams for presentation of his data[71] (with no reference to Lilienthal). One of Eiffel's polar diagrams is shown in Figure 7.5, with data for a flat plate and three circular-arc airfoils with cambers of $\frac{1}{27}$, $\frac{1}{13.5}$, and $\frac{1}{7}$. The wing planforms were rectangular, 90×15 cm; hence the aspect ratio was 6 for all cases. K_x and K_y were the "unit reactions" parallel to and perpendicular to the relative wind. Eiffel defined a unit resistance K_i as

$$R = K_i S V^2$$

where R was the resultant aerodynamic force on the wing or body in kilograms force, S was the planform area in square meters, and V was the free-stream velocity in meters per second. Hence K_i was the "unit resistance" in kilograms force per unit area and per unit velocity at an angle of attack i. The components of that unit resistance parallel to and perpendicular to the relative wind were K_x and K_y, essentially unit drag and unit lift, respectively. (Eiffel, like others during that period, used the term "head resistance" for drag.) Eiffel described the value and convenience of the polar diagram as follows:

> In the diagram, the abscissae are the components of the unit reaction parallel to the wind K_x (head resistance for an aeroplane), and the ordinates are the components perpendicular to the wind K_y (sustentation for an aeroplane). A vector drawn from the origin, therefore, represents the total unit reaction $K_i = \sqrt{K_x^2 + K_y^2}$. The angle Θ of any vector with the

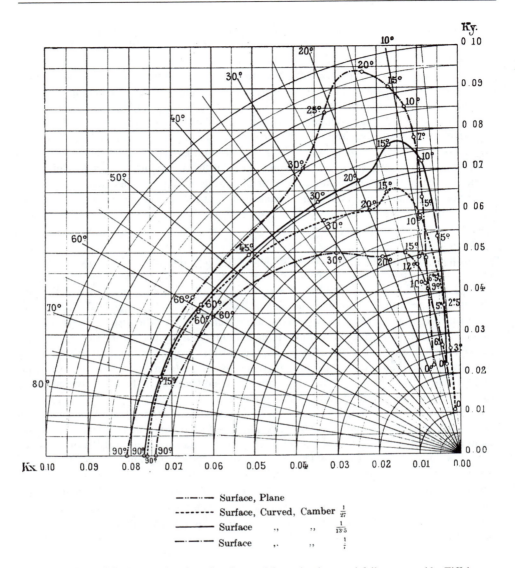

Figure 7.5   Drag polars for a flat plate and three circular-arc airfoils measured by Eiffel.

vertical is the angle between the resultant reaction and the vertical. In addition, on each curve, the corresponding angles of inclination of the chord to the wind are marked. In this way, a single polar curve represents the variations in the five quantities K_i, K_x, K_y, Θ, and i [p. 47].[71]

In addition to the seven substantial contributions we have discussed, Eiffel's wind-tunnel work led to other interesting findings. Like Wenham, Langley, and the Wright brothers, Eiffel recognized the importance of the aspect ratio in determining the aerodynamic properties of wings and airplanes. He ran numerous tests on wings with different aspect ratios and cambers and clearly demonstrated that higher aspect ratios resulted in higher lift-to-drag ratios. For flat-plate wings, Eiffel correlated the effects of aspect ratio and angle of attack in the following empirical formula:

$$\frac{K_i}{K_{90}} = \left[3.2 + \frac{n}{2} \right] \frac{i}{100}$$

where K_{90} was the resultant aerodynamic force on the plate at a $90°$ angle of attack, n was the aspect ratio, and i was the angle of attack in degrees. Eiffel stated that the formula applied to aspect ratios from 1 to 9. As before, we interpret the ratio K_i/K_{90} to be the modern-day resultant-force coefficient C_R. Hence, we can compare data obtained with Eiffel's formula and the data of Langley and Lilienthal shown in Figure 4.49. We pick a low angle of attack germane to airplane flight, say $i = 5°$. From Eiffel's formula we obtain $K_i/K_{90} = C_R = 0.185$ and 0.316 for aspect ratios of 1 and 6.25, respectively. Langley's data from Figure 4.49 yield 0.17 and 0.295, respectively, for the same two aspect ratios, agreeing to within 8% – again, surprisingly close agreement considering the differences in technique and conditions of testing for the two sets of data. Eiffel's formula showed the variation of K_i/K_{90} to be linear with the angle of attack i – a finding we know to be correct for low angles of attack. It also shows K_i/K_{90} to vary linearly with the aspect ratio – a finding that is not precisely correct analytically, but qualitatively it is in the right direction. Eiffel gave similar empirically based formulas applicable to wings with cambered airfoils.[71]

Eiffel's wind-tunnel experiments were aimed directly at aeronautical applications. His work was mainly in applied aerodynamics, rather than basic aerodynamics. For example, he gave findings for model wings that were scaled replicas of wings used on real airplanes.[71] Of particular interest are his tests of a wing model for the Wright Type A, as flown by Wilbur at Le Mans. Eiffel's wind-tunnel data for the Wright wing are shown in Figure 7.6 as a polar diagram. The solid curve shows the data for the Wright wing; for comparison, the dashed curve shows the data for a wing with a circular-arc airfoil and a camber of $\frac{1}{13.5}$. The top and bottom chordwise surface pressure distributions at the center line of the Wright wing at a $6°$ angle of attack are shown in Figure 7.7. The pressure distribution over the top surface (dashed curve) drops to a low value just behind the leading edge and then adjusts to a plateau region until almost reaching the $\frac{1}{2}$-chord point. Such a plateau region (region of constant pressure) is a sure sign of flow separation. Eiffel's data showed that a substantial region of separated flow existed over the top of the Wright wing. That is no surprise – such flow separation is usual for the flow over a very thin airfoil at an angle of attack. Most likely the Wright airplanes (as well as many others using very thin airfoils at that time) flew with substantial regions of separated flow over the top surface just downstream of the leading edge. Also, there may have been some low-Reynolds-number effects reflected in Eiffel's data in Figure 7.7. For the model size and free-stream conditions in Eiffel's tunnel, the Reynolds number (based on chord length) was 121,000 (that compares with a Reynolds number of 2,370,000 for the full-size Wright Type A in flight at 40 mph). Today we know that at low Reynolds numbers like 100,000, a laminar-separation bubble is easily formed on the top surface of the airfoil just downstream of the leading edge. Such a separation bubble can cause the pressure plateau shown in Figure 7.7.

Figure 7.8 shows one of the propellers tested by Eiffel, a design by the propeller expert M. Drzewiecki that Eiffel labeled the *normale* propeller. The twist of the propeller was designed such that each local airfoil section of the propeller would be at the same angle of incidence to the local relative wind. That incidence angle "is taken as constant, and is as close as possible to the angle of maximum efficiency for the profile adopted, where K_x/K_y

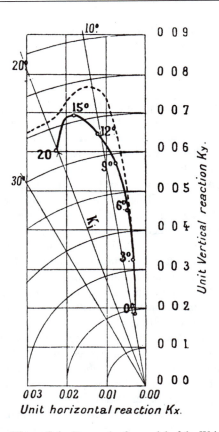

Figure 7.6   Drag polar for model of the Wright Type A wing measured by Eiffel.

is a minimum." That is, each airfoil section was to operate near its minimum drag-to-lift ratio. Eiffel's analysis[71] clearly showed that propeller efficiency (power produced by the propeller divided by the engine shaft power transmitted to the propeller) varied with the parameter V/nD, where n was the propeller's revolutions per second, D was its diameter, and V was the free-stream velocity. That parameter today is called the *advance ratio* and is clearly shown by dimensional analysis to be a major similarity parameter for propeller performance. Eiffel's analysis bordered on dynamic similarity (which is the reason we wondered earlier why he did not understand that his wind-tunnel data for lift and drag were Reynolds-number-dependent). Figure 7.9 shows one of Eiffel's plots of propeller efficiency versus V/nD. A modern aeronautical engineer will see here the familiar shape of the propeller efficiency curve plotted versus advance ratio – it is remarkable that such curves were being plotted by Eiffel in 1910. At the time of his wind-tunnel experiments in 1910, Eiffel was 78 years old – and starting a new career in the relatively unknown and very challenging area of aerodynamics.

Well before the turn of the century, Eiffel had become relatively wealthy from his work in structural engineering. He was one of the most famous civil engineers of his time, having been awarded the Legion of Honor in France and many other medals and awards from around the world. His reputation was at its peak when he started work on the tower

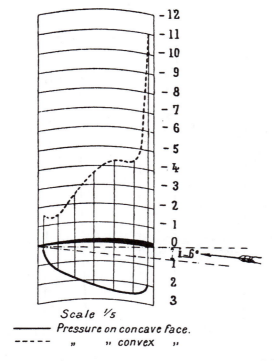

Scale ⅕
——— Pressure on concave face.
- - - - -   „   „ convex   „

Figure 7.7 Pressure distribution on the center line of the Wright wing at a 6° angle of attack as measured by Eiffel.

in January 1887, but in that same year he had the misfortune to become involved in the French effort to build a canal across the Panama isthmus.[72] Ferdinand de Lesseps, famous promoter of the Suez Canal, led that effort. The French consortium, which included Eiffel, went bankrupt in February 1889. Thinking that the French government would not allow so prestigious a national project to fail, Eiffel put another 8 million francs into the project. The project still failed. To make matters worse, Eiffel and de Lesseps were charged with swindling and breach of contract. The scandal widened when it was revealed that deputies of the National Assembly had been bribed in connection with the project. Eiffel denied those charges, claiming that he had carried out his contracted duties, and in an exemplary fashion at that. The courts found him guilty of only one charge: misusing funds. He was fined and sentenced to prison for two years, though he never served that time; his conviction was reversed by a higher court, but only because of a legal technicality. After a massive investigation by Legion of Honor authorities, Eiffel was cleared of all charges. The whole process played out over five years, and Eiffel was permanently intimidated. He retired from his engineering company and withdrew to pursue more scientific interests. He was still a wealthy man, owning a steam-powered yacht and mansions in both Paris and Bordeaux.

Eiffel continued his aerodynamics experiments throughout World War I, focusing on gathering wind-tunnel data on wings, fuselages, propellers, and models of the latest airplanes. In 1921, at age 89, Eiffel gave his laboratory at Auteuil to the French Ministry of Air, but not before carrying out a last series of wind-tunnel experiments that represented

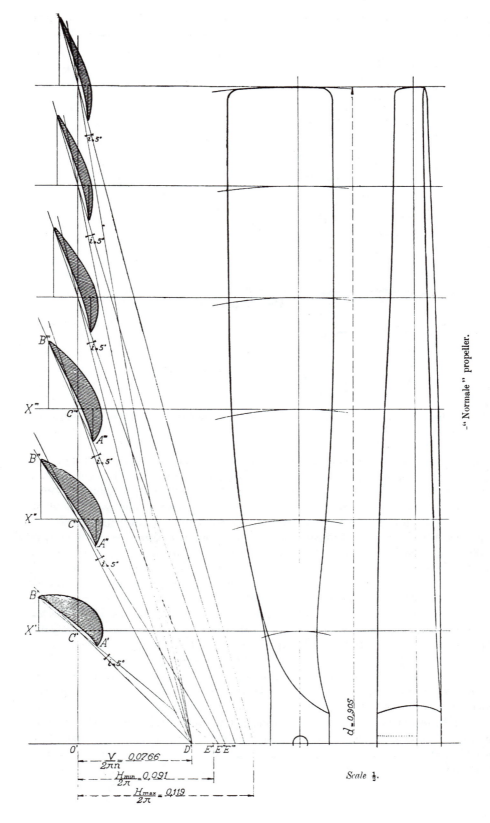

$$\frac{V}{2\pi n} = 0.0766$$

$$\frac{H_{min}}{2\pi} = 0.091$$

$$\frac{H_{max}}{2\pi} = 0.119$$

$d = 0.905$

Scale ½.

-" Normale " propeller.

Figure 7.8 One of the propellers tested in Eiffel's wind tunnel (propeller designed by M. Drzewiecki).

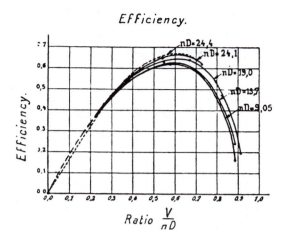

Figure 7.9   Propeller efficiency versus advance ratio, as measured by Eiffel.

another important contribution to aerodynamics. Over the course of his earlier work, Eiffel had measured the drag on spheres, for which he had found $K = 0.011$, but Prandtl's experiments at Göttingen showed sphere drag to be more than twice as high as Eiffel's measurement. During the exchange of information between the two laboratories, one of Prandtl's young engineers suggested that "M. Eiffel forgot a factor of two. He calculated the coefficient referred to ρV^2, not $1/2 \rho V^2$."[68] That remark somehow reached Paris, and "the elderly M. Eiffel became very angry."[68] As a result, in 1914 Eiffel carried out a definitive series of drag measurements on spheres, covering large ranges of diameters and velocities. He found that for each size there was a velocity above which the drag decreased markedly – by slightly more than a factor of 2 (Figure 7.10). The sudden decrease in drag is associated with a transition of the boundary layer from laminar to turbulent at a Reynolds-number value on the order of 300,000. That Eiffel was the first person to measure and publish the drag decrease on a sphere is a fact not known by most students of aerodynamics today. Following Eiffel's experiments, Prandtl eventually was able to explain why the drag decrease occurs.

Eiffel was well respected by his contemporaries in aerodynamics. By 1912, some of Eiffel's experiments were being considered "classic" and were included in new courses being taught at French universities. When he patented his wind-tunnel design in 1912, almost immediately licenses to build such tunnels were obtained by a number of agencies and universities in Rome, Moscow, Tokyo, Amsterdam, and Stanford. In 1913, Eiffel was awarded the second Langley Medal by the Smithsonian Institution; the first had gone to the Wright brothers in 1910. Alexander G. Bell, a regent of the Smithsonian and a longtime aeronautics enthusiast, presented the medal to the French ambassador acting on behalf of Eiffel.

Eiffel died peacefully in his mansion in Paris on December 27, 1923. In all the obituaries, the role of Eiffel as a pioneer in aeronautical research was recognized by his contemporaries. He led the way into the unknown territory of experimental aerodynamics during the first decade of the twentieth century. Building on the contributions to applied aerodynamics by the Wright brothers described in Chapter 5 and the circulation theory of lift discussed in Chapter 6, Eiffel's work provided a powerful springboard for aerodynamics in the twentieth century.

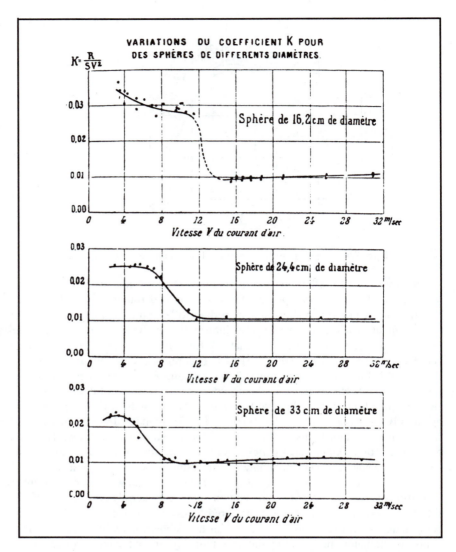

Figure 7.10 Drag on a sphere as a function of velocity and sphere diameter, as measured by Eiffel in 1914.

Wing and Airfoil Theory: Prandtl, Betz, and Munk

The circulation theory of lift, one of two developments that energized theoretical aerodynamics at the beginning of the twentieth century (the other being boundary-layer theory), was rather quick to take hold. Within 15 years it had yielded several models for the calculation of lift on finite wings and airfoils and was within the grasp of practicing aerodynamicists. The "discovery" of the induced drag on finite wings, which could be calculated by means of one of those models, was another by-product of the circulation theory. The basic elements of wing and airfoil theory were articulated during the age of strut-and-wire biplanes, largely at the Göttingen University laboratory run by Ludwig Prandtl. In contrast, most of the practical applications for boundary-layer theory came later, in the era of advanced propeller-driven airplanes, as discussed in Chapter 8.

Just as Gustave Eiffel's pioneering research and applications in the field of experimental aerodynamics in the early twentieth century were directly motivated by interest in the airplane, the most important work in theoretical aerodynamics was being conducted by Ludwig Prandtl, Albert Betz, and Max Munk at Göttingen, also motivated by aeronautical concerns. They took the pillar of the circulation theory of lift (the Kutta-Joukowski law, $L = \rho V \Gamma$) and combined it with appropriate models of the vortex flows over wings and airfoils in order to derive useful engineering formulas for the lift and induced drag on a finite wing and the lift and moments on airfoils of various shapes. We discussed the evolution of the Kutta-Joukowski law in Chapter 6, and by 1906 it was firmly in place and was about to bring some important changes in theoretical aerodynamics. Its apparent simplicity is disarming; it relates lift per unit span, L, to the circulation, Γ, around an airfoil in a straightforward equation, but it is not easy to find the value of Γ appropriate to a given situation. Once Γ is determind, then L is readily obtained from the Kutta-Joukowski law. The important contributions of Prandtl and his colleagues to an understanding of wing and airfoil aerodynamics concerned the development of appropriate models for wing and airfoil vortices, allowing determination of the circulation for a given case (and thus the lift from the Kutta-Joukowski law).

Chronologically, an appropriate engineering formula for a *finite wing* of arbitrary plan-form shape was in place before that for an *airfoil* of arbitrary shape, although the former might seem to have been an inherently more difficult problem. Let us first discuss how Prandtl approached the finite-wing problem and developed his *lifting-line theory* (called the Lanchester-Prandtl theory in England).

By 1911, Prandtl was studying the idea of modeling the effects of the airflow over a finite wing by simply replacing the wing with a single vortex line that would run from one wing tip to the other and then would trail downstream from the two tips (Figure 7.11),[73] earning it the name "horseshoe vortex." The problem was to calculate the strength of that horseshoe vortex such that for a given wing of given shape and angle of attack, the circulation Γ produced by the vortex in a flow of velocity V would properly predict the lift. However, there was a flaw in that model: The value calculated for the induced flow velocity at the wing tip went to infinity – a nonphysical result. Prandtl was quite perplexed; the single-horseshoe-vortex model was not the answer.

A hint that might suggest an improved model had already been published in Lanchester's *Aerodynamics* (1907). In Chapter 6 we discussed Lanchester's idea, dating back to 1891, of replacing the airplane-wing model with a model of several vortex filaments aligned along the span and then trailing downstream from the wing tips as "vortex trunks" (Figure 6.3).

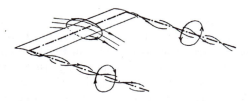

Figure 7.11 Prandtl's early idea for replacing a finite-wing model with a single vortex line.

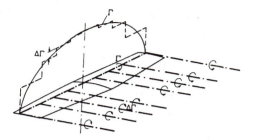

Figure 7.12   Prandtl's lifting-line model for a finite wing.

Moreover, Lanchester had in mind a multiple system, with much smaller vortices trailing downstream of the wing's trailing edge: "We may suppose that the air skirting the upper surface of the airfoil has a component motion imparted towards the axis of flight, and that skirting the under surface in the opposite direction, so that when the airfoil has passed, there exists a Helmholtz surface of gyration. This surface of gyration will, owing to viscosity, break up into a number of vortex filaments or vortices" (p. 177).[60] Lanchester was saying that there was a spanwise component of airflow on the top surface of the wing in the direction away from the wing tips and a spanwise component of airflow on the bottom surface of the wing in the opposite direction, toward the wing tip. When the airflow left the trailing edge of the wing, those spanwise components of velocity from the top and bottom surfaces, being in opposite directions, would form a vortex sheet (a Helmholtz surface of gyration) that would trail downstream of the wing.

Six years after publication of Lanchester's book, Prandtl began working with an analogous theoretical model. Instead of the single horseshoe vortex of finite strength, he expanded that model to include an infinite number of infinitesimally weak horseshoe vortices[73] (Figure 7.12). That was Prandtl's lifting-line model, consisting of a large number of very weak horseshoe vortices ranging along a single line across the wingspan (the lifting line) and trailing downstream of the wing in a continuous vortex sheet. That would allow the circulation Γ to vary continuously along the lifting line, starting with zero at the wing tips and reaching a maximum at the midspan location. As to what the variation of Γ along the lifting line should be, Prandtl was still having a problem. Theodore von Kármán related the following complaint by Prandtl:

> Now look here, I am calculating these damned vortices and can't get a reasonable result for the induced drag. I tried to make the lift suddenly drop to zero at the wing tips, but the induced velocity becomes infinite. All right, I thought, Nature does not like such a discontinuity, so I made the lift increase linearly with distance from the wing tip. That did not work either. This distribution of lift also does not produce finite induced velocity at the tip [p. 66].[68]

That was van Kármán's paraphrase of Prandtl's complaint as he later recalled that conversation, but certainly the frustration of Prandtl made an impression (also, von Kármán had Prandtl mentioning "induced drag," but that term was not coined until 1917).

A technical comment is in order concerning the source of induced drag. The vortices that are generated at the tips of a finite wing (the wing-tip vortices) and trail downstream from the wing are analogous to mini-tornadoes at the wing tips. Clearly, the effects of those mini-tornadoes will reach throughout the flow field over the finite wing, inducing some auxiliary components of velocity in the flow that would not otherwise be there. They compose the "induced velocity" referred to by Prandtl. In turn, the pressure distribution exerted over the surface of the wing by the airflow will be influenced, indeed perturbated, by the wing-tip vortices. The change in the pressure distribution will be such as to create an unbalanced pressure force on the wing acting in the drag direction, and that unbalanced pressure force is called induced drag. An alternative way of looking at the source of induced drag concerns the energy contained in the circulatory motions inside the vortices themselves. That energy has to come from somewhere; it comes from the engine propelling the airplane, and the engine has to put out more power to overcome the increase in total drag due to the presence of induced drag.

Lanchester first recognized the existence of induced drag. In *Aerodynamics* (1907) he discussed how the creation of the two wing-tip vortices required a continuous expenditure of energy, stating that "a source of power is consequently necessary to maintain the airfoil in horizontal flight." He recognized that drag would be created by the presence of the vortices – a drag that would not be dependent on friction, but rather would be a pressure effect somehow connected with the concurrent generation of lift. Thus, in addition to his seminal ideas in the 1890s that presaged the circulation theory of lift (as discussed in Chapter 6), Lanchester was responsible for the idea that tip vortices caused extra drag, which today we call induced drag. However, Lanchester was unable to develop a useful theory for the calculation of induced drag; that was left to Prandtl and his colleagues at Göttingen.

By 1911, four years after Lanchester's book was published, Prandtl was working with the same concept of induced drag expressed by Lanchester. For Prandtl, there was no doubt about the existence of induced drag, and, like Lanchester, he saw linkage between the energy in wing-tip vortices and the induced drag. In a chapter entitled "Fluid Motion," in the *Encyclopedia of Natural Sciences,* first published in 1913,[74] Prandtl described the physical aspects of the wing-tip vortices, stating that there was "a resistance which corresponds to the energy left in the vortex system" (p. 376).[10] Furthermore, Prandtl had concluded that wing-tip vortices also reduced lift. Aerodynamicists today are quite familiar with the *downwash* (a downward vertical component of airflow induced by wing-tip vortices) that when combined with the free-stream velocity decreases the effective angle of attack for the local airfoil section of the wing. Hence the lift produced by each section of the wing will be lower than it would be if there were no wing-tip vortices (i.e., a wing with infinite span). In his 1913 encyclopedia article, Prandtl discussed that phenomenon: "The pair of vortices react on the form of the current flowing over the wing, determining a descending current and thereby the developed lift diminishes in comparison with that deduced by means of the Kutta-Joukowski calculation for the infinite span wing" (p. 376).[10] That was the first published mention of the concept of reduced lift due to the tip vortices.

Before proceeding further with the evolution of the lifting-line theory, let us consider the extent to which Prandtl's work was influenced by Lanchester's thinking. Lanchester's ideas dated to 1891, but were not published at that time. It was not until there began to be rumors that the Wright brothers were on the verge of success, and the public mood began to shift toward seeing flying machines as a real possibility, that Lanchester began to find a more receptive attitude toward his ideas. Publication of his books *Aerodynamics* (1907) and *Aerodonetics* (1908) finally brought recognition of the importance of his work. *Aerodynamics* was expanded in three later editions and was translated into German in 1909 and French in 1914. It is clear that Prandtl read the German translation (but there is no direct evidence as to precisely when) and that his colleagues were also familiar with it. A 1911

Göttingen report written by Otto Foppl mentioned Lanchester's description of wing-tip vortices in *Aerodynamics,* and Lanchester made two visits to Göttingen (1908 and 1909):

> Lanchester visited Göttengen in 1908 and in 1909 and explained his ideas to us. I recall that Prandtl had trouble understanding Lanchester, partly because of his unconventional terms and mathematics, and partly because Lanchester spoke no German and Prandtl spoke no English. On the other hand, Carl Runge, whose mother was an English-woman, spoke perfect English, and he and Lanchester were able to converse fluently and profitably. Since I was a close friend of Runge, I received the benefit of these ideas, and from that early date I grew to appreciate the importance of Lanchester [p. 60].[67]

In spite of that information exchange, Prandtl's publications on the lifting-line theory gave little or no recognition to Lanchester, though von Kármán attempted to explain that away:

> It is hard for an active and creative brain to remember from what reading or from what conversation the first inspiration arose. So I am sure Prandtl never felt that he did not give full recognition to Lanchester's work. It was probably not quite clear to him how many elements of the theory that he worked out with such great success were already contained in Lanchester's work [p. 52].[68]

When the importance and usefulness of Prandtl's lifting-line theory became fully appreciated in the 1920s, a great debate broke out. Were the basic ideas Prandtl's? Or were they Lanchester's? Prandtl squarely faced the issue in 1927 when he was invited by the Royal Aeronautical Society to give the annual Wilbur Wright Memorial Lecture (Lanchester had given the 1926 lecture):

> In England you refer to it as the Lanchester-Prandtl theory, and quite rightly so, because Lanchester obtained independently an important part of the results. He commenced working on the subject before I did, and this no doubt led people to believe that Lanchester's investigations, as set out in 1907 in his "Aerodynamics," led me to the ideas upon which the aerofoil theory was based. But this was not the case. The necessary ideas upon which to build up that theory, so far as these ideas are comprised in Lanchester's book, had already occurred to me before I saw the book. In support of this statement, I should like to point out that as a matter of fact we in Germany were better able to understand Lanchester's book when it appeared than you in England. English scientific men, indeed, have been reproached for the fact that they paid no attention to the theories expounded by their own countryman, whereas the Germans studied them closely and derived considerable benefit therefrom. The truth of the matter, however, is that Lanchester's treatment is difficult to follow, since it makes a very great demand on the reader's intuitive perceptions, and only because we had been working on similar lines were we able to grasp Lanchester's meaning at once. At the same time, however, I wish to be distinctly understood that in many particular respects Lanchester worked on different lines than we did, lines which were new to us, and that we were able to draw many useful ideas from his book [pp. 720–1].[75]

In modern aerodynamics, the lifting-line theory is referred to as *Prandtl's* lifting-line theory in the United States and throughout most of the world, except in England, where, as noted by Prandtl himself, it is referred to as the Lanchester-Prandtl theory. The English custom in this regard has much justification. I believe that Lanchester's work must have had at least a reinforcing effect on Prandtl's thinking (if not providing the original germ of the idea). On the whole, it appears that Lanchester got somewhat of a raw deal in the early days of theoretical aerodynamics, and the effects of that persist to the present. However, it was Prandtl and his colleagues, not Lanchester, who developed the quantitative applications of lifting-line theory and worked it into the form that has made it a valuable engineering tool for calculating the distribution of lift over a finite wing and the induced drag. From that viewpoint, the label "Prandtl's lifting-line theory" is perhaps justified. Prandtl put the meat and the muscle on the bones.

That brings us back to Prandtl's frustration in the summer of 1914. Although convinced of the validity of the lifting-line model (Figure 7.12), he was having trouble finding a proper mathematical representation for the distribution of lift values along the span of the wing. As mentioned earlier, his calculations indicated that the induced velocity would become infinite at the wing tips (a nonphysical result) for every distribution that Prandtl tried. The problem was solved later that year by choosing lift distributions that increased as the square root of distance from the wing tip (such as an elliptical distribution). However, even for that quantitative success it is not clear to whom the credit belongs. In addition to his students, Prandtl had a very able assistant in Albert Betz. Born in Schweinfurt in 1885, Betz had studied naval architecture at the technical university in Berlin, graduating in 1911, before joining Prandtl's research laboratory at Göttingen. According to Klaus Oswatitsch, who joined Prandtl's laboratory in 1938 after receiving a doctorate in theoretical physics at Graz University, "Prandtl and Betz complemented each other ideally. Betz was not only a scientist of high rank but also an efficient administrator – quite in contrast to Prandtl. Very soon he became the right hand of Prandtl" (p. 5).[65]

In 1914 Betz published an article in the *Zeitschrift für Flugtechnik und Motorluftschiffahrt* in which he mentioned an equation obtained by Prandtl for the calculation of induced drag, as well as a more extensive article that presented the first equation for the *minimum* induced drag for a given lift and wingspan.[76] Moreover, the minimum induced drag corresponded to an elliptical distribution of lift along the span – a finding that was attributed to Prandtl. In August 1914, World War I broke out, and all research at Prandtl's laboratory was kept secret; there were virtually no publications until 1917. During that time, the lifting-line theory was further refined. In 1917, in a confidential technical report for the German military authorities, Betz first reported that the lifting-line theory could be properly applied for a more general, nonelliptical lift distribution. Finally, in 1918–19, Prandtl published the definitive publications that described in detail their past decade of work on the lifting-line theory, essentially enshrining the theory for posterity.[77,78] When later investigators cited the basis for the lifting-line theory, it was those two papers that were most frequently referenced.

By the end of World War I, Prandtl's lifting-line theory was solidly in place, and there was no doubt that Prandtl had been responsible for the seminal ideas, but considerable credit must also go to his colleagues, principally Albert Betz and, toward the later stages, Max Munk (to be discussed later). Sorting out the details of attribution for various steps in the development of the theory is somewhat difficult, because of three factors: (1) the normal interchange of ideas that takes place among a group of researchers working together, obscuring the matter of who had which idea and when, (2) the secrecy imposed on wartime research, and (3) the custom in European universities of crediting the findings from a group effort to the senior professor. Nevertheless, by 1918 Prandtl and his colleagues had developed a rational, engineering-oriented theory for calculating the aerodynamic lift and induced drag for finite wings that was to find widespread use. It is still taught in most college aerodynamics classes and is still used by design engineers to make a first estimate of the aerodynamic behavior of a wing. Airplane designers now have sophisticated and highly detailed computer programs for accurate calculations of the aerodynamic properties of wings, but Prandtl's lifting-line theory still has its uses.

Before we leave this topic, a few additional points should be mentioned. First, the term "induced drag" was not introduced until the development of lifting-line theory was almost complete. The pressure drag due to the wing-tip vortices was first called "edge drag" by the Prandtl group (the wing tips were called the "edges" of the wing). Sometimes it was simply referred to as "additional drag." "Induced drag" was coined by Max Munk in 1917 in a confidential technical report, and Prandtl gradually warmed to its use. In his definitive 1918 and 1919 papers[77,78] he used the term only parenthetically, and only at the end of the first paper. In 1921, in a lengthy paper for the NACA in the United States, Prandtl stated that "the part of the drag obtained from [lifting-line] theory is called 'edge drag', since it depends upon the phenomena at the edges of the wings. More justifiably the expression 'induced drag' is used, since in fact the phenomena with the wings are to a high degree analogous to the induction phenomena observed with electric conductors" (p. 192).[79] With that endorsement by Prandtl, the term "induced drag" took hold in the aerodynamics literature. In 1926, Hermann Glauert in England published the first English-language textbook that detailed Prandtl's lifting-line theory, and he used the term "induced drag" throughout.[80]

Second, we note that Prandtl's work on lifting-line theory up to 1919 had involved calculations of finite wing properties for a specified distribution of lift along the span (e.g., an elliptical distribution). In 1919 Albert Betz turned that around and developed a more general approach wherein the distribution of lift was calculated for a specified wing shape and angle of attack. In discussing three major problems associated with the development of lifting-line theory, Prandtl had this to say about Betz's work: "A 'third problem' consists in determining the lift distribution for a definite wing having a given shape and a given angle of attack. This problem, . . . leads to an integral which is awkward to handle. Dr. Betz in 1919 succeeded after very great efforts in solving it for the case of a square-cornered wing having everywhere the same profile and the same angle of attack" (p. 195).[79] Those "very great efforts" earned Betz his Ph.D. from Göttingen in 1919. In 1921, another of Prandtl's students, Erich Trefftz, whom von Kármán described as "tall and sporty,"[67] and who was the nephew of the distinguished mathematician Carl Runge, made the awkward integral much easier to handle by introducing Fourier analysis. Trefftz's Fourier-series representation of the generalized lift distribution for a specified wing shape is the form in which lifting-line theory is used today.

Finally, we note a confusing difference in nomenclature between the early twentieth century and today involving the terms "wing" and "airfoil." Today a wing is a wing, and an airfoil is a cross section of a wing, sometimes called an airfoil section, a wing section, or an airfoil profile. That seems logical and clear-cut. In the early twentieth century the word "airfoil" was frequently used to denote a wing. Much earlier, Octave Chanute[45] used "wings" in conjunction with the flapping wings of birds; he used "artificial wings" for the man-made flapping wings of an ornithopter, as well as for the wooden, fabric, or feather wings that some people had attached to their arms and/or legs before jumping from a roof or tower, flapping wildly in an attempt to fly like a bird. In contrast, a fixed wing on a flying machine was called an "aeroplane." Chanute most likely was influenced by George Cayley, who used the word "wings" in describing the flapping wings of birds, but referred to "surfaces" or "planes" for the fixed lifting surfaces on flying machines ("planes" was a literal use, because Cayley tested only flat, planar surfaces). Samuel Langley also referred to the lifting surfaces tested by his whirling arm as "planes" or "inclined planes," most likely because they also were planar surfaces.[49] Elsewhere[50] he clearly identified the wings of his aerodromes as "wings." As for the Wright brothers, in their wind-tunnel tests they consistently referred to wings as "surfaces" and to airfoils as "profiles."[55] In distinguishing between their upper and lower wings, they frequently referred to the "upper surface" and "lower surface." Thus it is no surprise that the aerodynamic literature of the early twentieth century was somewhat ambiguous compared with today's nomenclature. In the German literature, the words *Tragflügel* and *Tragfläche* meant "wing" and "airfoil." In many of the papers by Prandtl and his colleagues, the lifting-line theory was called, in general, *Tragflächentheorie* ("airfoil theory"). Even Glauert's book *The Elements of Aerofoil and Airscrew Theory*[80] consistently referred to a finite wing as an "airfoil of finite span" or an "airfoil in three dimensions." In the 1921 paper that revealed Prandtl's work to the English-speaking world, wings were called "finite wings," and airfoils were called "wing sections," "wing profiles," or "aerofoils." That is essentially the nomenclature in use today, though at least one modern British textbook[81] refers to a finite wing as a "finite aerofoil."

Because Prandtl's lifting-line theory was conceived and developed during the era of the strut-and-wire biplane, much emphasis was placed on its application to biplanes. The first such work was reported by Betz in 1913,[82] and Prandtl, Betz, and Munk continued to refine the biplane theory throughout World War I. One of the more important contributions to biplane theory was the Ph.D. dissertation of Max Munk in 1919, in which he proved the "stagger theorem" that the total drag on any multiple-plane lifting system remains unchanged if the individual wings are displaced in the direction of flight without changing their lift forces.

Mention of the work of Max Munk brings us to another related and equally important theoretical development: thin-airfoil theory (by "airfoil" we mean the modern usage, a cross section of a wing). A solution for the stream function (from which all other flow properties can be obtained) of the flow over a lifting airfoil was presented by Wilhelm Kutta in his Ph.D. dissertation in 1902. As discussed in Chapter 6, Kutta treated the special case of a thin, circular-arc airfoil at a zero angle of attack, obviously motivated by Lilienthal's earlier experiments with airfoils of the same shape. Kutta's paper presented the first theoretical solution for lift; however, it was severely limited because of the specialized geometry it treated, as well as the complexity of the mathematics. It was in no way a practical, engineering-oriented method. In 1910, Joukowski, in Moscow, using the method of conformal transformation from the mathematical theory of complex variables, determined the flow fields and lift values for a series of reasonable-looking airfoils. Starting with the known, exact analytical solution for the potential flow over a circular cylinder, Joukowski transformed that solution to apply to a series of streamlined shapes with sharp trailing edges – airfoil shapes that came to be known as *Joukowski airfoils* (Figure 7.13). Joukowski's solutions for those airfoils remained the only exact analytical solutions for cambered airfoils

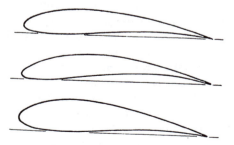

Figure 7.13   Airfoil sections generated analytically by Joukowski (1910).

of reasonable thicknesses until 1931, when Theodore Theodorsen published an analytical method for exact solution of the general problem of flow over a thick airfoil of arbitrary shape.[83] However, none of those techniques had led to simple, engineering-oriented solutions for airfoil properties. It was left to Max Munk to provide such a method, known today simply as thin-airfoil theory.

Michael Max Munk was born in Hamburg, Germany, October 22, 1890, to a lower-middle-class Jewish family. Gifted in mathematics and science, he received an engineering degree from the Hannover polytechnical school in 1914 and moved on to Göttingen, where he became one of Prandtl's most gifted students. As mentioned earlier, Munk's dissertation research involved an extension of Prandtl's lifting-line theory, and he was responsible for coining the term "induced drag." In 1919 he received two doctorates from Göttingen (engineering and physics). Munk was adept at both theory and experiment, as were most of Prandtl's early students. Although his contributions to the lifting-line theory were clearly theoretical, he spent much of his time at Göttingen assisting Prandtl with analyses of wind-tunnel experiments. Immediately after the war, Munk went to work briefly for the Luftschiffbau Zeppelin, where he designed a small wind tunnel for testing airship models. Munk's uncle, Adolph Lewisohn, had gone to America in 1868 and made a fortune in mining, becoming president of three mining companies and a director of Mt. Sinai Hospital in New York City. In the United States, people like Jerome Hunsaker were looking for German aeronautical engineering and science talent. Hunsaker, while visiting Eiffel in 1913, also toured Prandtl's Göttingen laboratory and was most impressed. When Max Munk expressed interest in going to the United States, Prandtl contacted Hunsaker about finding a suitable position for Munk. Hunsaker persuaded Joseph Ames, chairman of the NACA executive committee, to hire Munk. In 1920 Munk moved to America and for the next six years worked at NACA Headquarters in Washington. Thus, at about the same time NACA was publishing Prandtl's first English account of his aerodynamic research,[79] it was also importing one of his brightest graduates.

Munk's relationships with other NACA engineers and technicians ranged from strained to disastrous, as described in James Hansen's history of the Langley Memorial Laboratory.[84] Munk was considered arrogant and overbearing by co-workers, and that caused problems when Munk was put in charge of monitoring the construction of the NACA variable-density wind tunnel at Langley in 1921. In spite of his prolific output of more than 40 technical papers for the NACA from 1920 to 1926, and in spite of chairman Joseph Ames's high regard for the quality of Munk's theoretical work, Munk's reputation among workers at Langley went from bad to worse. Not fully aware of what had been building, and apparently misjudging Munk's ability to manage people, George Lewis, then director of research at NACA Headquarters in Washington, named Munk to be chief of aeronautics at Langley in January 1926. Over the next year there was a mass revolt against Munk by Langley professionals; in early 1927, all of the section heads of the Aerodynamics Division at Langley resigned in protest against Munk's supervision. Munk's pride was hurt, and he promptly resigned from the NACA. "Peace, and the section heads, returned to Langley, but only at the cost of losing one of the best theorists ever to work there."[84]

During his first few years with the NACA, Munk developed an engineering-oriented method for theoretical prediction of airfoil lift and moments, a method still in use today: thin-airfoil theory. His idea was novel but straightforward: Consider an airfoil of arbitrary shape. If the airfoil is reasonably thin (say the maximum thickness is about 12% or less of the chord), then the overall lift and moments on the airfoil ought to be approximately those for the flow over an infinitely thin surface in the shape of the mean camber line of the airfoil (i.e., for the calculation, the thin airfoil is replaced by its mean camber line). In his 1922 report,[85] Munk derived expressions for the zero-lift angle of attack and moment coefficient in terms of integrals that depended only on the geometric shape of the camber line – integrals that could be conveniently evaluated analytically (if the camber line was given as an analytic curve) or numerically (if the camber line was specified as a series of points). The tremendous advantage of the method is that airfoils of *arbitrary shape* (limited only to being thin) can be analyzed. It is also assumed that the angle of attack is small enough to allow the trigonometric approximation that $\tan \alpha \approx \alpha$ (not a severe restriction, because the angles of attack for conventional airplanes in normal flight are indeed small). Munk's thin-airfoil theory was a major contribution to the state of the art of theoretical aerodynamics during the era of strut-and-wire biplanes. Its long-term impact on aerodynamics has been only slightly less than that of Prandtl's lifting-line theory.

Munk derived his results for thin airfoils by using the idea of conformal mapping, from the theory of complex variables. One year later, Birnbaum, in Germany, approached the same problem by replacing the mean camber line with a vortex sheet, which led to a simpler derivation of the equations for thin-airfoil theory. Finally, in 1926, Glauert[80] applied the method of Fourier series to the solution of those equations. It is Glauert's formulation that is still used today. However, Munk was clearly responsible for the seminal ideas behind thin-airfoil theory, further extending the legacy of Prandtl and the Aerodynamics Laboratory at Göttingen.

Figure 7.14 shows Munk in his office at Langley in 1926. After his resignation from the NACA, Munk worked for Westinghouse, Brown Boveri, and the Alexander Airplane Company. During the Depression he became a consulting editor for *Aero Digest*. He also taught part-time in the Mechanical Engineering Department at the Catholic University of America, Washington, D.C. There were various overtures to the NACA about returning to work there or doing consulting work, but they were all rejected. Later he returned to Catholic University for three years, retiring in 1961. Munk moved to the eastern shore of Delaware, where he continued to be intellectually active. In 1977 he published a book at

Figure 7.14 Max Munk in his office at the NACA Langley Memorial Laboratory (1926).

his own expense in which he claimed to have proved Fermat's "last theorem," a proof that mathematicians had been attempting for 300 years, but his "proof" did not convince the mathematics community. Munk died in 1986 at the age of 96 – the last of that select group of Prandtl's students from the World War I period who led the way in wing and airfoil applications of the circulation theory of lift.

Aerodynamics Culture Shock, 1920–6: Theory versus Empiricism

The important advances in theoretical aerodynamics produced by Prandtl and his colleagues before and during World War I threw the aeronautics community into a kind of culture shock, producing the reverse of the sociotechnological alignments that had existed in the nineteenth century. The Aeronautical Society of Great Britain had been created in 1866 to advance the cause of aeronautics, to assemble a group of knowledgeable people with common interests, and to publish a journal through which rational ideas on flying machines could be disseminated. The early members of that society generally were not university-educated, and most had little or no formal advanced education in mathematics or the sciences. They were self-educated, practical in outlook, and steeped in the tradition of empiricism. They were part of a rising group of technologists who represented the beginnings of the profession of engineering. At that time, the scholarly community had only disdain for any research pertaining to flying by human beings; such work certainly was not fashionable and was thought by many to be the business of simpletons and madmen. Although there had been major advances in theoretical fluid dynamics during the nineteenth century because of the work of university-based scientists, there had been no effort by those scientists to apply such findings to the development of flying machines. In short, there was a wide gulf between the scientific and aeronautical communities, mainly because of the attitude of the scientists.

By the end of World War I, that situation had been completely reversed. Driven by a newfound intellectual interest in the science of flight, as well as the practical need to learn how to design better airplanes, the university community, as represented by Prandtl's group at Göttingen and to some extent by Joukowski's research at the University of Moscow, made substantial contributions to theoretical aerodynamics. However, by that time it was the practicing aeronautical community that had become recalcitrant, not readily making use of the new theoretical developments. The attitude of that community was reinforced by its long commitment to empiricism; moreover, most aeronautical engineers at that time did not have sufficient technical education to be able to understand such theories, let alone know how to apply them. In the United States at that time, most B.S. graduates in engineering barely understood the elements of differential and integral calculus; to understand the theoretical aerodynamics developed by Prandtl and Joukowski it was also necessary to understand complex variables and differential equations. Hence, in 1921, when Prandtl's work became known to the English-speaking world,[79] and when Max Munk began to work for the NACA, ready to expand and disseminate the legacy of Göttingen, there was major culture shock within the American aeronautical-engineering community. Munk must have recognized that culture shock immediately, because he was compelled to defend and virtually teach the importance of theoretical work in aerodynamics:

> A theoretical investigation at least gives the limit of what to expect. It enables the investigator to survey and keep in mind a great number of isolated experiences, whether the agreement between theory and experience be more or less close. It induces him to reflect on the

> phenomenon and thus becomes a source of progress by guiding him to new observations and experiments. It has often occurred even that some relation was thought to be confirmed by experience till the progress of theory made the relation improbable. And only then the experiments confirmed the improved relation, contrary to what they were supposed to do before.... But is it really necessary to plead for the usefulness of theoretical work? This is nothing but systematical thinking and is not useless as sometimes supposed, but the difficulty of theoretical investigation makes many people dislike it [p. 245].[85]

The fact that Munk had to emphasize such basic matters in an official NACA publication attests to the severity of the culture shock. Keep in mind that Munk was not just an excellent theoretician; he had had extensive practical experience with wind tunnels. Indeed, Munk was mainly responsible for the design and construction of the NACA variable-density wind tunnel – a story to be told in a subsequent section. So his comments were from the perspective of one whose feet were planted in both the world of theory and the world of experiment.

Following World War I, the snail's pace at which the U.S. aeronautical-engineering community was able to absorb, appreciate, and apply the aerodynamic theory developed by Prandtl and his students was to be expected, given the differences in educational philosophy between Göttingen and the typical engineering college in the United States at that time. Göttingen was a research university with major emphasis on the sciences; Prandtl's research group was the first major collection of *engineering scientists* to work in aeronautics. In contrast, U.S. engineering colleges produced *engineers,* who were taught to focus on the empirical methods required to build something. It is no wonder that the early NACA engineers were slow to understand and embrace the new aerodynamics theory, and Munk's attitude certainly did not help the situation:

> Having internalized the social relations of German academic life, Munk considered himself the absolute master of the division he directed. He intended to set the research goals, and like a German university professor, himself receive whatever credit was forthcoming. A proud genius, Munk was frequently autocratic and arrogant in his dealings with people, treating his men at Langley as graduate students and obliging some of them to attend a seminar on theoretical aerodynamics he conducted in a way that at least two talented young men, Elliott Reid and Paul Henke, a Ph.D. in physics from Johns Hopkins University, found rude and condescending [p. 93].[84]

Acceptance of airfoil and finite-wing theory in Britain was also slow at first, but interest picked up when the wind-tunnel tests carried out under the auspices of the British Aeronautical Research Committee verified Prandtl's lifting-line theory.[86,87] Of course, Prandtl had already verified his theory in wind-tunnel tests carried out as early as 1915.[79]

Prandtl's lifting-line theory for a finite wing and Munk's thin-airfoil theory were enshrined in the English-language aeronautical-engineering literature by Hermann Glauert in 1926.[80] Glauert, born in Sheffield, England, October 4, 1892, was educated at Cambridge, where he excelled in science and mathematics. In 1916 he joined the Royal Aircraft Establishment in Farnborough as an aerodynamicist. After the war, Glauert visited Göttingen and became familiar with the details of Prandtl's research. Glauert's *The Elements of Aerofoil and Airscrew Theory,*[80] in which the work of Prandtl and Munk was clearly explained and even improved upon, became the standard text for aerodynamicists in Britain and the United States; it is a classic in the aeronautical literature and is still in print more than 70 years after its first appearance. An example of its impact can be seen in comparing the first and second editions of another classic, Leonard Bairstow's *Applied Aerodynamics.*[88]

Bairstow was a professor of aeronautical engineering at the University of London, and he had been a highly respected and influential force in British aeronautical-engineering circles since before World War I. The first edition of his book (1920), published six years before Glauert's book, had only a brief mention of Prandtl's work:

> The chief aerodynamics laboratory, prior to 1914, in Germany was the property of the Parsebal Airship Company, but was housed in the Göttingen University under the control of Professor Prandtl. Some particularly good work on balloon models was carried out and the results published in 1911, but in 1914 the German Government started a National laboratory in Berlin under the direction of Prandtl, of which no results have been obtained in this country [p. 7].[88]

There was no further mention of Prandtl or any aspects of the lifting-line and airfoil theories, despite the fact that by that time Prandtl had completed all the important features of the lifting-line theory. There clearly had been no technology transfer to the English-speaking world at that time. In contrast, in Bairstow's second edition (1939), published 13 years after Glauert's book, there were 12 separate entries in the index dealing with Prandtl and a complete chapter (55 pages) on lifting-line theory.

In the United States, in 1928 Captain Walter Diehl of the U.S. Navy Bureau of Aeronautics published an aeronautical-engineering textbook that became the standard for students and professionals in the United States during the 1930s and 1940s: *Engineering Aerodynamics*[89] had a chapter on finite wings that described Prandtl's lifting-line theory and a chapter on airfoils that mentioned Munk's thin-airfoil findings. Diehl had worked closely with the NACA during the time that Munk was there, and he knew Munk, but in his book, Diehl avoided the rigorous mathematical theory and gave only the important engineering applications derived from the theory. In 1941, Clark Millikan, at the California Institute of Technology, published *Aerodynamics of the Airplane*,[90] an introductory text for aeronautical engineering that presented Prandtl's lifting-line theory in more detail than had Diehl's treatment, although Millikan also sidestepped a lot of the mathematical theory. Clark Millikan was the son of Robert Millikan, the Nobel physicist who had carried out a famous oil-drop experiment to measure the electrical charge on an electron, and who was the head of Cal Tech at that time. Clark Millikan became head of the aeronautics program at Cal Tech, and his book enjoyed great popularity. With Bairstow's second edition in 1939 in England and Millikan's book in 1941 in the United States, the Göttingen legacy became a permanent part of aeronautical-engineering education in the English-speaking world.

The Reawakening of Aerodynamics in America: Creation of the NACA

When Max Munk came to America in 1920, he went to work for the NACA, the only U.S. organization conducting research and development in aerodynamics at that time. It was still a fledgling agency, having been created only five years earlier in response to a critical need. To understand the situation, let us go back to the status of aerodynamics in America at the turn of the century.

By the middle of 1903, six months before the Wrights flew, two important series of aerodynamics investigations had been completed in the United States: (1) the whirling-arm and model-aircraft tests carried out by Samuel Langley[49,50] at the Smithsonian Institution between 1887 and 1903; (2) the wind-tunnel tests carried out by the Wright brothers during 1902–3 wherein the lift and drag characteristics for more than 200 different wings and airfoil shapes were measured and tabulated. Lilienthal's work in Germany had produced the only significant European data on the aerodynamic properties of wings and airfoils. In addition to the two important series of investigations carried out in the United States, research on the friction drag on dirigible hulls and aircraft surfaces had been carried out by Albert Zahm in a 6- × 6-ft wind tunnel at the Catholic University in Washington, D.C., during 1901 (to be discussed later). So in 1903, the center of gravity of the significant activity in experimental aerodynamics was squarely in the United States – but not for long. After Langley's two well-publicized failures in 1903, the Smithsonian board of regents summarily terminated his aeronautical research. And after their successful flights on December 17, 1903, the Wrights drastically reduced their wind-tunnel testing and concentrated on mechanical improvements to their flying machine. Progress in basic aerodynamics was suddenly reduced to a crawl in the United States.

That was not the case in Europe. In a farsighted move, the British began in-house government research in aerodynamics. At the National Physical Laboratory in London, a 2-ft-diameter wind tunnel was built and used by Thomas Stanton in 1903. In 1905, her Majesty's Balloon Factory was moved to Farnborough, and by 1908 it was the site of experimentation with flying machines designed and flown by Samuel F. Cody, an American who took British citizenship. In 1909 the British prime minister appointed an Advisory Committee for Aeronautics, with the famous Lord Rayleigh, at age 67, as president. In 1910 the National Physical Laboratory built a new wind tunnel with a 4- × 4-ft test section, and in 1912 a larger 7- × 7-ft tunnel. In 1911 the balloon factory at Farnborough was renamed the Army Aircraft Factory to better reflect the type of work being carried out, and in 1912 it became the Royal Aircraft Factory, and in 1918 the Royal Aircraft Establishment (RAE). Elsewhere in Europe, Eiffel began his experimental aerodynamics research in 1902 with drop tests from the Eiffel Tower, in 1909 turning to wind-tunnel testing. Prandtl's laboratory at Göttingen became the leading center of theoretical aerodynamics research during the first decade of the twentieth century, and experimental work in aerodynamics began at Göttingen in 1908.

After 1903, the center of gravity for experimental aerodynamics research shifted from the United States to Europe. The center of gravity for theoretical aerodynamics had always been in Europe. By 1915, American aerodynamics research and aeronautics in general had descended to a wretched state in comparison with the progress in Europe. The only aerodynamics research funded by the U.S. government was a small program at the Washington Navy Yard run by Albert Zahm, using a large 8- × 8-ft wind tunnel constructed in 1913 to gather data for future naval aircraft. There was even less activity in aerodynamics in private industry. In 1915 the U.S. aeronautical industry was still suffering from a barrage of patent-infringement suits filed by the Wright brothers beginning in 1909.[52,53,91,92] In American universities there were no aerodynamics research programs that could hold a candle to the work at Göttingen; the only glimmers of hope for developing such programs were Hunsaker's new 4- × 4-ft wind tunnel, built at the Massachusetts Institute of Technology in 1914, and William Durand's propeller research at Stanford University, which did not get well under way until 1916. The few people in the United States who understood the imminent importance of aeronautics and who were aware of the rapid progress being made in Europe had a lot to be concerned about.

It was that group of concerned people who finally succeeded in bringing about the formation of the National Advisory Committee for Aeronautics (NACA) in 1915. As early as 1911, ideas had begun to be floated concerning some kind of national aeronautical organization. For example, it was suggested that Langley's old aeronautical laboratory, unused at the Smithsonian since 1903, be rejuvenated and expanded, with the Smithsonian in charge. Such ideas were opposed on two fronts. First, the U.S. Navy, in the person of Rear Admiral David W. Taylor, chief constructor for the navy, claimed that the model basin at the Washington Navy Yard was already conducting aeronautical research, and a civilian laboratory would be needless duplication. Second, the location of such a laboratory at the Smithsonian was opposed by Richard Maclaurin, president of the Massachusetts Institute of Technology, and Samuel Stratton, director of the National Bureau of Standards, who thought that the aeronautical laboratory should be located at and run by their respective institutions. In spite of that opposition, the Smithsonian board of regents authorized the reopening of Langley's laboratory in 1913, with a distinguished group (including Orville Wright) serving as an advisory committee. The entire project had to be abandoned almost immediately because it was found to be in violation of a 1910 law concerning such activities at the Smithsonian. However, relentless pressure from many people, including Charles Walcott, who had taken over as secretary of the Smithsonian after Langley's death, finally resulted in the creation of an advisory committee for aeronautics. On March 3, 1915, Congress passed a resolution creating the committee. The Smithsonian would not dominate the program; rather, the act created a broad-based NACA committee, called the National Advisory Committee for Aeronautics, composed of seven government and five private members that would meet twice a year to identify the major research needs. In the end, a European influence was present – the committee was partly modeled after the British Advisory Committee for Aeronautics.

Public Law 271, which created the NACA, stated the duty of the committee:

> It shall be the duty of the Advisory Committee for Aeronautics to supervise and direct the scientific study of the problems of flight, with a view to their practical solution, and to determine the problems which should be experimentally attacked, and to discuss their solution and their application to practical questions. In the event of a laboratory or laboratories, either in whole or in part, being placed under the direction of the committee, the committee may direct and conduct research and experiment in aeronautics in such laboratory or laboratories.

An appropriation of $5,000 per year was given to the committee, whose members were unpaid.

Creation of the NACA was a turning point that led to a program of research in aerodynamics that over the next 43 years would prove to be instrumental in advancing the state of the art in aerodynamics, not only for the United States but also for worldwide applications and spin-offs. The members of the NACA immediately urged the creation of a major research laboratory, and with U.S. involvement in World War I seeming imminent, in 1916 Congress appropriated $53,580 for construction of such a laboratory. Four years later, on June 11, 1920, the Langley Memorial Aeronautical Laboratory in Hampton, Virginia, was formally dedicated, and its aeronautical research began. The NACA would play an important role in the further development of aerodynamics, and throughout the remainder of our discussion of twentieth-century aerodynamics, we shall see how the NACA (and then NASA after 1958) carried out that role.

The Evolution of Wind Tunnels: The First 20 Years after the Wright Flyer

The era of strut-and-wire biplanes was also the period in which the wind tunnel came into its own. After the Wright brothers' wind-tunnel experiments in 1901–2 and the use of their findings to design their successful 1902 glider (and the Wright Flyer after that), the wind tunnel became the testing device of choice in aerodynamics. The Whirling arm was passé. For the next 20 years, wind tunnels were improved and developed to such an extent that by the middle 1920s, almost all new airplane designs were based directly on a set of wind-tunnel data, or else complete scale models of the designs were tested in wind tunnels. Because wind tunnels played an important role in the evolution of the state of the art of aerodynamics, it is appropriate to examine their development and impact. We have already discussed the development of the Eiffel-type wind tunnel, but there were several other notable wind-tunnel facilities in Europe, and there was even some limited wind-tunnel research in the United States.

At the time that Orville and Wilbur Wright were carrying out their first wind-tunnel experiments, a much larger wind tunnel was being put into operation at the Catholic University of America, in Washington, D.C., by Albert Zahm, a professor in the Department of Mechanics. The test section was 6 × 6 ft in cross section and 40 ft long, with an airspeed of 27 mph. Housed in its own building on the campus, it was by far the largest wind tunnel built to that time (1901). Zahm was an experienced scientist with a Ph.D. in physics from Johns Hopkins University, and the tunnel he designed was technically more advanced than the Wright brothers' tunnel. Zahm's facility had several screens of cheesecloth and wire mesh to straighten the airflow and decrease the turbulence level – a technique still used today. The blower was downstream of the test section, sucking the air into the tunnel, whereas the Wrights had their fan upstream of the test section. The tunnel was heavily instrumented for that day. For example, the test-section airspeed was measured with a Pitot-static tube connected to an extremely sensitive manometer. Zahm developed several balances for measuring aerodynamic force, including a wire suspension balance much like those used in more modern tunnels well into the twentieth century.

Zahm's tunnel was operational beginning in 1901. He measured the drag on dirigible hulls, and, of most importance, he made detailed measurements of the friction drag on flat plates and other lifting surfaces. Zahm's tests were the first to show that skin friction was a major contributor to drag; that was in direct contrast to Samuel Langley's assumption that friction drag was negligible. Zahm and Langley were friends and colleagues, and some of Zahm's experiments were funded by small grants from the Smithsonian, brokered by Langley.

Although the wind tunnel had been invented in England by Francis Wenham and used by Horatio Phillips for the design and testing of the first cambered airfoils well before the turn of the century, America momentarily became the center of serious wind-tunnel research with the Wright brothers' work in 1901–2 and Zahm's research, also beginning in 1901. Zahm's facility was not sponsored by the government nor the university; funding for constructing and operating Zahm's tunnel came from Hugo Mattullath, a wealthy industrialist with a strong interest in the commercial future of the airplane, but Mattullath died before he could realize any useful gain from the facility. Without funding after Mattullath's death, Zahm's research at the Catholic University withered on the vine, and the facility was shut down in 1908.

By that time, the center of wind-tunnel activities had already shifted back to Europe. In England, Thomas Stanton at the National Physical Laboratory (NPL) had built a tunnel in 1903 that had a 2-ft-diameter test section and a test-stream velocity of 20 mph. At that time, the purpose of the NPL tunnel was to measure wind pressures on surfaces and structures. However, following Wilbur's successful flights at Le Mans in 1908, the NPL became interested in aeronautics. A larger tunnel was built in 1910. With a square test

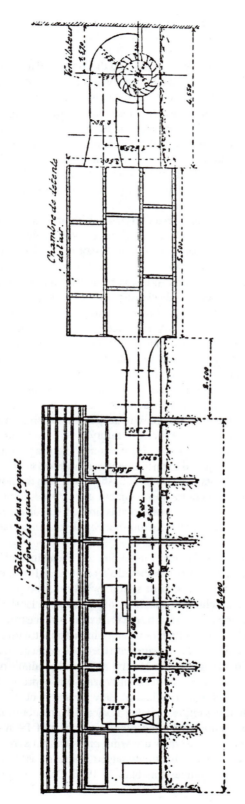

Figure 7.15 Crocco's wind tunnel near Rome (1903).

section 4 × 4 ft in cross section and a flow velocity of 30 mph, that facility put the NPL on the map of aeronautical research. However, by that time there were several other European wind tunnels on the map. Also in 1903, Lieutenant Arturo Crocco constructed a tunnel on the grounds of an army unit near Rome. With a square test section 1 × 1 m in cross section, it was capable of the stunning speed of 65 mph (Figure 7.15) and was used to study the drag and stability of dirigibles and the aerodynamic characteristics of propellers. Crocco went on to become a general in the Italian Air Force and was a strong proponent of aerodynamic research in the quest for high-speed flight in the 1930s.

In Russia, Dimitri Pavlovitch Riabouchinsky founded the Aerodynamic Institute at Koutchino, near Moscow, in 1904. Riabouchinsky, only 22 years old at the time, and the son of a wealthy merchant, supported the institute, which was situated on the grounds of his father's estate. The focal point of the institute was a wind tunnel with a cylindrical duct 14.5 m long and 1.2 m in diameter. The air was sucked into the tunnel by a fan at the downstream end, and the air velocity in the test section was about 15 mph. The test section was equipped with windows for observing the test models. It can be argued that Riabouchinsky's tunnel was not the first in Russia, because Konstantin Tsiolkovsky, the Russian rocket pioneer, had built a crude blower in 1897 at Kaluga, which he used to carry out aerodynamic experiments on models of airfoils, dirigibles, and various geometric shapes.[8] But Riabouchinsky's work was the first Russian wind-tunnel research to gain acceptance in scientific circles. With the revolution of 1917, Riabouchinsky nationalized the institute in order to preserve its staff and activities.

Finally, European dominance in wind tunnels was solidified in 1908 with Prandtl's construction of a wind tunnel at Göttingen 2 × 2 m in cross section. It introduced a novel design feature: The air passing through the test section was not exhausted directly into the room, but was ducted back to the entrance of the tunnel – the first closed-circuit, return-flow wind tunnel. The increased efficiency obtained with the closed-circuit design, as well as the use of strategically placed screens and honeycombs that reduced the turbulence of the airflow, made Prandtl's tunnel the model for future subsonic wind tunnels. That was reinforced in 1916 when Prandtl constructed a second-generation closed-circuit tunnel (Figure 7.16) with a square test section 2.2 × 2.2 m in cross section and a phenomenal airspeed of 116 mph. Like Eiffel's tunnels in France, Prandtl's tunnel had a convergent nozzle leading into the test section, providing relatively smooth, low-turbulence airflow. The 1916 Göttingen tunnel "launched a new era in tunnel design."[93] With Eiffel's wind tunnel at Auteuil and Prandtl's tunnel at Göttingen, by the middle of World War I Europe was the undisputed center of experimental aerodynamics.

The United States was not totally out of business in regard to wind tunnels at that time. Albert Zahm used his earlier experience to design a large wind tunnel at the Washington Navy Yard. With a square test section 8 × 8 ft in cross section, it went into operation in 1913. Zahm designed a square insert (4 × 4 ft in cross section) that reduced the test section's cross-sectional area by a factor of 4 and thus increased the airspeed to 160 mph – well beyond the maximum cruising speed of any airplane at that time. At Stanford University, William Durand established a propeller research program in 1916 for which he constructed a wind tunnel with a 5.5-ft-diameter test section. At the National Bureau of Standards in Washington, D.C., a 4.5-ft-diameter octagonal wind tunnel was built in 1918 to carry out research on basic fluid dynamics, with emphasis on turbulence and boundary layers. However, those efforts were not enough to seriously challenge the European dominance. But the situation was about to change, and change dramatically.

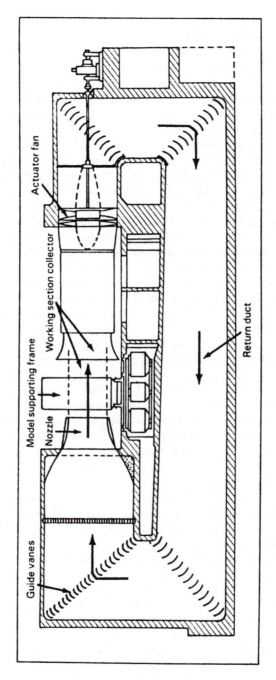

Figure 7.16 Prandtl's wind tunnel at Göttingen, a closed-circuit wind tunnel (1916).

When Max Munk moved to the United States in 1920, he took with him a number of innovative ideas in theoretical aerodynamics and a concept for a revolutionary new wind tunnel in which the air pressure in the test section would be considerably higher than the surrounding ambient pressure outside the tunnel. There was a compelling reason for such thinking: Extrapolations from wind-tunnel findings generally were limited because the test model usually was so much smaller than the actual flight vehicle. For example, the wingspan of the Curtiss JN-4 Jenny, a popular biplane in the United States at the end of World War I, was 43.6 ft. A $\frac{1}{20}$-scale model of the Jenny would have a wingspan of 2.18 ft, a typical size for a wind-tunnel model. As a result, for such a model tested in a conventional wind tunnel, where the test-section pressure and density would be essentially the same as ambient conditions, the Reynolds number would be smaller by a factor of 20 than that for actual flight by the real airplane. That was not a satisfactory situation. The measured drag coefficient and the value of the maximum lift coefficient were sensitive to the value of the Reynolds number. Those coefficients, measured in a wind tunnel where the Reynolds number was smaller by a factor of 20, would not be equal to the corresponding data in actual flight. By 1920 that problem had been recognized. It was called the *Reynolds-number scaling problem*. As mentioned earlier, Eiffel understood that his wind-tunnel measurements for a small-scale model would have to be modified before they would be applicable to the actual airplane; however, his modifications to the wind-tunnel data were strictly intuitive, by rule of thumb. Because a model had to be much smaller than the actual airplane in order to fit into most wind tunnels, the only way to obtain a true simulation of the Reynolds number for full-scale flight would be to adjust the flow properties in the test section. Because $Re = \rho V c/\mu$, Re could be increased by increasing the velocity V, increasing the density ρ, or decreasing the viscosity coefficient μ. The viscosity coefficient was a function of temperature; it could be reduced by cooling the air. Also, when the air was cooled, the density would be increased. Both the lower μ and higher ρ would serve to increase Re. (That is the idea behind the few modern cryogenically cooled high-Reynolds-number facilities existing today.) Another approach would be to increase ρ simply by increasing the pressure in the test section. That was Munk's idea.

Munk had suggested the idea of a pressurized wind tunnel to the Zeppelin company before 1920, but the idea was never pursued. After arriving in the United States, he campaigned for the NACA to build such a facility. The benefit of being able to simulate the true Reynolds number was fully appreciated by George Lewis, director of research for the NACA, and construction of the NACA Variable Density Tunnel (VDT) was authorized on March 4, 1921, with Munk as the chief designer (Figure 7.17). The method used to increase the pressure of the test-section airflow was straightforward: The entire wind tunnel was placed inside a larger pressure vessel, and the vessel was pressurized to 20 atm., thus raising the pressure in the test section to 20 atm. Thus full-scale Reynolds numbers could be achieved during tests of $\frac{1}{20}$-scale models. The VDT was simply a closed-circuit wind tunnel placed inside a pressure vessel; Figure 7.18 shows an external view of the pressure vessel (with the tunnel inside).

It took only 16 months to design and build the VDT, which became operational at the NACA Langley Memorial Laboratory on October 19, 1922. This was remarkable considering the poor working relationships between Munk (the de facto chief designer of the VDT, though residing at NACA Headquarters in Washington) and the Langley engineers who were constructing the tunnel. Munk visited Langley on a sporadic basis during the early construction, which only exacerbated the problems. Frederick Norton was head of

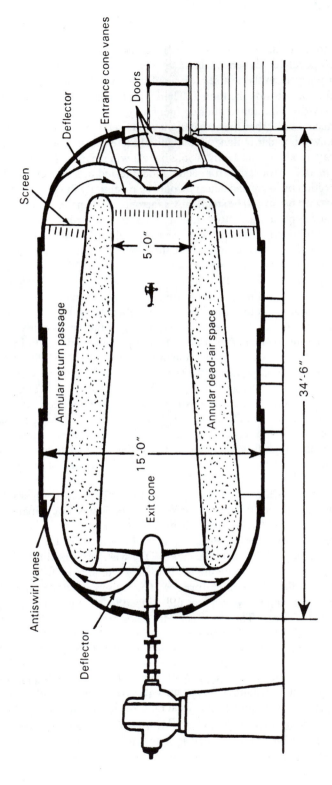

Figure 7.17 The NACA Variable Density tunnel (1922).

Figure 7.18   External view of the NACA Variable Density Tunnel.

the aerodynamics section at Langley and bore the full brunt of the day-to-day construction problems. Munk was a problem for Norton:

> Frederick Norton detected in Munk a stubborn unwillingness to take personal responsibility for transforming the idea of the compressed air tunnel into reality. In 1921 Norton complained to Washington about the chaos brought on by Munk's vague yet overbearing direction of the construction of the Variable-Density Tunnel. He reported that the work of designing the interior and balance system of the VDT was being carried out very inefficiently, due chiefly, he believed, to the lack of sympathy between Munk and the Langley draftsmen and engineers. "Dr. Munk does not seem to have any clear idea as to what he wishes in the engineering design," Norton reported, "excepting that he is sure that he does not want anything that [I or my men] suggest." According to Norton, many portions of Munk's design were quite unsatisfactory [p. 85].[84]

NACA Headquarters did not listen to Norton; indeed, they compounded the situation in late 1922 by detailing Munk to Langley for long periods (weeks at a time) to take charge of the tunnel. Officially, Munk was "responsible for the preparation of the research program, the control of the operation of the apparatus and the preparation of reports." In 1923 Norton resigned from the NACA.

Norton's successor, David Bacon, was even more strongly opposed to Munk's interference. Bacon refused a direct order from Washington to let Munk take charge of the VDT for a period of four weeks; after Munk pressured NACA Headquarters to do something, Bacon relented, though a month later he resigned from the NACA.

The personality conflicts notwithstanding, the record shows that the VDT was a tremendous success. When it became operational in 1922, it was the first wind tunnel capable of testing airfoils, wings, and complete airplane configurations under the conditions of full-scale Reynolds numbers. With that facility the NACA leapfrogged over all other existing wind tunnels, and the experimental data obtained with the VDT made the United States the undisputed leader in applied aerodynamics for the next 15 years. Aircraft companies, universities, and foreign agencies, including the National Physical Laboratory in Great Britain, sent teams of people to Langley to learn about the VDT and, in some cases, to make plans for building their own high-density facilities. Much later, in 1956, Jerome Hunsaker commented that the VDT "represented the first bold step by the NACA to provide its research personnel with the novel, often complicated, and usually expensive equipment necessary to press forward the frontiers of aeronautical science."[94] The impact of the VDT was nicely summarized by Baals and Corliss: "It was the VDT above all that established NACA as a technically competent research organization. It was a technological quantum jump that rejuvenated American aerodynamic research and, in time, led to some of the best aircraft in the world."[93] The research-and-development work in aerodynamics carried out using the VDT clearly vindicated all of the efforts by that small group of visionaries who worked so hard before 1915 for the creation of the NACA and a national aerodynamics laboratory.

The VDT was used for testing until the early 1940s, though well before that time it had been superseded by other wind tunnels, including a large full-scale tunnel with a 30- × 60-ft test section constructed at Langley in 1931. From the early 1940s until 1983, the VDT, minus its internal wind-tunnel plumbing, served as a high-pressure storage tank. In 1983, the Langley Research Center Pressure Systems Committee recommended that the pressure vessel be decommissioned because of its advanced age and riveted construction. The outside shell of the VDT still exists, designated a National Historic Landmark, and it is on display at the NASA Langley Research Center.

Among the most important findings from VDT testing were the data from an extensive series of NACA airfoil tests carried out in the 1920s and 1930s – perhaps the most important product of the NACA in its early years, and used by every nation with a role to play in aeronautics.

The Evolution of Airfoil Design: The First 25 Years after the Wright Flyer

Interest in the "proper" shape for an airfoil dates as far back as the late fifteenth century and Leonardo da Vinci's ornithopter designs. However, da Vinci was not influenced by the science of fluid dynamics, which at that time was in an embryonic stage; he had only the natural shapes of birds' wings for guidance. Engineering inquiry about the proper shapes for airfoils began with George Cayley during the period 1799–1810, when he theorized that a cambered airfoil shape might be better than a flat surface and that lift was produced by a difference in pressure between the bottom and top surfaces of an airfoil. Actual engineering design of cambered airfoils began with the work of Otto Lilienthal in Germany between 1866 and 1889, based on whirling-arm tests, and continued with the work of Horatio Phillips in England between 1884 and 1891, based on wind-tunnel tests. Lilienthal's data were used extensively by the Wright brothers in their early work, although the circular-arc airfoils studied and used by Lilienthal were never employed by the Wrights on any of their flying machines. Instead, the Wrights used airfoils with the maximum camber much closer to the leading edge, and they eventually generated their own wind-tunnel data to guide their airfoil designs. In none of that early work, however, was there any semblance of a rational approach to airfoil design. Everything was completely ad hoc, and each flying-machine enthusiast had his own approach to the problem.

Such was the state of the art of airfoil design at the time of the Wright Flyer (1903). In this section we shall examine the evolution of airfoil understanding and design during the 25 years after the Wright Flyer, spanning the era of the strut-and-wire biplane. Figure 7.19 shows various airfoil shapes from that era.[90]

The ad hoc approach that characterized airfoil design prior to the Wright Flyer continued for the next 25 years; airfoil design continued to be very personalized, customized, and basically random. Although the state of the art of theoretical aerodynamics had been

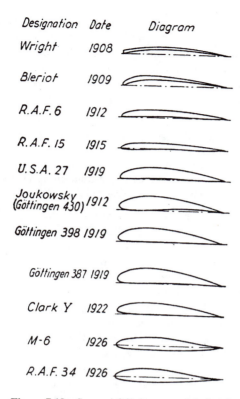

Designation	Date	Diagram
Wright	1908	
Bleriot	1909	
R.A.F. 6	1912	
R.A.F. 15	1915	
U.S.A. 27	1919	
Joukowsky (Göttingen 430)	1912	
Göttingen 398	1919	
Göttingen 387	1919	
Clark Y	1922	
M-6	1926	
R.A.F. 34	1926	

Figure 7.19 Some airfoil shapes used during the era of the strut-and-wire biplane.

advancing, there were virtually no applications of that theory to the design of efficient airfoils. On the experimental side, a great deal of wind-tunnel testing had been conducted on airfoils, but there was no unifying thread running through that empirical work that could have led to the development of efficient, standardized airfoil sections. That situation was recognized in the first annual report of the NACA in 1915 in a discussion of the important technical problems to be tackled by the fledgling agency:

> Of the many problems now engaging general attention, the following are considered of immediate importance and will be considered by the committee as rapidly as funds can be secured for the purpose. [A list of 17 items followed, the third being airfoils.]
> Wing Sections – The evolution of more efficient wing sections of practical form, embodying suitable dimensions for an economical structure, with moderate travel of the center of pressure and still affording a large range of angle of attack combined with efficient action.

The NACA's first effort to address the airfoil problem was to sponsor a series of wind-tunnel tests by Edgar S. Gorrell and H. S. Martin at the Massachusetts Institute of Technology in 1917. By that time, two families of airfoil shapes were in common use: a series tested by Eiffel in France, and the RAF series of airfoils designed and tested by the Royal Aircraft Factory (part of the National Physical Laboratory) in England. Gorrell and Martin concluded that "although a great many airfoils have been tested, many are useless from a practical point of view. It seems safe to assert that in this country [the United States] nearly every airfoil used is either one of the best five or six tested by M. Eiffel near Paris or by the National Physical Laboratory at Teddington, England, or based upon them, with some slight modifications."[95] They went on to comment that the findings from their experiments showed that even slight variations in the shape of an airfoil could result in large differences in aerodynamic performance – an important observation at that early stage of airfoil development. They also emphasized the importance of the structural design of an airfoil and the practical compromise that imposed on aerodynamic performance:

> We are thus limited to a few aerofoils, and some of these lack certain desirable characteristics as to the depth of wing spars combined with aerodynamical efficiency. It would seem of advantage to have the following results of tests made upon the six structurally excellent and heretofore aerodynamically unknown aerofoils designed by the Aviation Section, Signal Corps, United States Army. This constitutes the largest single group of aerofoils, excepting those of the N.P.L. and M. Eiffel, which has been tested and published.[95]

The "aerodynamically unknown aerofoils" were the U.S.A. series developed by the U.S. Army. The data on those airfoils were the NACA's first published airfoil results. The tests carried out by Gorrell and Martin reflected the state of the art in airfoil experiments in 1917, and thus it will be instructive to summarize their techniques and findings. The wind tunnel they used was at the Massachusetts Institute of Technology, with a test-stream velocity of 30 mph. The test models were rectangular finite wings, each with an 18-in. span and a 3-in. chord (aspect ratio = 6), carefully machined from solid brass. For testing, each wing model would be mounted in the wind tunnel on a vertical spindle attached to a three-component balance that measured lift, drag, and pitching moment. The wing model would be mounted at a given angle of attack, and the forces and moment measured. Then the model would be set to a new angle of attack, and the forces and moment measured again, and so forth. Using their notation, which was common at the time, the lift and drift (drag) coefficients K_y and K_x were calculated from the measured lift and drag forces using

$$L = K_y A V_\infty^2$$

and

$$D = K_x A V_\infty^2$$

where V_∞ was in miles per hour, and A was the wing planform area in square feet. Of the six airfoils tested, the U.S.A. 1 was found to have the highest maximum lift-to-drag ratio: $(L/D)_{max} = 17.8$. However, the range of $(L/D)_{max}$ values for the other five airfoil shapes was not much different, the lowest value being 15.9 for the wing with the U.S.A. 4 section. The shape of the U.S.A. 1 airfoil and the measured variations of its lift coefficient K_y, drift (drag) coefficient K_x, and lift-to-drag ratio L/D versus angle of attack are shown in Figure 7.20.

The values for K_y and K_x in Figure 7.20 are in units of pounds force per square foot per (miles per hour)[2]. The relationship between K_y and the standard lift coefficient C_L used today is as follows: The modern lift coefficient is defined by

$$L = \tfrac{1}{2}\rho_\infty V_\infty^2 A C_L$$

where V_∞, A, and ρ_∞ are in consistent units (i.e., V_∞ in ft/s, A in ft^2, and ρ_∞ in slugs/ft^3, or V_∞ in m/s, A in m^2, and ρ_∞ in kg/m^3). Hence, comparing lift in terms of K_y and lift in terms of C_L, both given by the foregoing equations, we have

$$K_y A[V_\infty \quad (\text{in mph})]^2 = \tfrac{1}{2}\rho_\infty [V_\infty \quad (\text{in ft/s})]^2 A C_L$$

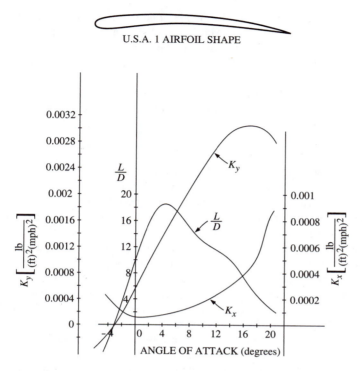

Figure 7.20 Lift drag coefficients and lift-to-drag ratio as functions of the angle of attack for the U.S.A. 1 airfoil.

Using the appropriate conversion between feet per second and miles per hour, we have

$$C_L = (0.9298/\rho_\infty)K_y$$

Assuming the standard sea-level value of $\rho_\infty = 0.002377$ slug/ft^3, that becomes

$$C_L = 391K_y$$

The same relation holds between the modern drag coefficients C_D and K_x, namely,

$$C_D = 391K_x$$

Thus, in Figure 7.20, the maximum lift coefficient is shown as $K_y = 0.00318$. In modern terms, that is equivalent to $C_L = (391)(0.00318) = 1.24$, a number more familiar to modern aerodynamicists.

The findings of Gorrell and Martin were representative of the state of the art in 1917 in at least two respects:

(1) The variations of the lift and drag coefficients and of L/D with the angle of attack, as seen in Figure 7.20, are familiar to a modern aerodynamicist; the trends of the data are consistent with most airfoils. For example, for angles of attack above 2°, the lift curve is essentially linear until stall is reached at about 15°. The drag curve is relatively flat over most of the range of low angles of attack. And the maximum

value of L/D occurs at a fairly low angle of attack. In putting those three curves on the same graph, Gorrell and Martin established a mode for presenting airfoil and wing data that has carried through to the present.

(2)   The data suffered from low-Reynolds-number effects. For the 3-in. chord in a 30-mph test stream at standard atmospheric pressure, the Reynolds number was only 69,960, but that low-Reynolds-number environment was typical for all airfoil and wing testing at that time. It was not until the VDT became operational at the NACA Langley Memorial Laboratory in 1922 that airfoils and wings were tested at true flight Reynolds numbers.

Thus, in 1917, experiments in airfoil aerodynamics were yielding lift and drag data that qualitatively resemble those obtained today; however, they were quantitatively misleading because of the very low Reynolds numbers during the experiments.

That situation had an impact on airfoil design that was far more serious than simply the magnitude of some experimentally obtained numbers being somewhat off because of the low Reynolds number. It pointed the early airfoil designers in the *wrong general direction,* toward the design of very thin airfoils, instead of the more proper consideration of airfoils with some reasonable thickness. There was a belief and tradition, going as far back as Geroge Cayley's time, that airfoils should be thin, because of an intuitive, but mistaken, notion that thick airfoils would produce more drag. That notion was reinforced by the early wind-tunnel findings (e.g., by Phillips and the Wright brothers), which indicated that thick airfoils did experience more drag than thin airfoils. But we know today that that finding is valid only at low Reynolds numbers. Low-Reynolds-number flow over an airfoil is characterized by the formation of a laminar-separation bubble on the top surface – a bubble that can burst and cause the flow to separate over almost the entire upper surface, leading to the undesirable situation of low lift and high drag. Low-Reynolds-number laminar-separation bubbles are most likely to occur over thick airfoils, because the pressure gradients along the surface of a thick airfoil are stronger than those along the surface of a thin airfoil, and the formation of a laminar-separation bubble is promoted in regions of strong adverse pressure gradients. In contrast, the pressure gradients along the top surface of a very thin airfoil are much weaker, and a low-Reynolds-number laminar-separation bubble is much less likely to form. In a low-Reynolds-number flow, a thin airfoil is a better performer than a thick airfoil, but the opposite is true in a high-Reynolds-number flow. In a high-Reynolds-number flow, the rapid expansion around the rather sharp leading edge of a very thin airfoil will promote flow separation at the leading edge, much like that which occurs for a thin flat plate at an angle of attack. In contrast, the flow over the larger, rounded leading edge of a thick airfoil will tend to remain attached, and thus the thick airfoil is a better performer. Therefore, when very thin airfoils (which had looked good in low-Reynolds-number wind-tunnel tests) were used on real airplanes, for which the flight Reynolds numbers were orders of magnitude larger, the wing lift performances were not all they could have been. The wind-tunnel data of Gorrell and Martin provide a good example of that situation. Figure 7.20 shows that the lift coefficients for the very thin U.S.A. airfoils were quite respectable, but they were applicable only in the low-Reynolds-number flow conditions of the wind-tunnel tests. It is no wonder that the airfoils used for most airplanes through 1917 were very thin. Of course, the fundamentals of low-Reynolds-number airfoil aerodynamics as we know them today simply were not sufficiently understood by the designers at that time. They had no way of knowing what they were missing.

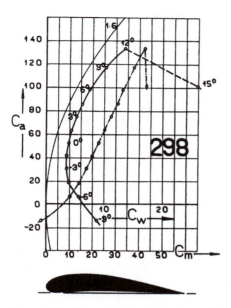

Figure 7.21   Drag polar for the Göttingen 298 airfoil measured by Prandtl in 1917.

Figure 7.22   Fokker Dr-1 triplane, the first airplane to use a thick airfoil section.

That situation changed dramatically in 1917. Work carried out by Prandtl at Göttingen finally demonstrated the superiority of a thick airfoil section, which quickly led to the family of Göttingen airfoils, two of which, the 398 and 387, are shown in Figure 7.19. Prandtl's data[73] for the Göttingen 298 airfoil are shown in Figure 7.21, where we see the drag polar (the middle curve): the lift coefficient versus the drag coefficient, or, in the German notation, C_a versus C_w. (Note that in the early German notation, the values for C_a and C_w were 100 times larger than the values we quote today. For example, in Figure 7.21, a value of $C_a = 100$ corresponds to a lift coefficient $C_L = 1.0$ today.) The Reynolds number was 2.1×10^6, well within the high-Reynolds-number regime.

That revolutionary discovery was immediately picked up by the famous designer Anthony Fokker, who incorporated the 13%-thick Göttingen 298 profile in his new Fokker Dr-1, the famous triplane flown by the "Red Baron," Manfred Freiherr von Richthofen (Figure 7.22).

There were two major benefits in Fokker's use of the thick airfoil:

(1)   There was sufficient room for the wing structure to be made completely internal. That is, the wings of the Dr-1 involved a cantilever design, which obviated the need for the conventional wire bracing used in other aircraft. In turn, that eliminated the high drag associated with such wires. For that reason the Dr-1 had a zero-lift drag coefficient of 0.032, among the lowest for World War I airplanes (for comparison, the zero-lift drag coefficient for the French SPAD XIII was 0.037).

(2)   The thick airfoil gave the Fokker Dr-1 a high maximum lift coefficient, which gave it an exceptionally high rate of climb, as well as enhanced maneuverability – characteristics that were extremely useful in close combat.

Anthony Fokker continued to use thick airfoils as in his design of the D-VII, and that gave the D-VII a much faster rate of climb than its two principal opponents toward the end of the war, the English Sopwith Camel and the French SPAD XIII, both of which still used very thin airfoil sections. That rate-of-climb performance, as well as its excellent handling characteristics, made the Fokker D-VII the most feared of all German World War I fighters. As mentioned earlier, in the armistice agreement the Fokker D-VII was the only airplane named to be turned over to the victorious Allies.

The ad hoc, individualized nature of airfoil design at that time is illustrated by the compilation of existing airfoil data published by the NACA in 1920. In NACA technical report 93, "Aerodynamic Characteristics of Aerofoils," the lift and drag coefficients, lift-to-drag ratios, and center-of-pressure locations for 214 different airfoils are plotted as functions of the angle of attack. For instructional purposes, the complete list of airfoils is given in Figure 7.23. Such a large number of airfoils, many of them obscure, is testimony to the proliferation of custom-designed airfoils by 1920, and the list in Figure 7.23 was not complete (e.g., it did not include any of the thick Göttingen airfoils discussed earlier). However, the Göttingen airfoils were discussed in NACA reports 124, 182, and 244, published in 1921, 1923, and 1926, respectively, which were simply continuations of the review that began with NACA report 93 in 1920. Those four NACA reports listed the characteristics of over 600 airfoil shapes from the United States, Britain, France, Germany, and Italy.

One of the first thick airfoils designed in the United States was the Clark Y airfoil, listed in Figure 7.19. Conceived in 1922 by Colonel Virginius Clark, the shape of the Clark Y section was arrived at simply by adding the thickness distribution of a Göttingen airfoil to a flat undersurface. The flat bottom was chosen for convenience of manufacture. The series of Clark Y airfoils became very popular in the 1920s.

By 1925, toward the end of the era of the strut-and-wire biplane, there were hints of the forthcoming revolution in NACA airfoil designs. Max Munk designed a systematic series of "Munk airfoils" using the design approach of adding the thickness distributions of three symmetrical airfoils to various camber lines, thus obtaining a systematic family of relatively thick cambered airfoils. The shapes of the camber lines were calculated so as to minimize the movement of the center of pressure as the angle of attack was changed.[96] The shape and drag polar for one of those, the M12 airfoil, are shown in Figure 7.24. All of the Munk airfoils were relatively thick, and by 1925 the advantages of relatively thick airfoils over the very thin shapes characteristic of most World War I airplanes were widely recognized. The series of 27 Munk airfoils[96] was the first systematic family of thick airfoils designed in the United States. Munk's airfoil data[96] were pioneering in another sense, having been obtained using the NACA Variable Density Tunnel (VDT), which allowed testing under

A. D. No. 1
A. D. No. 4
Albatross
Avro
B. I. R. 1a
B. I. R. 3
B. I. R. 33a
Bleriot Triplane
Bristol
Clark, V. E.
Cowley & Levy—A. 1
Cowley & Levy—A. 2
Cowley & Levy—A. 3
Cowley & Levy—A. 4
Cowley & Levy—A. 5
Cowley & Levy—A. 6
Cowley & Levy—A. 7
Cowley & Levy—B. 1
Cowley & Levy—B. 2
Cowley & Levy—B. 3
Cowley & Levy—B. 4
Cowley & Levy—B. 5
Cowley & Levy—B. 6
Cowley & Levy—B. 7
Curtiss
DeH-2
DeH-3
Dorand
Durand Propeller No. 4
Durand Propeller No. 7
Durand Propeller No. 10
Durand Propeller No. 13
Durand Propeller No. 16
Eiffel 8
Eiffel 9
Eiffel 10 (Wright)
Eiffel 11 (Voisin)
Eiffel 12 (M. Farman)
Eiffel 13 (Bleriot 11)
Eiffel 13 bis (Bleriot 11a)
Eiffel 14 (Breguet)
Eiffel 15 (M. Ernoult)
Eiffel 16 (M. Drzewiecki)
Eiffel 16a
Eiffel 16b
Eiffel 16c
Eiffel 16d
Eiffel 17 (M. Drzewiecki)
Eiffel 18 (M. Drzewiecki)
Eiffel 30
Eiffel 31
Eiffel 32 (Lanier-Lawrance)
Eiffel 33 (Breguet)
Eiffel 34 (Colliex)
Eiffel 35 (Dorand)
Eiffel 36 (Odier)
Eiffel 37 (Kauffmann)
Eiffel 38 (Coanda)
Eiffel 39 (16b Modified)
Eiffel 40
Eiffel 41
Eiffel 42
Eiffel 43
Eiffel 44 (Voisin)
Eiffel 45 (Buch)

Eiffel 46 (Buch)
Eiffel 47 (Howard-Wright)
Eiffel 48 (Howard-Wright)
Eiffel 49 (Howard-Wright)
Eiffel 52 (Nieuport)
Eiffel 53 (Nieuport)
Eiffel 54 (Deperdussin)
Eiffel 55 (Deperdussin)
Eiffel 56 (Deperdussin)
Eiffel 57
Eiffel 58
Eiffel 59
Eiffel 60
Eiffel 61
Eiffel 62
F. 2 B
Fairey
Halbronn 2
Halbronn 3
Handley Page 166
Handley Page 166a
Handley Page 166b
Handley Page 166c
Italian 1
Italian 2
Italian 3
N. P. L. 4
N. P. L. 4a
N. P. L. 4b
N. P. L. 4c
N. P. L. 4cα
N. P. L. 4cβ
N. P. L. 4cγ
N. P. L. 64
N. P. L. 73
N. P. L. 214
Naylor & Griffiths 1
Naylor & Griffiths 2
Naylor & Griffiths 3
Naylor & Griffiths 4
Naylor & Griffiths 5
Naylor & Griffiths 6
Naylor & Griffiths 7
Naylor & Griffiths 8
Naylor & Griffiths 9
Naylor & Griffiths 10
Naylor & Griffiths 11
Naylor & Griffiths 12
Naylor & Griffiths 13
Naylor & Griffiths 14
Naylor & Griffiths 15
Naylor & Griffiths 16
Naylor & Griffiths 17
Naylor & Griffiths 18
Naylor & Griffiths 19
Naylor & Griffiths 20
Naylor & Griffiths 21
Naylor & Griffiths 22
Naylor & Griffiths 23
Naylor & Griffiths 24
Naylor & Griffiths 25
Naylor & Griffiths 26
Naylor & Griffiths 27
Naylor & Griffiths 28

Naylor & Griffiths 29
Naylor & Griffiths 30
Naylor & Griffiths 31
Offenstein
Offenstein (modified)
Portholme
R. A. F. 3
R. A. F. 4
R. A. F. 5
R. A. F. 6
R. A. F. 6 (modified)
R. A. F. 6a
R. A. F. 6c
R. A. F. 6c (both surfaces)
R. A. F. 8
R. A. F. 9
R. A. F. 12
R. A. F. 13
R. A. F. 14
R. A. F. 14 (modified)
R. A. F. 15
R. A. F. 15 (modified)
R. A. F. 16
R. A. F. 17
R. A. F. 18
R. A. F. 19
R. A. F. 20
S. E. A.
St.-Cyr 1
St.-Cyr 2
St.-Cyr 3
Scout E
Sloane
Sopwith
Spad
Standard Aircraft Corp. 48
Turin 1
Turin 2
U. S. A. 1
U. S. A. 2
U. S. A. 3
U. S. A. 4
U. S. A. 5

U. S. A. 6
U. S. A. 7
U. S. A. 8
U. S. A. 9
U. S. A. 10
U. S. A. 11
U. S. A. 12
U. S. A. 14
U. S. A. 15
U. S. A. 16
U. S. A. 18
U. S. A. 19
U. S. A. 20
U. S. A. 21
U. S. A. 23
U. S. A. 24
U. S. A. T. S. 1
U. S. A. T. S. 2
U. S. A. T. S. 3
U. S. A. T. S. 4
U. S. A. T. S. 5
U. S. A. T. S. 6
U. S. A. T. S. 7
U. S. A. T. S. 8
U. S. A. T. S. 9
U. S. A. T. S. 10
U. S. A. T. S. 11
U. S. A. T. S. 12
U. S. A. T. S. 13
U. S. A. T. S. 14
U. S. A. T. S. 15
U. S. A. T. S. 16
U. S. A. T. S. 17
U. S. A. T. S. 18
U. S. D. 9A
W-1
Washington Navy Yard 1
Washington Navy Yard 2
Washington Navy Yard 3
Washington Navy Yard 4
White

Figure 7.23   A partial list of airfoil sections used in 1920.

conditions that could simulate full-scale Reynolds numbers. The Reynolds number for the data shown in Figure 7.24 was 3.6×10^6. Munk commented on some of the differences between data obtained in the VDT and data obtained in conventional wind tunnels: "The test charts show that at full size Reynolds Number, the minimum drag is much smaller than we are accustomed to obtain in the ordinary atmospheric wind tunnel. The maximum lift is not necessarily larger at a larger Reynolds Number."[96]

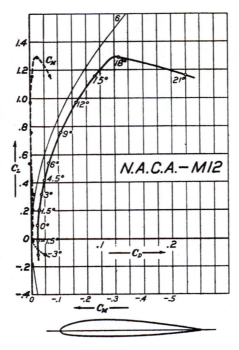

Figure 7.24   Drag polar for the NACA M12 airfoil designed by Max Munk (1925).

By 1925 there had been a confluence of two factors that would later establish the NACA as the leader in airfoil design and the premier source of data on airfoils: (1) the beginnings of a systematic procedure for designing effective airfoil shapes and (2) the testing of those shapes in the VDT, which at that time was the only wind tunnel in the world capable of testing at full-scale Reynolds numbers.

Aerodynamic Coefficients: Evolution of the Modern Nomenclature

The modern nomenclature for lift and drag coefficients came into use during the era of the strut-and-wire biplane. Otto Lilienthal was the first to express aerodynamic lift and drag data in terms of coefficients. He tabulated normal-force and axial-force coefficients as functions of the angle of attack (the famous Lilienthal table first used and then discarded by the Wright brothers, as discussed in Chapters 4 and 5). Because Lilienthal presented his data in a form ratioed to the findings for a flat plate at a 90° angle of attack, his normal-force and axial-force coefficients, η and θ, respectively, were not influenced by the prevailing uncertainty in Smeaton's coefficient (as discussed in Chapter 4). However, Lilienthal did use the following equations[46] for the normal and axial forces, N and T, respectively:

$$N = 0.13\eta F V^2$$
$$T = 0.13\theta F V^2$$

where, in Lilienthal's notation, F was the reference area, and the number 0.13 was the classic, original value for Smeaton's coefficient in metric units. Lilienthal suspected that the original value for Smeaton's coefficient was inaccurate; he probably included the preceding equations in his book as a formality. In any event, those two equations marked the first time that expressions for aerodynamic force had been expressed in terms of *force coefficients*.

Only a few years later, Samuel Langley followed Lilienthal's lead by presenting his data in a form ratioed to the findings for a flat plate at a 90° angle of attack, thus obtaining aerodynamic-force coefficients that were independent of any uncertainty in Smeaton's coefficient. Langley went a step further: He *measured* a value for Smeaton's coefficient that was within 3% of today's accepted true value. Langley employed the concept of an aerodynamic-force coefficient by using the following equation:

$$R = kSV_\infty^2 F(\alpha)$$

where R was the resultant aerodynamic force, k was the *more accurate value* of Smeaton's coefficient, and $F(\alpha)$ was the corresponding force coefficient, a function of the angle of attack α. Note that Langley and Lilienthal used the same form of the equation, namely, expressing the aerodynamic force *explicitly* in terms of Smeaton's coefficient, a reference area, the square of the free-stream velocity, and the force coefficient.

The Wright brothers followed suit. Their novel and ingenious wind-tunnel force balance measured the lift *coefficient* directly; nowhere was Smeaton's coefficient used in the reduction of their data. However, in calculating the lift and drag forces, the Wrights used their own (reasonably accurate) measurement of Smeaton's coefficient in a force equation identical with the one used by Lilienthal and Langley, namely,

$$L = kSV^2 C_L$$
$$D = kSV^2 C_D$$

where C_L and C_D were the lift and drag coefficients from their wind-tunnel measurements.

The Wrights were among the last to use expressions for aerodynamic coefficients explicitly written in terms of Smeaton's coefficient. Eiffel, in 1909, defined a "unit-force coefficient" K_i as

$$R = K_i SV^2$$

where R was the resultant aerodynamic force. Smeaton's coefficient was nowhere to be seen – it was buried in the measured value of K_i. After Eiffel's work became known, Smeaton's coefficient was never used in the aerodynamics literature – it was totally passé.

Gorrell and Martin,[95] for their 1917 wind-tunnel tests on various U.S.A. airfoils, adopted Eiffel's approach, giving the following expressions for lift and drag:

$$L = K_y AV_\infty^2$$
$$D = K_x AV_\infty^2$$

where K_y and K_x were the lift and drag coefficients, respectively.

By 1917, the density ρ was appearing in the expressions for force coefficients. One of the early objectives of the NACA was to try to standardize nomenclature. In 1917, Joseph Ames, president of Johns Hopkins University, and later to be chairman of NACA (1927–39), in NACA report 20, "Aerodynamic Coefficients and Transformation Tables," led off with the following equation:

$$F = C\rho SV^2$$

where F was the total force acting on the body, S was the reference area, V was the free-stream velocity, and C was the force coefficient, which Ames described as "an abstract number, varying for a given airfoil with its angle of incidence, independent of the choice of units, provided these are consistently used for all four quantities $(F, \rho, S, \text{and } V)$." The expression given by Ames was very close to the standard usage today, differing only by a factor of $\frac{1}{2}$. In Ames's expression, the quantity C was one-half the modern resultant force coefficient C_R (i.e., $C = \frac{1}{2}C_R$).

At about the time that Ames's report was published, Prandtl, at Göttingen, was already using the nomenclature that is standard today. In his 1921 English-language review[79] of the work performed at Göttingen before and during World War I, Prandtl defined the dynamic pressure (he called it "dynamical pressure") as

$$q \equiv \tfrac{1}{2}\rho V_\infty^2$$

He went on to say that

> the dynamical pressure is also well suited to express the laws of air resistance. It is known that this resistance is proportional to the square of the velocity and to the density of the medium; but $q = (\rho/2)V_\infty^2$; so the law of air resistance may also be expressed by the formula
>
> $W = cFq$
>
> where F is the area of the surface and c is a pure number.[79]

As with so many other developments in aerodynamics, Prandtl led the way in defining our modern terms for aerodynamic-force coefficients. Today, for lift and drag we write

$$L = q_\infty S C_L$$
$$D = q_\infty S C_D$$

where C_L and C_D are the "pure numbers" referred to by Prandtl (i.e., the lift and drag coefficients).

In the early German usage of lift and drag coefficients, C_a and C_w, respectively, the values cited were always 100 times larger than C_L and C_D, as we saw in Figure 7.21. There was no physical reason for such usage, and "the only advantage seems to be that the coefficients can be expressed as whole numbers and large decimals."[97]

The word "drift" was used to denote aerodynamic drag in the days of Langley, Chanute, and the Wright brothers, but it fell out of favor during World War I. By 1917, the NACA was using the word "drag" in all its literature.

The State of the Art as Reflected in World War I Aircraft

The era of the strut-and-wire biplane was a transition period in the evolution of aerodynamics, with the fundamental advances in experimental aerodynamics (Chapter 5) and theoretical aerodynamics (Chapter 6) at the turn of the century leading to the relatively mature understanding of aerodynamics that characterized the era of advanced propeller-driven airplanes, as discussed in Chapter 8. Some of the most important developments in that transition period have been discussed in this chapter: (1) the development of lifting-line theory for finite wings, which led to an understanding of induced drag, (2) the subsequent appreciation that increasing the aspect ratio would decrease the induced drag, (3) the development of thin-airfoil theory, which allowed the calculation of lift and moment coefficients for a thin airfoil of specified shape, (4) the rapid development of the wind tunnel as the primary laboratory tool for aerodynamics, (5) the awareness of Reynolds-number effects, followed by construction of the VDT for testing at full-scale Reynolds numbers. Along with those advances there was a general rectification process whereby old, archaic ideas and terminology were weeded out, to be replaced by new terms, new notations, and new conceptions that have defined the language of today's aerodynamics.

To what extent were those advances in the state of the art reflected in airplane design during the era of strut-and-wire biplanes, particularly in World War I aircraft? The answer is a mixed bag. On one hand, the development of the thick airfoil was promptly used by Anthony Fokker for the design of the Dr-1 triplane and the D-VII. That was the first time in the history of aeronautics that an aerodynamic breakthrough was immediately applied in the design of a practical airplane. On the other hand, the new understanding of the aspect ratio, and how significantly the induced drag could be reduced by increasing the aspect ratio, was not immediately reflected in airplane design. Figure 7.25 shows the wing planforms for four airplanes typical of the World War I era, with aspect ratios listed at the right, ranging from 3.88 to 5.51. Those were, by any standard, relatively low aspect ratios, and in comparison with the Wright Flyer, with its aspect ratio of 6, those World War I aircraft seem retrograde. One consequence of those low aspect ratios was that they had relatively low lift-to-drag ratios L/D; the maximum L/D values for the Albatross, Nieuport, De Havilland, and Fokker airplanes in Figure 7.25 were 7.5, 7.9, 7.0, and 8.0, respectively (whereas the L/D

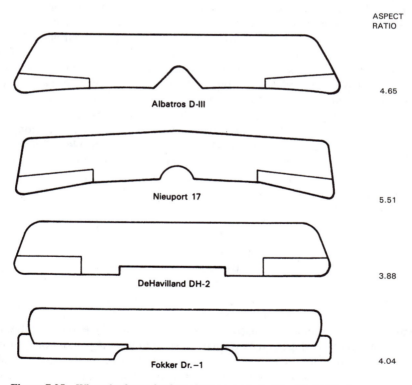

Figure 7.25   Wing planforms for four airplanes typical of the World War I era.

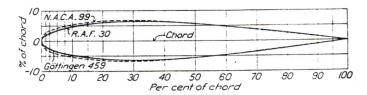

Figure 7.26   Comparison of the RAF 30, the NACA 99, and the Göttingen 459 airfoil shapes.

values are 15–20 for modern, low-speed airplanes). Of course, there are many factors to be considered in airplane design in addition to aerodynamics, an important one being the structural integrity of the wings, which is more of a problem for high-aspect-ratio wings. The higher the aspect ratio, the heavier the wing structure required for strength and stiffness. However, I am inclined to give the designers of World War I aircraft low marks for their inadequate appreciation of the aerodynamic benefits of high-aspect-ratio wings, especially the German airplane designers, because they were closest to Prandtl's innovative ideas and research data at Göttingen. Not all airplanes during that era had low aspect ratios. The Curtiss JN-4 Jenny, for example, one of the most popular U.S. airplanes at the end of the war and well into the 1920s, had an aspect ratio of 7.76, with a maximum L/D of 9.24.

By 1927, state-of-the-art techniques in experimental aerodynamics were being used, in hindsight, to assess World War I airplanes. The best wind tunnel in existence at that time, the NACA's VDT, was being used to measure the aerodynamic properties of earlier aircraft to gain a better understanding of their overall aerodynamic nature. Of particular note were the measurements obtained from models of three British airplanes by Higgins and associates.[98] Their work systematically demonstrated the effects on C_L and C_D of increasing the Reynolds number. Let us examine their data for a $\frac{1}{15}$-scale model of a Bristol fighter with an RAF 30 airfoil section (Figure 7.26). It was a relatively thick, symmetric section almost identical with the Göttingen 459 airfoil. Also shown in Figure 7.26, for comparison, is the shape of the NACA 99 airfoil (dashed curve). Some of the data obtained in the VDT for the Bristol fighter are shown in Figures 7.27–7.29. The variation of the lift coefficient C_L with the angle of attack is shown in Figure 7.27 for five different pressures in the VDT. The Reynolds numbers, based on wing chord, for those pressures are as follows:

Tank pressure (atm)	Reynolds number
1	152,000
2.5	404,000
5	760,000
10	1,500,000
20	3,050,000

The presentation of the findings in Figure 7.27 is classic: (1) The lift curve is linear up to the point of stall, and the slope of the lift curve is essentially not dependent on the Reynolds number. (2) The value for the maximum lift coefficient $(C_L)_{max}$ is strongly dependent on the Reynolds number, with higher values of $(C_L)_{max}$ at higher Reynolds numbers. The drag polars in Figure 7.28 show the classic finding that the minimum drag coefficient is also

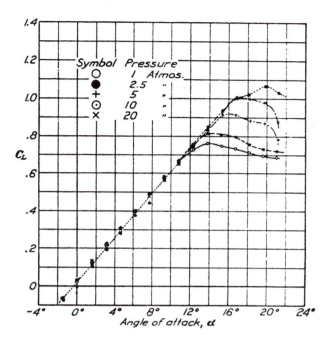

Figure 7.27 Lift curve for the Bristol fighter measured in the VDT (1927).

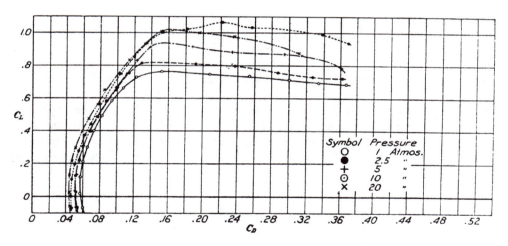

Figure 7.28 Drag polars for the Bristol fighter.

a strong function of the Reynolds number, decreasing as the Reynolds number increases. The variation of L/D with C_L (and hence with the angle of attack) is shown in Figure 7.29. Note that the maximum value for L/D increases as the Reynolds number increases – another classic result.

By the middle of the 1920s the VDT was paying off. In no other wind tunnel in the world could one conduct tests over such a wide range of Reynolds numbers, and tests in

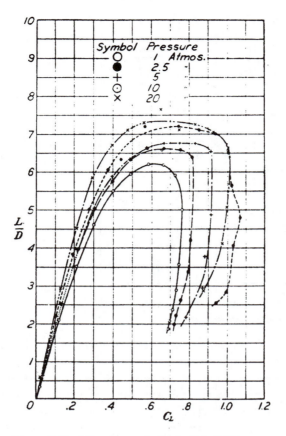

Figure 7.29   Lift-to-drag ratio for the Bristol fighter.

the VDT on whole-airplane configurations were clearly establishing the powerful effect of the Reynolds number on airplane aerodynamics. Airplane designers were beginning to pay attention to those findings, and in that sense the state of the art in experimental aerodynamics was beginning to be reflected in flying machines, a trend that would be greatly accelerated in the next decade.

CHAPTER 8

Aerodynamics in the Age of Advanced Propeller-Driven Airplanes

Every advance in aeronautics is the result of long and painstaking research. More often than not the development of a particular kind of airplane cannot take place because the time is not yet ripe, and revolutionary ideas must be laid aside until the materials and techniques, i.e., the technology have caught up.

Darrol Stinton (1966)

To fly seems to have been man's dream from the earliest recorded days. There have always been "scientists" who wanted to find out how flying was done, and there have always been "engineers" who wanted to create the tools to do it with.

Dietrich Kuchemann (1975)

The major thrust of aerodynamics in the age of the advanced propeller-driven airplane can be summarized in a word: *streamlining*. Research in aerodynamics in that age, generally covering the period from about 1930 to 1945, was driven by two practical concerns: (1) the need to reduce the drag on aircraft so they could fly faster and more efficiently, and (2) the problems associated with fluid compressibility as new, high-performance propeller-driven aircraft began to reach speeds approaching the speed of sound. That age was characterized by intensive efforts to better understand the mechanisms of drag production, to find ways to better predict drag, to meticulously refine a configuration in order to reduce drag, and to begin to sneak up on the speed of sound without disastrous consequences. The outcome of all that effort was the trend toward more streamlining of aerodynamic bodies.

It was also during that age that the fundamental innovative ideas in aerodynamics that had been articulated at the turn of the century finally began to yield great rewards. Prandtl's boundary-layer theory, first presented in 1904, spread throughout the world of aerodynamics in the late 1920s, allowing airplane designers to make the first intelligent (and sometimes reasonably accurate) predictions of skin-friction drag. The boundary-layer concept also helped to explain how flow could separate from a surface; such flow separation is the cause of form drag (pressure drag due to flow separation). Lanchester's vortex theories, finally published in 1907, and Prandtl's lifting-line theory, developed during World War I, became widely known in the 1920s, allowing the first predictions of induced drag. It was the *applications* of those fundamental breakthroughs that led to the development of advanced propeller-driven airplanes. In this chapter we shall examine the development of the new thinking in aerodynamics that encouraged its maturation process and led to improved designs for airfoils, airplanes, and wind tunnels. And we shall examine the impact of all those factors on the evolution of the flying machine between 1930 and 1945. Before we proceed further, we shall discuss very briefly some of the technical aspects of the production of aerodynamic drag, in order to clarify the practical concerns that drove the evolution of aerodynamics during that period.

319

Friction Drag, Form Drag, and Induced Drag

In Chapter 1 we emphasized that an airflow over a body creates (1) a pressure distribution that is exerted over the surface of the body and (2) a shear-stress distribution, due to friction, that is exerted over the surface of the body. The pressure acts locally *perpendicular* at each point on the surface, and the shear stress acts *tangentially* at each point on the surface. The pressure and shear-stress distributions are the two hands by which Nature grabs hold of the body and exerts an aerodynamic force on it. The component of that force that acts in the direction of the free stream is, by definition, the drag. Hence drag is produced both by friction and by pressure.

For a purely subsonic airflow over an airplane, the total drag is made up of contributions from three sources:

(1) Skin-friction drag: Shear stress, which is the tugging action of friction between the air and the surface of the airplane, creates a force in the drag direction (Figure 8.1a). This force is called *skin-friction drag* (or sometimes simply friction drag).

(2) Form drag: There are situations in which the airflow separates from the surface of an aerodynamic body (Figure 8.1b). The pressure in the low-energy, recirculating, separated region is relatively low, and thus the surface pressure

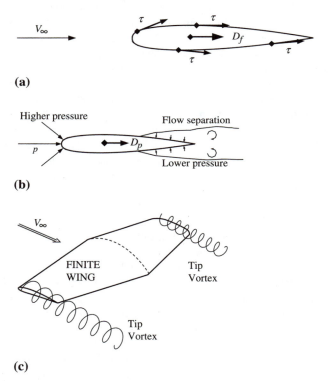

Figure 8.1   Sketches showing the physical sources of (a) skin-friction drag (τ, shear stress due to friction; D_f, resultant friction drag), (b) form drag, and (c) induced drag.

exerted on the body in that region is less than it would be if the flow were still attached. In turn, that creates a pressure imbalance between the front and back surfaces of the body; the lower pressure acting on the back of the body trying to push the body forward is overwhelmed by the higher pressure acting on the front of the body trying to push the body backward. That pressure imbalance creates a force in the drag direction, known as "pressure drag due to flow separation," also called *form drag*. If on an airplane there were absolutely no regions of separated flow, then the form drag would be zero. Unfortunately, aeronautical engineers have not yet been able to design that ideal airplane.

(3) Induced drag: An airplane wing produces lift because the pressure on the bottom of the wing is higher than the pressure on the top of the wing. A by-product of that pressure difference between the top and bottom surfaces of the wing is that at the tips of the wing, the flow is pushed from the high-pressure region on the bottom surface to the low-pressure region on the top surface (i.e., the flow tends to curl around the wing tip from bottom to top). The curling action, superimposed on the main flow over the wing, produces a vortex at each wing tip that flows downstream (Figure 8.1c). Those wing-tip vortices are like mini-tornadoes. Imagine a small tornado swirling around next to you. You would certainly feel a change in air pressure due to the presence of the tornado. The wing also experiences a change in pressure, for the same reason. That change in pressure acting on the surface of the wing always creates an extra pressure imbalance in the drag direction, thus increasing the drag on the wing. That increase in drag is called *induced drag*. It is strictly a pressure drag, and it is caused by the presence of the wing-tip vortices. Clearly, the high pressure on the bottom of the wing and the low pressure on the top of the wing combine to cause both the lift and the wing-tip vortices. Hence, induced drag is directly related to lift. Indeed, Prandtl's lifting-line theory shows that $C_{D,i} \propto C_L^2 / \text{AR}$, where $C_{D,i}$ is the induced-drag coefficient, C_L is the lift coefficient, and AR is the wing aspect ratio. Designing to achieve lift on an airplane is not without its price; induced drag is the cost paid for the production of lift. The induced drag can be decreased by increasing the aspect ratio.

It will be helpful to keep these three sources of drag in mind in our subsequent discussions of aerodynamics; in the age of the advanced propeller-driven airplane, they became the fundamental drivers of aerodynamic research and development.

Streamlining: An Idea Whose Time Had Come

On the evening of April 6, 1922, Louis-Charles Breguet read a paper before the Royal Aeronautical Society in London: "Aerodynamical Efficiency and the Reduction of Air Transport Costs."[99] Breguet was already famous as a successful pilot and airplane designer. Born in Paris in 1880, he had been educated in electrical engineering at French technical universities and had joined the family electrical-engineering firm, Maison Breguet. However, motivated by the spectacular flying demonstrations by Wilbur Wright in 1908 in France, Breguet built and flew his first airplane. After that, he plunged headlong into aviation. He immediately opened an airplane assembly factory in Douai, and by 1912 he had an assembly line turning out a biplane powered by a Renault 80-hp engine. During

World War I he manufactured the Breguet 14 bomber in large numbers for the French forces. In 1919 he founded a commercial airline company that later grew into Air France. Thus on that evening in London as Louis Breguet addressed the Royal Aeronautical Society, the audience listened attentively to the famous French aviation pioneer.

What the audience heard was one of the first important calls for major improvements in the aerodynamic efficiency of airplanes. Breguet's measure of aerodynamic efficiency was the drag-to-lift ratio. (In the United States, it is conventional to work with the reciprocal of that number, namely, the lift-to-drag ratio. However, in Europe, even today, the drag-to-lift ratio is frequently quoted.) Breguet called the drag-to-lift ratio the "fineness" of the airplane; the smaller the fineness, the more aerodynamically efficient the aircraft. Moreover, he referred to the equation for the range of an airplane, which shows that the range is directly proportional to the lift-to-drag ratio, or inversely proportional to the fineness. That equation was first used by Breguet during World War I and today is known worldwide as the Breguet range equation. About that equation, Breguet stated that "one at once realizes the very great importance of the fineness which in that formula is the only term depending upon the aerodynamic qualities of the aeroplane."[99] Later in his talk he elaborated as follows: "The conclusion is that one must bring to the minimum the value of the fineness. It can be obtained by choosing the best possible profile for the wings, the best designs for the body, empennage, etc. Moreover, the undercarriage should be made to disappear inside the body or the wings when the airplane is in flight, etc."[99] Breguet was emphasizing that the aerodynamics of the airplane should be such as to minimize the fineness (i.e., maximize the L/D ratio), and his suggestions to achieve minimum fineness all centered on the reduction of drag. His recommendation for "choosing the best possible profile for the wings, the best designs for the body, empennage, etc.," implied streamlining those geometric shapes so as to reduce the pressure drag due to flow separation (form drag). That was especially true of his idea for retractable landing gear. The fixed landing gear used on airplanes during that period were simply blunt bodies exposed to the flow, with consequent massive flow separation on their back surfaces, and thus high form drag. He knew that a substantial reduction in drag could be achieved by retracting the landing gear out of the flow. (The first ideas for retractable landing gear can be traced to some of da Vinci's sketches for flying machines and to Alphonse Pénaud in France in 1876, who patented a design for his "airplane of the future," which included a retractable undercarriage with compressed-air shock absorbers. It was not until 1920 that the first practical retractable landing gear was used – on the Dayton-Wright R. B. high-wing monoplane built for the Gordon-Bennett air races in France. As a regular feature of airplane designs, retractable landing gear did not become common until the 1930s.)

Breguet went on to note that "an airplane of high standard quality now has a fineness $[D/L]$ equal to 0.12." That was an L/D of 8.3, in keeping with the values shown in Chapter 7 for typical strut-and-wire biplanes. He gave an example of a typical transport airplane with that fineness and calculated a payload cost of 35 francs per ton per kilometer. If the aerodynamics could be improved to give that airplane a fineness of 0.065 ($L/D = 15.4$), then Breguet calculated a cost of 7.4 francs per ton per kilometer – a cost reduction by almost a factor of 5. Clearly, streamlining, with its consequent reduction in form drag and hence increase in L/D, would pay off financially for civil air transport, which was a major focus of Breguet's remarks to the Royal Aeronautical Society that night. His "improved" fineness value of 0.065, pulled out of thin air in 1922 during the age of the strut-and-wire biplane, would not be achieved until the mid-1930s and the development of the Douglas

DC-3, which had an L/D of 14.7, or a fineness of 0.068. It was the insistence of Breguet and others like him who campaigned for improvements in aerodynamic efficiency that finally led to significant efforts to achieve drag reductions through streamlining; Breguet's 1922 paper was an important precursor to the aerodynamics of the age of the advanced propeller-driven airplane.

Breguet went on to practice what he preached. He designed a number of airplanes during the 1920s and 1930s that set long-range records, including the first nonstop crossing of the South Atlantic in 1927. Breguet was active in running his airplane company until his death in 1955, and his influence permeates a substantial part of French aviation history.

The progress in drag reduction during the age of advanced propeller-driven airplanes was facilitated by the spread of Prandtl's boundary-layer theory in the 1920s and the widespread applications of various aspects of that theory in the 1930s. Boundary-layer theory made it possible to calculate the skin-friction drag on an aerodynamic surface, and for laminar flow, such calculations were quite accurate, especially for flow over a flat plate. That case had been treated as early as 1908 by Blasius, one of Prandtl's students at Göttingen.[63] Blasius derived the classic equation for the coefficient for laminar skin-friction drag on a flat plate:

$$C_f = \frac{1.328}{\sqrt{Re}}$$

where $C_f = D_f / q_\infty S$, D_f was the total friction drag on one side of a plate of area S, and Re was the Reynolds number based on the chord length of the plate. Right away that formula explicitly showed the strong Reynolds-number scaling effect on skin-friction drag. Unfortunately, most boundary layers in aeronautical applications are turbulent rather than laminar, and our knowledge of turbulent flows is much less precise than for laminar flows. Calculations to describe turbulent boundary layers became a major focus in theoretical aerodynamics during the 1930s, driven primarily by the mixing-length theory proposed by Prandtl in 1925. The effort to understand turbulent flows and derive equations adequate to describe them has consumed the entire careers of legions of aerodynamicists during the twentieth century, and continues to do so. And after all that effort, we still cannot calculate turbulent skin-friction drag with anything like the accuracy we can achieve for laminar skin-friction drag. Today, just as in the 1930s, some empirical data must always be used to adjust the theoretical formulas for turbulent skin-friction drag. Such work has led to *approximate* formulas for the calculation of turbulent skin-friction drag – half a loaf is better than nothing. That work has also highlighted the role of Reynolds numbers in turbulent boundary layers. For example, for turbulent flow over a flat plate, the skin-friction coefficient is less sensitive to the Reynolds number than in the laminar case. One of the many approximate Reynolds-number variations for the friction drag on a flat plate in a turbulent flow is

$$C_f \propto \frac{1}{Re^{\frac{1}{5}}}.$$

Clearly, the explicit dependence of C_f on the Reynolds number for both laminar flow and turbulent flow is one of the reasons for concern with scale effects in wind-tunnel testing. Related to the matter of scale effects is the problem of determining the location on an aerodynamic surface where the transition from a laminar boundary layer to a turbulent boundary layer takes place – a transition point that is still a poorly understood function of the Reynolds number. An enlightened appreciation of the effects of the Reynolds number on skin-friction and transition measurements (a direct result of Prandtl's boundary-layer theory) was instrumental in Max Munk's advocacy of the variable-density wind tunnel, which allowed testing at full-scale Reynolds numbers, as discussed in Chapter 7.

Whereas boundary-layer theory led to reasonably accurate estimations of skin-friction drag, the same cannot be said in regard to flow separation and the resulting form drag. The best that the theory could do was to allow one to estimate the *location* on the surface where separation would occur, and such an estimate would have some degree of accuracy only for laminar flow; where separation would occur in a turbulent flow was much more in doubt. The only means for obtaining aerodynamic information on form drag were empirical – principally wind-tunnel testing and, to a lesser extent, actual flight tests.

Against that background, in 1929 there came a second resounding call for streamlining – in that case from the famous British aeronautical engineer Sir B. Melvill Jones. Addressing the Royal Aeronautical Society, as had Breguet seven years earlier, Jones entitled his lecture "The Streamline Airplane."[100] Like Breguet, Jones was highly respected, but whereas Breguet was an aeronautical industrialist, Jones was a professor of aeronautical engineering at Cambridge University. Jones's analysis of the advantages of streamlining was so compelling that it was said that "designers were shocked into greater awareness of the value of streamlining."[101] Jones's paper marked a turning point in the practice of aerodynamics during the age of the advanced propeller-driven airplane.

Jones led off his discussion with the following thought:

> Ever since I first began to study aeronautics, I have been annoyed by the vast gap which has existed between the power actually expended on mechanical flight and the power ultimately necessary for flight in a correctly shaped aeroplane. Every year, during my summer holiday, this annoyance is aggravated by contemplating the effortless flight of the sea birds and the correlated phenomena of the beauty and grace of their forms.[100]

Jones went on to underscore the importance of drag reduction, pointing out that such reduction for an airplane with a given power output from the engine would result in a higher cruising velocity, or a lower fuel consumption. Taking a page from Breguet's analysis, this would result in increased range and/or payload, which in Jones's words are "both factors of the first importance in aeronautical development."[100]

Jones clearly identified the kind of drag that was most in need of reduction: *form drag*. He correctly pointed out that induced drag was important at low speeds (because the airplane flies at high C_L at low speeds, hence high $C_{D,i}$), but that its importance diminished as the speed increased. Also, he noted that significant reductions in induced drag could not be achieved without using much larger wingspans (i.e., higher aspect ratios). Thus, Jones suggested that the major area in which drag reduction could be achieved was "head resistance," a term deriving from as far back as Chanute's *Progress in Flying Machines* – simply the sum of skin-friction drag and form drag for the airplane. Jones quoted a characteristic typical of airplanes in the 1920s, namely, that the power required by an airplane to overcome head resistance was 75–95% of the total power used. Because little could be done to reduce skin-friction drag, except to reduce the exposed surface area of the airplane, the primary target for drag reduction had to be form drag – the pressure drag due to flow separation: "We all realize that the way to reduce this item in the power account is to attend very carefully to *streamlining*."[100]

With that in mind, Jones defined what he called the "perfectly streamlined airplane" as one that

(1)   generates a flow identical (except in a very thin boundary layer) with the flow of an inviscid fluid (a fluid with no friction),

(2)   experiences a pressure distribution identical with that due theoretically to the
      inviscid fluid (i.e., no flow separation), and therefore
(3)   experiences a drag that is the sum of the induced drag and the tangential skin-
      friction forces resolved in the downwind direction.

Thus Jones's ideal airplane was simply one with no form drag. He went on to describe what
would be necessary to achieve that lofty goal:

> Unless bodies are "carefully shaped", they do not necessarily generate streamline flow, but
> shed streams of eddies from various parts of their surface . . . . The power absorbed by these
> eddies may be, and often is, many times greater than the sum of the powers absorbed by
> skin friction and induced drag. The drag of a real aeroplane therefore exceeds the sum of
> the induced power and skin friction drag by an amount which is a measure of *defective
> streamlining*.[100]

Jones went on at length about the importance of designing the perfectly streamlined airplane, but he did not specify how it should be shaped. To underscore that importance, he
estimated the power required to overcome skin-friction drag on several generic airplanes
and then compared that with the power required for various real airplanes. His skin-friction
calculations assumed that the friction drag on an airplane was the same as that exerted by
a turbulent boundary layer on a flat plate of equal exposed area; he called that assumption
"convenient and safe." In fact, he suggested that flat-plate turbulent skin friction would be
a good estimate for the drag coefficient for any "good streamlined body."

The aspect of Jones's paper that most "shocked" airplane designers into greater awareness was his plot of horsepower required versus velocity, which compared Jones's ideal,
"perfectly streamlined" airplane with various real airplanes of that time (Figure 8.2). The
solid curves at the bottom show the power required for the ideal airplane (they take into
account only skin-friction drag and induced drag). Four different curves are shown for four
different combinations of span loading (W/b^2) and wing loading (W/S), where W is the
weight of the airplane, and b and S are the span and area of the wing, respectively. The
solid symbols are data points for real airplanes; Jones obtained those data from the 1927
edition of *Jane's*, the annual compendium of aircraft performance and design characteristics. Jones pointed out that the vertical distance between any of those data points and
the solid "ideal" curve was the power expended by the real airplane in "the generation of
unnecessary eddies" (i.e., the power expended in overcoming form drag), and that "unnecessary" power consumption was considerable for all the airplanes listed. That is no surprise.
Consider the data point for the Argosy. (The Armstrong-Whitworth Argosy was one of the
first multiengine airplanes to be designed for a specific buyer: Imperial Airways in Britain.
Introduced in 1926, it was widely used on the routes from London to Paris, Basel, Salonika,
Brussels, and Cologne. Only seven were built, but they were popular airliners for that day.
The last Argosy was retired from service in 1935.) A three-view of the Argosy (Figure 8.3)
shows a rather boxy configuration, with fixed, protruding landing gear and the wires and
struts typical of the early 1920s designs. Clearly the Argosy was a long way from Jones's
ideal of the perfectly streamlined airplane. The vertical distance between the Argosy data
point and the ideal curve in Figure 8.2 is intuitively explainable simply from the three-view.
The Argosy was a perfect example of the conservative design approach common in the late
1920s and early 1930s, especially in Europe: "Designers had acquired the attitude of the
practical man, who knew how airplanes should be designed, because that was how they had
designed them for the previous 20 years."[101] No wonder the designers were "shocked" by
what Jones had to say.

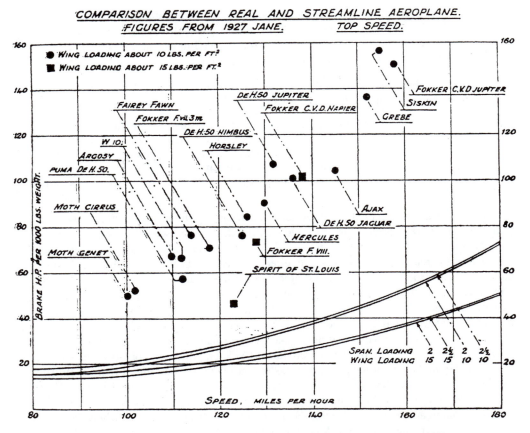

Figure 8.2 Power required for various airplanes, as compiled by Jones (1929).

Another way of interpreting Jones's graph (Figure 8.2) is to examine the horizontal distance between a given data point for a real airplane and the ideal curve. That represents the *increase in velocity* that could be achieved at the given power available to the airplane if there were no form drag. For example, with no form drag, the top speed of the Argosy would have been a blistering 175 mph rather than the actual value of 110 mph. Any way they looked at it, Jones's graph made a strong case for streamlining. It is interesting that of all the late-1920s aircraft listed in Figure 8.2, the *Spirit of St. Louis* came closest to Jones's ideal airplane.

After Jones finished his presentation, one very impressed member of the audience, identified only as Mr. Bramson, declared Jones's findings to be as important as the statement of the Carnot cycle in thermodynamics. Jones was more modest than that in response, noting that his findings could not be considered on the same plane as the Carnot theory of heat engines, though their practical outcomes were similar: the provision of an ideal toward which to work. Jones noted that whereas "the Carnot cycle is a precise theorem, my paper is more in the nature of an exercise in approximations." Jones was referring primarily to the approximate formula he used for flat-plate turbulent skin-friction drag, for lack of a more precise equation for the turbulent case. In any event, by the late 1920s, streamlining was an idea whose time had come. It meant drag reduction, which would be the primary concern in applied aerodynamics during the age of the advanced propeller-driven airplane.

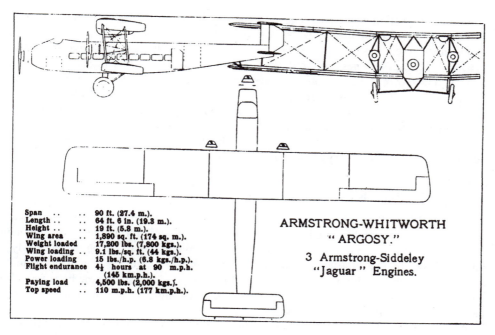

Span	..	..	90 ft. (27.4 m.).
Length ..	..		64 ft. 6 in. (19.3 m.).
Height ..	..		19 ft. (5.8 m.).
Wing area	..		1,890 sq. ft. (174 sq. m.).
Weight loaded			17,200 lbs. (7,800 kgs.).
Wing loading ..			9.1 lbs./sq. ft. (44 kgs.).
Power loading			15 lbs./h.p. (6.8 kgs./h.p.).
Flight endurance			4½ hours at 90 m.p.h.
			(145 km.p.h.).
Paying load	..		4,500 lbs. (2,000 kgs.).
Top speed	..		110 m.p.h. (177 km.p.h.).

ARMSTRONG-WHITWORTH

" ARGOSY."

3 Armstrong-Siddeley
"Jaguar " Engines.

Figure 8.3　Three-view of the Armstrong-Whitworth Argosy.

Figure 8.4　Sir B. Melvill Jones.

Sir B. Melvill Jones (1887–1975) (Figure 8.4) was the first occupant of a new chair of aeronautical engineering at Cambridge University, endowed by Emile and Angela Mond in memory of their son Francis Mond who had been killed during World War I while in action with the Royal Air Force. The focus of that position was to be research, but it was attached to an engineering laboratory with little tradition of research. With few resources and no track record, Jones soon completed a small wind tunnel in a wooden hangar used by the university air squadron, where he carried out a long series of aerodynamic experiments on wing stall, streamlining, and drag reduction. He later built a low-turbulence tunnel to study laminar and turbulent boundary layers. His research team was always small, involving only two or three research students at any one time: "It was unusual for a research student to become a Ph.D., for Jones preferred men who were untrammeled by degree regulations and could take part in the laboratory's investigations as equals. Consequently some of the published papers gave Cambridge Aeronautics Laboratory as the author."[102] Jones, one of England's most respected professors of aeronautical engineering, remained interested in undergraduate teaching, in addition to his research, delivering at least two lectures each week. When he died in 1975 at the age of 89, his life had spanned virtually all of the major advances in twentieth-century aerodynamics.

Wind Tunnels: A Maturing Species

In 1930, "Streamline!" was the call to action in applied aerodynamics. The objective was to design aerodynamic bodies with the lowest possible form drag – to approach Melvill Jones's ideal airplane. Although there had been considerable progress in aerodynamics theory by that time, it offered only an *understanding* of the mechanisms of flow separation and the resulting form drag; it could not yet provide accurate methods for calculating such phenomena, especially in a turbulent flow. Thus, to meet that call to action, aerodynamicists had to turn to the wind tunnel.

As discussed in Chapter 7, the commissioning of the NACA Variable Density Tunnel (VDT) in October 1922 opened the door to wind-tunnel testing at full-scale Reynolds numbers – an important first in the history of aerodynamics. Because of that, the NACA took a place in the forefront of aerodynamics testing, and its worldwide prominence was reinforced by its design and operation of several new wind tunnels in the late 1920s. Of those new facilities, two would prove to be of historic significance: the Propeller Research Tunnel (PRT) and the Full-Scale Tunnel (FST). Both of those new wind tunnels were designed for testing at full-scale Reynolds numbers, but the approaches used were different from that for the VDT. In the case of the VDT, full-scale Reynolds numbers were achieved in a relatively small wind tunnel by pressurizing the tunnel as high as 20 atm, thus increasing the air density up to 20-fold. Both the PRT and the FST achieved full-scale Reynolds numbers at ordinary atmospheric conditions because of their size; their test sections were large enough to hold full-scale airplanes or components of airplanes.

There was good reason for the new designs of those large tunnels: There was nagging doubt about the precision of the data obtained in the VDT, because of the unusual amount of turbulence generated in the pressurized circuit. That was caused by the small contraction ratio of the nozzle, the double-return design (Figure 7.17), and the inexpensive synchronous-drive motor that introduced small, high-frequency fluctuations into the flow. Indeed, the chief physicist at Langley, Fred Norton, had seen that as a problem right from the beginning. In a letter to George Lewis at NACA Headquarters, dated April 30, 1921, Norton stated that "the probability that the steadiness of flow in the compressed-air tunnel because of the small room required [to turn the airstream] would be inferior to that in the usual type tunnel, thus considerably decreasing the accuracy of the test." As useful as it was, "the VDT was far from the total aerodynamic triumph trumpeted in the NACA brochures."[84] Therefore, a group of engineers at Langley believed that testing at full-scale Reynolds numbers could best be achieved by using large tunnels at atmospheric conditions.

The first of those large wind tunnels, the PRT, became operational in July 1927, with a

Figure 8.5   The NACA PRT (1927).

test section 20 ft in diameter (Figure 8.5) and a maximum airspeed of 110 mph. Initially the PRT was intended only for full-scale propeller testing. It had been recognized that there were discrepancies between the propeller data obtained using models in small tunnels and the actual performance data for real propellers on airplanes in flight. William F. Durand at Stanford University had carried out an extensive series of propeller tests in a 5.5-ft wind tunnel beginning in 1916 – considered state-of-the-art research at that time.[36] However, his data were not consistent with the data from later flight tests carried out by the NACA. Fully aware of the importance of propeller aerodynamics and of the current inadequate knowledge in that area, George Lewis at NACA headquarters approved the construction of a special wind tunnel for propeller research. The tunnel would have to be large enough to accommodate actual airplane fuselages with installed engines and propellers, and that dictated the 20-ft diameter of the test section. The PRT would prove to have an impact that reached well beyond the field of propeller testing. For example, it was the facility in which the famous NACA engine cowling was developed – an important development in the story of streamlining to be discussed in the next section.

The obvious next step in the evolution of wind tunnels was a facility large enough to accommodate a complete airplane – wings, fuselage, and tail structure. Thus the FST was authorized in February 1929, and it became operational on May 27, 1931. Its open-throat test section was 30 ft by 60 ft, and its maximum airspeed was 118 mph. Figure 8.6 shows the first complete airplane to be tested in the FST, a Vought O3U-1. The exit of the wind-tunnel nozzle is at the left, clearly showing the gigantic size of the tunnel. From a financial point of view, the timing of the construction of the FST was fortuitous for the NACA. The congressional appropriation for the tunnel passed in February 1929, eight months before the Wall Street crash, so its construction enjoyed considerably reduced prices for materials and labor during the first two years of the Depression. The cost of the FST was about $1 million. At the time it was finished, the FST was the world's largest wind tunnel. The

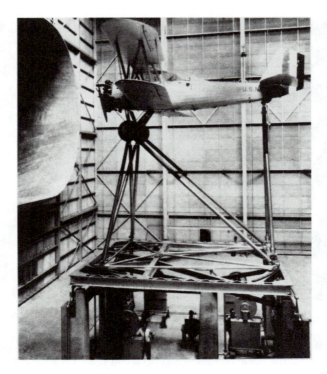

Figure 8.6   The NACA FST (1931).

engineers at Langley found the turbulence levels in the FST to be extremely low – so low that Smith J. DeFrance, a key member of the design team, stated in 1932 that its effects could "be neglected in applying the data to design." The low turbulence levels in the FST made it an ideal facility for the extensive studies of drag cleanup carried out in the late 1930s; those studies were instrumental in the aerodynamic streamlining process that led to advanced propeller-driven airplanes. In that era the FST was clearly the world's premier wind tunnel, and it is still in operation today, in this age of supersonic flight, more than 65 years after its dual 8,000-hp engines were first turned on.

A Success in Streamlining: The NACA Cowling

With adequate wind tunnels in operation, the developing technology in aerodynamics soon led to the design of advanced propeller-driven airplanes. The decade of the 1930s was the heyday of that maturation process in aeronautical engineering, beginning almost immediately after the NACA PRT came on-line. One of the most important drag-reduction programs of that era was initiated at the NACA Langley Memorial Laboratory: the NACA cowling program, an important first step toward Melvill Jones's ideal airplane. We shall examine the NACA cowling program as an example of the developmental process in both experimental and theoretical aerodynamics in the 1930s; it provides a representative case history in the evolution of twentieth-century aerodynamics.

In 1926, airplanes could be divided into two general categories on the basis of the type of piston engine used: the liquid-cooled in-line engine or the air-cooled radial engine. The former was generally enclosed within the fuselage and did not present much of a problem in regard to streamlining. The latter relied on the airflow over the cylinders to cool the engine, and thus the cylinders, arrayed like the spokes on a wheel, were directly exposed to the airstream. As a consequence, air-cooled radial engines created a lot of drag – just how much was yet to be realized. However, radial engines had several advantages that led to their use in many airplane designs: lower weight per horsepower, fewer moving parts, and lower maintenance costs. The U.S. Navy was partial to air-cooled radial engines because they continued to perform well despite the jarring impacts of carrier landings. In June 1926 the chief of the U.S. Navy's Bureau of Aeronautics requested that the NACA study how a cowling could be wrapped around the cylinders of radial engines so as to reduce drag without interfering with cooling capacity.

The idea of cowlings was not new. For example, the French Déperdussin racing airplane in 1913 had a rounded and streamlined shroud wrapped around a Gnome 14-cylinder two-row rotary engine. Also, many of the rotary engines used on World War I airplanes were housed inside curved metal cowlings. Those cowling designs were based more on art than on science or any sound knowledge of aerodynamics. Fortunately, cooling was not a problem for those cowled rotary engines, because the cylinders were always rotating through the air behind the cowling. The problem arose with the stationary radial engines that became prevalent in the 1920s.

On May 24, 1927, representatives of the major U.S. aircraft manufacturers met at Langley to become more familiar with the NACA's work and facilities and to make suggestions regarding future NACA research that would benefit the industry. That second meeting in what was to become an annual series at the NACA played an important role in guiding the development of aerodynamic research at Langley, and the cowling program in particular, as recounted in the 1927 NACA annual report:

> At a preliminary meeting held in the morning the functions and work of the committee were briefly outlined, following which the representatives of the industry were conducted on a tour of inspection of the laboratory and the investigations under way were explained. This occasion marked the formal opening of the committee's new propeller research equipment. In the afternoon the conference proper convened and after a brief statement by the chairman as to the purpose of the meeting, there was general discussion of the problems of commercial aviation in which the representatives of the industry participated. Among the problems which were mentioned as of importance to commercial aviation were the various factors relating to the comfort and convenience of passengers in airplanes and particularly the elimination of noise; the question of controllability at low speeds; and the effect of protuberances on an otherwise faired stream-line body. One of the problems suggested, the study of the effect of cowling and fuselage shape on the resistance and cooling characteristics of air-cooled engines, was promptly incorporated in the committee's research program.

Although the U.S. Navy had asked the NACA to begin research on cowlings a year earlier, it took the political clout of the collective aircraft industry to have such work "promptly incorporated in the committee's research program."

The NACA cowling research was the first major test program to be carried out in the PRT at Langley. Fred Weick,[103] a relatively young aeronautical engineer from the University of Illinois, had just become director of the PRT after the sudden departure of Max Munk in 1927. At the end of the May 24 conference, Weick was given responsibility for the NACA cowling program, because the PRT was the logical place to carry out the research.

For the next 10 years the NACA carried out research on cowlings, most of it experimental; it was not until 1935 that any emphasis was placed on an analytical understanding of the aerodynamic processes associated with cowlings. Within a year after beginning the program, Weick and his associates had demonstrated that a properly designed cowling could dramatically reduce the form drag associated with radial engines without adversely affecting engine cooling – findings that were immediately snapped up by industry and incorporated into new airplane designs. The research method used by Weick was an experimental variation of parameters, and that approach in experimental aerodynamics was to take root at Langley and elsewhere during the 1930s.[84]

The method of "experimental parameter variation" was "the procedure of repeatedly determining the performance of some material, process, or device while systematically varying the parameters that define the object or its conditions of operation."[36] Weick described his approach in the first NACA publication on the cowling research:

> The program as finally arranged included ten main forms of cowling to be tested on a J-5 engine in connection with two fuselages, three on an open cockpit fuselage and seven on a closed cabin type. The seven forms of cowling on the cabin fuselage range from the one extreme of an engine entirely exposed except for the rear crank case, to the other extreme of a totally enclosed engine. One of the cowlings with the open cockpit fuselage includes individual fairings behind each cylinder. Three forms of cowling, two of which are on the cabin fuselage, afford direct comparisons with and without a propeller spinner. The program involves the measurement of the engine cylinder temperatures, each cowling being modified, if necessary, until the cooling is satisfactory. The cowling is then tested for its effect on drag and propulsive efficiency.[104]

The key to the success of the NACA cowling program was that as they experimented with configurations that would reduce form drag through external streamlining, they were careful to maintain effective cooling of the engine by internal ducting of the flow. The cowling found most successful by Weick was cowling no. 10 in the NACA series, which was really a cowling within a cowling. Cowling no. 10 was devised by taking cowling no. 5 (Figure 8.7) (consisting of a smooth, rounded fairing enclosing only part of the cylinders) and wrapping around it an exterior rounded and streamlined shroud that totally enclosed the cylinders (Figure 8.8). The aerodynamically tailored internal flow in the passage formed by those two walls allowed effective cooling of the engine.

The most important graph in NACA technical report 313[104] was a plot of the measured drag versus the dynamic pressure (or the velocity) for various cowling designs (Figure 8.9), which clearly showed the outstanding performance achieved with the "NACA cowling" (i.e., cowling no 10). The bottom line was for the bare fuselage with no engine, and the top line was for the fuselage with the engine installed, but with the cylinders totally exposed. In comparing those two extremes, it was seen that addition of the uncowled engine increased the drag by an absolutely stunning factor of 4.76. Until that measurement, no one had ever understood the devastating extent to which the exposed cylinders were increasing the drag on such airplanes. The lines labeled "No. 10" in Figure 8.9 were for the no. 10 cowling ("No. 10-0" was the cowling shown in Figure 8.8, and "No. 10-M" had a slight modification of the cowling inlet to improve the engine cooling). Compared with the case of totally exposed cylinders, cowling no. 10-M reduced the drag by a factor of 0.41 (almost 60%). Indeed, with the cowling, the drag was reduced almost back to the value measured without the engine installed.

That was a dramatic finding, and the normally staid NACA was in a state of euphoria.

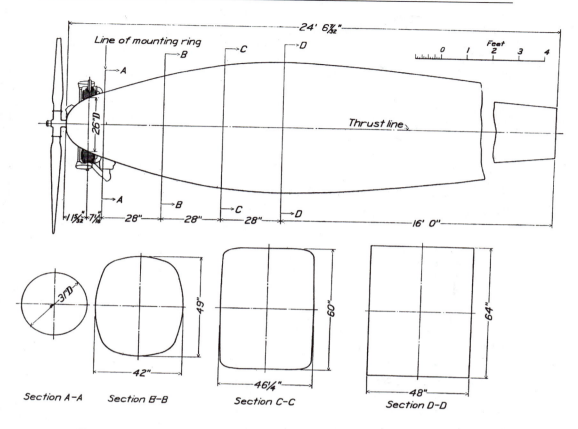

Figure 8.7 Cowling no. 5 in the series of NACA cowling tests.

It lost no time in getting the word out. The U.S. aircraft industry was given advance notice of the findings well before the public announcement (that policy is still followed today, giving the U.S. aerospace industry first access to critical NASA-generated data to enhance its position relative to foreign competitors). In 1928 Weick wrote an article for the weekly periodical *Aviation* (forerunner of *Aviation Week and Space Technology*) on the dramatic performance of the NACA cowling: "In conclusion, it would seem from the tests made to date that a very substantial increase in high speed and all-round performance can be obtained on practically all radial engined aircraft by the use of the new NACA complete cowling."[105] In view of the normal policy of the NACA at that time to be very conservative and guarded in any communications to the public (data and analyses had to be authenticated beyond question before publication), Weick's statement in *Aviation* must have been quite revealing of the excitement at the NACA.

Actually, the NACA had remained conservative in one respect: Before releasing the cowling data, NACA engineers had flight-tested the cowling, as reported in an appendix to technical report 313[104] by Thomas Carroll, who had been responsible for the flight tests:

> In order that the practical value of the information in the foregoing report might be demonstrated, simple flight tests have been made of the Number 10 cowling.
> Through the courtesy of the Army Air Corps at Langley Field, VA, a Curtiss AT-5A

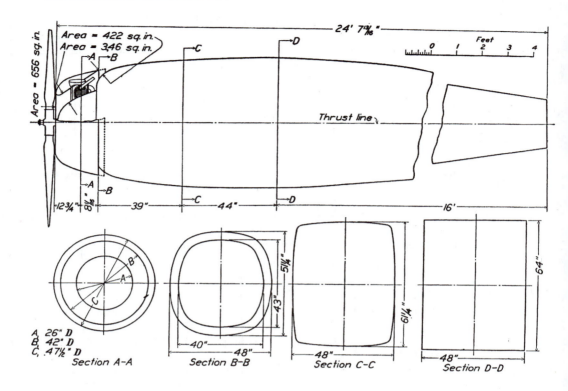

Figure 8.8   Cowling no. 10, the most successful of the early NACA cowlings (1928).

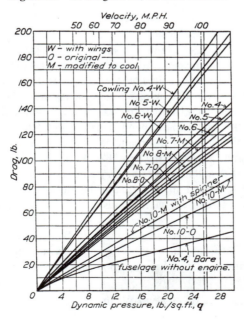

Figure 8.9   Drag versus velocity for various cowlings.

airplane was obtained on which an adaptation of the Number 10 cowling was installed . . . . A series of flights was made by the three pilots of the laboratory.

The maximum speed of this type airplane as in use at Langley Field had been reported at 118 miles per hour. This was checked by making a series of level runs with a Curtiss AT-5A airplane at low altitude over the water at full power. The maximum speed was found to be 118 miles per hour at 1,900 R.P.M., both air speed and R.P.M. being measured on calibrated instruments. Similar high speed runs made with the modified AT-5A showed a performance of 137 miles per hour at 1,900 R.P.M., an increase of 19 miles per hour. The original speed of 118 miles per hour was attained at 1,720 R.P.M. on the modified airplane.

While the type of cowling as normally installed on an AT-5 is not particularly adaptable to speed, the increase is considered remarkable. Furthermore, the improvement of flying qualities in smoothness of operation was also very favorably commented upon by all pilots who have flown it. The air flow over the fuselage and over the tail surfaces is very obviously improved.

The cooling of the engine was found to be normal in these tests. The oil temperature reached 58° and was fairly constant, and there was no other indication of overheating. Likewise, there was no interference to the pilot's vision in any useful field.[104]

Given the data, the aircraft industry was quick to adopt the NACA cowling, and the expense of the conversion was almost trivial; the NACA estimated that it would cost about $25 to build and install a cowling on an existing aircraft. The Lockheed Vega was the first production-line airplane to use the NACA cowling. The Vega first flew in 1927, with its cylinders exposed to the airflow, as was conventional at that time. In 1929, with the NACA cowling added, the maximum speed of the Vega was increased from 165 mph to 190 mph.[69] The cowling-equipped Vega (Figure 8.10) became one of the most famous airplanes of the early 1930s, used by pilots such as Wiley Post and Amelia Earhart. With its cowling and aerodynamically streamlined wheel pants, its zero-lift drag coefficient was 0.0278, quite low for that day.

In 1929 the NACA cowling won the Collier Trophy, an annual award commemorating the most important achievement in American aviation. It would be the first of many Colliers to be won by NACA and NASA in the years to come.

After the initial euphoria of 1928 and 1929, the NACA cowling research program settled down to a series of tests intended to further improve the design, and perhaps to provide an understanding of why the cowling worked the way it did. That latter concern was particularly important, for the cowling program had been and continued to be totally empirical. It may have been a bit of an embarrassment that there was no fundamental understanding of the detailed aerodynamic processes involved, particularly in regard to the internal flow used to

Figure 8.10   Lockheed Vega with the NACA cowling.

cool the engine. There certainly was no accompanying aerodynamic analysis, for Weick's report (TR 313) had not a single equation, though he hinted that the performance of the cowling was sensitive to small changes in its design: "it must be carefully designed, however, to cool properly."[104]

In April 1929, Weick left Langley to take a job with the Hamilton Aero Manufacturing Company (part of the United Aircraft companies) in Milwaukee to design propellers. Donald H. Wood, his colleague and assistant from the start of the PRT program, took charge of the wind tunnel and the NACA cowling research program. For the next few years, Wood faced a post-euphoria period of cowling tests, during which questions were raised about the propriety of the NACA claims. Worse yet, additional testing led more to confusion than understanding of the fundamental aerodynamics of the cowling.

The propriety question stemmed from work carried out at the National Physical Laboratory in England by Hubert C. Townend, beginning in 1927. Townend developed a ring with an airfoil cross section that wrapped around the outside of the exposed cylinders of a radial engine for the purpose of reducing drag. His work was published in 1929 by the British Aeronautical Research Council,[106] a few months before publication of Weick's report. Neither man was aware of the work by the other. Because the "Townend ring" left the cylinders more or less exposed, it did not interfere with cooling,[101] which was reassuring to airplane designers at the time. As a result, a number of airplanes designed in the early 1930s used Townend rings rather than the NACA cowling. Boeing particularly favored the ring, using it on several fighter and bomber designs of the period; the three-view of the 1932 Boeing P-26A single-seat fighter in Figure 8.11 clearly shows the Townend ring. To assess the competition, Wood compared the aerodynamic performances of the Townend ring and the NACA cowling in the PRT.[107] A typical ring configuration examined by Wood is shown in Figure 8.12. The nacelles were mounted on stub wings, and each enclosed a $\frac{4}{9}$-scale model of the J-5 radial engine. Wood presented his findings in the drag polars shown in Figure 8.13. The wing alone had the lowest C_D for a given C_L, but the wing with the NACA cowling was a close second. The ring produced a considerably larger C_D and was clearly inferior to the NACA cowling. On the strength of those data, George Lewis at NACA Headquarters in Washington convinced Glenn Martin to replace the Townend ring on the Martin B-10 bomber with the NACA cowling. Equipped with the NACA cowling, the B-10's maximum speed increased from 195 mph to 225 mph; also, its landing speed was significantly reduced. As a result, in 1933 and 1934 the U.S. Army bought more than 100 B-10s, which kept the Martin Company solvent during the worst of the Depression. Hansen[84] theorized that Martin's use of the NACA cowling may have been why Martin won the U.S. Army contract over the Boeing B-9, which used Townend rings.

A less well known by-product of the NACA cowling research program, but of almost as much importance for aircraft with wing-mounted engines, was its study of the proper placements for engine nacelles on a wing. Shortly after Weick's initial tests of the NACA cowling, and as a complement to those tests, Wood performed a series of parametric studies of nacelle placement. Using the PRT, Wood examined 21 possible positions for a nacelle on a thick wing (Figure 8.14). As in the tests of the Townend ring, a $\frac{4}{9}$-scale model of the Wright J-5 radial engine was installed in a nacelle with the NACA cowling. The crosses in Figure 8.14 indicate the different positions of the propeller hub. From his experimental data,[107,108] Wood reached the following conclusions:

> Taking into account the lift, interference, and propulsive efficiency, the best location of the nacelle, with tractor propeller on a monoplane wing, for high speed and cruising, is with

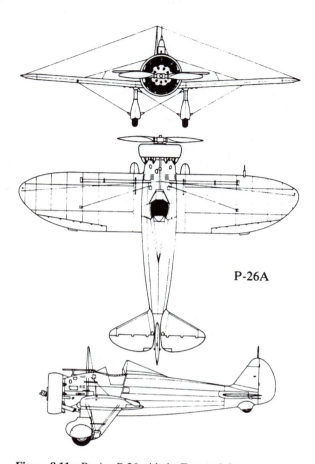

Figure 8.11 Boeing P-26 with the Townend ring.

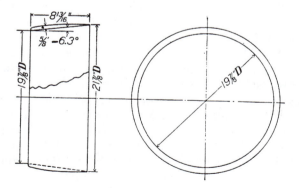

Figure 8.12 A typical ring configuration tested by the NACA (1932).

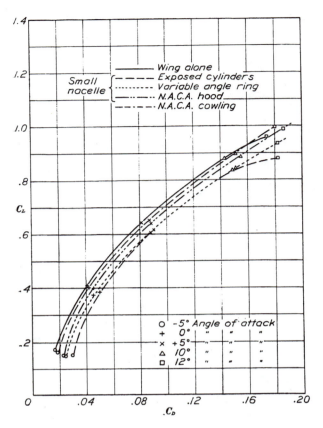

Figure 8.13   Drag polars for the engine ring and cowlings, as measured by the NACA in 1932.

the thrust axis in line with the center of the wing and with the propeller about 25 percent of the chord ahead of the leading edge. This same location also appears to be the best in climb and landing, therefore excels in all conditions of flight.[108]

In Figure 8.14, the best location was found to be point B. The engine locations for airplanes such as the Douglas DC-3, Boeing B-17, and Consolidated B-24 were derived from Wood's data obtained in 1932.

In the post-euphoria period, the NACA cowling program continued its experimental testing using the parameter-variation method that had been initiated by Weick in 1928, but the euphoria occasioned by the early experiments faded over the next seven years of testing. Weick had been lucky in choosing his test conditions and parameters, which were very favorable for highlighting the advantages of the NACA cowling. Essentially, Weick had skimmed the cream off the top. As time went on, new data gathered over a wider range of parameters yielded mixed findings. For example, when the NACA cowling was tested on a Fokker trimotor with its wing engines mounted below the wings, virtually no improvement in performance was achieved. It was quickly recognized that the configuration of the airplane downstream of the cowling had some effect on the drag. It had been fortuitous that the NACA had chosen a Curtiss AT-5A for the first series of flight tests with the cowling, for which the findings were spectacular, as discussed earlier. However, all

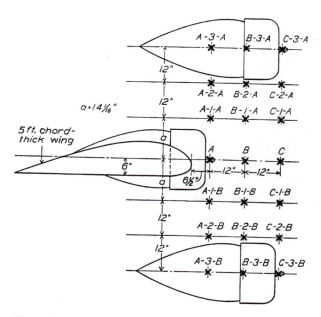

Figure 8.14 Different engine nacelle locations on a wing, as tested by the NACA in 1932.

the experimental data on the cowling generated by the NACA during the period 1928–35 were obtained without a fundamental understanding of the aerodynamics of the cowling. Of course, such a situation was common in aerodynamics, even at that time. We have seen that the early researchers such as Lilienthal, Langley, the Wright brothers, and Eiffel amassed large bodies of useful aerodynamic data and effectively applied them, without ever understanding the fundamental aerodynamic principles involved. However, a hallmark of the empirical advances in aerodynamics during the era of the advanced propeller-driven airplane was that they were accompanied by an understanding of the basic principles involved, but the NACA cowling program was not meeting that standard. Hansen[84] described that period of cowling research as a time of "paralyzing confusion" leading to an "experimental impasse."

That situation was soon to change. For its first eight years, the cowling program had been run by engineers who were pure experimentalists, and that worked well in the beginning, because the initial state of streamlining was so abysmal that even approximate solutions from applied aerodynamics led to great improvements, and the wind tunnel and actual flight testing were the most readily available tools with which to attack the problem. But in the summer of 1935, Henry Reid, Langley's "Engineer in Charge" (the title formerly held by the director of Langley), transferred most of the responsibility for cowling research to Theodore Theodorsen, the most respected theoretician in the NACA at that time. Theodorsen (Figure 8.15) had been born in Norway in 1897. He earned an engineering degree from the Norwegian Institute of Technology at Trondheim in 1922 and received his Ph.D. in physics from Johns Hopkins University in 1929, joining the NACA Langley Memorial Laboratory as an associate physicist that same year. Theodorsen soon became head of the Physical Research Division, the smallest of the three research divisions at Langley at that time, the others being Engine Research and Aerodynamics. He quickly made his mark in a variety of research areas: airfoil theory, propeller theory, icing problems, wind-tunnel

Figure 8.15   Theodore Theodorsen.

theory, aircraft flutter, and aircraft noise, to name only a few. Theodorsen believed in a balance between theoretical work and experimental research, much as had Max Munk nine years earlier:

> A science can develop on a purely empirical basis for only a certain time. Theory is a process of systematic arrangement and simplification of known facts. As long as the facts are few and obvious no theory is necessary, but when they become many and less simple theory is needed. Although the experimenting itself may require little effort, it is, however, often exceedingly difficult to analyze the results of even simple experiments. There exists, therefore, always a tendency to produce more test results than can be digested by theory or applied by industry.[83]

By 1935 the NACA leadership at Langley had decided that Theodorsen's approach was just what was needed to correct the problems in cowling research, and he was given responsibility for the cowling research program and free rein to use the PRT to delve further into the aerodynamic fundamentals underlying the empirically successful cowlings.

Theodorsen, like Munk before him, believed that many NACA engineers were deficient in mathematics. He was critical of some of the work, stating that "a large number of investigations are carried out with little regard for the theory." But unlike Munk, Theodorsen was liked and respected by most of the people who worked for him. Much later, I. Edward Garrick, a talented mathematician who worked closely with Theodorsen in the 1930s and 1940s, commented on Theodorsen's style:

> When a problem captured his attention, he would work on it during relatively short periods of intense concentrated activity, almost incommunicado, followed by periods of apparent desultory inactivity. Often, some of us would walk with him among trees and orchards then still existing at Langley, and he would discuss foibles of mankind. As head of a division, he had the major virtue, now rare, of protecting his staff from routine and time-draining demands of government, allowing a person to develop his own talents and resources. When one had a finished or semifinished product, Theodorsen would then be a helpful though severe critic.[109]

When Theodorsen took charge of the cowling research program in 1935, it entered its third and final phase, with the aim of achieving fundamental understanding of the aerodynamics of cowlings. Whereas Fred Weick's reports on the NACA cowlings in 1928–9 had been devoid of analysis and contained no equations,[104,105] by 1937 Theodorsen had studied the aerodynamic details of the internal and external flows through and over the cowling and had developed an approximate engineering analysis for the aerodynamic processes. On January 26, 1938, he presented a paper at the sixth annual meeting of the Institute of Aeronautical Sciences (IAS) in New York in which he discussed the latest NACA findings – the first rational analysis of the aerodynamics of cowlings. That analysis focused on the balance between the internal airflow that would effectively cool the engine and yet would keep drag (called "cooling drag") to a minimum, and it examined the nature of the external flow and how it could be directed to achieve minimum form drag. His presentation was an excellent example of engineering analysis – making appropriate assumptions that would simplify the mathematics, but retaining sufficient rigor to arrive at findings that would be adequate for engineering design. He artfully used the continuity equation, Bernoulli's equation (the flow velocity was low enough to assume incompressible flow), and his experimental data on the friction losses for the flow through the cylinder fins to obtain formulas that related the pressure drop across the cowling, the volume flow rate through the cowling, and the relative inlet and exit areas for the internal flow. Part of the appeal of his analysis was that it was all algebraic – not a partial differential equation in sight. The findings from his analysis were quickly published by the IAS[110] (the analysis and the supporting experimental data had been published a year earlier by the NACA).[111]

Theodorsen also carried out a series of cowling experiments in the PRT to study the fundamental aerodynamic characteristics of the flow field, such as pressure distributions, streamline patterns, and heat-transfer distributions. One finding from those more detailed experiments was that the impact of the air on the front of the cowling produced massive turbulence, and that was a major contributor to total nacelle drag – a disadvantage of using a cowling. However, it greatly increased the cooling at the front of the engine – a definite advantage that outweighed the disadvantage. That was an example of the increased understanding of cowling aerodynamics achieved at the NACA during Theodorsen's watch.

The NACA was quick to package the new analysis and data into a form useful to airplane designers: "Inasmuch as the designer of an airplane has neither the time nor the opportunity to acquire a detailed knowledge of every part of the airplane, he wants a simple method of obtaining the optimum cowling dimensions and, perhaps, some of the more important reasons for selecting these dimensions. It is the purpose of this report to present such a method and to illustrate the method with a discussion of practical examples."[112] That report went on to present design-oriented calculations based on Theodorsen's analysis, treating the geometric design of proper cowlings for given flight conditions and emphasizing how cooling drag and external form drag could be minimized. With that report, the NACA cowling research program reached its high point. Eleven years after Fred Weick's initial breakthrough in empirical design, a better theoretical understanding of the aerodynamics of cowlings had finally become an integral part of the state of the art in aeronautics and was having a major impact on the design of flying machines.

Theodore Theodorsen left the NACA in 1947 to help organize and administer an aeronautical institute in Brazil. From 1950 to 1954 he was "Chief Scientist" for the U.S. Air Force. In 1955 he joined the Republic Aviation Corporation as chief of research, retiring in 1962. In retirement he was an active consultant to the United Aircraft Corporation, specializing in ducted propeller work. His health began to fail in 1974, and he died in Centerport, New York, November 6, 1978.

Airfoil Aerodynamics: Systematic Progress

Experimental and theoretical research on the aerodynamics of airfoils led to significant progress in the early 1930s, and again the NACA was a leader in that work. As discussed in Chapter 7, from the beginning of aeronautics to the end of the 1920s, the design of airfoils was essentially ad hoc, leading to a great proliferation of different airfoil shapes, but only fragmentary understanding of basic airfoil aerodynamics. That situation changed dramatically in the early 1930s, primarily because of the work of Eastman Jacobs on the experimental side, Theodore Theodorsen on the theoretical side, and the Langley Variable Density Tunnel (VDT).

Eastman N. Jacobs (Figure 8.16) joined the NACA Langley Memorial Laboratory in 1925, one year after graduating with honors from the University of California at Berkeley. He was soon recognized as an outstanding addition to the Langley staff, often taking innovative approaches to challenging problems. Assigned to the VDT, Jacobs played an important role in the early aerodynamic research at high Reynolds numbers. By the time of the NACA experimental airfoil program in the early 1930s, Jacobs had become head of the VDT section, a position he held for the next decade.

From April 1931 to February 1932 Jacobs and his colleagues carried out a series of airfoil measurements that provided a standard for the era of the advanced propeller-driven airplane, arguably the most important measurements made in the VDT. Expanding on the idea that Max Munk used to design his M-series airfoil shapes in the 1920s, Jacobs used a systematic approach to obtain what was to become the family of NACA "four-digit" airfoils. The scheme was simplicity itself: Construct a single curved line, called the mean camber line, and wrap a mathematically defined thickness distribution around the camber line: "The major shape variables then become two, the thickness form and the mean-line form. The thickness form is of particular importance from a structural standpoint. On the other hand, the form of the [mean line] determines almost independently some of the most important aerodynamic properties of the airfoil section, e.g., the angle of zero lift and the pitching-moment characteristics."[58] The thickness distribution chosen by the NACA

Figure 8.16 Eastman N. Jacobs.

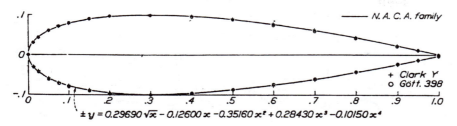

$$\pm y = 0.29690 \sqrt{x} - 0.12600\, x - 0.35160\, x^2 + 0.28430\, x^3 - 0.10150\, x^4$$

Figure 8.17 The thickness distribution used by the NACA for the four-digit airfoil series (1931).

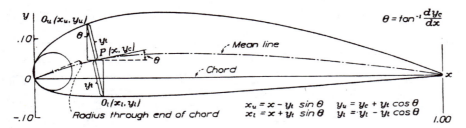

Figure 8.18 Resulting NACA airfoil shape obtained by wrapping a given thickness distribution around a given camber-line shape (1931).

was patterned after that of "well-known airfoils of a certain class including the Göttingen 398 and the Clark Y" (Figure 8.17). Airfoils of different thickness-to-chord ratios were obtained simply by multiplying the thickness distribution by a constant factor. When the prescribed thickness distribution was wrapped around the prescribed mean camber line, the resulting airfoil shape was as shown in Figure 8.18. The family of airfoils designed by the NACA in 1931 using that simple technique (Figure 8.19) was the famous NACA four-digit series of airfoils, where the first digit gave the maximum camber in hundredths of the chord length, the second digit gave the location of the maximum camber in tenths of the chord length measured from the leading edge, and the last two digits gave the maximum thickness of the airfoil in hundredths of the chord length. For example, the NACA 2412 had a maximum camber of 0.02 of the chord length, located at 0.4 of the chord length from the leading edge, with a maximum thickness of 0.12 of the chord length. The lift, drag, and moment coefficients for that entire family of airfoils were carefully measured in the VDT at Langley. The models used in the wind tunnel were finite rectangular wings with the aspect ratio of 6, and the data were modified and plotted[58] for an infinite aspect ratio using the appropriate formulas from Prandtl's lifting-line theory, as discussed in Chapter 7. Because the measurements were made in the VDT, the Reynolds numbers were on the order of 3 million, well within the range encountered in practical flight at that time.

The airfoil data from those studies[58] were used by aircraft manufacturers in the United States, Europe, and Japan during the 1930s. The combination of Jacobs's engineering talent, the rational simplicity of the NACA design process, and the high-Reynolds-number conditions of the VDT had finally produced a useful data base on the aerodynamic properties of airfoils – "a classic, a designer's bible." That contribution to applied aerodynamics in the early 1930s was a major step toward the development of advanced propeller-driven airplanes.

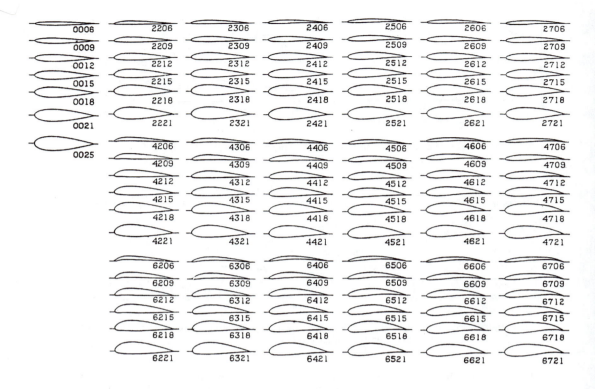

Figure 8.19 Family of NACA four-digit airfoils (1931).

Concurrent with Jacobs's experimental work on airfoils, a major theoretical advance in the calculation of airfoil properties was reported by Theodore Theodorsen. Theodorsen, educated in the European tradition, an engineer with a strong background in advanced mathematics, brought a large dose of theory to counterbalance the massive experimental emphasis at the NACA during the 1930s. Considered the best theoretical aerodynamicist in the United States at the time, Theodorsen was responsible for the next major advance in airfoil design after Munk's thin-airfoil theory during the early 1920s. In 1931 Theodorsen published the first general analysis for airfoils of any arbitrary shape and thickness, and he took that opportunity to criticize the fact that the major emphasis in airfoil design at that time was on empirical experimentation: "Investigations are carried on with little regard for the theory and much testing of airfoils is done with insufficient knowledge of the ultimate possibilities. This state of affairs is due largely to the very common belief that the theory of the actual airfoil necessarily would be approximate, clumsy, and awkward, and therefore useless for nearly all purposes."[83] Theodorsen's efforts went a long way toward correcting the imbalance between the experimental and theoretical approaches to airfoil design. His work was based on the theory of complex variables, which had been used by Joukowski to determine, by conformal transformation, the flow over an airfoil from the known solution for the flow over a circular cylinder. However, the resulting Joukowski airfoils, obtained at the turn of the century (as discussed in Chapter 6), had not been particularly practical airfoils – the empirically derived airfoil shapes generated during the era of the strut-and-wire biplane

(Chapter 7) were consistently better. The Joukowski airfoils were simply airfoil-like shapes that came out of the transformation, starting with the circle (they were, in essence, "slaves" of the circle and the particular transformation used). Lamb, in the most authoritative fluid-dynamics textbook at the turn of the century, stated that the problems associated with that approach were "now so great as to render this method of very limited application."[54] In the 1930s, what was needed was a comprehensive theoretical method, a direct approach that would allow accurate calculations of the aerodynamic properties of an airfoil of *any given shape*. Max Munk's thin-airfoil theory met that description, but it was applicable only for airfoils that were thin and at small angles of attack. Theodorsen's new method eliminated those restrictions; it was the first theoretical analysis for airfoils of any arbitrary shape and thickness at any angle of attack, and it proved to be the most important advance in airfoil theory during the age of the advanced propeller-driven airplane.

Despite the importance of Theodorsen's new method, the problem of theoretical calculation of airfoil properties was far from being solved. A case in point involves the comparisons between experimental and theoretical pressure distributions over the surface of an airfoil presented by Theodorsen in 1932. Figure 8.20 shows the variation of the pressure coefficient over the top and bottom surfaces of a Clark Y airfoil at an angle of attack of $5.3°$; the solid curve shows Theodorsen's calculations, and the dashed curve plots experimental data from the VDT. Although the theoretical calculation yielded appropriate qualitative trends, the quantitative agreement was not precise, most likely because of viscosity; Theodorsen's analysis was for an inviscid, incompressible flow, and it did not account for the effects of a viscous boundary layer. (Proper calculation of viscous effects on airfoil properties is still a problem today; even with the power of modern computational fluid dynamics, there remains an uncertainty in such calculations, especially for turbulent, separated flows.) Nevertheless, Theodorsen's method allowed the calculation of pressure distributions and lift and moment coefficients for airfoils of arbitrary shape and thickness, and those calculations usually were within 10% of the measured values – a tremendous accomplishment in the early 1930s.

Theodorsen's reaction to such comparisons between his theory and the experimental data was to suggest that perhaps the accuracy of the experimental data should be reexamined. His only comment about one such comparison with experimental data was a single sentence: "The experimental values are from original data sheets for N.A.C.A. Technical Report No. 353, and are not entirely consistent due to difficulties experienced in these experiments."[83] A year later, Theodorsen and his colleague Edward Garrick published a more thorough discussion of his airfoil theory, with extensive comparisons between theory and experiment for an M6 airfoil.[113] The agreement was essentially no better or worse than that obtained a year earlier, and again there was virtually no discussion about the comparison. Indeed, the only hint as to Theodorsen's opinions was relegated to a footnote that listed the effects of finite span (the wind-tunnel data had been taken using a finite wing and had been modified to apply to an infinite wing – an airfoil), tunnel-wall interference, and viscosity as possible sources of error. Given the importance of Theodorsen's work and his background in mathematics, the fact that virtually no effort was made to explain the discrepancies between theory and experiment, no matter how slight, is very curious. On the basis of my experience with theoreticians and experimentalists over the past 36 years, my guess is that a lot more was being said behind the scenes, but that the normal NACA conservatism, especially during the editorial process for NACA technical reports, was a very strong filter. Keep in mind that the experimental data were being produced by Eastman Jacobs and his colleagues in the VDT. By that time, Jacobs was the leading experimentalist in the NACA, and Theodorsen was

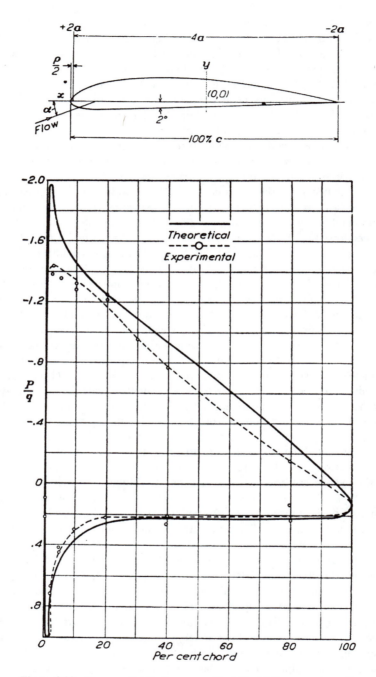

Figure 8.20 Pressure distribution over a Clark Y airfoil: comparison of experiment with Theodorsen's theory (1932).

the leading theoretician in the NACA, and there was no love lost between them: "Beneath the basic difference in their approaches to gaining aeronautical knowledge, there existed a strong personal rivalry and mutual dislike that moved most of their confrontations beyond more objective disagreement. At Langley both men controlled fiefdoms, and because both men were so valuable to the NACA, George Lewis [director of research at NACA Headquarters in Washington] had permitted the feudal arrangement to flourish."[84] Jacobs had joined the NACA in 1925, and Theodorsen arrived in 1929. Within two years after his arrival, Theodorsen had published his airfoil theory and had made known his views that too many experiments were being conducted with too little regard for theory. Clearly, the Jacobs–Theodorsen enmity had early roots, and the fact that there was virtually no discussion of the discrepancy between experiment and theory in cases like that shown in Figure 8.20 suggests that their hostility, filtered by the NACA editorial process, was the reason why.

Theodorsen's airfoil theory was accorded almost instant respect. Within a year the U.S. Navy's Bureau of Aeronautics asked the NACA to carry out a series of calculations of the pressure distributions on airfoils for use in determining the structural loads on wings. Edward Garrick applied Theodorsen's theory to calculate the pressure distributions and lift coefficients for 20 different airfoils, ranging from the earlier U.S.A. 27 and Göttingen 398 airfoils to the most recent NACA four-digit series.[114] The production-line nature of Garrick's calculations is typified in Figure 8.21, which shows his pressure-distribution findings for the Göttingen 398 airfoil at four different angles of attack. Garrick's prolific calculations were reminiscent of the earlier NACA compilations of empirical airfoil data in the 1920s and the contemporary experiments on the NACA four-digit series of airfoils by Jacobs.[58] Finally the overwhelming tendency for airfoil data to be determined experimentally was being redressed.

Jacobs continued to design and develop improved airfoil shapes through the 1930s. The second popular NACA series was the five-digit series of airfoils, developed by Jacobs in 1935 – a family of related airfoils having the position of maximum camber unusually far forward, within 5–15% of the chord length from the leading edge. Following the NACA

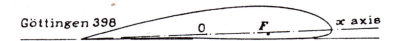

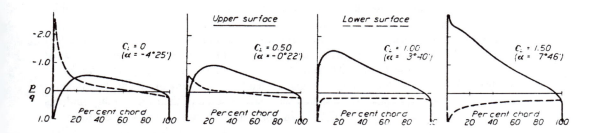

Figure 8.21 Theoretical pressure distributions over a Göttingen 398 airfoil: calculations by Garrick using Theodorsen's theory (1933). (From Garrick.)[114]

N.A.C.A. 23012

Figure 8.22   The NACA 23012 airfoil (1935). (From Jacobs.)[115]

tradition of parameter variation, exhaustive tests were carried out in the VDT, with the amount and location of the camber and the thickness ratio being systematically varied. With the forward location of the maximum camber, those new airfoils had higher maximum lift coefficients and lower pitching moments than the NACA four-digit series – attributes that were appealing to aircraft designers.[115,116] Jacobs found that the best airfoil in that series was the NACA 23012 airfoil (Figure 8.22); note the mean camber line, with maximum camber near the leading edge. The measured variations of the lift and drag coefficients, the lift-to-drag ratio, and the center of pressure for the NACA 23012 airfoil are shown in Figure 8.23. Although the five-digit airfoils had larger maximum lift coefficients than the previous airfoils, the decrease in lift at the point of stall was abrupt and dramatic (Figure 8.23). That was a distinct disadvantage compared with the earlier four-digit series, which had smoother and more gradual decreases in lift at the point of stall. Nevertheless, the NACA five-digit airfoils were widely used by the aircraft industry. For example, the Douglas DC-4, a four-engine transport that saw service in World War II as the C-54 and was a mainstay of airlines in the United States immediately after the war, used an NACA 23012 airfoil section. The NACA four-digit airfoils have been used for a number of general-aviation airplanes over the past 40 years, such as those of Cessna (the Caravan, the model 310, and the Citation II jet) and Beechcraft (Bonanza, Baron, and King Air).

The final chapter in airfoil research and design in the era of advanced propeller-driven airplanes was also written by Jacobs: the development of the *laminar-flow airfoil*. The impetus for that work can be traced back to Melvill Jones's paper on the ideal airplane,[100] one in which perfect streamlining would eliminate pressure drag due to flow separation, leaving only induced drag and skin-friction drag to be dealt with by the designer. During the 1930s the progress in streamlining was such that airplanes began to approach Jones's ideal, and aerodynamicists then began to turn their attention to reduction of skin-friction drag. That was what motivated Jacobs to begin thinking in terms of the laminar-flow airfoil. From Prandtl's early boundary-layer research it was well known that the skin-friction drag in a laminar flow was less (often considerably less) than that in a turbulent flow. Unfortunately, Nature prefers turbulent flows, and therefore it is extremely difficult (sometimes impossible) to maintain a laminar flow over a surface.

In late 1935, Jacobs attended the Fifth Volta Congress in Rome (whose significance will be discussed in Chapter 9). During that trip, he visited the major European aeronautical laboratories and spent some time at Cambridge University, where he had conversations with Geoffrey I. Taylor and Melvill Jones, England's leading fluid dynamicist and aerodynamicist, respectively. Taylor and Jones shared with Jacobs some of their preliminary findings that a laminar boundary layer would remain laminar if the surface pressure continued to decrease in the flow direction (a favorable pressure gradient), and a transition to turbulent flow would occur at about the location where the pressure began to increase in the flow direction (an adverse pressure gradient). Jones showed Jacobs his findings from actual flight experiments in which large regions of laminar flow over a wing were observed in areas where

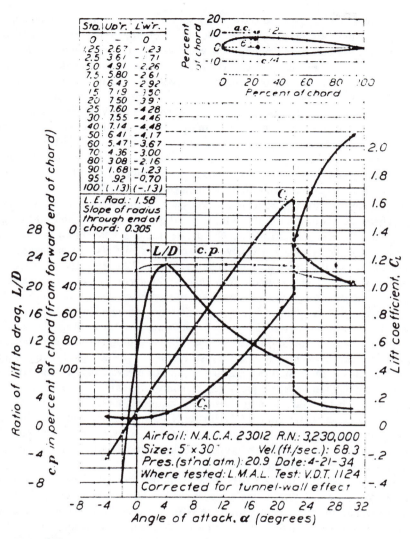

Figure 8.23   Measured variations of the lift and drag coefficients, the lift-to-drag ratio, and the center of pressure for the NACA 23012 airfoil (1935). (From Jacobs.)[115]

there were favorable pressure gradients. Jacobs returned to the United States convinced that airfoils could be designed to maintain laminar flow simply by shaping them to have large running lengths of decreasing pressure along the surface. It was a nice idea, but it would be difficult to implement. At that time, the recent advances in airfoil theory had been oriented toward calculating the pressure distribution for a given airfoil shape. Jacobs needed to turn that theory inside out and design an airfoil for a given pressure distribution. By background and nature, Jacobs, an experimentalist, would not appear the ideal person to take on that theoretical challenge. One of his colleagues and close friends at Langley, Robert T. Jones, who later would be recognized as the NACA's leading theoretician in the postwar period, said that Jacobs, "one of the most skillful and innovative American aerodynamicists," "had a wide appreciation of science but did not devote much time to theoretical studies. Rather, he used his theoretical understanding to devise intelligent experiments."[117] Nevertheless, Jacobs ultimately took on that theoretical challenge. He turned to Theodorsen's 1931 airfoil theory and began to examine how it could be reversed to design an airfoil shape from a given pressure distribution. He received no help from Theodorsen, who was quite negative about the whole idea. Later, one of Jacobs's engineers in the VDT, Ira Abbott, stated that "we were told that even the statement of the problem was mathematical nonsense with the implication that it was only our ignorance that encourages us."[118] Hansen stated that "encouraged now by hearing this negative peer response, Jacobs stubbornly persisted in directing an all-out effort to devise a satisfactory inversion of the Theodorsen method."[84] The Jacobs–Theodorsen enmity was surfacing again.

Jacobs, becoming the theoretician by necessity, in the peace and quiet of his home, studied Theodorsen's theory carefully for a few days. Finally he managed to modify the theory to allow the design of airfoil shapes with large regions of favorable pressure gradients. From that he designed a completely new family of NACA airfoils: the laminar-flow airfoils. Figure 8.24a shows the shape of the standard NACA 0012 airfoil and its surface pressure distribution at a zero angle of attack. Note that the favorable pressure gradient (decreasing pressure) would exist only over the most forward portion of the airfoil; over the remaining 90% of the airfoil there would be an adverse pressure gradient (increasing pressure). For that airfoil, the transition to turbulent flow would occur near the leading edge, and virtually the entire remaining airfoil surface would experience a turbulent boundary layer, with its attendant high skin-friction drag. In contrast, Figure 8.24b shows the shape of an NACA laminar-flow airfoil, the NACA 66-012, and its surface pressure distribution at a zero angle of attack. Note that the favorable pressure gradient would exist over more than 60% of the airfoil surface. That would encourage laminar flow over at least the first 60% of the airfoil surface – a dramatic change compared with the NACA 0012 airfoil. Both of those airfoils were symmetric airfoils with 12% thickness, but their shapes were completely different, the laminar-flow airfoil having its maximum thickness much farther back from the leading edge than the conventional airfoil.

From one perspective, Jacobs's persistence with the concept of the laminar-flow airfoil yielded a great success: Wind-tunnel tests showed a considerable decrease in drag for the new airfoils, and the excitement within the NACA was much like that generated by the NACA cowling in 1928. But in 1938, with war clouds on the horizon, security restrictions prevented the NACA from going totally public with Jacobs's important new findings. Nevertheless, some of that excitement was evident in the 1939 annual report of the NACA, in the following cryptic statement having to do with experiments in the new low-turbulence wind tunnel at Langley: "These preliminary investigations were started by the development of new airfoil forms that, when tested in the new equipment, immediately gave drag coefficients of one-third to one-half the values obtained for conventional sections." The actual data on the laminar-flow airfoils were not publicly released until after the war, but it was clear from the wind-tunnel data that the laminar-flow airfoils worked, at least in the laboratory, and those findings were quickly put to use by North American Aircraft in the design of the P-51 Mustang wing, the first airplane to use the NACA's laminar-flow airfoil.

But from another perspective – that of the real world of airplane manufacture and operation – the laminar-flow airfoils did not work. The NACA wind-tunnel models were like finely polished jewels with very smooth surfaces. Real airplanes were not. The realities

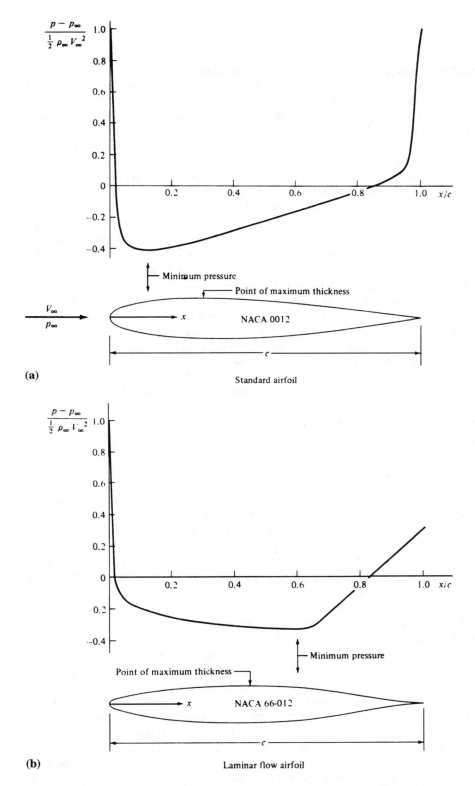

Figure 8.24 Comparison of the pressure distributions over two NACA airfoils: (a) NACA 0012, (b) NACA 66-012 laminar-flow airfoil.

of manufacturing introduced surface roughness and nonuniformities. When used in the field, bug splatters and other foreign-object impacts added to the surface roughness. The net result was that such surface roughness, which led to turbulent flow, won out over the effect of the favorable pressure gradients. In the field, the NACA laminar-flow airfoils experienced almost totally turbulent flow, like any other standard airfoil.

But from a final perspective, the NACA laminar-flow airfoils would enjoy success: Those airfoil shapes, with the maximum thickness far back from the leading edge, and with the resulting large regions of favorable pressure gradients, were found to have excellent *high-speed* characteristics; they had higher critical Mach numbers than conventional airfoils (high-speed effects will be discussed in Chapter 9). That was almost a fluke – one of those rare instances in the history of technology in which a system becomes a success because it unexpectedly excels at something for which it was not originally designed. The most successful of the NACA laminar-flow airfoils was the "six series," of which the airfoil in Figure 8.24b is an example. Because of their desirable high-speed characteristics, the NACA six-series laminar-flow airfoils were used on almost all high-speed airplanes in the 1940s and 1950s and are still in use today. (However, most aircraft manufacturers now have sophisticated computer programs to design their airfoil shapes for their own purposes, which in a sense is a return to the custom-made product and the ad hoc approach of the early twentieth century, but certainly no longer in ignorance.)

In the end, the development of the laminar-flow airfoil series was the crowning achievement from a decade of important airfoil research by the NACA led by Eastman Jacobs. The decade of the 1930s brought an increased understanding of airfoil aerodynamics, the derivation of an airfoil theory for arbitrary shapes, and the compilation of a massive collection of substantive wind-tunnel data on airfoils – all substantial factors in the development of advanced propeller-driven airplanes.

Drag Cleanup

The aerodynamicists put the final touches on their evocations of Melvill Jones's ideal airplane in the late 1930s and early 1940s, when the concept of streamlining was pushed to the maximum, when every effort was made to reduce or eliminate even the slightest sources of local flow separation on an airplane. In the laboratory, there could be no better way to locate small regions of drag production than to dispense with small wind-tunnel models and instead to put a real airplane in a wind tunnel. During the 1930s, the only wind-tunnel facility in which that could be done was the 30- × 60-ft Full-Scale Tunnel (FST) at NACA Langley. So the NACA began a series of detailed, laborious wind-tunnel tests whose purpose was to reduce the drag coefficients for conventional airplanes as much as possible without interfering with their practical operation. Within the NACA those early wind-tunnel tests were collectively referred to as the drag-cleanup program, which started in 1938 and lasted essentially through the end of World War II.

The typical drag-cleanup process was one of parameter variation: The airplane was first put in its most faired and sealed condition (protuberances removed, gaps sealed, etc.) and mounted in the wind tunnel, and the drag was measured. Then, one by one, each element was restored to its service condition, and the drag was measured each time. In that fashion the increment in drag due to each element was measured. Although the drag increment for each element usually was small, the total accumulation due to all the drag-producing elements usually was large. For example, the drag-cleanup series began in 1938

Figure 8.25   Brewster XF2A Buffalo mounted in the NACA FST (1938).

with the testing of a Brewster XF2A Buffalo single-seat navy pursuit airplane; the navy had become concerned when the experimental prototype had been unable to fly faster than about 250 mph. The airplane was flown to Langley and mounted in the FST (Figure 8.25). After a detailed series of tests, a number of drag-producing protuberances were identified (landing gear, exhaust stacks, machine-gun installation, gunsight, etc.). That led to some modifications of the airplane, after which the maximum speed was found to be 281 mph, a 31-mph increase over the original prototype. Referring to Jones's Figure 8.2, that drag cleanup was a push toward the right-hand side of the graph.

The drag cleanup for the Brewster Buffalo was such a success that within 18 months 18 different military prototypes were tested in the FST. A quantitative example of the drag-cleanup technique, that for the XP-41, is shown in Figure 8.26. Starting with the most streamlined configuration (condition 1), for which the drag coefficient was 0.0166, the airplane was restored to its original configuration through 17 different steps. The drag coefficient for the fully restored configuration was 0.0275, a 66% increase over the most stream-lined condition. Many of the sources of drag appear rather pedestrian (e.g., Sanded walkway added, and Oil cooler installed), but collectively they accounted for considerable drag.

The drag-cleanup procedures begun toward the end of the 1930s represented an important step in the evolution of advanced propeller-driven airplanes. Although the tests were mainly for military aircraft because of wartime priorities, they provided an educational experience and a massive aerodynamic data base that would later be used to design aircraft of all types.

Closure

Completing something of an international full circle, on December 17, 1937, eight years after his famous paper on streamlining to the Royal Aeronautical Society, B. Melvill

Airplane Condition

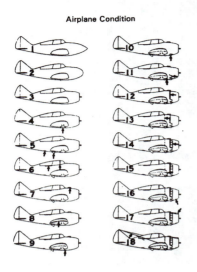

Condition number	Description	C_D (C_L = 0.15)	ΔC_D	ΔC_D, percent[a]
1	Completely faired condition, long nose fairing	0.0166		
2	Completely faired condition, blunt nose fairing	.0169		
3	Original cowling added, no airflow through cowling	.0186	0.0020	12.0
4	Landing-gear seals and fairing removed	.0188	.0002	1.2
5	Oil cooler installed	.0205	.0017	10.2
6	Canopy fairing removed	.0203	−.0002	−1.2
7	Carburetor air scoop added	.0209	.0006	3.6
8	Sanded walkway added	.0216	.0007	4.2
9	Ejector chute added	.0219	.0003	1.8
10	Exhaust stacks added	.0225	.0006	3.6
11	Intercooler added	.0236	.0011	6.6
12	Cowling exit opened	.0247	.0011	6.6
13	Accessory exit opened	.0252	.0005	3.0
14	Cowling fairing and seals removed	.0261	.0009	5.4
15	Cockpit ventilator opened	.0262	.0001	.6
16	Cowling venturi installed	.0264	.0002	1.2
17	Blast tubes added	.0267	.0003	1.8
18	Antenna installed	.0275	.0008	4.8
	Total		0.0109	

[a]Percentages based on completely faired condition with long nose fairing.

Figure 8.26 Progression of the drag cleanup for the Republic XP-41 carried out in the NACA FST (1939).

Jones arrived at Columbia University in New York to deliver the first Wright Brothers Lecture to the Institute of Aeronautical Sciences, addressing 300 members and guests of the institute, including Orville Wright. Jones took that opportunity to discuss some new data on boundary-layer behavior, obtained during flight experiments at Cambridge University using a Hawker Hart military biplane. Boundary-layer measurements, including velocity profiles, the location of the transition from laminar flow to turbulent flow, and boundary-layer thickness, had been obtained on the lower wing of the airplane at airspeeds from 60 to 120 mph in level flight, and 240 mph in long, steep dives.[119] Jones's lecture was a fitting closure to one phase in the development of applied aerodynamics in the era of the advanced propeller-driven airplane. Much had been accomplished in the decade since his original call for streamlining, and he was directing his attention to one of the remaining major sources of drag, namely, friction drag. Jones repeated his Wright Brothers Lecture at the California Institute of Technology on December 21, described in the January 1938 issue of the *Journal of the Aeronautical Sciences* as "the most outstanding meeting of the Institute of the Aeronautical Sciences ever held on the Pacific Coast." One member of the audience, Francis Clauser, at that time with the Douglas Aircraft Company, recognized the historical full-circle significance of Jones's talk: "It was a pleasure to hear from the man who provided the stimulation some years ago which has led to the practical elimination of unnecessary form drag in modern airplanes and it is reassuring that this same man is now engaged in research which may conceivably reduce the remaining skin friction to some fraction of its present value." Today, at the end of the twentieth century, aerodynamicists are still striving to "reduce the remaining skin friction to some fraction of its present value."

Six years later, the achievements in drag reduction due to streamlining were nicely summarized by another English aeronautical engineer, William S. Farren, who delivered the seventh Wright Brothers Lecture to the Institute of Aeronautical Sciences in New York.[120] Farren, who at that time was director of the Royal Aircraft Establishment, had been a member of Melvill Jones's research group at Cambridge during the 1920s and 1930s.[102] Farren was an experimentalist who specialized in instrumentation. He built the first wind tunnel at Cambridge and later designed the instrumentation for the in-flight boundary-layer measurements described earlier. Although his remarks to the Institute of Aeronautical Sciences centered on the role of research in aeronautics, he singled out drag reduction as a major example of the progress in aerodynamic research. His presentation involved some excellent illustrations of the metamorphosis of the airplane over the period from 1918 to 1944: At the top in Figure 8.27a is the British S.E.5 single-seat fighter from 1917. The middle figure shows how aerodynamic research during the 1920s and early 1930s transformed the 1917 biplane shape into the streamlined monoplane configuration typical of the Supermarine racers that finally won the Schneider Trophy for Britain. At the bottom is the famous Spitfire from World War II, which incorporated the latest advances in aerodynamic streamlining. A similar progression for multiengine bombers is shown in Figure 8.27b. At the top is the Handley Page 0/400 twin-engine bomber from 1917. The middle figure shows the transformation of the old biplane configuration into the streamlined airliner shape of the late 1930s. At the bottom is the Lancaster bomber from World War II, a product of the advances in streamlining. These figures provide graphic testimony to the advances in applied aerodynamics during the era of the advanced propeller-driven airplane.

Impact on Flying Machines

During the era of advanced propeller-driven airplanes, the state of the art in aerodynamics was directly reflected in the flying machines to a much greater extent than in previous eras in the history of aerodynamics. During the 1920s and 1930s, airplane designers became acutely aware of the need for aerodynamic improvements: Airplanes were continually flying faster and farther, and new technology had to be developed to keep pace with the steadily increasing challenges. Effective channels for communication between the two worlds of aerodynamics research and airplane design had to be developed during the period between the two world wars. Prandtl's research findings spread to countries outside of Germany through translations of his publications, as well as the export of some of his students, such as the emigration of Max Munk and Theodore von Kármán to the United States. During the 1920s and 1930s the research findings of government agencies (particularly the Royal Aeronautical Establishment in Britain and the NACA in America) were published in technical reports that were widely disseminated across the aeronautics community. Indeed, the NACA technical reports from that period became classics – data from careful research presented in terms readily understood by both researchers and airplane designers. The annual industry conferences hosted by the NACA served to expedite the flow of government research findings to industry. Certainly the airplane designers were still conservative by nature, and it took time for them to assimilate and develop trust in the state-of-the-art applications flowing from research in aerodynamics. By the end of the 1930s the propeller-driven airplane had become a sophisticated flying machine that clearly reflected such assimilation and trust.

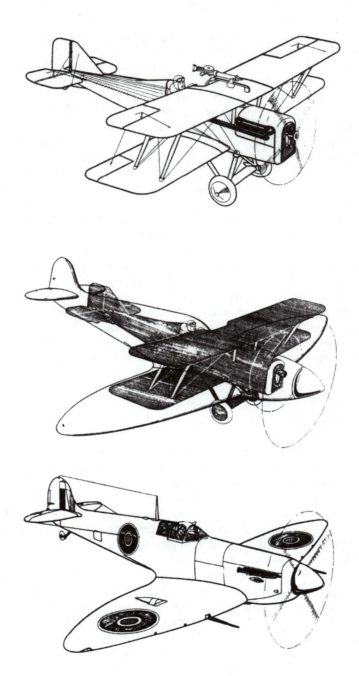

a)

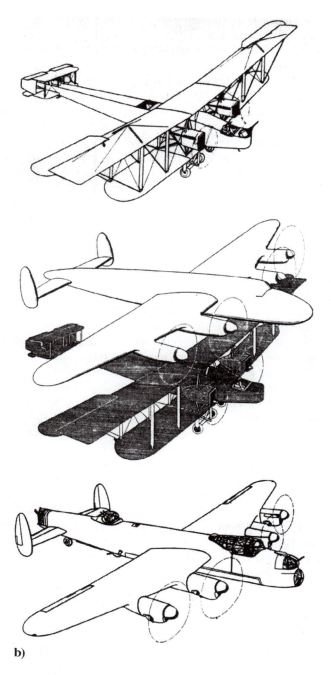

b)

Figure 8.27 Examples of streamlining: (a) evolution from the S.E.5 of World War I to the Spitfire of World War II. (b) evolution from the Handley Page 0/400 bomber of World War I to the Avro Lancaster of World War II.

The Boeing 247D twin-engine airliner (Figure 8.28) was the first important airplane to embody the state of the art in contemporary aerodynamics: retractable landing gear (reflecting Breguet's statement in 1922 that "the undercarriage should be made to disappear"), an NACA cowling (reflecting the work of Fred Weick at the NACA and the agency's rapid dissemination of the cowling research to industry), and a controllable-pitch propeller (reflecting the latest understanding of propeller aerodynamics). The Boeing 247D was widely used by United Airlines in the middle 1930s; it was a substantial improvement in commercial airliners,[121] and it helped to convince the public that airplanes could provide rapid and relatively safe transportation. A Boeing 247D hangs in the National Air and Space Museum, testimonial to its importance in the progression of airplane design.

Loftin[69] singled out three aircraft as epitomizing the advanced propeller-driven airplanes of the 1930s: the Douglas DC-3, the Boeing B-17, and the Seversky P-35, a commercial transport, a bomber, and a fighter, respectively. The Douglas DC-3 (Figure 8.29) incorporated all the advanced technology featured in the Boeing 247D, but was larger and had a higher cruising speed. It also had fillets at the junctures of the wings and fuselage to maintain attached flow at those locations, thus reducing form drag. One does not have to be an aerodynamicist to intuitively appreciate the beauty and excellent subsonic aerodynamic features of the airplane. The DC-3 became the most popular airliner of the late 1930s and 1940s; between 1936 and 1945, 10,926 were built, and many are still flying today. The DC-3 "surely must be considered as one of the truly outstanding aircraft developments of all time."[69] A DC-3 also hangs in the National Air and Space Museum. On the military side, the Boeing B-17 Flying fortress (Figure 8.30) was an outgrowth of Boeing's experience with the 247D. It embodied in a four-engine bomber the features of retractable landing gear, the NACA cowling, engine nacelles on the wings at the proper centerline locations (per the NACA research discussed earlier), and constant-speed propellers. Nearly 13,000 Fortresses were built; they were the mainstay of the U.S. strategic bomber force during World War II. Representative of fighter planes in the late 1930s was the Seversky P-35 (Figure 8.31), with an NACA cowling, retractable landing gear, a constant-speed propeller, and wing fillets. Because of the army's attachment to fighter planes with open cockpits (a World War I feature preferred by most fighter pilots at that time), open-cockpit designs had been perpetuated far too long – simply a cavity in the fuselage that increased the drag. The Seversky P-35, with its enclosed cockpit, was a marked innovation in fighter design and a welcomed aerodynamic improvement. First flown in 1936, the P-35 was underpowered, and only about 75 were built. However, after the Seversky company changed its name to Republic Aviation in 1939, the P-35 became the basis for the design of the famous P-47 Thunderbolt of World War II.

An excellent example of the use of state-of-the-art aerodynamic features in general aviation in the 1930s was the staggered-wing Beechcraft D-17 (Figure 8.32), sporting the same features as the latest commercial and military aircraft: retractable landing gear, NACA cowling, wing fillets (on the lower wing), and variable-pitch propeller. Because of its importance in the development of general-aviation airplane design, a Beech Staggerwing hangs in the Golden Age of Flight gallery at the National Air and Space Museum.

To illustrate the improvements in aerodynamics between the two world wars, the following table[69] shows some technical data for eight different airplanes. The first three were representative of the era of the strut-and-wire biplane, and the remaining five were from the age of the advanced propeller-driven airplane. Tabulated are the zero-lift drag coefficient, $C_{D,0}$, the maximum lift-to-drag ratio, $(L/D)_{max}$, and the wing loading, W/S, where W is

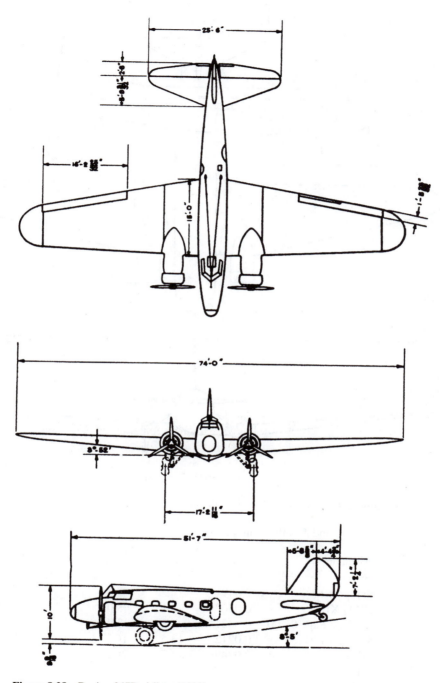

Figure 8.28 Boeing 247D airliner (1933).

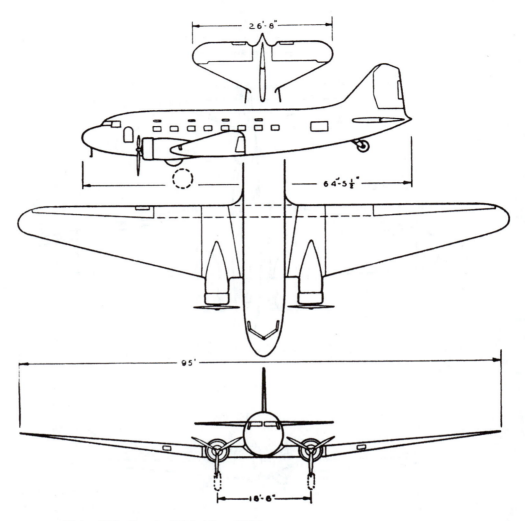

Figure 8.29   Douglas DC-3 airliner (1936).

the airplane weight, and S is the wing planform area:

Airplane	$C_{D,0}$	$(L/D)_{max}$	W/S
Nieuport 17	0.0491	7.9	7.8
SPAD XIII	0.0367	7.4	8.0
Handley-Page 0/400	0.0427	9.7	8.7
Boeing 247D	0.0212	13.5	16.3
DC-3	0.0249	14.7	25.3
B-17	0.0302	12.7	38.7
P-35	0.0251	11.8	25.5
Beech D-17S	0.0182	11.7	14.2

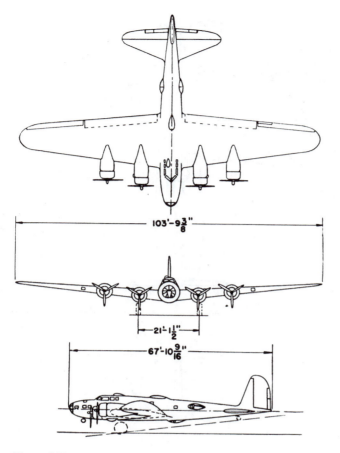

Figure 8.30   Boeing B-17 (1935).

The marked effect of aerodynamic streamlining during the 1930s is clearly seen in the table. On average, there was almost a factor-of-2 decrease in the zero-lift drag coefficient and a factor-of-2 increase in the maximum lift-to-drag ratio from the older strut-and-wire biplanes to the more advanced propeller-driven airplanes of the 1930s.

The wing loading almost quadrupled over that time span, because 1930s airplanes simply flew faster and thus generated more lift from the increased dynamic pressure, $\frac{1}{2}\rho_\infty V_\infty^2$. In steady, level flight, the lift equals the weight, and we can write

$$L = W = \frac{1}{2}\rho_\infty V_\infty^2 S C_L$$

where C_L is the lift coefficient. As V_∞ increases, the same lift (to balance the same weight) can be generated with a smaller wing area, and therefore the design wing loadings greatly increased during the 1930s. That was not without consequence, however: The higher the wing loading, the higher the required takeoff and landing speeds. That was handled in the 1930s by the use of "flaps," high-lift devices that increase the maximum lift coefficient and thus allow the airplane to fly slower while still generating enough lift to balance the weight. For example, the DC-3 used a simple split flap at the wing's trailing edge, a device that

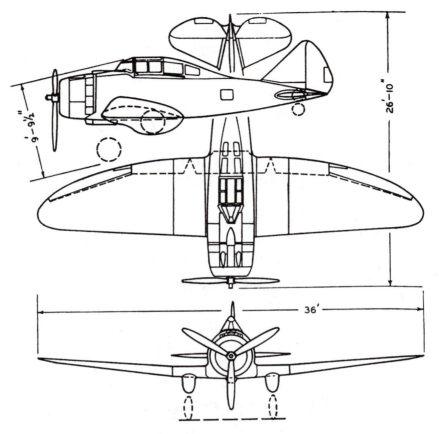

Figure 8.31   Seversky P-35 (1936).

was developed by Orville Wright in 1920. Also, airport runways were lengthened during the 1930s to allow for higher takeoff and landing speeds. The combination of those two factors allowed airplane designers to increase the wing loading. With higher wing loadings, they got more "bang for their buck" (i.e., more lift per square foot of wing area). For a given weight, if the wing loading can be increased, then the surface area of the wing can be decreased, and that will reduce the surface skin-friction drag. Thus the higher wing loading for the more advanced propeller-driven airplanes (as shown toward the bottom of the table) was a direct contribution to improved aerodynamics. Note that the wing loading for the Boeing 247D was almost half that for the DC-3; the 247D had no flaps. Also note that the DC-3 had a slightly higher $(L/D)_{max}$ than the 247D, although the zero-lift drag coefficient for the DC-3 was about 17% higher. That was an aspect-ratio effect; the aspect ratio for the DC-3 was 9.14, whereas that for the Boeing 247D was 6.55. The higher aspect ratio for the DC-3 resulted in considerably lower induced drag, as Prandtl had shown years earlier, and that gave the DC-3 a higher lift-to-drag ratio than the 247D.

  Wing flaps had an interesting history of development during the age of the advanced propeller-driven airplane. The aerodynamic function of a flap is simply to increase the lift of the airplane. Various high-lift devices are shown in Figure 8.33, ranging from the

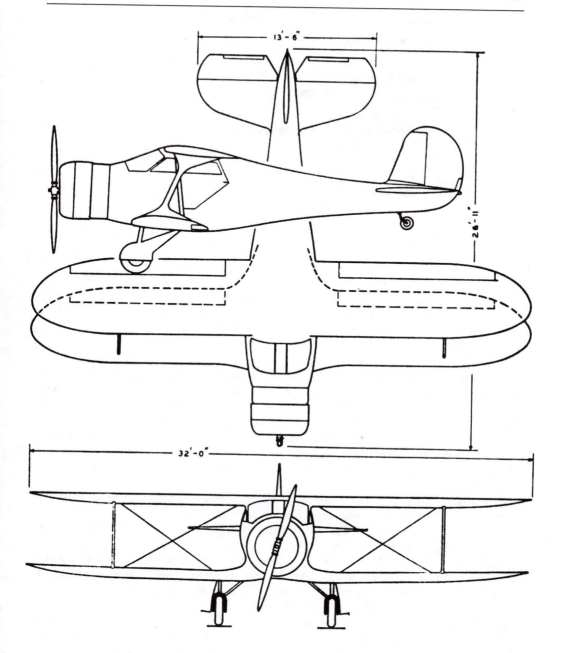

Figure 8.32 Beechcraft D-17 (1938).

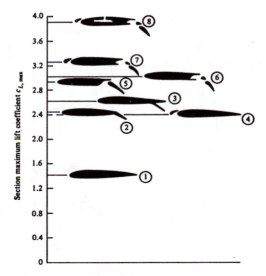

Figure 8.33   Typical values of airfoil maximum lift coefficients for various types of high-lift devices: (1) airfoil only, (2) plain flap, (3) split flap, (4) leading-edge slat, (5) single-slotted flap, (6) double-slotted flap, (7) double-slotted flap in combination with a leading-edge slat, (8) addition of boundary-layer suction at the top of the airfoil. (From Loftin.)[69]

simple plain flap (item 2) to the complicated double-slotted flap with a leading-edge slat and boundary-layer suction over the top of the airfoil (item 8). Typical values for the maximum lift coefficients for the devices are also shown. The aerodynamic action of a plain flap is illustrated in Figure 8.34, which shows the variation of the lift coefficient with the angle of attack for wings with and without flaps. At a given angle of attack below the point of stall, deflecting the flap downward will increase the lift coefficient – it is as though the original lift curve for the unflapped wing is raised and moved to the left in Figure 8.34 when the flap is deflected.

There is a detailed discussion of the historical development of flaps in the book by Miller and Sawers,[101] on which the following account is based. Flaps evolved directly from the aileron that Henry Farman first used in the autumn of 1908. Farman was the son of British parents living in France. From 1907 to well into 1920 he was recognized as Europe's most outstanding pilot, and he was a prominent airplane manufacturer in France until 1937. After witnessing Wilbur's flights in August 1908, Farman set out to achieve better lateral control over his aircraft. Not wishing to copy the Wrights' wing-warping idea (the Wrights were ever alert to any possible infringement of their patent for lateral control), Farman modified the trailing edges of his wings with flush-mounted sections near the wing tips that could be controlled to flap up and down independently of the wings. When those flap-like surfaces were deflected, the effective camber of the airfoil section of the wing was changed, and thus the lift was changed. Although ailerons had been used earlier by various inventors, they had always been separate rotatable surfaces, sometimes placed between biplane wings, or in front of the wings. Farman's ailerons in 1908 were the first that were integral parts of the wing, in the manner that we design them today. Farman used them as ailerons in the conventional sense. That is, one aileron would be deflected upward to decrease the lift on that wing, and the other aileron would be deflected downward to increase the lift on the

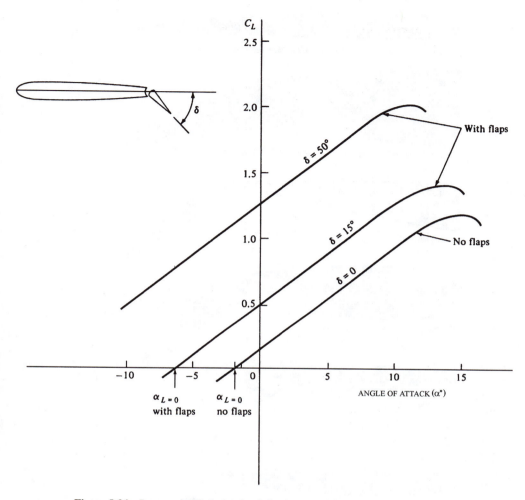

Figure 8.34   Increase in lift due to the deflection of a plain flap.

other wing, thus creating a lift imbalance between the right and left wings that would cause the airplane to roll about its longitudinal axis. The idea of simultaneously deflecting both flap surfaces in the downward direction to gain a balanced increase in lift came from the Royal Aircraft Factory in 1914, which used them on the S.E.4 biplane. After 1916, flaps were routinely used on airplanes built by Fairey Aircraft. However, the maximum speeds of World War I aircraft were so low that flaps were not particularly useful; the low wing loadings made flaps essentially redundant, and most pilots rarely bothered to use them.

Nevertheless, engineers continued to develop ideas for high-lift mechanisms. The slotted flap, with a gap (slot) between the main wing and the leading edge of the flap (Figure 8.35), was developed more or less independently in three places and in different ways. G. V. Lachmann, a young German pilot, conceived the idea of the slotted wing – simply a wing with a long, spanwise slot located near the leading edge. The idea was that the pressure differential between the lower and upper surfaces of the wing would force a

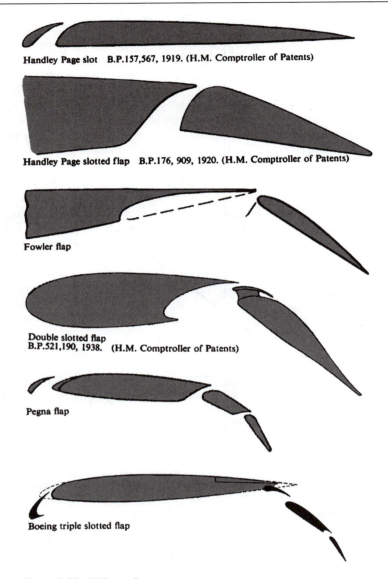

Handley Page slot   B.P.157,567, 1919. (H.M. Comptroller of Patents)

Handley Page slotted flap   B.P.176, 909, 1920. (H.M. Comptroller of Patents)

Fowler flap

Double slotted flap
B.P.521,190, 1938.   (H.M. Comptroller of Patents)

Pegna flap

Boeing triple slotted flap

Figure 8.35   Different flap concepts.

high-energy jet of air through the slot, which would flow tangentially over the top surface of the wing, energizing the boundary layer and delaying flow separation (hence stall) to much higher angles of attack. Lachmann ran some primitive smoke-tunnel tests in 1917 to verify his idea. However, his first application for a patent was turned down on the grounds that slots would destroy lift, not enhance it. Meanwhile, in England, Handley Page ran wind-tunnel tests on slotted wings in 1920. His data showed an increase in lift by 60% because of the slot. Lachmann read about Handley Page's work and promptly asked Ludwig Prandtl to run similar wind-tunnel tests at Göttingen. Prandtl was skeptical at first, but ran the tests anyway. The data showed that lift was increased by 63%! Lachmann got his patent and pooled rights with Handley Page in 1921. Handley Page, in the meantime, combined a slot and a flap to get the slotted flap, which he tested in a wind tunnel at the National Physical Laboratory during 1920–2. The data showed that the slotted flap produced its greatest lift increase when used on thick airfoils (that might explain in part why flaps were rarely used on the very thin airfoils of most World War I aircraft). Ultimately, Lachmann joined Handley Page's company in 1929. A third person to work independently on the slotted wing was O. Mader, working for Junkers in Germany. Mader tested a crude slotted wing in a wind tunnel and on an airplane during 1919–21. However, when the Junkers company applied for a patent in 1921, it was denied as an infringement on Lachmann's earlier patent.

Airplane designers during the 1920s were reluctant to use flaps, and most of the airplanes equipped with flaps during that time were designed by Handley Page or Lachmann.

In the United States, the split flap (Figure 8.33) was invented in 1920 by Orville Wright and J. M. H. Jacobs, working in Orville's small laboratory provided by the U.S. Army at McCook Field in Dayton. The split flap was found to increase both lift and drag; the increase in drag was useful for increasing the glide angle of an airplane on its landing approach, and that is still one of the primary uses for flaps.

In 1924, Harlan D. Fowler, an engineer working for the U.S. Army, invented a new kind of flap as a private venture using his own money. The Fowler flap not only deflected downward at the trailing edge of the wing but also deployed downstream of the wing (Figure 8.35). By that mechanical extension of the flap, the wing area was effectively increased. Thus the Fowler flap achieved an increase in lift by a combination of two effects: the increased camber due to flap deflection, and the increased surface area due to flap extension. For years afterward, Fowler was unable to obtain financial support to develop his idea. Finally, in 1932 the NACA ran some limited wind-tunnel tests that suggested that Fowler's flap might have some value. Fowler, then working as a salesman in California, traveled to Baltimore in 1933 and convinced the noted airplane manufacturer Glenn Martin of the value of the Fowler flap. Martin hired Fowler to help design flaps for several new Martin airplanes.

Finally, in the early 1930s, airplane designers, faced with the increasing speeds and wing loadings of the new airplanes, embraced the idea of flaps. In the United States, the first use of flaps was on the Northrop Gamma and the Northrop-designed Lockheed Orion; whether or not in deference to Orville Wright, Northrop used split flaps. When Douglas designed the DC-1 in 1932, split flaps were included. The first mass-produced airplane to use split flaps was the Douglas DC-3 (Figure 8.29). Throughout the middle and late 1930s, split flaps were used on the majority of civil and military airplanes, and even longer on fighter aircraft. The Fowler flap was part of the design for Martin's 146 bomber in 1935, but the airplane was never produced. The first production-line airplane to use Fowler flaps was the Lockheed 14 twin-engine airliner in 1937. After that, both slotted flaps and Fowler flaps were widely used, resulting in great improvements in lifting capacity and allowing airplane designers to use even higher wing loadings.

In 1937, G. Pegna of Piaggio Aircraft in Italy designed the double-slotted flap (Figure 8.35), first used on the Italian M-32 bomber in 1937 and widely used on airliners in the 1940s. Douglas first used that flap on its A-26 bomber, designed in 1941, and then on the DC-6 and on all later Douglas airliners. The most recent step in that line of evolution was a triple-slotted flap (Figure 8.35), first used on the Boeing 727 jet airliner in the early 1960s.

The airplanes shown in Figures 8.28–8.32 were streamlined, even coming close to Melvill Jones's ideal streamlined airplane. Clearly, the state of the art in aerodynamics during the 1930s had a substantial impact on flying machines during that time.

In the United States during World War II, improvements to the existing conventional airplanes continued, but there were no radical new developments. That was intentional. The government made a management decision to concentrate totally on mass production of the existing aircraft, rather than emphasizing the development of new types. That situation was later reviewed by J. H. "Dutch" Kindleburger, president of North American Aircraft, in 1953: "As far as United States military aviation is concerned, the Second World War may be characterized as a period of intensive design improvements and refinement rather than as a period of innovation."[122] To some extent the same can be said about developments in England. In Germany, however, during the same period, research in aerodynamics was thriving. Even though the Nazi government exhibited some anti-scientific attitudes and had a tendency to interfere with scientific and engineering work, German scientific research prospered. That was because several well-placed, forward-looking individuals had worked to set up the proper atmosphere in which research engineers and scientists could work.[101] German aerodynamics research during the war produced a great wealth of data on swept-back wings and delta wings for high-speed airplanes, data that helped to move aerodynamics into the age of the jet-propelled airplane. Miller and Sawers[101] saw a certain irony in that the German long-term research in aerodynamics did not help Germany greatly during the war, but rather provided handsome gifts for the victors.

In general, when considering the impact of the contemporary state of the art of aerodynamics on flying machines over the centuries, we have made the case in this book that there was virtually none, at least not until well after World War I. However, that pattern changed during the 1930s. Advanced propeller-driven airplanes would not have been developed without significant transfers of technology from the research laboratory to the airplane designers. Sometimes the designers, usually conservative by nature, were careful about accepting fundamentally new ideas, but when faced with ever increasing demands for planes that would fly faster, higher, and farther, the designers of the 1930s were driven to the research literature for new solutions. That was made possible, of course, by the ready availability of research findings from a variety of sources: technical meetings, journals, reports, and, by that time, the telephone. But in spite of that transfer of technology, there remained a certain divide between the acquisitions of research findings and their implementation in industry. That problem had existed from the very beginnings of the aeronautical industry, but it became particularly evident during the 1930s. The problem was that whereas industry used the findings from aerodynamics research, industry performed very little of that research. Miller and Sawers, in studying the technical development of modern aviation, stated that "invention with the aircraft industry is most noticeable for its absence . . . . The institutions that have been most productive of inventions are the universities and government-financed research institutes – especially in Germany . . . . The aircraft industry has done surprisingly little inventive research" (p. 246).[101]

The first concern of the aircraft industry, and of all industries in a capitalist society, is to earn a profit. Research is usually an expensive activity whose returns are realized only over the long term. Moreover, money and labor resources allocated for research are resources that will not be used in the industry's main activity: manufacturing. Hugo Junkers, a professor of mechanics at the Aachen Technische Hochschule, and head of his own aircraft company, was in a position to see the dilemma clearly: "Increased production results in research being sidetracked when it is a function of the same factory. The giant mechanism of mass production has its own laws, which run counter to the principles which apply to the handling of pioneer problems in the differently-ordered world of research. That is what happened at the Junkers factory" (p. 55).[101] Even today, when acquisition of new technology has never been more crucial in aeronautics, support for a research laboratory in an aircraft company is an expensive luxury that most companies forgo.

Although the mechanisms for transferring state-of-the-art developments in aerodynamics to airplanes were reasonably well in place during the evolution of propeller-driven airplanes (ready availability of current research findings in journals and technical reports, technical meetings, etc.), there still remained a kind of cultural divide between the design-and-manufacturing entity and the research laboratory. That situation was summarized by Miller and Sawers:

> Neglect of the lessons that could be learned from the available theoretical knowledge in the years before 1930 is a characteristic feature of aeronautical history; as in other industries, technical progress has not followed any clear and logical path. Even in Germany, where the research of Prandtl and Lanchester was better-known and understood than it was in Britain and the United States, designers indicated oddly little awareness of the gains that their theoretical work showed were possible from the use of better streamlining. They left it to the younger glider designers, who had often studied under Prandtl at Göttingen; but the impetus to better streamlining came as much from the intuitive engineering approach of men like Northrop in the United States or Mitchell [designer of the Supermarine Schneider Trophy racers and the World War II Spitfire] in Britain as from the applications of theory, though Northrop and Douglas were later aided by scientists like Prandtl's pupil von Kármán. German designers followed the American example on streamlining in the early 1930's. But later in the decade a new generation of designers came to the fore in Germany, products of Göttingen and the gliding movement. These men better appreciated the lessons they could learn from science than their predecessors or designers in Britain and the United States, who remained practical men, in the tradition of the industry. So Germany quickly attained a lead in design when understanding of recent aerodynamic research – mostly done in Germany – became essential to the designer, as it did in the development of transonic and supersonic airplanes during the war. Only the military defeat of Germany in 1945 prevented the German industry from becoming as dominant as the American industry is today [p. 247].[101]

With the advent of high-speed flight – transonic and supersonic airplanes – after World War II, a new age in the development of the airplane began – the age of the jet-propelled airplane. That age would require the application of some different aerodynamic principles to the flying machine and mandate even closer connections between science and airplane design. Some of those different aerodynamic principles would be old, and some new.

CHAPTER 9

Aerodynamics in the Age of the Jet Airplane

We call the speed range just below and just above the sonic speed – Mach number nearly equal to 1 – the transonic range. Dryden [Hugh Dryden, well-known fluid dynamicist and past administrator of the National Advisory Committee for Aeronautics] and I invented the word "transonic." We had found that a word was needed to denote the critical speed range of which we were talking. We could not agree whether it should be written with one *s* or two. Dryden was logical and wanted two *s*'s. I thought it wasn't necessary always to be logical in aeronautics, so I wrote it with one *s*. I introduced the term in this form in a report to the Air Force. I am not sure whether the general who read it knew what it meant, but his answer contained the word, so it seemed to be officially accepted. . . . I will remember this period [about 1941] when designers were rather frantic because of the unexpected difficulties of transonic flight. They thought the troubles indicated a failure in aerodynamic theory.

Theodore von Kármán (*Aerodynamics,* 1954, p. 116)

The morning of Tuesday, October 14, 1947, dawned bright and beautiful over Muroc Dry Lake, a large expanse of flat, hard surface in the Mojave Desert in California. At 6:00 a.m., teams of engineers and technicians at the Muroc Army Air Field began to prepare a small rocket-powered airplane for flight. Painted orange, and resembling a 50-caliber machine-gun bullet mated to a pair of straight, stubby wings, the Bell X-1 research vehicle was carefully installed in the bomb bay of a four-engine B-29 bomber of World War II vintage. At 10:00 a.m. the B-29 took off and climbed to an altitude of 20,000 ft. As it rose past 5,000 ft, Captain Charles ("Chuck") Yeager, a veteran P-51 pilot from the European theater during World War II, struggled into the cockpit of the X-1. Yeager was in pain from two broken ribs incurred during a horseback accident the previous weekend, but not wishing to disrupt the events of the day, he informed no one except his close friend Captain Jack Ridley, who was helping him to squeeze into the X-1 cockpit. At 10:26 a.m., at a speed of 250 mph, the brightly painted X-1 dropped free from the B-29. Yeager fired his Reaction Motors XLR-11 rocket engine, and powered by 6,000 lb of thrust, the sleek airplane accelerated and climbed rapidly. Trailing an exhaust jet of shock diamonds from the four convergent-divergent rocket nozzles of the engine, the X-1 was soon flying faster than Mach 0.85, the speed beyond which there were no wind-tunnel data in 1947 – and beyond which no one knew what problems might be encountered in transonic flight. Entering that unknown realm, Yeager momentarily shut down two of the four rocket chambers and carefully tested the controls of the X-1 as the Mach meter in the cockpit registered 0.95 and still increasing. Small invisible shock waves were dancing back and forth over the top surface of the wings. At an altitude of 40,000 ft, the X-1 began to level off, and Yeager fired one of the two shutdown rocket chambers. The Mach meter moved smoothly through 0.98 and 0.99 to reach 1.02. There the meter hesitated, and then jumped to 1.06. A stronger bow shock wave was formed in the air ahead of the needle nose of the X-1 as Yeager reached a velocity of 700 mph, Mach 1.06, at 43,000 ft. The flight was smooth; there was no violent buffeting of the airplane and no loss of control, as had been feared by some engineers. At

Figure 9.1 Bell X-1.

that moment, Yeager became the first pilot to fly faster than the speed of sound, and the small, streamlined Bell X-1 (Figure 9.1) became the first supersonic airplane in the history of flight.

As the sonic boom from the X-1 propagated across the California desert, that flight became the most significant milestone in human flight since the Wright brothers' first flight at Kill Devil Hills 44 years earlier. That flight was equally significant in the history of scientific accomplishments, for it was one of the high points in 260 years of research into the mysteries of high-speed gas dynamics and aerodynamics. In particular, it was the climax of 23 years of outstanding research in high-speed aerodynamics carried out by the NACA, research that is one of the most important stories in the history of aeronautical engineering.

In this chapter we move into the age of the jet-propelled airplane, and the history of aerodynamics emphasizes research on high-speed flows. There is a clear physical dividing line in the aerodynamics of high-speed flows: the speed of sound, which is 1,117 ft/s (340 m/s) at standard sea-level conditions. When the local velocity at a point in a flow is less than the local speed of sound, the flow is *subsonic* at that point; when the local velocity is greater than the local speed of sound, the flow is *supersonic*. For the aerodynamic flow field around an airplane in flight, there are local regions in the flow that are at higher or lower velocities than the overall flight velocity of the airplane. Hence the flow field around an airplane flying at a speed slightly less than the speed of sound may have pockets of locally supersonic flow. Similarly, an airplane flying at a speed slightly greater than the speed of sound may have pockets of locally subsonic flow. In such cases the flow fields are called *transonic* – flow fields with mixed regions of locally subsonic and supersonic flows. The transonic flight regime for airplanes is that narrow band of flight speeds ranging from just below to just above the speed of sound.

For historical assessment of high-speed aerodynamics, it would be too artificial to divide our discussion into separate treatments of subsonic, transonic, and supersonic speeds. During the evolution of the state of the art in high-speed aerodynamics, those three regimes were (and still are) inexorably intertwined. Here we shall consider various aspects of high-speed subsonic, transonic, and supersonic aerodynamics in a manner that parallels their evolution, integrating the stories of their development when appropriate.

The Speed of Sound

Most golfers know the following rule of thumb: When you see a flash of lightning in the distance, start counting at a rate of one count per second. For every count of 5 before you hear the thunder, the lightning bolt will have struck a mile away. Clearly, sound travels through air at a definite speed, much slower than the speed of light. Indeed, the standard sea-level speed of sound is 1,117 ft/s, and in 5 s a sound wave will travel 5,585 ft, slightly more than a mile. That is the basis for the golfer's count-of-5 rule of thumb.

The speed of sound is one of the most important parameters in aerodynamics; it is the dividing line between subsonic flight (speeds less than that of sound) and supersonic flight (speeds greater than that of sound). The Mach number is the ratio of the speed of a fluid flow to the speed of sound in that flow. If the Mach number is 0.5, the velocity of the fluid flow is one-half the speed of sound; a Mach number of 2.0 means that the flow velocity is twice that of sound. The rules of physics for a subsonic flow are totally different from those for a supersonic flow – a contrast as striking as that between day and night. That was why there was such drama and anxiety at the first supersonic flight of the X-1, and it is why the precise value of the speed of sound is so important in aerodynamics.

By the seventeenth century it was understood that sound propagates through the air at some finite velocity. By the time Isaac Newton published his *Principia* in 1687, artillery tests had already shown that the speed of sound was approximately 1,140 ft/s. The seventeenth-century gunner had prefigured the modern golfer's experience: The tests were performed by standing a known large distance away from a cannon and noting the elapsed time between the flash from the cannon and the sound of its firing. In Proposition 50, Book II, of the *Principia,* Newton correctly theorized that the speed of sound was related to the "elasticity" of the air. (The elasticity is the reciprocal of the compressibility τ, a thermodynamic quantity that is a measure of the fractional change in the volume of a gas per unit change in pressure.) However, he made the erroneous assumption that a sound wave was an isothermal process (i.e., he assumed that the air temperature inside the sound wave was constant) and consequently proposed the following incorrect expression for the speed of sound:

$$a = \sqrt{1/\rho\tau_T}$$

where τ_T was the isothermal compressibility. Much to his dismay, he calculated a value of 979 ft/s from that expression – 15% lower than the value indicated by gunshot data. Undaunted, he resorted to a familiar ploy of theoreticians: He proceeded to try to explain away the difference on the basis of the presence of solid dust particles and water vapor in the atmosphere. That misconception was corrected a century later by the French mathematician Laplace, who properly assumed that a sound wave was adiabatic, not isothermal. Laplace went on to derive the proper expression

$$a = \sqrt{1/\rho\tau_s}$$

where τ_s was the isentropic compressibility (an isentropic process is one that is both adiabatic and frictionless). Therefore, by the 1820s the process and relationship for the propagation of sound in a gas were fully understood.

That is not to say that the issue of the precise value for the speed of sound was totally settled; that debate lasted well into the twentieth century. Indeed, although this event is little known today, the NACA was an arbiter in setting the standard sea-level value for the speed of sound. On October 12, 1943, a group of 27 distinguished U.S. leaders in aerodynamics were at NACA Headquarters in Washington, D.C., for a meeting of the

Committee on Aerodynamics, one of the adjunct committees set up by the main NACA, Among those present were Hugh L. Dryden from the Bureau of Standards, John Stack from the NACA Langley Memorial Laboratory, and Theodore von Kármán, director of the Guggenheim Aeronautical Laboratory at the California Institute of Technology, representing the legacy of Prandtl's aerodynamic research at Göttingen. After subcommittee reports on the progress in helicopter aerodynamics and recent aerodynamic problems with wing flutter and vibration, the matter of the speed of sound was brought up as new business by John Stack, who stated that "the problem of establishing a standard speed of sound was raised by an aircraft manufacturer."[123] Stack reported that the committee's laboratory staff had surveyed the available information on the specific heats of air (thermodynamic information that goes into the calculation of the speed of sound) and had calculated a value for the speed of sound of 1,116.2 ft/s. Recently measured values had given weighted means of 1,116.8 to 1,116.16 ft/s. Dryden noted that the specific heats were " not necessarily the same for all conditions" and suggested that the committee select 1,117 ft/s as a round figure to serve as the standard value for the speed of sound for sea-level conditions for aeronautical purposes. Today, the accepted standard speed of sound depends on which standard-atmosphere table is being used, ranging from a value of 1,116.4 ft/s in the 1959 ARDC (Aeronautical Research and Development Command, U.S. Air Force) model atmosphere to 1,116.9 ft/s in the 1954 International Civil Aviation Organization (ICAO) model atmosphere. However, for engineering purposes that is splitting hairs, and Dryden's suggestion of a round value of 1,117 ft/s is used today for many engineering calculations.

The Early History of High-Speed Aerodynamics

Some of the fundamental phenomena of supersonic flows were discovered and understood in the nineteenth century, the most important being the physics of shock waves. A shock wave is a phenomenon of nature that occurs only at supersonic speeds. Shock waves are very thin regions (much thinner than the paper of this page) across which pressure and temperature increase abruptly and severely. Imagine that you are a small fluid element moving in a flow. As you pass through the shock wave, you will feel almost instantaneous increases in pressure and temperature, much as an explosion would create – hence the name *shock* wave.

The existence of shock waves was recognized in the early nineteenth century. Following the successful approach of Laplace in calculating the speed of sound, in 1858 the German mathematician G. F. B. Riemann attempted to calculate shock properties by also assuming isentropic conditions. That effort failed, because a shock wave is, in thermodynamic language, an irreversible process, caused by viscosity and thermal-conduction effects inside the shock wave. The measure of the degree of irreversibility is a thermodynamic variable called *entropy,* which, according to the second law of thermodynamics, always increases in any process involving such irreversibilities. The entropy of a gas always increases as it passes through a shock wave. Unfortunately, Riemann made the incorrect assumption that the entropy remained constant across a shock. However, 12 years later, the first major breakthrough in shock-wave theory was reported by a Scottish engineer, William John Macquorn Rankine (1820–72) (Figure 9.2), one of the founders of the science of thermodynamics. At the age of 25 he was offered the Queen Victoria Chair of Civil Engineering and Mechanics at the University of Glasgow, a post he occupied until his death on December 24, 1872. During that period, Rankine worked as an engineer in the true sense,

Figure 9.2   W. J. M. Rankine.

applying scientific principles to study metal fatigue in rail-car axles, to develop new meth-ods of mechanical construction, and to test soil mechanics (dealing with earth pressures and the stability of retaining walls). His best-known contributions were in the field of steam engines: the Rankine cycle, a thermodynamic cycle used as a standard of efficiency for steam power, and an engineering scale for absolute temperature based on the Fahrenheit scale.

Rankine's contribution to shock-wave theory came late in life, two years before his death. In 1870, in the *Philosophical Transactions of the Royal Society,* Rankine clearly presented the proper normal-shock equations for continuity, momentum, and energy in much the same form as used today. (In those equations, Rankine defined a quantity he called "bulkiness," identical with what we now define as "specific volume." Apparently the use of the term "bulkiness" died of its own cumbersomeness.) Moreover, Rankine properly assumed that the internal structure of the shock wave was not isentropic; rather, it was a region of dissipation. He was thinking about thermal conduction, not the companion effect of viscosity within the shock, but he was able to derive relationships for the thermodynamic changes across a shock wave.

The equations obtained by Rankine were subsequently rediscovered by the French ballis-tician Pierre Henry Hugoniot. Not aware of Rankine's work, in 1887 Hugoniot published a paper in the *Journal de 1' Ecole Polytechnique* in which the equations for normal-shock ther-modynamic properties were presented. As a result of the pioneering work by Hugoniot and by Rankine before him, all equations dealing with changes across shock waves are known as the *Rankine-Hugoniot relations,* a label that appears frequently in modern gas-dynamics literature.

However, the work of Rankine and Hugoniot did not establish the directions of changes across a shock wave. Noted in both works was the mathematical possibility of either compression shocks (pressure increases) or rarefaction shocks (pressure decreases). It was not until 1910 that the ambiguity was resolved. In two almost simultaneous and independent papers, first Lord Rayleigh and then G. I. Taylor invoked the second law of thermodynamics to show that only compression shocks are physically possible (i.e., the Rankine-Hugoniot relations apply physically only to the case in which the pressure behind the shock is greater than the pressure in front of the shock). Rayleigh's paper, published in *Proceedings of the Royal Society* (vol. 84), September 15, 1910, summarized his findings as follows:

But here a question arises which Rankine does not seem to have considered. In order to secure the necessary transfers of heat by means of conduction it is an indispensable condition that the heat should pass from the hotter to the colder body. If maintenance of type be possible in a particular wave as a result of conduction, a reversal of the motion will give a wave whose type cannot be so maintained. We have seen reason already for the conclusion that a dissipative agency can serve to maintain the type only when the gas passes from a less to a more condensed state.

In addition to applying the second law of thermodynamics, Rayleigh showed that viscosity played a role in the structure of a shock wave as essential as the role of conduction. (Recall that Rankine considered conduction only; also, Hugoniot's work was without reference to any dissipative mechanism.)

One month later, in the same journal, a young Geoffrey I. Taylor (who was to become one of the leading fluid dynamicists of the twentieth century) published a short paper that supported Rayleigh's conclusions. Finally, over a course of 40 years, ending in the second decade of this century, the theory of normal shock waves was fully established.

It should be noted that the shock-wave studies by Rankine, Hugoniot, Rayleigh, and Taylor were viewed at the time as interesting research into the basic mechanics of a relatively academic problem. The rush to apply the theory did not begin until 30 years later, with the sudden interest in supersonic vehicles during World War II. That was a classic example of the importance of an ongoing program of basic research, even when such work appears irrelevant at the time. The rapid advances in supersonic flight during the 1940s were clearly expedited because shock-wave theory was sitting there, fully developed and ready for application.

The first person to observe and record the nature of a supersonic flow in the laboratory was Ernst Mach, a famous nineteenth-century physicist and philosopher. In the early history of high-speed aerodynamics, Mach deserves special attention.

Ernst Mach was born at Turas, Moravia, in Austria, on February 18, 1838. His father and mother were extremely private and introspective intellectuals. His father was a student of philosophy and classical literature; his mother was a poet and musician. The family lived on an isolated farm, where Mach's father was pioneering the beginning of silkworm culture in Europe. At an early age, Mach was not a particularly successful student. Later, Mach described himself as a "weak pitiful child who developed very slowly." Tutored by his father at home, Mach learned Latin, Greek, history, algebra, and geometry. After marginal performances in grade school and high school (not for lack of intellectual ability, but for lack of interest in the material usually taught by rote), Mach entered the University of Vienna, where he excelled, spurred by interest in mathematics, physics, philosophy, and history. In 1860 he received a Ph.D. in physics, with a thesis entitled "On Electrical Discharge and Induction." By 1864 he was a professor of physics at the University of Graz. (The variety and depth of his intellectual interests were attested by the fact that he was offered, but turned down, a chair in *surgery* at the University of Salzburg, preferring to go to Graz.) In 1867, Mach became a professor of experimental physics at the University of Prague, a position he would occupy for the next 28 years.

In the modern technological world, engineers and scientists are virtually forced to concentrate their efforts in narrow areas of specialization, but in Mach's time one could still contemplate the Renaissance man, and Mach was a supreme generalist. A listing of Mach's contributions and writings would include works on physical optics, the history of science, mechanics, philosophy, the origins of relativity theory, supersonic flow, thermodynamics,

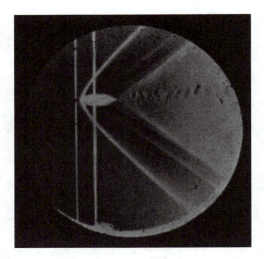

Figure 9.3   The first photograph of a shock wave from a body (a bullet) moving at supersonic speed.

the sugar cycle in grapes, the physics of music, and classical literature. He even wrote on world affairs: one of Mach's papers commented on the "absurdity committed by the states-man who regards the individual as existing solely for the sake of the state," which provoked strong criticism from Lenin. We can only regard him with awe and envy, for Mach, in the words of the American philosopher William James, knew "everything about everything."

Mach's contributions to supersonic aerodynamics were highlighted in his paper "Pho-tographische Fixierung der durch Projektile in der Luft eingeleiten Vorgange," presented before the Academy of Sciences in Vienna in 1887, which showed the first photograph of a shock wave in front of a bullet moving at supersonic speed (Figure 9.3). Also visible are weaker waves at the rear of the projectile and the structure of the turbulent wake downstream of the base region. The two vertical lines were made by trip wires designed to time the photographic light source (a spark) with the passing of the projectile. Mach was a precise and careful experimentalist; the quality of the photograph and the fact that he was able to make the shock waves visible in the first place (he used an innovative technique called the shadowgraph) attest to his exceptional experimental abilities. Note that Mach was able to carry out such experiments involving split-second timing without the benefit of electronics – indeed, the vacuum tube had not yet been invented.

Mach was the first researcher to understand the basic physical characteristics of a super-sonic flow and to point out the importance of the flow velocity V relative to the speed of sound a and to note the discontinuous and marked changes in a flow field as the ratio V/a changed from below 1 to above 1. He did not, however, call that ratio the Mach number, as we do today. The term "Mach number" was introduced in 1929 by the Swiss engineer Jakob Ackeret during a lecture in Zürich and did not reach the English literature until the late 1930s.

Ernst Mach (Figure 9.4) was an active thinker, lecturer, and writer up to the time of his death, February 19, 1916, near Munich, one day after his 78th birthday. Today Germany honors him via the Ernst Mach Institute, which conducts research in experimental gas dynamics, ballistics, high-speed photography, and cinematography.

Figure 9.4   Ernst Mach.

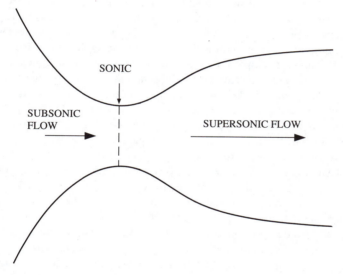

Figure 9.5   Convergent-divergent nozzle.

A continuous supersonic flow can be produced by expanding a gas through a nozzle that first contracts and then expands in area. Such a convergent-divergent nozzle is sketched in Figure 9.5; this is the shape of the nozzles used for supersonic wind tunnels and the exhaust nozzles for rocket engines. Today the gas-dynamic phenomena of convergent-divergent nozzles compose part of the classic subject matter in a first course in compressible flow, but the first practical use of the convergent-divergent supersonic nozzle was reported before the twentieth century. The Swedish engineer Carl G. P. de Laval designed a steam turbine in the late 1880s that incorporated supersonic expansion nozzles upstream of the turbine blades (Figure 9.6), and such convergent-divergent nozzles are often called "Laval nozzles" in the literature.

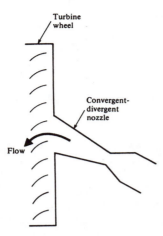

Figure 9.6   Schematic of de Laval's supersonic nozzle for driving a turbine wheel.

Carl Gustaf Patrik de Laval was born at Blasenborg, Sweden, May 9, 1845. The son of a Swedish army captain, de Laval showed an early interest in mechanical mechanisms, disassembling and then reassembling such devices as watches and gun locks. His parents encouraged his development along those lines, and at the age of 18 de Laval entered the University of Uppsala, graduating in 1866 with high honors in engineering. He was then employed by a Swedish mining company, the Stora Kopparberg, where he quickly realized that he needed more education. He returned to Uppsala and studied chemistry, physics, and mathematics, earning a Ph.D. in 1872. From there, he returned to the Stora company for three years and then joined the Kloster Iron Works in Germany in 1875. By that time his inventive genius was beginning to surface; he developed a sieve for improving the distribution of air in Bessemer converters and a new apparatus for galvanizing processes. During his time with Kloster, he was experimenting with centrifugal machines to separate the cream in milk. Unable to convince Kloster to manufacture his cream separator, he resigned in 1877, moved to Stockholm, and started his own company. Within 30 years he had sold more than a million de Laval cream separators, and today he is better known in Europe for cream separators than for steam turbines.

However, it was with his steam-turbine designs that de Laval made an important contribution to an understanding of compressible flows. In 1882 he constructed his first turbine using conventional nozzles. Such nozzles had convergent shapes – nothing more than orifices in some designs. Thus the kinetic energy of the steam hitting the rotor blades was low, resulting in low rotational speeds for the turbines. The cause of that deficiency was recognized: The pressure ratio across such nozzles was never less than one-half. Today we know that such nozzles were choked and that the flow exhausted from the nozzle exit at a velocity that was not greater than sonic, but in 1882 engineers did not fully understand such phenomena. Finally, in 1888, de Laval hit upon the idea of further expanding the gas by adding a divergent section to the original convergent shape. Immediately his steam turbines turned at incredible rotational speeds: more than 30,000 revolutions per minute (rpm). Overcoming the many mechanical problems introduced by such as improvement in rotational speed, de Laval developed his turbine business into a large corporation in Stockholm and quickly signed up a number of international affiliates in France, Germany,

England, The Netherlands, Austria-Hungary, Russia, and the United States. His design was demonstrated at the World Columbian Exposition in Chicago in 1893.

In addition to his successes as an engineer and businessman, de Laval was adept at social relations, respected and liked by his social peers and employees. He was elected to the Swedish Parliament (1888–90) and later became a member of the Senate. He was awarded numerous honors and decorations and was a member of the Swedish Royal Academy of Science. After a full and productive life, de Laval died in Stockholm in 1912, at the age of 67. However, his influence and his company have lasted to the present day.

In 1888, de Laval and his contemporary engineers were not quite certain that supersonic flow actually existed in the "Laval nozzle," a point of contention that was not properly resolved until the experiments of Stodola in 1903.

The innovative de Laval design for steam-turbine nozzles sparked interest in the fluid mechanics of flows through convergent-divergent nozzles at the turn of the century, and an important figure in such studies was a Hungarian-born engineer, Aurel Boleslav Stodola, who would eventually become the leading European expert on steam turbines. Whereas de Laval was an idea and design man, Stodola was a scholarly professor. He tied up the loose ends of the theory and scientific phenomena associated with Laval nozzles and was a major force in developing an understanding of compressible flows, thermodynamics, and steam turbines.

Stodola was born in 1859 in Hungary, the second son of a leather manufacturer. He attended the Budapest Technical University for one year and was an exceptional student, but in 1877 he transferred to the University of Zürich in Switzerland, and then in 1878 to the Eidgenossische Technische Hochschule, also in Zürich, from where he graduated in 1880 with a mechanical-engineering degree. Subsequently he worked briefly for Ruston & Co. in Prague, where he was responsible for the design of several types of steam engines. However, his superb record as a student soon earned him a position as a professor of thermal machinery back at the Eidgenossische Technische Hochschule in Zürich, a position he held until his retirement in 1929. There Stodola enjoyed a distinguished academic career that included teaching, industrial consultation, and engineering design. However, his main contributions were in applied research, fueled by his synergistic combination of considerable mathematical competence and an intense devotion to practical applications. He understood the importance of engineering research at a time when it was virtually nonexistent anywhere the world. In 1903, the same year as the Wright brothers' first powered flight, Stodola wrote that

> we engineers of course know that machine building, through widely extended practical experimenting, has solved problems, with the utmost ease, which baffled scientific investigation for years. But this "cut and try method," as engineers ironically term it, is often extremely costly; and one of the most important questions of all technical activity, that of efficiency, should lead us not to underestimate the results of scientific technical work [p. iii].[124]

That commentary on the neglected importance of basic scientific research was directed primarily at the field of steam-turbine design, but it was prophetic of the massive and varied research programs to come during the latter half of the twentieth century.

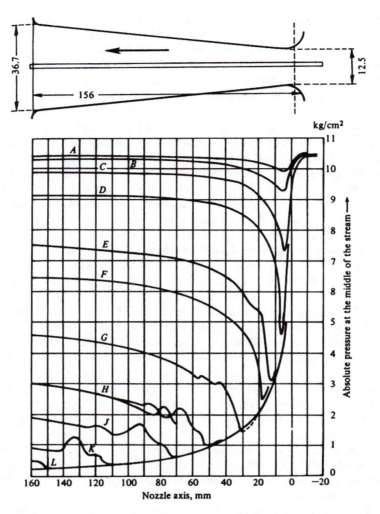

Figure 9.7 Pressure distributions through a supersonic nozzle measured by Stodola. (From Stodola.)[124]

The importance of Stodola in the early history of high-speed aerodynamics derives from his pioneering work on the flow of steam through Laval nozzles. As mentioned earlier, the possibility of supersonic flow in such nozzles, although theoretically possible, had not been demonstrated experimentally. To study that problem, Stodola constructed a convergent-divergent nozzle with the shape illustrated at the top of Figure 9.7. He could vary the back-pressure over any desired range by closing a valve downstream of the nozzle exit. With pressure taps in a long, thin tube extended through the nozzle along its centerline Stodola measured the axial-pressure distributions associated with different back-pressures. His data are shown below the nozzle sketch in Figure 9.7, and they provided the first experimental confirmation of the characteristics of a supersonic flow through a nozzle. In Figure 9.7, the lowest curve corresponds to a complete isentropic expansion. The curves D through

L correspond to a shock wave inside the nozzle, induced by higher back-pressures. The curves *A*, *B*, and *C* correspond to completely subsonic flow induced by high back-pressures. Regarding the large jumps in pressure shown by some of the data, Stodola commented that "I see in these extraordinary heavy increases of pressure, a realization of the 'compression shock' theoretically derived by von Riemann; because steam particles possessed of great velocity strike against a slower moving steam mass and are therefore compressed to a higher degree" (p. 63).[124] Stodola referred to the work of Riemann, but historically he would have been more accurate to have referred to Rankine and Hugoniot. Stodola's nozzle experiments and his data (Figure 9.7) represented a quantum jump in the understanding of supersonic nozzle flows. In conjunction with de Laval's contributions, Stodola's work provided the foundations for the aerodynamic developments discussed in this chapter.

Stodola died in Zürich in 1942 at the age of 83. He had become the world's leading authority on steam turbines, and his students were found throughout the Swiss companies that made steam turbines, the international leaders in that field. He had exceptional personal charm, and his students composed an almost disciple-like group during his long life in Zürich. Clearly, Stodola left a permanent mark in the history of compressible flow.

The final important figure in the early history of high-speed aerodynamics was the ubiquitous Ludwig Prandtl. We have already discussed Prandtl's creative influence in many areas of twentieth-century aerodynamics: boundary-layer theory, low-speed airfoil theory, and finite-wing lifting-line theory, to name only a few. It is not widely recognized by many students of aerodynamics that Prandtl made major contributions to the theory and understanding of compressible flows. In 1905 he built a small Mach-1.5 supersonic nozzle to study steam-turbine flows and the movement of sawdust in sawmills. For the next three years he continued to study the flow patterns associated with such supersonic nozzles. Figure 9.8 shows some striking photographs made in Prandtl's laboratory during that period that clearly illustrate a progression of expansion and oblique shock waves emanating from the exit of a supersonic nozzle. The dramatic contrast is that Prandtl was learning about

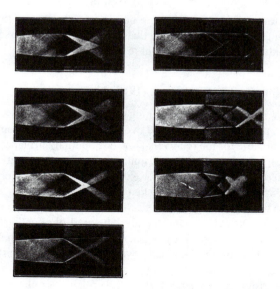

Figure 9.8   Various wave patterns in a supersonic nozzle, photographed by Prandtl (1908).

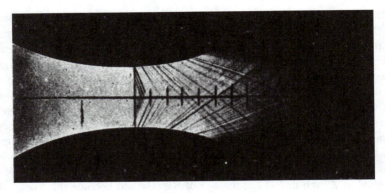

Figure 9.9   Flow in a supersonic nozzle, with Mach waves generated by an intentionally roughened surface. (From Meyer.)[125]

supersonic flows at the same time that the Wright brothers were introducing the world to powered flight at maximum velocities no greater than 40 mph.

Those observations of shock and expansion waves naturally prompted Prandtl to explore the theory they represented, and in 1908 in a doctoral dissertation,[125] Theodor Meyer, one of Prandtl's students at Göttingen, presented the first practical development of the relations for expansion waves and oblique shock waves – essentially the same theory taught in modern classes on compressible flows. Without fanfare, Meyer ended his paper with an impressive photograph of the internal flow within a supersonic nozzle (Figure 9.9). The walls of the nozzle had been intentionally roughened so that weak waves (essentially Mach waves) would be visible in the schlieren photograph – an amazing photograph for 1908, worthy of a modern supersonic laboratory.

The work of Prandtl and Meyer on expansion and oblique shock waves was contemporary with the normal-shock studies of Rayleigh and Taylor in 1910, as discussed earlier. So once again we are reminded of the importance of conducting basic research in areas that may appear unprofitable or irrelevant at the time. The practical value of Meyer's dissertation was not fully known until the advent of supersonic flight in the 1940s.

Compressibility Problems: The First Hints (1918–23)

From the time of the Wright Flyer to the beginning of World War II it was assumed that changes in air density were negligible as air flowed over an airplane. That assumption of *incompressible flow* was reasonable for the relatively slow (< 350 mph) airplanes of that era. In dealing with the theory, it was a considerable advantage to assume constant density, and physically the low-speed aerodynamic flows usually exhibited smooth variations, with no sudden changes or surprises. All of that changed when flight speeds began to edge up close to the speed of sound. At that point, aerodynamics theory had to begin to account for changes in air density in the flow field around the airplane, and physically the flow field sometimes acted erratically, frequently springing surprises that greatly challenged aerodynamicists. In the 1930s, all of those phenomena were tossed into one basket and called "compressibility problems."

The first hints of the compressibility problems that faster airplanes would face came during the era of the strut-and-wire biplane, and they came from only one part of the

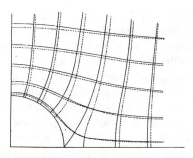

Figure 9.10 Compressibility effects on streamlines and equipotential lines for the flow over a circular cylinder. (From Bryan.)[126]

airplane: the propeller. Although typical flight speeds for World War I airplanes were less than 125 mph, the tip speeds of their propellers, because of their combined rotational and translational motions through the air, were quite large, sometimes exceeding the speed of sound. That was understood by aeronautical engineers at the time, and it prompted the British Advisory Committee for Aeronautics to take an interest in the theory of compressible flows. (The Advisory Committee for Aeronautics was created in 1909 by the British government to define the important problems in aeronautics and "to seek their solution by the application of both theoretical and experimental methods of research." Lord Rayleigh was appointed the first president. When the NACA was created in 1915, the British Advisory Committee for Aeronautics was used as a model.) In 1918 and 1919, G. H. Bryan, working for the committee at the Royal Aircraft Establishment, carried out theoretical analyses of subsonic and supersonic flows over a circular cylinder, a simple geometric shape chosen for convenience.[126,127] He was able to show that in a subsonic flow, the effect of compressibility was to displace adjacent streamlines farther apart. His data (Figure 9.10) revealed the streamlines and equipotential lines for the flow over a circular cylinder, where the flow was from right to left (only the first quadrant of his figure is shown, because the flow was symmetric about the horizontal and vertical axes through the center of the cylinder). The streamlines are the lines that essentially flow across the diagram from right to left; the equipotential lines (lines of constant velocity potential) are locally perpendicular to the streamlines. The dashed lines are for the incompressible-flow situation, and the solid lines are for the subsonic compressible-flow situation. The compressible-flow streamlines are displaced farther apart than those for the incompressible flow. Such displacement of the streamlines due to compressibility has subsequently been confirmed. In 1918, Bryan's findings represented new information, albeit of limited practical value. Also, his analysis was necessarily cumbersome and complex – a harbinger of the difficulties to come in dealing with compressible flows. Nevertheless, it was a beginning toward dealing with the problems of compressibility in propeller performance. Bryan was quite enthusiastic about his theory: "The methods of this Report thus appear to open up a wide field for future research and one in which there is every prospect of bringing theoretical investigations into close relationship with practical applications."[127] He was partly wrong, because nobody chose to continue to pursue the details of his cumbersome analysis. However, he was partly right, because his general approach – solving for an incompressible flow in the complex plane, and then modifying the incompressible streamlines to apply to a compressible flow – provided the seed for the later theoretical development of "compressibility corrections."

At the same time, Frank Caldwell and Elisha Fales, at the U.S. Army Air Service Engineering Division, McCook Field, Dayton, Ohio, were taking a purely experimental approach to the problem. That was the beginning of a divide between the British and American approaches to research on compressibility effects; over the next two decades the major experimental contributions to an understanding of compressibility effects would be made in the United States, principally by the NACA, and the major theoretical contributions would come from England. In 1918 Caldwell and Fales designed and built the first high-speed wind tunnel in the United States, solely to investigate the problems associated with propellers. Its velocity range was from 25 mph to a stunning 465 mph. It had a length of almost 19 ft, and the test section was 14 in. in diameter – a big, powerful machine for its day. They tested six different airfoils with thickness ratios (ratio of the maximum thickness to the chord length) from 0.08 to 0.2. At the higher speeds, the data showed "a decreased lift coefficient and an increased drag coefficient, so that the lift–drag ratio is enormously decreased."[128] They denoted the airspeed at which those dramatic departures took place as the "critical speed" (that would appear to have been the origin for the term "critical Mach number" that was to come into wide use in aerodynamics beginning in the late 1930s). The critical Mach number is defined as the free-stream Mach number at which sonic flow is first encountered on the surface of a body. The large drag rise due to compressibility effects normally occurs at a free-stream Mach number slightly above the critical Mach number; that is called the drag-divergence Mach number. Caldwell and Fales had reached and exceeded the drag-divergence Mach number in their experiments, and their introduction of the word "critical" in conjunction with that speed was the inspiration for its use in later coining the term "critical Mach number." Some of their data are shown in Figure 9.11, where the lift coefficient for an airfoil at an 8° angle of attack is plotted versus airstream velocity. Note the dramatic drop in lift coefficient at the "critical speed" of 350 mph – that is the compressibility effect. That plot and others like it for other angles of attack were the first published data on the adverse effects of compressibility on airfoils. Figure 9.11 shows a gradual decrease in the lift coefficient (denoted as K_y by Caldwell and Fales) as

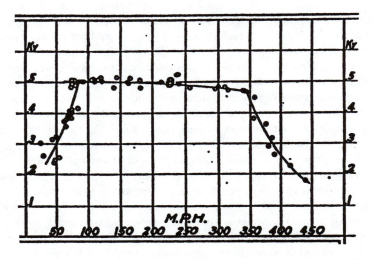

Figure 9.11   Lift coefficient versus velocity. (From Caldwell and Fales.)[128]

the velocity increased in the moderate speed regime, just before the precipitous drop at the critical speed. We know today that their graph was not correct; the effect of compressibility is to *increase* the value of the lift coefficient for increasing velocities below the critical speed (i.e., the middle part of the curve should have shown a gradual increase rather than a gradual decrease). That prompted me to make a closer study of their data-reduction method, from which it was clear that Caldwell and Fales made an error (an understandable error, because no one had experience in dealing with compressible-flow conditions in 1919) that caused their reported lift and drag coefficients to be about 10% too low at the higher speeds (the details of their error are discussed in Appendix I). However, their error did not compromise the dramatic discovery of the large increases in drag and decreases in lift when the airfoil sections were tested above the critical speed. Moreover, they were the first to show that the critical speed for thin airfoils was higher than that for thick airfoils (and thus by making the airfoil section thinner, one could delay the adverse compressibility effects to higher Mach numbers). That was an important finding that would have a lasting impact on designs for high-speed vehicles.

The fledgling NACA got into the act at that time as the publisher of the Caldwell and Fales data, carrying out its duty as stated in Public Law 271 "to supervise and direct the scientific study of the problems of flight, with a view to their practical solution, and to determine the problems which should be experimentally attacked, and to discuss their solution and their application to practical questions." The NACA was earmarking compressibility effects as a problem "which should be experimentally attacked."

The British were next to examine the effects of compressibility on propellers. In 1923, G. P. Douglas and R. M. Wood, two aerodynamicists at the Royal Aircraft Establishment, tested model propellers at high rotational speeds in a 7-ft low-speed wind tunnel (100-mph airstream) at the National Physical Laboratory in London.[129] They also carried out flight tests on a De Havilland D. H. 9A biplane, taking global measurements of the thrust and torque generated by the whole propeller, and so the details of the compressibility effects affecting the airfoil sections at the tip of the propeller were somewhat obscured. However, one of their conclusions anticipated the adverse effects of compressibility: "higher tip speeds than at present used will probably involve a serious loss of efficiency."[129]

Those first hints of compressibility effects on airfoils at high speeds should not have been any surprise, for a suggestion that aerodynamic forces could do strange things near Mach 1 had first been advanced by Benjamin Robins in 1742 (as in Chapter 3). On the basis of his ballistic-pendulum measurements, Robins observed that projectiles experienced a large increase in drag when the speeds approached the speed of sound. Specifically, he reported that the aerodynamic force began to vary as the velocity cubed (not squared as in the lower-speed case). At the beginning of the twentieth century, ballistics measurements on projectiles carried out by Bensberg and Cranz[130] in Germany yielded the variations of the drag coefficient with velocity shown in Figure 9.12, for the transonic and supersonic regimes. That curve shows the large rise in the drag coefficient near Mach 1 and the gradual decrease in the drag coefficient in the supersonic region. Although those data were obtained by ballistics engineers for use with artillery, they should have alerted aerodynamicists to the problems they should have expected to encounter with high-speed airplanes later in the century. But there was no reference to that work in the literature on high-speed aerodynamics early in the twentieth century; the aerodynamicists who carried out the pioneering work on compressibility effects acted as if they were unaware of such data.

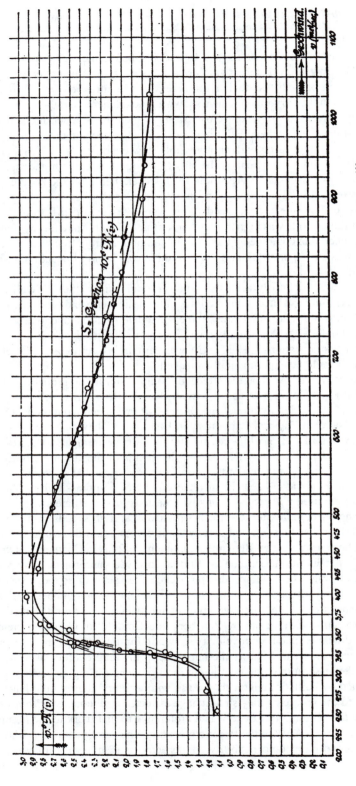

Figure 9.12 Projectile drag coefficient versus velocity in the transonic and supersonic regimes. (From Bensberg and Cranz.)[130]

The Compressibility Burble: NACA's Seminal Research (1924–9)

During the 1920s the NACA sponsored a series of fundamental experiments in high-speed aerodynamics at the Bureau of Standards, conducted by Lyman J. Briggs and Hugh L. Dryden. In 1919, at age 20, Dryden had a Ph.D. in physics from Johns Hopkins University (he would later become director of research for the NACA, 1947–58). That work progressed in three stages that covered the period from 1924 to 1929, the primary motivation being to understand the compressibility effects at the tips of propellers.

The first stage simply confirmed the trends already observed by Caldwell and Fales four years earlier. Briggs and Dryden, with the help of G. F. Hull of the U.S. Army Ordnance Department, jury-rigged a high-speed wind tunnel by connecting a vertical standpipe 30 in. in diameter and 30 ft high to a large centrifugal compressor at the Lynn works of the General Electric Company in Massachusetts.[131] At the other end of the pipe was a cylindrical orifice that served as a nozzle, 12.24 in. in diameter. With that device, "air speeds approaching the speed of sound were obtained." Unlike Caldwell and Fales, Briggs and Dryden used the proper equations for compressible flow to calculate the air velocity. Although not yet in the standard textbooks, those equations were known to Dryden as a result of his Ph.D. studies in physics. (The first engineering textbook in English to focus on compressible flow was by Liepmann and Puckett,[132] published in 1947.) Rectangular planform models, with a span of 17.2 in. and a chord length of 3 in., were placed in the high-speed airstream, and the lift, drag, and center of pressure were measured. The findings supported the earlier trends observed by Caldwell and Fales. In Particular, Briggs et al.[131] found that

(1) the lift coefficient for a fixed angle of attack decreased very rapidly as the speed increased,
(2) the drag coefficient increased rapidly,
(3) the center of pressure moved back toward the trailing edge, and
(4) the critical speed at which those changes occurred decreased as the angle of attack was increased and the airfoil thickness was increased.

In 1924, the effect of that work, as well as the research that had gone before, was to raise a red flag: Compressibility effects were a nasty lot, and they markedly degraded airfoil performance. But nobody had a fundamental understanding of the physical features of the flow field that were causing those adverse effects – nor would anyone for another decade.

An important step toward such a fundamental understanding came with the second stage of the work by Briggs and Dryden[133] in 1926. Because the compressor at the Lynn works was no longer available to them, they moved their experimental activity to the U.S. Army's Edgewood Arsenal, where they constructed another high-speed wind tunnel, much smaller, with an airstream only 2 in. in diameter. By careful design of the small airfoil models, two pressure taps could be placed in each model. Seven identical models were used, all with different locations of the pressure taps. Thirteen pressure-tap locations, seven on the upper surface and six on the lower surface, were employed (for the reader who is counting, the seventh model had only one tap). With that technique, Briggs and Dryden measured the pressure distributions over the airfoil at Mach numbers from 0.5 to 1.08. The findings were dramatic: Beyond the critical speed, the pressure distributions over the top of the airfoil exhibited a sudden pressure jump at about one-third to one-half the distance from the leading edge, followed by a rather long plateau toward the trailing edge. Such a pressure plateau was familiar (it was similar to that over the top surface of an airfoil in a low-speed flow when the airfoil stalls at a high angle of attack), and it was well known at that time that airfoil stall was caused by separation of the flow off the top surface of the airfoil. Briggs and Dryden concluded that the adverse effects of compressibility were caused by flow separation over the top surface, even though the airfoil was at low (even zero) angles of attack. To substantiate that, they conducted oil-flow tests: An oil, with pigment added to make it visible, was painted on the model surface, and when the model was placed in the high-speed airstream, the telltale line of flow separation was revealed by the oil pattern. Clearly, beyond the critical speed, flow separation was occurring on the top surface of the airfoil. But what was causing the flow to separate? The answer to that question was eight years in the future.

The third stage of the work by Briggs and Dryden was utilitarian, in keeping with the stated duty of the NACA to work toward practical solutions. At the end of the 1920s they carried out a large number of detailed measurements of the aerodynamic properties of 24 different airfoils at Mach numbers from 0.5 to 1.08.[134] The airfoils tested were those conventionally used by the army and navy for propellers: the standard family of British-designed RAF airfoils and the American-designed Clark Y family. Their data were the first definitive measurements to show compressibility effects on the standard series of airfoils.

By the time of World War I, aerodynamicists were well aware that an airfoil would stall at a high angle of attack because the flow would separate from the top surface. The resulting drastic loss of lift was termed "lift burble." When Briggs and Dryden found that the drastic loss of lift at high speeds, beyond the critical speed, was also due to flow separation, it was natural to call that effect the "compressibility burble," and that NACA terminology was used in the literature on high-speed aerodynamics throughout the 1930s.

The First Theoretical Compressibility Correction: The Prandtl-Glauert Rule

Progress toward theoretical solutions for the compressibility effects in a high-speed subsonic flow was virtually nonexistent during the 1920s. The only major contribution was that by the British aerodynamicist Hermann Glauert, who rigorously derived a correction to be applied to the lift coefficient for low-speed, incompressible flows in order to correct it for compressibility effects.[135] That was the first in a series of theoretical rules called "compressibility corrections." As discussed in Chapter 7, Glauert learned the details of the circulation theory of lift during visits with Prandtl in the early 1920s and then wrote the first book in English to clearly describe that theory.[80]

Hermann Glauert was born in Sheffield, England, October 4, 1892. He was well educated, first at the King Edward VII School at Sheffield, and then later at Trinity College, Cambridge, where he won many honors for his leadership in the classroom. For example, he was awarded the Ryson Medal for astronomy in 1913, an Isaac Newton Scholarship in 1914, and the Rayleigh Prize in 1915. In 1916, as World War I dragged on, Glauert joined the staff of the Royal Aircraft Establishment (RAE) at Farnborough, where he quickly grasped the fundamentals of aerodynamics and wrote numerous reports and memoranda dealing with airfoil and propeller theory, the theory of the autogyro, and the performance, stability, and control of airplanes. In 1926 he published *The Elements of Aerofoil and Airscrew Theory*,[80] the single most important instrument for spreading Prandtl's airfoil and wing theories in the English-speaking world, and still used as a reference in courses dealing with incompressible flows. By the early 1930s, Glauert was one of the leading theoretical aerodynamicists in England, having become the principal scientific officer of the RAE and head of its Aerodynamics Department. When he died in an accident in 1934, the world lost one of its best aerodynamicists.

After Bryan initiated the theoretical study of subsonic compressible flows in 1918–19 at the Royal Aircraft Establishment, with little practical effect, Glauert took up the initiative eight years later and brought in some very practical findings. To understand the nature of Glauert's contribution, consider that by the late 1920s a relatively large body of theoretical and experimental data on low-speed, incompressible flows had already been accumulated. When faced with the emerging reality of high-speed flight, where compressibility would be an important concern – first for the tips of propeller blades and later for the airplane itself – aerodynamicists had two choices: (1) scrap all the existing data on incompressible flows, and compile a new data base for compressible flows, or (2) try to modify the existing incompressible-flow data to take into account the effects of compressibility. Glauert's work was in the latter category. Deriving a transformation that related a subsonic compressible flow to a corresponding incompressible flow over the same airfoil shape, in 1927 Glauert obtained the following amazingly simple equation:[135]

$$C_L = \frac{C_{L,0}}{\left(1 - M_\infty^2\right)^{1/2}}$$

where $C_{L,0}$ is the incompressible-flow lift coefficient for an airfoil, and C_L is the corresponding lift coefficient for compressible flow over the same airfoil in a free stream with a Mach number M_∞. That derivation was based on an approximation that reduced the governing nonlinear Euler equations for inviscid fluid motion to a much simpler linear equation. That approximation assumed that the airfoil would introduce only relatively small disturbances in an otherwise uniform free stream, and thus the applications were limited to slender bodies at small to moderate angles of attack. However, most standard airfoils at normal cruise conditions for airplanes satisfied that constraint, and the foregoing equation applied to such airfoils. Also, the approximate, linearized form of that equation was reasonably valid only up to M_∞ values of about 0.7. Nevertheless, in 1927, that equation, today called the Prandtl-Glauert rule, was a major advance in the theoretical analysis of compressible subsonic flows.

The attachment of Prandtl's name to that rule is worth a comment. Students at Göttingen reported that Prandtl had somehow obtained that equation in the early 1920s and had discussed it in a few of his lectures. However, Prandtl never published it nor its derivation. Glauert was not aware of any work by Prandtl on that problem and certainly did not collaborate with Prandtl on the development of the Prandtl-Glauert rule. Rather, Glauert worked totally independently and in 1927 was the first to publish the rule and its logical derivation from the established governing equations for fluid flow, first in an RAE report[135] in September 1927 and then in the *Proceedings of the Royal Society* in the following year. In any event, to recognize Prandtl's association with that equation, it has been handed down in the aerodynamic literature as the Prandtl-Glauert rule.

The Prandtl-Glauert rule is a "compressibility correction" – a means to modify the existing incompressible-flow data for compressibility effects. In the 1920s and well into the 1930s it was the only such means. It is also an example of how the development of aerodynamics theory for high-speed, subsonic compressible flows was centered in Europe, specifically in Germany and England. In contrast, compressibility research in the United States during the same period was completely experimental. That is not to say that experimental research was totally lacking in Europe.

Early British Experimental Work on Compressibility Effects

Caldwell and Fales, in the United States, prompted by the propeller problem, carried out the first experiments to clearly delineate the adverse effects of compressibility on airfoils. The British faced the same problem, and the Royal Aircraft Establishment began compressibility experiments in 1922, when G. P. Douglas and R. M. Wood[129] carried out propeller tests in still air, with propeller tip speeds slightly beyond the speed of sound. From their measurements of propeller performance (thrust and torque) they extracted values for the lift and drag coefficients for their propeller airfoil sections. Some of their data, published in 1923, are shown in Figure 9.13, where the lift coefficient k_L is plotted versus the quantity $2\pi nrD/a$, where n is the propeller rotation in revolutions per second, D is the propeller diameter, r is the location of the airfoil section from the propeller hub nondimensionalized by $D(r = 0.4$, which corresponds to a location 80% of the distance from the hub to the tip), and a is the speed of sound. In Figure 9.13, when the quantity $2\pi nrD/a = 0.8$, the speed of the propeller tip was at the speed of sound. Note that k_L for a fixed propeller-blade angle α increased with speed up to a certain value, namely, that speed where the tip went supersonic, and then decreased with further increases in speed. The increase in k_L with velocity below the critical speed was the trend caused by subsonic compressibility effects; the Prandtl-Glauert rule predicts the same trend of an increasing lift coefficient with an increase in Mach number. Recall that the Caldwell and Fales date (Figure 9.11) did not show that trend, because of the error in the reduction of their data. Thus the British tests in 1923 yielded the first experimental data to show the increase in the lift coefficient below the critical speed. The same trend was measured one year later by Briggs et al.[131] Of more importance, of course, was the precipitous drop in the lift coefficient beyond some critical speed, a finding first reported in 1920 by Caldwell and Fales,[128] reinforced by Douglas and Wood[129] in 1923 at the RAE and by Briggs et al.[131] in 1924 with the NACA-sponsored work at the Bureau

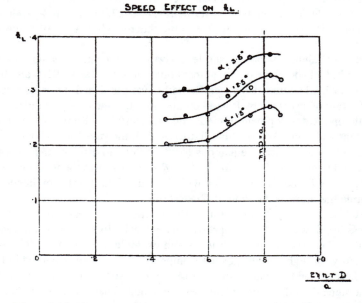

Figure 9.13 Compressibility effects on the lift coefficient. (From Douglas and Wood.)[129]

of Standards. In all of that early compressibility research, more attention was given to the precipitous drop in the lift coefficient beyond the critical speed than to the corresponding increase in the drag coefficient, most likely because of the focus on propeller performance, where the drop in the lift coefficient meant a precipitous drop in propeller thrust.

The RAE at Farnborough expanded its compressibility testing with a series of propeller tests in its 7-ft wind tunnel: Two-blade propellers, 2 ft in diameter, were tested in a 100-mph test stream in the tunnel, with the model propellers being driven at high rotational velocities by a high-speed air turbine. Those wind-tunnel tests were conducted by G. P. Douglas and W. G. A. Perring,[136–138] who carried out an exhaustive series to measure the variations of the lift and drag coefficients for each propeller airfoil section as functions of velocity. Some typical data (Figure 9.14) show the variations of the lift coefficient (upper graph) and the drag coefficient (lower graph) with section speed, with the shape of the biconvex airfoil section at the bottom. Those data provided the clearest picture up to that time of the dramatic decrease in the lift coefficient and the precipitous increase in the drag coefficient due to high-speed compressibility effects (i.e., due to the "compressibility burble").

The lengths to which people went to obtain data in the compressible-flow regime were exemplified, in the extreme, by the novel and relatively forgotten approach taken in 1928 by Geoffrey I. Taylor, the leading fluid dynamicist in England for the first half of the twentieth century. We mentioned him earlier in conjunction with shock-wave theory. A professor at Cambridge University, Taylor was described by Rouse and Ince[2] as follows: "England's vast series of 19th-century contributions to the analysis of fluid motion were matched in quality in the early 20th century by the work of Geoffrey Ingram Taylor, a meteorologist at the University of Cambridge." Taylor's work covered many topics in fluid flows, from compressible flows (including supersonic flows) to meteorology. He is perhaps best known for his papers over two decades on the fundamental analysis of turbulence by statistical methods.

For our purposes here, we refer to the work done by Taylor for the RAE toward solving compressible subsonic flows. Taylor combined theory and experiment, using an electrical analogy to correlate the equations for compressible, irrotational flow with the equations for "electric flow" in a conservative electromagnetic field. He saw an analogy "between the flow of a fluid of variable density in two dimensions and the flow of electric current in a conducting sheet of variable thickness."[139] Taylor proceeded to set up an experiment consisting of a copper sulfate solution in a tank of shallow but variable depth; the solution was therefore a conducting sheet of variable thickness. The bottom of the tank was made of paraffin wax that could be easily carved into the desired shape. A two-dimensional body (an airfoil and a circular cylinder made of nonconducting material were used in various experiments) was inserted in the conducting solution, and an electric field was imposed across the conducting sheet. When the local thickness of the conducting sheet was made proportional to the local density in the analogous compressible flow, the intensity of the electric force in the conducting sheet would be proportional to the velocity in the analogous compressible flow, and the direction of the greatest electric intensity (the lines of flow of the electric current) would be analogous to the streamlines in the compressible flow. To arrive at the proper variation of the depth of the copper sulfate solution so that the analogy would hold, Taylor set up a series of iterative measurements. Starting with a flat bottom for the tank, he measured the field of electric flow, which in that case corresponded to an incompressible (constant-density) flow. A measurement of the electric-field intensity gave the corresponding velocity field in the flow. Assuming irrotational flow, that velocity field corresponded to a certain pressure field, which, assuming adiabatic conditions, also corresponded to a certain density

ANALYSIS FOR LIFT COEFFICIENT FINE AND COARSE PITCH AIRSCREWS.

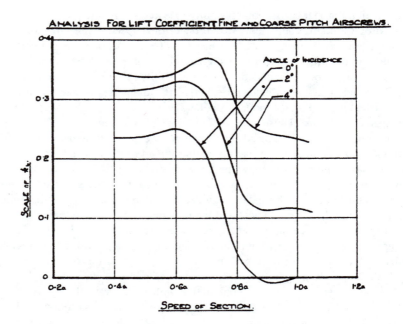

ANALYSIS FOR DRAG COEFFICIENT FINE AND COARSE PITCH AIRSCREWS.

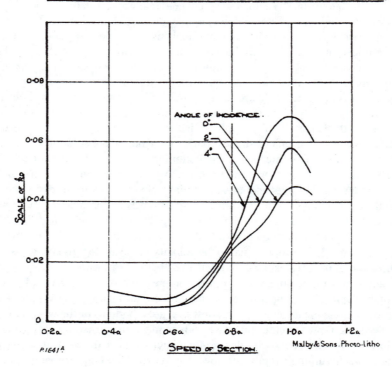

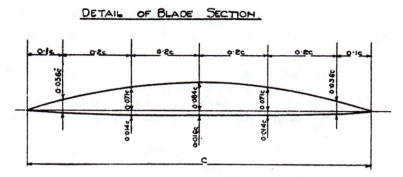

Figure 9.14   Compressibility effects on the lift and drag coefficients. (From Douglas and Perring.)[137]

field. However, that density field would not be quantitatively accurate, because the bottom of the tank was flat (a conducting sheet of constant thickness). Thus, the paraffin-wax bottom was carved to a contour such that the local depth would be proportional to the density obtained earlier. The electric field would then be imposed again, and a new field of electric intensity measured. That would yield an analogous new density field, which would then allow the wax bottom to be carved again. That process was repeated until convergence was achieved. The final result gave the proper density field for the flow (as reflected by the final local variations in the depth of the tank) and the streamline directions and velocity field for the flow (as reflected by the electric-field direction and intensity).

It appears that Taylor's technique was never used by anyone else. It was too limited in scope and was soon supplanted by more conventional testing in high-speed wind tunnels. However, there was one important result from Taylor's experiments: an approximate measurement of the critical Mach number for a circular cylinder. The critical Mach number is the free-stream Mach number at which sonic flow is first encountered at some point on the body surface:

> One of the most important results of the present work is the discovery that the failure of convergence occurs not when the speed of the main body of the stream reaches that of sound but at some lower speed. In the case of a circular cylinder convergence fails *when the maximum velocity in the field reaches the speed of sound in the air at that point.* This first occurs when the speed of the stream is between 0.4 and 0.5 of that of sound. *At a speed of 0.5 of that of sound it appears that no continuous irrotational motion is possible past a circular cylinder* [Taylor's italics].[139]

Taylor was nearly right. Today we know that the critical Mach number for a circular cylinder is 0.404, and for free-stream Mach numbers above that value, shock waves will appear in the locally supersonic pockets at the top and bottom of the cylinder, destroying the irrotationality of the flow. Coming in 1928, Taylor's finding was an important step toward a fundamental understanding of the compressible flow over an aerodynamic body.

Despite those experiments in Britain, the center of gravity for experimental investigations of subsonic compressibility effects was in the United States, with the work sponsored by the NACA. That became particularly clear in the early 1930s, when the NACA began testing in a specially designed high-speed wind tunnel at the Langley Memorial Laboratory.

John Stack and the NACA Compressible-Flow Research in the 1930s

By 1928, two of the three elements essential for a fundamental understanding of the high-speed compressible flow over an airfoil were in hand:

(1) From the work of Caldwell and Fales in 1920, and the investigations that followed, it was clear that something dramatic happened to the aerodynamics of an airfoil when the free-stream velocity approached the speed of sound: Beyond some "critical speed," the lift would drop precipitously, and the drag would suddenly and rapidly increase.

(2) The work of Briggs and Dryden in 1926 showed that those precipitous changes corresponded to a sudden separation of the flow over the airfoil surface, even at low angles of attack.

But what caused the flow to separate? The answer to that question would provide the third element, but the breakthrough to that understanding would take another six years.

In July 1928, John Stack, born and raised in Lowell, Massachusetts, began his career with the NACA Langley Memorial Laboratory. Having just graduated from the Massachusetts Institute of Technology (M.I.T.) with a B.S. degree in aeronautical engineering, he was assigned to the Variable Density Tunnel (VDT), the world's premier wind tunnel at that time. Stack had long been dedicated to aeronautical interests. While in high school, he worked to earn enough money for a few hours of flight instruction in a Canuck biplane, and he helped out with the maintenance of a Boeing biplane owned by one of his part-time employers. He had made up his mind to study aeronautical engineering, but his father, a carpenter who was also very successful in real estate, wanted him to study architecture at M.I.T.[140] When Stack entered M.I.T., he enrolled in aeronautical engineering, keeping it a secret from his father for the first year, but with the approval of his mother. Much later, Stack commented that "when Dad heard about it, it was too late to protest."

When John Stack arrived at the NACA Langley Memorial Laboratory in 1928, a year's worth of design work had already been done on Langley's first high-speed wind tunnel, and the facility was already operational, with an open-throat test section. Because of the success of Briggs and Dryden in studying compressibility effects, and because a few visionaries recognized the importance of research into high-speed aerodynamics, in 1927 Joseph S. Ames, president of Johns Hopkins University and new chairman of the NACA, assigned a higher priority to research with high-speed wind tunnels.[93] Eastman Jacobs was the chief designer of the open-throat 11-in. High-Speed Tunnel at Langley. (Jacobs, as discussed in Chapter 8, designed the NACA airfoil sections in the 1930s and the NACA laminar-flow airfoils just before the beginning of World War II.) An innovative feature of the 11-in. High-Speed Tunnel was that it was driven from the 20-atm pressure tank of the Langley Variable Density Tunnel (VDT). To change a model in the VDT, the 20-atm tank that encased the entire tunnel had to be blown down to 1 atm, which was a waste of an energy source that the Langley engineers realized could be used for the 11-in. High-Speed Tunnel. The 5,200-ft^3 capacity of the high-pressure tank allowed about 1 min of operation for the 11-in. tunnel. Stack was given the responsibility for upgrading the High-Speed Tunnel by designing a closed throat, and that improved facility (Figure 9.15) became operational in 1932. It was his participation in the design and development of the 11-in. High-Speed Tunnel that launched Stack's career in high-speed aerodynamics.

While Stack was working on the High-Speed Tunnel, he was impressed by an event in England that would lead to a rapid refocusing of the NACA high-speed research program:

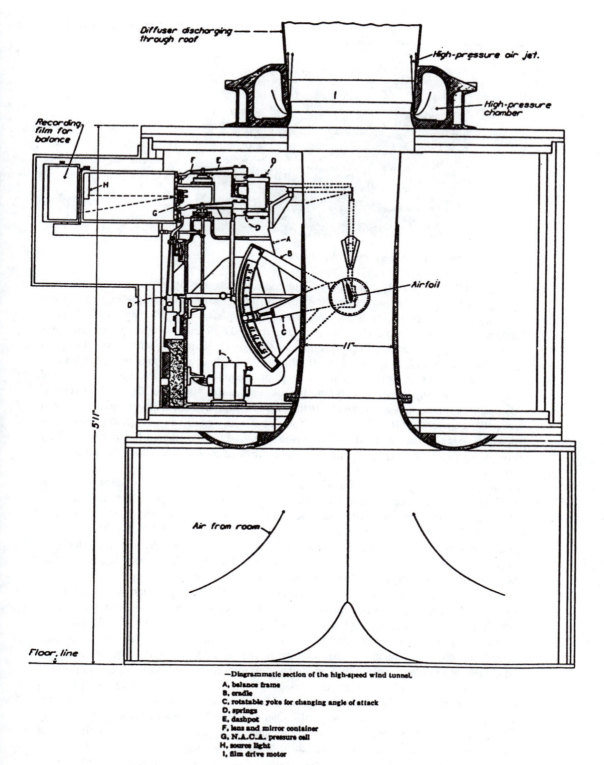

Diffuser discharging through roof

High-pressure air jet.

High-pressure chamber

Recording film for balance

Airfoil

Air from room

5'11"

Floor line

—Diagrammatic section of the high-speed wind tunnel.

A, balance frame
B, cradle
C, rotatable yoke for changing angle of attack
D, springs
E, dashpot
F, lens and mirror container
G, N.A.C.A. pressure cell
H, source light
I, film drive motor

Figure 9.15   Closed-throat High-Speed Tunnel at NACA Langley (1932).

Figure 9.16 Supermarine S.6B (1931).

On September 13, 1931, a highly streamlined airplane, the Supermarine S.6B, flashed through the clear afternoon sky near Portsmouth, along the southern English coast. Piloted by Flight Lieutenant John N. Boothman, that racing airplane averaged a speed of 340.1 mph around a long, seven-lap course, winning the coveted Schneider Trophy permanently for Britain. Later that month, Flight Lieutenant George H. Stainforth set the world's speed record of 401.5 mph in the same S.6B (Figure 9.16). It does not take an aerodynamics expert to appreciate that by 1931 the concept of streamlining in order to reduce drag had taken root: The Supermarine S.6B simply *looks* fast, and at 400 mph (Mach 0.53, over half the speed of sound), it was. Suddenly, in the face of that kind of speed, the prior concern over propeller compressibility effects, which for propeller tips posed an important but tolerable problem, became transformed into an absolutely vital concern about the compressibility effects on the entire airplane, and the complexities of those effects raised a problem of showstopping proportions.

Stack was acutely aware of the new compressibility challenge. In 1933 he published the first data to come from the newly modified, closed-throat High-Speed Tunnel. Although the airfoils tested were propeller sections, Stack obviously had the Schneider Trophy racer in mind: "A knowledge of the compressibility phenomenon is essential, however, because the tip speeds of propellers now in use are commonly in the neighborhood of the velocity of sound. Further, the speeds that have been attained by racing airplanes are as high as half the velocity of sound. Even at ordinary airplane speeds the effects of compressibility should not be disregarded if accurate measurements are desired."[141] For the most part, Stack's data in 1933 confirmed the trends observed earlier. His measurements of the variations of the lift, drag, and moment coefficients with Mach number for a Clark Y airfoil 10% thick are shown in Figure 9.17; the precipitous drop in lift and the large increase in drag at high speeds are clearly evident. He also confirmed that when there were increases in airfoil thickness or angle of attack or both, the adverse compressibility effects began to appear at lower Mach numbers. One of his conclusions reflected the theory of the Prandtl-Glauert compressibility correction discussed earlier: His measurements indicated that "the limited theory available may be applied with sufficient accuracy for most practical purposes only for speeds below the compressibility burble," presaging almost 40 years of a theoretical void. The aerodynamic equations applicable to the transonic flight regime (Mach numbers between about 0.8 and 1.2) were nonlinear partial differential equations that defied solution until the 1970s, and even then the solution was by brute force: numerical solutions using

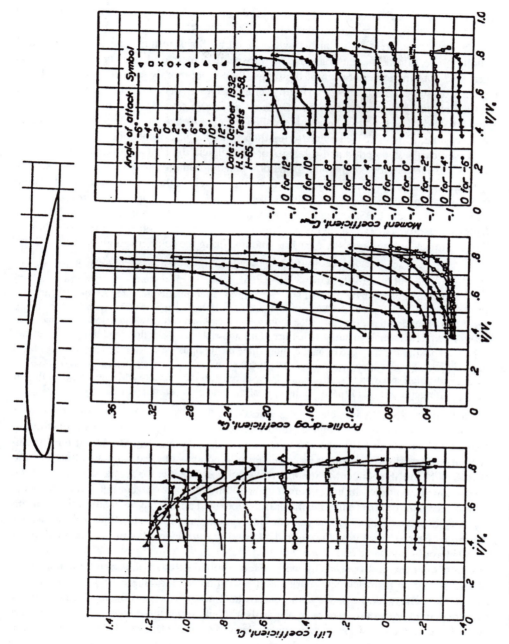

Figure 9.17 Compressibility effects on the lift, drag, and moment coefficients for a Clark Y airfoil 10% thick, measured by Stack (1933).

the power of the newly developed discipline of computational fluid dynamics carried out on high-speed digital supercomputers.

The term "compressibility burble" was coined by Stack in the same NACA technical report: "The lift coefficients increase as the speed is increased, slowly as the speed is increased over the lower portion of the range, then more rapidly as speeds above half the velocity of sound are exceeded, and finally at higher speeds, depending on the airfoil section and the angle of attack, the flow breaks down as shown by a drop in the lift coefficient. This breakdown of the flow, hereinafter called the *compressibility burble,* occurs at lower speeds as the lift is increased by changing the angle of attack of the model."[141]

Driven by the conviction and foresight of John Stack, the NACA continued to alert the worldwide aeronautics community to the problems of compressibility effects. In January 1934, the first significant professional aeronautical society in the United States, the Institute of Aeronautical Sciences, began publishing its *Journal of the Aeronautical Sciences,* which contained an article by Stack emphasizing what would be an important NACA theme for the next several decades:

> The effects of compressibility have commonly been neglected because until the relatively recent development of the last Schneider trophy aircraft the speeds have been low as compared with the velocity of sound, and the consequent local pressures over the surfaces of high speed airplanes have differed but slightly from atmospheric pressure. At the present time, however, the speeds associated with the fastest airplanes approach 60 percent of the velocity of sound, and the induced velocities over their exposed surfaces lead to local pressures that differ appreciably from the pressure of the atmosphere. When this condition exists, air can no longer be regarded as an incompressible medium. The effects of compressibility on the aerodynamic characteristics of airfoils have been under investigation by the N.A.C.A. in the high speed wind tunnel, and it is the purpose of this paper to examine the possibility of further increases in speeds in the light of this relatively recent research.[142]

By that time the NACA was clearly the world's leading research institution in the area of compressibility effects. Through its influence and sponsorship of the experiments in the 1920s by Caldwell and Fales at McCook Field and by Briggs and Dryden at the Bureau of Standards, and more recently its own carefully conducted experiments at Langley, the NACA had been able to identify the first two elements essential for an understanding of the basic nature of compressibility effects: (1) the finding that above a certain "critical speed," the lift decreased dramatically, and the drag skyrocketed almost beyond comprehension and (2) the finding that such behavior was caused by a sudden, precipitous flow separation over the top surface of the wing or airfoil. The missing third element was an explanation for such behavior.

In 1934, Stack and the NACA were able to provide that explanation. By that time, Stack had a new instrument to work with: a schlieren photographic system, an optical arrangement that made the density gradients in the flow visible. One of nature's mechanisms for producing very strong density gradients is a shock wave, and hence such waves would be visible with the schlieren system. Stack's boss, Eastman Jacobs, was familiar with such optical systems through his hobby of astronomy, and he suggested to Stack that the use of a schlieren system might make some of the unknown features of the compressible flow over an airfoil visible and might shed some light on the nature of the compressibility burble. It did just that, and more. With the 11-in. High-Speed Tunnel running above the critical speed for the NACA 0012 symmetric airfoil mounted in the test section, and with the aid of the schlieren system, Stack and Jacobs recorded the first observation of a shock wave

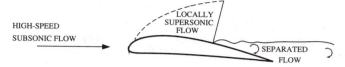

Figure 9.18 The physical nature of the transonic flow over an airfoil.

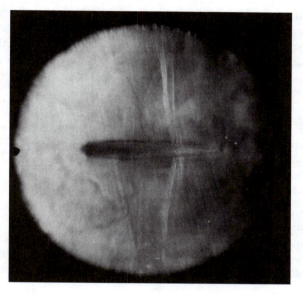

Figure 9.19 A schlieren photograph showing the existence of shock waves in the transonic flow over an airfoil. (From John Stack Files.)[143]

in the flow over the top surface of the airfoil. It was immediately clear that the separated flow over the top surface of the airfoil and the resulting compressibility burble, with all its adverse consequences, were caused by the presence of a shock wave. The nature of that flow is sketched in Figure 9.18: When the free-stream velocity is high enough, the rapid expansion of the flow over the airfoil creates a local pocket of supersonic flow over the top surface of the airfoil. That pocket is terminated by a shock wave. In turn, the shock wave interacts with the thin, friction-dominated boundary layer adjacent to the surface of the airfoil and causes the boundary layer to separate in the region where the shock impinges on the surface. A massive region of separated flow trails downstream, greatly increasing the drag and decreasing the lift. One of the pioneering schlieren pictures of the flow over the NACA 0012 airfoil taken by Stack in 1934 is shown in Figure 9.19; the quality is poor by modern standards, but it is certainly sufficient for identifying the phenomena. For that case, which was flow over a symmetric airfoil at a zero angle of attack, shocks appeared on both the upper and lower surfaces of the airfoil. That photograph reflected an important discovery in the history of high-speed aerodynamics, one that led to a complete understanding of the physics of the compressibility burble – a breakthrough of enormous theoretical and practical importance. It was due to the work of two innovative aerodynamicists at the Langley Memorial Laboratory, John Stack and Eastman Jacobs, operating in a creative atmosphere promoted throughout the NACA, made possible by the foresight of Joseph Ames and George Lewis at NACA Headquarters in Washington, who accorded the NACA high-speed research program a high priority at a time when most airplanes were lumbering along at 200 mph or less.

Almost any new scientific discovery will encounter some initial skepticism, and Theodore Theodorsen was skeptical about Stack's findings. As discussed in Chapter 8, Theodorsen was the leading theoretical aerodynamicist in the NACA, with a worldwide reputation for his pioneering papers on airfoil theory. John Becker, who joined the NACA in 1936 and went on to become one of the most respected high-speed aerodynamicists at Langley, reported Theodorsen's reaction to the schlieren photographs taken by Stack, illustrating that the new findings constituted a radical departure from the expected norm:

> The first tests were made on a circular cylinder about 1/2 inch in diameter, and the results were spectacular in spite of the poor quality of the optics. Shock waves and attendant flow separations were seen for the first time starting at subsonic stream speeds of about 0.6 times the speed of sound. Visitors from all over the Laboratory, from Engineer-in-Charge H. J. E. Reid on down, came to view the phenomena. Langley's ranking theorist, Theodore Theodorsen, viewed the results skeptically, proclaiming that since the stream flow was subsonic, what appeared to be shock waves was an "optical illusion," an error in judgement which he was never allowed to forget.[144]

The fifth Volta conference was held in Italy in 1935, affording the NACA a timely opportunity to inform the international research community of the breakthrough to a fundamental understanding of compressibility effects and the compressibility burble. That conference was the most important assemblage of aerodynamicists in the early years of high-speed aerodynamics.

The 1935 Volta Conference: Threshold to Modern High-Speed Aerodynamics and the Concept of the Swept Wing

Because of the rapidly growing interest in high-speed flight, by 1935 it was time for an international meeting of those few fluid mechanicians dealing with compressible flows. The time was right, and in Italy the circumstances were right: Since 1931 the Royal Academy of Science in Rome had been conducting a series of important scientific conferences sponsored by the Alessandro Volta Foundation. (Alessandro Volta was the Italian physicist who invented the electric battery in 1800, and the unit of electromotive force, the volt, was named in his honor.) The first conference dealt with nuclear physics, and subsequent conferences dealt with the sciences and the humanities in alternate years. The second Volta conference carried the title "Europe," and in 1933 the third conference dealt with immunology, followed by "The Dramatic Theater" in 1934. During that period, Italian aeronautics was gaining momentum, led by General Arturo Crocco, an aeronautical engineer and the father of Luigi Crocco, a leading aeronautical scientist in the middle of the twentieth century. (Luigi was responsible for Crocco's theorem, an important correlation of the entropy change, rotationality, and energy loss or gain in a compressible flow, and an integral part of modern compressible-flow analysis.) General Crocco had become interested in ramjet engines in 1931 and therefore was well aware of the importance of compressible-flow theory and experiment in the future of aviation. The topic chosen for the fifth Volta conference was "High Velocities in Aviation." Participation was by invitation only, and because of the prestige of the conference and the rapidly increasing interest in the subject matter, the participants were careful in preparing their papers. On September 30, 1935, the major figures in the development of compressible-flow theory gathered in Rome (Theodore von Kármán and Eastman Jacobs from the United States, Ludwig Prandtl and Adolf Busemann from Germany, Jakob Ackeret from Switzerland, Geoffrey I. Taylor from England, Crocco and Enrico Pistolesi from Italy, and many more) for the fifth Volta conference, which would open up the newly established theory of compressible flows to practical applications in the decades to come.

The technical content at that Volta conference ranged from subsonic flows to supersonic flows and from experimental testing to theoretical considerations. Prandtl gave a general introduction and survey paper on compressible flows, showing many schlieren pictures for illustration. Taylor discussed the theory of supersonic conical flows, and von Kármán presented research on minimum wave-drag shapes for axisymmetric bodies. The linearized Prandtl-Glauert relation was once again derived and presented by Enrico Pistolesi, along with several higher-order calculations for compressibility corrections. Jakob Ackeret read a paper on different designs for subsonic and supersonic wind tunnels. There were also presentations on propulsion techniques for high-speed flight, including rockets and ramjets. There was a field trip to the new Italian aerodynamics research center at Guidonia, near Rome. Guidonia was equipped with several high-speed wind tunnels, subsonic and supersonic, all designed in consultation with Ackeret. That laboratory would produce a large body of experimental supersonic data before and during World War II and would produce from its ranks a leading aerodynamicist, Antonio Ferri.[145]

Because of his work in the design and testing of the NACA four-digit airfoil series, and the fact that he was the section head for the NACA VDT, which had put the NACA on the international aerodynamics map in the 1920s, Eastman Jacobs also received an invitation to the Volta conference. He took the opportunity to present a paper on the new NACA compressibility research carried out under his supervision by John Stack.[146] Eastman Jacobs ably represented the NACA. His paper derived and presented the basic equations for compressible flow assuming no friction and no thermal conduction. Then he described the NACA High-Speed Tunnel, the schlieren system, and the airfoil experiments carried out in the tunnel. Then he showed, for the first time in a technical meeting, some of the schlieren pictures that he and Stack had taken (e.g., Figure 9.19). In keeping with the NACA's penchant for perfection, especially in its publications, Jacobs apologized for the quality of the photographs, a fault that detracted little from their technical and historical importance: "Unfortunately the photographs were injured by the presence of bent celluloid windows forming the tunnel walls through which the light passed. The pictures nevertheless give fundamental information in regard to the nature of the flow associated with the compressibility burble."[146] With that, the NACA high-speed research program was not simply on the map, it was leading the pack.

One of the most farsighted and important papers at the fifth Volta conference was presented by Adolf Busemann (Figure 9.20): "Aerodynamischer Auftrieb bei Ueberschallgeschwindigkeit" (Aerodynamic Forces at Supersonic Speeds), which introduced the concept of the swept wing as a means to reduce the large drag increase encountered beyond the critical Mach number. Busemann reasoned that the flow over a wing was determined mainly by the component of velocity perpendicular to the leading edge. If the wing were swept, that component would decrease (Figure 9.21). Consequently, the free-stream Mach number at which the large rise in drag would be encountered would be increased. Therefore, airplanes

Figure 9.20   Adolf Busemann.

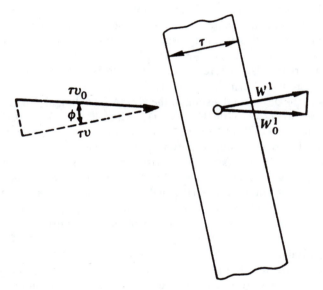

Figure 9.21   Busemann's diagram for a swept wing (1935).

with swept wings could fly faster before encountering the drag-divergence phenomenon. Busemann's swept-wing concept is reflected in the vast majority of high-speed aircraft in operation today.

The fifth Volta conference was accorded special treatment by the Italian government. Its prestige was reflected in its location: It was held in an impressive Renaissance building that had served as the city hall during the Holy Roman Empire. Moreover, the Italian dictator Benito Mussolini chose the conference as the place to make his announcement that Italy had invaded Ethiopia – a curious setting for such a political announcement.

The conference fueled the excitement about the future of high-speed flight and provided the first major international exchange of information on compressible flows, but in many areas its impact was delayed. For example, Busemann's ideas on swept wings were withdrawn from public view (the German Luftwaffe recognized their military significance and classified such work in 1936). The Germans went on to produce a large body of swept-wing research during World War II and the first operational jet airplane, the Me-262, which had a moderate degree of sweep (although the principal design reason was for center-of-gravity location). After the war, technical teams from England, Russia, and the United States descended on the German research laboratories at Peenemünde and Braunschweig and gathered all the data they could find. The United States also gathered Adolf Busemann, who was moved to the NACA Langley Memorial Aeronautical Laboratory. Later, Busemann became a professor at the University of Colorado. Virtually all modern high-speed airplanes can trace their lineage back to the original data obtained in Germany and ultimately to Busemann's paper at the fifth Volta conference.

On a more positive note, the Volta conference led to some increase in high-speed research in the United States. There were renewed efforts at the NACA to obtain data on the compressibility effects on high-speed subsonic airfoils, and shortly thereafter von Kármán and Tsien published a compressibility correction that improved on the older Prandtl-Glauert relation.[147,148] But in general, the United States reacted slowly to the stimulus provided by the Volta conference. On his return from Italy in late 1935, von Kármán urged both the army and the NACA to develop high-speed supersonic facilities, to no avail. Finally, when the war came to the United States in 1941, such urging got a more receptive hearing. In 1942, at the California Institute of Technology (Cal Tech), von Kármán established the first major university curriculum in compressible flow, a course of study that was heavily attended by military officers. Finally, in 1944, the first practical operational supersonic wind tunnel in the United States was built at the U.S. Army's Ballistics Research Laboratory in Aberdeen, Maryland, designed by von Kármán and his colleagues at Cal Tech. Twelve years after Busemann began to collect data in his supersonic tunnel in Germany, and nine years after the fifth Volta conference and the construction of supersonic tunnels at Guidonia in Italy, the United States finally was seriously into the business of supersonic research.

The High-Speed Research Airplane

By the time of the Volta conference, John Stack had a newer, larger facility at the NACA: a 24-in. high-speed tunnel equipped with an improved schlieren system, where basic testing of compressibility effects on flows over airfoils continued. In 1938, Stack published the most thorough report to that time on the nature of high-speed compressible flows over airfoils, including many detailed measurements of surface pressures.[149] With that, the NACA continued as the undisputed leader in studying the effects of compressibility and the consequences of the compressibility burble.

Jacobs's paper at the fifth Volta conference was very much a celebration of the second phase of the NACA research program on high-speed flight. The first phase had been the embryonic wind-tunnel compressibility work of the 1920s, clearly oriented toward applications to propellers. The second phase was a refocusing of that high-speed wind-tunnel research on the airplane itself, and that second phase would soon be augmented by a new initiative: the design and development of an actual research airplane.

The idea of a research airplane – an airplane designed and built strictly for the purposes of probing unknown flight regimes – can be traced to the thinking of John Stack in 1933. On his own initiative, Stack went through a very preliminary design analysis "for a hypothetical

Figure 9.22   Stack's sketch for a proposed high-speed experimental aircraft (1933).

airplane which, however, is not beyond the limits of possibility."[143] The purpose of the airplane, as presented in a 1933 article in the *Journal of the Aeronautical Science,* was to fly very fast, well into the compressibility regime. The airplane he designed (Figure 9.22) was highly streamlined for its time, with a straight, tapered wing having an NACA 0018 symmetric airfoil section at the center, thinning to an NACA 0009 airfoil 9% thick at the tip. Stack even went so far as to test a model of that design (without tail surfaces) in the Langley VDT. He estimated the drag coefficient for the airplane using the data he had measured in the 11-in. High-Speed Tunnel. Assuming a fuselage large enough to hold a 2,300-hp Rolls-Royce engine, Stack calculated that the propeller-driven airplane would have a maximum speed of 566 mph – far beyond that of any airplane flying at the time, and well into the regime of compressibility. Stack's enthusiasm about the possibilities for such an airplane was reflected in his hand-drawn graph (Figure 9.23) showing the horsepower required as a function of speed, and comparing the data with and without the effects of compressibility. His sketch of the airplane is at the top of the graph (along with the aged rust marks of two paper clips), which was found buried in the John Stack Files in the Langley archives. Barely distinguishable at the bottom, Stack had written "Sent to Committee Meeting, Oct. 1933." He was so convinced of the viability of his idea for a research airplane that he had sent that quickly prepared hand-drawn graph to the biannual meeting of the full committee of the NACA in Washington in October 1933. The NACA did not help Stack find a developer for the airplane, but "the optimistic results of his paper study convinced many people at Langley that the potential for flying at speeds far in excess of 500 miles per hour was there."[84]

The state of high-speed aerodynamics in 1939 can be illustrated by the trends shown in Figure 9.24, where the generic variation of the drag coefficient for an airplane is shown as a function of the free-stream Mach number. On the subsonic side, below Mach 1, wind-tunnel data had indicated the rapid increase in the drag coefficient as Mach 1 was approached. On the supersonic side, ballisticians had known for years (as supported by findings from the linearized supersonic theory developed by Jakob Ackeret[150] in Germany since 1925) how the drag coefficient would behave above Mach 1. Of course, all airplanes at that time were on the subsonic side of the curve shown in Figure 9.24. Stack summarized the situation in 1938:

> The development of the knowledge of compressible-flow phenomena, particularly as related to aeronautical applications, has been attended by considerable difficulty. The complicated nature of the phenomena has resulted in little theoretical progress, and, in general, recourse to experiment has been necessary. Until recently the most important experimental results have been obtained in connection with the science of ballistics, but this information has been of little value in aeronautical problems because the range of speeds for which most ballistic experiments have been made extends from the speed of sound upward; whereas the important region in aeronautics at the present time extends from the speed of sound downward.[149]

In essence, the flight regime just below and just above the speed of sound was unknown – a transonic gap, as shown schematically in Figure 9.24.

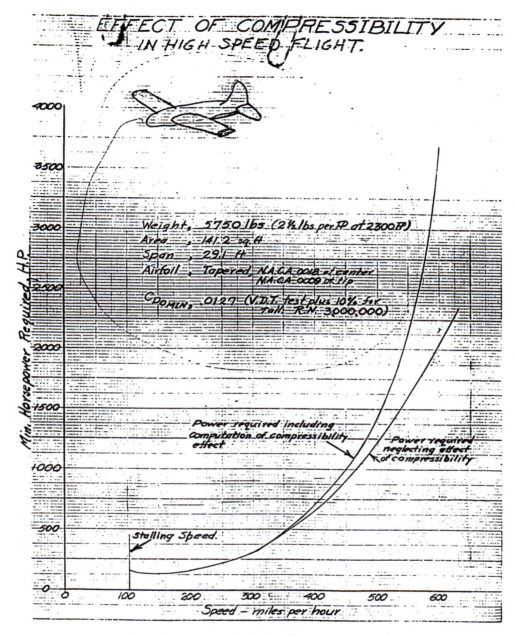

Figure 9.23 Stack's hand-drawn graph of the power required for a high-speed airplane, illustrating the effects of compressibility (1933). (From John Stack Files.)[143]

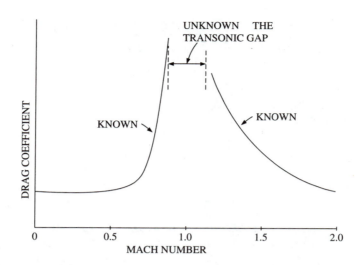

Figure 9.24   Generic variation of airplane drag coefficient versus Mach number for subsonic, transonic, and supersonic speeds.

The aeronautics community was suddenly wakened to the dangers of that unknown flight regime in November 1941, when Lockheed test pilot Ralph Virden, putting the new, high-performance P-38 through a high-speed dive, could not pull out in time, and crashed. That was the first fatality due to adverse compressibility effects, and the P-38 (Figure 9.25) was the first airplane to suffer from those effects. Virden's P-38 had exceeded its critical Mach number in an operational dive and had penetrated well into the regime of the compressibility burble at its terminal dive speed.[151] The problem encountered by Virden and many other P-38 pilots at that time was that beyond a certain speed in a dive, the elevator controls suddenly felt as if they were locked, and to make things worse the tail suddenly produced more lift, pulling the P-38 into an even steeper dive. That was called the tuck-under problem at the time.[17] Lockheed consulted various aerodynamicists, including von Kármán at Cal Tech, but it turned out that John Stack at NACA Langley, with his accumulated experience in compressibility effects, was the only one to properly diagnose the problem: The wing of the P-38 lost lift when it encountered the compressibility burble, and as a result the downwash angle of the flow behind the wing was reduced. That, in turn, increased the effective angle of attack at which the flow encountered the horizontal tail, increasing the lift on the tail and pitching the P-38 to a progressively steeper dive, totally beyond the control of the pilot. Stack's solution was to place a special flap under the wing, to be employed only when those compressibility effects were encountered. The flap was not a conventional dive flap intended to reduce the airplane's speed; rather, it was used to maintain lift in the face of the compressibility burble, thus eliminating the change in the downwash angle and therefore allowing the horizontal tail to function properly. That was a graphic example, from the early days of high-speed flight, of the vital importance of the NACA compressibility research as real airplanes began to sneak up on Mach 1.

By the late 1930s, it was time for real airplanes to be used to probe the mysteries of the unknown transonic gap illustrated in Figure 9.24, time for the high-speed research airplane to become a reality.[152] The earliest concrete proposal along those lines came from Ezra

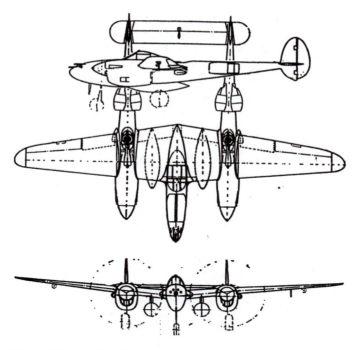

Figure 9.25   Lockheed P-38.

Kotcher, a senior instructor at the Army Air Corps Engineering School at Wright Field (a forerunner of today's Air Force Institute of Technology). Kotcher was a 1928 graduate of the University of California, Berkeley, with a B.S. degree in mechanical engineering. The same year that John Stack began work at Langley as a junior aeronautical engineer, Kotcher began work at Wright Field in a similar position; both interested in high-speed aerodynamics, they would later work together on the development of the Bell X-1. Kotcher's proposal, drafted in August 1939, was in response to Major General Henry ("Hap") Arnold's request for an investigation into the future of advanced military aircraft. The proposal contained a plan for a research program of actual high-speed flight. Kotcher pointed out the unknown aspects of the transonic gap and the problems associated with the compressibility burble, as elucidated by the NACA, and he concluded that the next important step should be a full-scale flight research program. The Army Air Corps did not immediately respond to his proposal.

At Langley, the idea of a high-speed research airplane was gaining momentum.[152] By the time the United States entered World War II in December 1941, Stack had studied the behavior of the flow in a wind tunnel when the flow in the test section was near or at Mach 1. He repeatedly found that when a model was mounted in the flow, the flow field in the test section essentially broke down, making any aerodynamic measurements worthless. He concluded that the development of a truly transonic wind tunnel would be an extremely difficult problem, probably far into the future. It appeared that the best way to learn about the aerodynamics of transonic flight would be to build a real airplane that would fly in that regime, and during several visits by George Lewis, NACA's director of aeronautical research, Stack continued to push that idea. Lewis, who appreciated the work that Stack had done at the NACA, was not immediately receptive to the idea of a research airplane, but in early 1942 he left the door open just a crack: "He left Stack with the idea, however, that some low-priority, back-of-the-envelope estimates to identify the most desirable design features of a transonic airplane could not hurt anyone, providing they did not distract from more pressing business."[84]

Given Stack's commitment to the idea, that was all that was needed. With the blessing of the local management at Langley, Stack immediately gathered a small group of engineers and began to work on the preliminary design aspects for a transonic research airplane, and by the summer of 1943 the group had produced such a design. That design established a mind-set that would guide the NACA's thinking on the transonic research airplane for the next five years, though it would be in conflict with some later ideas coming from Kotcher and the U.S. Army. The principal features of the preliminary NACA design were as follows:

(1)   It would be a small turbojet-powered airplane.
(2)   It would take off under its own power from the ground.
(3)   It would have a maximum speed of Mach 1, but the main concern would be its ability to fly safely at high subsonic speeds.
(4)   It would carry a large payload of scientific instruments to measure the aerodynamic parameters and its dynamic behavior in flight near Mach 1.
(5)   It would start its test program at the low end of the compressibility regime and progressively sneak up to Mach 1 in later flights. The important goal would be to gather aerodynamic data at high subsonic speeds, not necessarily to fly into the supersonic regime.

Those features came to be considered as almost inviolable in the mind-set of the Langley engineers, and Stack in particular.

The exigencies of wartime mandated accelerated research into high-speed aerodynamics, and compressibility problems finally had the attention not only of the NACA but also of the army and navy. Stack, who had become Eastman Jacobs's assistant section chief for the VDT in 1935, and head of the high-speed wind tunnels in 1937, was made chief of the newly formed Compressibility Research Division in 1943, giving him his most advantageous position thus far to push for the high-speed research airplane.

The Bell X-1: Point and Counterpoint

Although the NACA had the theoretical knowledge, the experience, and the data to deal with compressibility problems, the army and navy had the money that would be necessary to design and build a research airplane. The Bell X-1 was conceived on November 30, 1944, when Robert J. Woods of Bell Aircraft visited the office of Ezra Kotcher. Woods, who had NACA ties because of having worked at Langley during 1928–9 on the VDT, had joined with Lawrence D. Bell in 1935 to form the Bell Aircraft Corporation in Buffalo, New York. Woods had dropped by Kotcher's office simply to chat. During the conversation, Kotcher mentioned that the army, with the help of the NACA, wanted to build a special, non-military, high-speed research airplane. After detailing the army's specifications for the aircraft, Kotcher asked if Bell Aircraft would be interested in designing and building the airplane. Woods said yes, and the die was cast.[152]

When Kotcher was talking with Woods, he was not operating out of a vacuum. During 1944, army and NACA engineers were meeting to outline the nature of a joint-research airplane program, and by mid-1944 Kotcher had received the army's approval for the design and construction of such an airplane. However, the army's reasons for wanting the high-speed research airplane were somewhat in conflict with those of NACA. To understand that conflict, we need to consider two factors operative at that time.

The first factor was that there was a common public belief in a "sound barrier," a myth that had originated in 1935 when the British aerodynamicist W. F. Hilton had described to a journalist some of the high-speed experiments he was conducting at the National Physical Laboratory. Pointing to a plot of airfoil drag (similar to that shown at the left in Figure 9.24), Hilton described "how the resistance of a wing shoots up like a barrier against higher speed as we approach the speed of sound." The next morning, the leading British newspapers were misrepresenting Hilton's comment by referring to "the sound barrier."[153] The idea of a physical barrier to flight – that airplanes could never fly faster than the speed of sound – became widespread among the public. Furthermore, even though most engineers knew that that was not the case, they still had no idea how much the drag would increase in the transonic regime, and given the low levels of thrust produced by airplane powerplants at that time, dealing with the speed of sound certainly loomed as a tremendous challenge.

The second factor in the conflict between the army and the NACA was that Kotcher was convinced that the research airplane had to be powered by a rocket engine, rather than a turbojet. That stemmed from his experience in 1943 as project officer on the proposed Northrop XP-79 rocket-propelled flying-wing interceptor, as well as the army's knowledge of Germany's new rocket-propelled interceptor, the Me-163.

Therefore, the army viewed the high-speed research airplane as follows:

(1) It should be rocket-powered.
(2) Early in its flight schedule it should attempt to fly supersonically, to show everybody that the sound barrier could be broken.
(3) Later in the design process it was determined that it should be air-launched, rather than take off from the ground.

All of those requirements were in conflict with the NACA's more careful, more scientific approach, but the army was paying for the X-1, and the army's views prevailed.

Although John Stack and the NACA did not agree with the army's specifications, they nevertheless provided as much technical data as possible throughout the design of the X-1. Lacking appropriate wind-tunnel data and theoretical solutions for transonic aerodynamics, the NACA developed three stopgap methods for acquisition of transonic aerodynamic data. In 1944, Langley carried out tests using the *drop-body concept.* Wings were mounted on bomb-like missiles, which were dropped from a B-29 at an altitude of 30,000 ft. The terminal velocities of those models sometimes reached supersonic speeds. The data were limited, mainly consisting of estimates of the drag, but NACA engineers considered them reliable enough to estimate the power required for a transonic airplane. Also in 1944, Robert R. Gilruth, chief of the Flight Research Section, developed the *wing-flow method,* wherein a model wing would be mounted perpendicularly at just the right location on the wing of a P-51D. During a dive, the P-51 would pick up enough speed (to about Mach 0.81) that locally supersonic flow would occur over its wing (as sketched in Figure 9.18). The small model wing mounted perpendicularly on the P-51 wing would be totally immersed in that supersonic-flow region, providing a unique high-speed-flow environment for the model. Ultimately those wing-flow tests gave the NACA the most systematic and extensive plots of transonic data yet assembled. The third stopgap method was *rocket-model testing,* wherein wing models were mounted on rockets, which were fired from the NACA's facility at Wallops Island on the coast of Virginia's eastern shore. The data from all those methods, along with the existing body of compressibility data obtained by the NACA over the preceding 20 years, provided the scientific and engineering base from which Bell Aircraft designed the X-1.

Breaking the Sound Barrier

This chapter began by describing how Chuck Yeager flew the Bell X-1 through the sound barrier in 1947. The detailed events leading up to that flight (the design, construction, and early flight-test program by Bell, and the army's preparations to handle the X-1 at Muroc) have been described by Hallion[152] and Young.[154] That first supersonic flight of the Bell X-1 was the culmination of 260 years of research into the mysteries of high-speed aerodynamics, especially the 23 years of research by the NACA – one of the most important stories in the history of applied aerodynamics.

On December 17, 1948, President Harry S Truman presented the Collier Trophy for 1947 jointly to three men for "the greatest aeronautical achievement since the original flight of the Wright Brothers' airplane." That trophy was the highest possible official recognition for the accomplishments embodied in the X-1. John Stack was one of the three, recognized as the scientist, along with Lawrence D. Bell, the manufacturer, and Captain Charles E. Yeager, the pilot. The citation to Stack read "for pioneering research to determine the physical laws affecting supersonic flight and for his conception of transonic research airplanes," but an entire team of NACA researchers had worked to earn the 1947 Collier Trophy for the NACA high-speed research program.

At the time of that award, Stack (Figure 9.26) was assistant chief of research at NACA Langley, and in 1952 he was made assistant director of Langley. By that time he had been awarded the 1951 Collier Trophy for development of the slotted-throat wind tunnel. In 1961, three years after the NACA had been absorbed into the National Aeronautics and

Figure 9.26   John Stack.

Space Administration (NASA), Stack became director of aeronautical research at NASA Headquarters in Washington. Despairing at the deemphasis of aeronautics relative to space funding within NASA, Stack, after 34 years of government service with NACA/NASA, retired in 1962 and became vice-president for engineering at Republic Aircraft Corp. When Republic was absorbed by Fairchild-Hiller in 1965, Stack became a vice-president of that company, retiring in 1971. On June 18, 1972, Stack fell from a horse on his farm in Yorktown, Virginia, and was fatally injured. He was buried in the churchyard cemetery of Grace Episcopal Church in Yorktown, only a few miles from the NASA Langley Research Center. Today, F-15s from nearby Langley Air Force Base fly over the churchyard – airplanes that can routinely fly at almost three times the speed of sound, thanks to the legacy of John Stack and the NACA high-speed research program.

Transonic Aerodynamics: Probing the Mysteries

The flights of the Bell X-1 proved beyond any doubt that airplanes could fly safely in the mysterious aerodynamic region around Mach 1, in spite of the lack of theory and experimental data concerning the aerodynamic characteristics in that "transonic gap." In 1947, a great deal was known about the subsonic and supersonic regions that bracketed that gap, but very little about the transonic region of the gap itself.

In regard to wind-tunnel testing in the late 1940s, measurements of transonic flows below $M_\infty = 0.95$ and above $M_\infty = 1.1$ could be carried out with reasonable accuracy in the NACA high-speed wind tunnels, but the data obtained between 0.95 and 1.1 were of questionable accuracy. For Mach numbers very near 1, the flow was quite sensitive, and if a model of any reasonable cross-sectional area was placed in the tunnel, the flow would become choked ("choking" is the breakdown of flow in the test section when the proper mass flow cannot pass through). That choking phenomenon was one of the most difficult aspects of high-speed tunnel research; small models had to be used. Figure 9.27 shows a small model of the Bell X-1, with a wingspan slightly over 1 ft, whereas the test-section diameter was 8 ft. In spite of that small model size, valid data could not be obtained at free-stream Mach numbers above 0.92, because of choking in the tunnel at higher Mach numbers. Even when the flow did not choke, the shock waves from the model, which were nearly perpendicular to the flow at Mach numbers near 1, reflected off the tunnel walls and impinged back on the model itself. In either case, the aerodynamic data from the model were essentially worthless.

Figure 9.27   Wind-tunnel model of the Bell X-1 in the Langley 8-ft tunnel (1947).

The Mach-number gap between 0.95 and 1.1, in which valid data could not be obtained using the existing high-speed wind tunnels in the late 1940s, contributed greatly to the aerodynamic uncertainties that dominated the Bell X-1 program up to its first supersonic flight (that was why the Bell engineers made the X-1 fuselage precisely the shape of a 50-caliber machine-gun bullet, a shape whose attributes were well known to ballisticians by that time). Moreover, the advancement of basic aerodynamics in the transonic range was greatly hindered by that situation. Throughout the late 1930s and 1940s, NACA engineers attempted to alleviate the choking problem in their high-speed tunnels by using various test-section designs (closed test sections, totally open test sections, a bump on the test-section wall to tailor the flow constrictions), as well as various methods for supporting a model in the test section to minimize blockage. None of those ideas solved the problem. But the stage was being set for a technical breakthrough, which came in the late 1940s: the slotted-throat transonic tunnel.

In 1946, Ray H. Wright, a theoretician at NACA Langley, carried out an analysis that indicated that if the test section were given a series of long, thin, rectangular slots parallel to the flow direction that would leave about 12% of the test-section periphery open, then the blockage problem might be greatly alleviated. That idea met with some skepticism, but it was almost immediately accepted by John Stack, who by that time was a highly placed administrator at Langley. A decision was made to slot the test section of the small, 12-in. high-speed tunnel, which resulted in greatly improved performance in early 1947. However, that was simply an experiment, and much skepticism remained. On the surface, the NACA made no plans to further implement that idea. On the other hand, Stack confided privately to his colleagues that he favored slotting the large 16-ft high-speed tunnel. Without fanfare, that work began in the spring of 1948, buried in a larger project to increase the horsepower of the tunnel. Almost simultaneously, Stack made the decision to slot the 8-ft tunnel as well. The work on the 8-ft tunnel proceeded faster than that on the larger tunnel, and on October 6, 1950, it became operational for research. By December of that same year, the modified 16-ft tunnel also became operational. Subsequent operation of those facilities proved that the slotted-throat concept allowed smooth transition of the tunnel flow through Mach 1 simply because of the increase in tunnel power. With that, the problem of blockage was basically solved. Those tunnels became the first truly transonic wind tunnels, and for that accomplishment John Stack and his colleagues at NACA Langley were awarded the Collier Trophy in 1951. The problem of measuring transonic flows in the laboratory was well in hand.

The same could not be said for the problem of *computing* transonic flows. Such flows, like all aerodynamic flows, are governed by the continuity, momentum, and energy equations (as detailed in Appendixes A and B). If we neglect friction, the Euler equations (Appendix A) are the governing equations. They are nonlinear partial differential equations, and thus they are exceptionally difficult, if not impossible, to solve. However, if the flow field has only small changes from the free-stream flow (small perturbations in the flow), such as the flow over a thin body at small angles of attack, and if the free-stream Mach number either is below about 0.8 or is between 1.2 and 5, then the governing equations for such subsonic and supersonic flows reduce, in an approximate fashion, to linear partial differential equations, which can be solved. Unfortunately, in the transonic regime, with free-stream Mach numbers between 0.8 and 1.2, the governing equations cannot be linearized. Therefore, the analysis of such flows was exceptionally difficult in the period before the development of the high-speed digital computer. In 1951 there was virtually no useful aerodynamic method for the calculation of transonic flows. Clearly, in 1951 computational analysis of transonic flow fields was lagging greatly behind the experimental progress. That situation prevailed until the advent of modern computational fluid dynamics. By the 1980s, with a few exceptions, the problem of calculating transonic flows was reasonably well in hand, because of the power of computational fluid dynamics.

The Area Rule and the Supercritical Airfoil

There were two major developments having to do with configuration that made transonic flight practical: the area rule and the supercritical airfoil, both products of the transonic wind-tunnel research at Langley directed by Richard Whitcomb. On a technical basis, the area rule and the supercritical airfoil had the same objective: to reduce drag in the transonic regime. However, their drag reductions were accomplished in different ways. Consider the qualitative sketch of the drag coefficient versus Mach number in Figure 9.28 for a transonic body. The variation for a standard body shape without the area rule and without a supercritical airfoil is given by the solid curve.

Now let us consider the area rule by itself. The area rule is a simple statement that the cross-sectional area of the body should have a smooth variation with longitudinal distance along the body; there should be no rapid or discontinuous changes in the distribution of cross-sectional areas. For example, a conventional wing-body combination will have a sudden increase in cross-sectional area in the region where the wing cross section is added to the body cross section. The area rule says that to compensate, the body cross section should be decreased in the vicinity of the wing, producing a wasp-like or Coke-bottle shape for the body. The aerodynamic advantage of the area rule is shown in Figure 9.28, where the drag variation for the area-ruled body is given by the dashed curve. Simply stated, application of the area rule reduces the peak transonic drag by a considerable amount.

The supercritical airfoil, on the other hand, acts in a different fashion. A supercritical airfoil is shaped somewhat flat on the top surface, in order to reduce the local Mach number

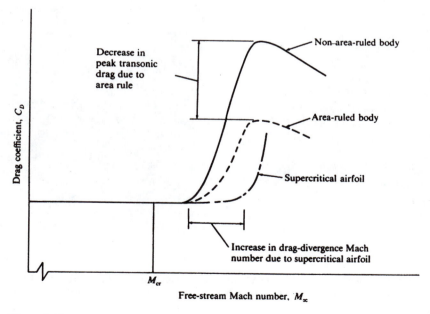

Figure 9.28   Illustration of the separate effects of the area rule and the supercritical airfoil.

inside the supersonic region below what it would be for a conventional airfoil under the same flight conditions. As a result, the strength of the shock wave is lower, and the boundary-layer separation is less severe, and thus a higher free-stream Mach number can be reached before the drag-divergence phenomenon sets in. The drag variation for a supercritical airfoil is shown in Figure 9.28 by the broken curve. The role of a supercritical airfoil is clearly evident: Although the supercritical airfoil and an equivalent standard airfoil may have the same critical Mach number, the drag-divergence Mach number for the supercritical airfoil will be much larger. That is, the supercritical airfoil can tolerate a much larger increase in the free-stream Mach number above the critical value before drag divergence is encountered. Such airfoils are designed to operate far above the critical Mach number – hence the term "supercritical" airfoils.

The area rule was introduced in a spectacular fashion in the early 1950s. Although there had been some analysis that had produced oblique hints in the direction of the area rule, and although workers in the field of ballistics had known for years that projectiles with sudden changes in their distributions of cross-sectional areas exhibited high drag at high speeds, the importance of the area rule was not fully appreciated until Richard Whitcomb conducted a series of wind-tunnel tests on various transonic bodies in the slotted-throat 8-ft wind tunnel. Those data, and an appreciation of the area rule, came just in time to save a new airplane program at Convair. In 1951, Convair was designing one of the new "century series" fighters intended to fly at supersonic speeds. Designated the YF-102 (Figure 9.29),

(a)   (b)

Figure 9.29   (a) The Convair YF-102, without application of the area rule. (b) The Convair YF-102A, with area-ruled fuselage. Note the wasp-like shape of the fuselage in comparison with the YF-102.

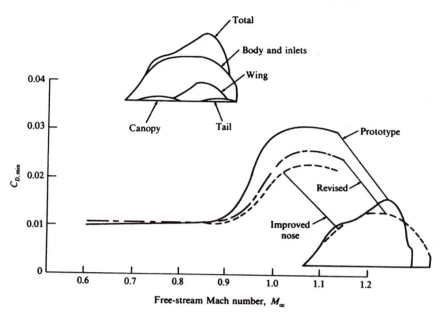

Figure 9.30   Variation of the drag coefficient with Mach number for the YF-102 and YF-102A. (From Loftin.)[69]

that aircraft had a delta-wing configuration and was powered by the Pratt & Whitney J-57 turbojet, the most powerful engine in the United States at that time. Aeronautical engineers at Convair expected the YF-102 to fly supersonically. On October 24, 1953, flight tests of the YF-102 began at Muroc Air Force Base (now Edwards), and a production line was being set up at Convair's San Diego plant. However, as the flight tests progressed, it became painfully clear that the YF-102 could not fly faster than sound – the transonic drag rise was simply too great for even the powerful J-57 engine to overcome. After consultation with NACA aerodynamicists and inspection of the area-rule findings that had been obtained in the Langley 8-ft tunnel, the Convair engineers modified the airplane to become the YF-102A, with an area-ruled fuselage. Wind-tunnel data on the YF-102A looked promising. Figure 9.30, produced from those data,[69] shows the variation of the drag coefficient with free-stream Mach number for the YF-102 and YF-102A. At the top is a sketch of the cross-sectional-area distribution for the YF-102, showing how it was built up from the different body components. Note the irregular, bumpy nature of the total cross-sectional-area distribution. At the bottom right, shown by the dashed line, is the cross-sectional-area distribution for the YF-102A – a much smoother variation than that for the YF-102. The comparison between the drag coefficients for the conventional YF-102 (solid curve) and the area-ruled YF-102A (dashed curve) dramatically illustrates the tremendous transonic drag reduction to be obtained with the use of the area rule. (Recall from Figure 9.28 that the function of the area rule is to decrease the peak transonic drag; Figure 9.30 quantifies that function.) Encouraged by those wind-tunnel findings, the Convair engineers began a flight-test program for the YF-102A. On December 20, 1954, the prototype YF-102A left the ground at Lindbergh Field, San Diego, and exceeded the speed of sound while still climbing. The use of the area rule had increased the top speed of the airplane by 25%. The

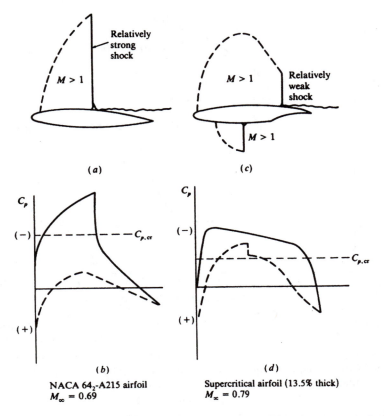

Figure 9.31   Flow-field and pressure-distribution comparison between a standard NACA 6-series airfoil and a supercritical airfoil. (From Whitcomb and Clark.)[155]

production line rolled, and 870 F-102As were built for the U.S. Air Force. The area rule made its debut in dramatic style.

The supercritical airfoil, also pioneered by Richard Whitcomb, based on data obtained in the 8-ft wind tunnel, was a development of the 1960s. Recall from Figure 9.28 that the function of the supercritical airfoil was to increase the increment between the critical Mach number and the drag-divergence Mach number. The data in the Langley tunnel indicated a possible 10% increase in cruise Mach number due to the use of a supercritical-airfoil wing. NASA introduced the technical community to the supercritical-airfoil data in a special conference in 1972. Since that time, the supercritical-airfoil concept has been employed on virtually all new commercial aircraft and some military airplanes. Physical data for a supercritical airfoil and for the standard NACA 64-A215 airfoil are compared in Figures 9.31 and 9.32 along with their shapes. At the top of Figure 9.31, the regions of supersonic flow over the airfoils are shown, and the corresponding variations of the pressure coefficients over the airfoil surfaces are shown at the bottom of the figure. Notice that for the supercritical airfoil, the region of supersonic flow is terminated by a weaker shock wave, and the pressure change is less severe. That delays the onset of the transonic drag rise to a higher free-stream Mach number, as shown in Figure 9.32. The performance advantage

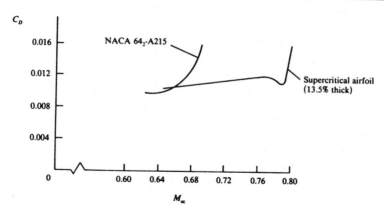

Figure 9.32 Drag coefficient versus Mach number for a standard NACA 6-series airfoil and a supercritical airfoil. (From Whitcomb and Clark.)[155]

of the supercritical airfoil is clearly evident. Whitcomb's original publication[155] on the supercritical airfoil was at first classified, but was released to the public in the early 1970s.

This section has provided a brief glimpse into what was one of the most exciting chapters in the history of aerodynamics and aeronautical engineering. We have seen how the secrets of transonic flow were slowly and painstakingly uncovered, how a concerted, intelligent attack on the problem eventually led to useful wind-tunnel data, as well as to modern methods of computation for transonic flows, and finally how those transonic data ultimately led to two of the major aerodynamic breakthroughs in the latter half of the twentieth century: the area rule and the supercritical airfoil.

Supersonic Aerodynamics Theory: Its Early Days

For more than a century, the speeds attained by some artillery shells and rifle bullets had been supersonic, but it was not until October 14, 1947, that a human being flew faster than the speed of sound in an airplane. That dramatic event opened a special era within the age of the jet-propelled airplane: the era of the supersonic airplane. We have already seen that the theory of supersonic shock waves was well in hand by that time; the work of Riemann, Rankine, Hugoniot, and Mach during the nineteenth century, and subsequent analyses by Rayleigh, Taylor, Prandtl, and Meyer in the first decade of the twentieth century, resulted in an understanding of supersonic shock and expansion waves and accurate methods for their calculation. That fundamental capability in supersonic aerodynamics had been achieved long before Chuck Yeager's historic flight in the Bell X-1 in 1947. However, the application of that theory to supersonic bodies, and an understanding of other aspects of the aerodynamic flow over such bodies, represented a challenge to aerodynamicists for most of the twentieth century, and is still a challenge in the design of certain supersonic military vehicles. In this section we shall focus on the early evolution of supersonic aerodynamics theory as applied to the flow over supersonic bodies. In the process, we shall examine the characteristics of *supersonic wave drag,* the most significant practical difference between the aerodynamics of supersonic flight and the aerodynamics of subsonic flight.

In Chapter 1 we emphasized that the aerodynamic force on a body is due to the net effect of the pressure distribution and the shear-stress distribution integrated over the exposed surface of the body. Wave drag is due to the increased pressure behind shock waves; that high pressure is exerted on the forward portions of the supersonic body, thus increasing the drag. The increase in drag that is due to the presence of shock waves is called *wave drag*. Wave drag is therefore a type of pressure drag.

If all supersonic vehicles were shaped as a series of planar surfaces, such as a wedge, or a flat plate, or a diamond airfoil, then the wave drag and lift could be readily calculated from direct application of oblique-shock and expansion theory, in hand since 1908. However, in most cases the body surface is curved, such that the leading-edge shock wave is curved, rather than straight. Furthermore, the flow field downstream of the curved shock wave cannot be calculated from straight-shock-wave theory. Question: How could the wave drag be predicted for a body with curved surfaces? Two researchers in the 1920s and early 1930s took giant strides toward answering that question: Jakob Ackeret in Switzerland and Theodore von Kármán in the United States.

Jakob Ackeret was born in Zürich, Switzerland, March 17, 1898. He received his diploma in mechanical engineering and a doctorate in science from the Swiss Institute of Technology in Zürich, where he served as an assistant to Aurel Stodola. Ackeret then joined Prandtl's laboratory at Göttingen for postdoctorate work, where he became a department head at the Kaiser Wilhelm Institute (Figure 9.33). From 1921 to 1928 Ackeret extended Prandtl's work on boundary-layer theory and was interested in practical applications of boundary-layer control through suction at the surface. Returning to Switzerland in 1928, he became chief engineer for the Escher Wyes company in Zürich, where he pioneered applications of new fluid-dynamics theories for the design of turbines and other rotating machinery. Four years later he returned to his alma mater, the Swiss Institute of Technology, as a professor and director of the Institute of Aerodynamics, where he remained for the rest of his professional

Figure 9.33   Some of the principal researchers at Göttingen in 1924, from left to right: Jakob Ackeret, Ludwig Prandtl, Albert Betz, and Reinhold Seiferth. (From Rotta,[73] with permission.)

life. In the early 1930s, Ackeret became active in the design and construction of supersonic wind tunnels, thus continuing the kind of work carried out by his advisor, Stodola, years before. Ackeret designed the first closed-circuit supersonic wind tunnel and participated in the design and construction of supersonic wind tunnels throughout Europe during the 1930s. Indeed, when attendees at the 1935 Volta conference toured the new Italian high-speed aerodynamics laboratory at Guidonia, a city near Rome, they saw a supersonic wind tunnel designed by Ackeret, who was acting as a consultant to the Italian Air Force. The tunnel was capable of Mach 4. By the beginning of World War II, Ackeret was the leading European authority on supersonic flows. At the end of the war, the U.S. Army had Ackeret's name on the list of important German scientists to be taken to the United States under a program known as "Operation Paperclip." Of course, Ackeret was not German, but it was only after Theodore von Kármán, a close friend of Ackeret, objected strenuously to that "U.S. raiding program" of academics in Germany that Ackeret was left alone. Ackeret retired as a professor emeritus in 1967. By that time he had been made an Honorary Fellow of the Institute of Aeronautical Science in New York and the Royal Aeronautical Society in London. Ackeret died after a long illness in Kusnault, Switzerland, March 26, 1981, at the age of 83.

Ackeret was an eminent teacher who showed a deep concern for his students. An interesting sidelight was his serious interest in the history of science, and in his later years he became an authority on the history of hydrodynamics. His library contained well over 2,500 volumes dating from 1850 to 1940, with another 500 volumes published after World War II, many signed by their authors. His library was a valuable resource on the historical evolution of mathematics and physics in the second half of the nineteenth century and the first half of the twentieth. In addition to collecting those books, Ackeret read many of them, writing notes and calculations on slips inserted between the pages.

Ackeret's contribution of relevance to this section was his pioneering work in the middle 1920s on the theory of supersonic flows. While at Göttingen he developed the essence of the theory of linearized supersonic flows.[150] By assuming small disturbances produced by an airfoil in an otherwise uniform supersonic stream, Ackeret was able to reduce the governing system of nonlinear Euler equations for an inviscid flow (Appendix A) to a much simpler linear equation that could be readily solved. Although restricted to use for thin airfoils at small angles of attack, Ackeret's calculations showed that the lift and drag coefficients in a supersonic flow are given by

$$C_l = \frac{4\alpha}{\left(M_\infty^2 - 1\right)^{1/2}}$$

and

$$C_d = \frac{f(\alpha, t)}{\left(M_\infty^2 - 1\right)^{1/2}}$$

respectively, where α is the angle of attack, and $f(\alpha, t)$ is some function of the angle of attack and the body thickness, depending on the shape of the body. Notice that the coefficient for supersonic wave drag decreases as the Mach number increases, a trend sketched at the right in Figure 9.24. Ackert's 1925 theory[150] predicted that trend, as reflected in Figure 9.34. Ackeret's work was the beginning of a whole spectrum of variations on the theory of linearized supersonic flows for the prediction of supersonic wave drag and lift on slender bodies at small angles of attack – a spectrum that is still being explored today.

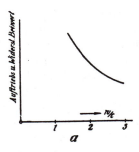

Figure 9.34   Calculation of the coefficient for supersonic wave
drag for a thin airfoil versus Mach number. (From Ackeret.)[150]

The second researcher who played an important role in the development of supersonic theory during the 1930s was Theodore von Kármán, a Hungarian-born scientist who moved to the United States. The name von Kármán has been accorded almost godlike importance in the history of aerodynamics in the United States, and that aura existed well before his death in 1963 and continues today because of the magnitude of his contributions to, and advocacy for, aerodynamic research and development. Also, his influence is seen today in the contributions of his students, many of whom have gone on to leadership roles in the field. To realize that von Kármán was a special individual, one need look no farther than three books that have been written about his life. Paul Hanle[156] emphasized that von Kármán's move to the California Institute of Technology from Aachen, Germany, brought to the United States a major infusion of mature aerodynamic thinking and sophisticated aerodynamic theory in the tradition of Prandtl's work at Göttingen. Michael Gorn[157] described von Kármán's massive influence in research-and-development policies in the United States during the 1940s and 1950s. Many of the features of our established research culture today are direct results of von Kármán's vision. von Kármán freely circulated from the United States to Europe and occasionally to the Orient, something of a universal man. He was a colorful individual, which perhaps accounted for the popularity of his autobiography, published after his death, *The Wind and Beyond,*[67] which best reveals his dynamic intellectual and personal nature. So far as I know, no other person in the history of aerodynamics has been the subject of two published biographies and one autobiography. Because of the extensive literature available concerning von Kármán, only a short biographical sketch will be given here.

Theodore von Kármán was born May 11, 1881, in Budapest, Hungary, to a middle-class Jewish family. His father, Maurice, had risen from a modest beginning to become a distinguished professor of education at the Pazmany Peter University in Budapest. Theodore's mother, Helen Konn, came from a long line of distinguished scholars, beginning with a sixteenth-century rabbi and mathematician at the imperial court in Prague who invented the world's first mechanical robot, called the "Golem" of Prague. In his autobiography, von Kármán commented that "as I look back on this impressive parental merger, it seems very logical to me that my career should have been directed toward science from the beginning" (p. 15).[67] Theodore was a child prodigy; he enrolled at the age of 9 in an open educational laboratory called the Minta ("Model Gymnasium"), founded by his father, and known as a "nursery for the elite." Its graduates included Edward Teller, Leo Szilard, and John von Neumann. At the age of 22, von Kármán graduated with highest honors from the Royal Joseph University of Polytechnics and Economics in Budapest with a degree in mechanical engineering. After brief periods of compulsory military service and employment in industry, he enrolled as an advanced student at Göttingen in 1906, receiving his doctorate under Prandtl in 1908. He remained at Prandtl's laboratory for the next four years as a *Privat Dozent* (the lowest rung on the academic ladder in Germany). During that period at Göttingen, von Kármán was at the center of the advances being made in boundary-layer theory and the development of airfoil and wing theory. In 1913, anxious to get away from the overpowering presence of Prandtl, he accepted a professorship in aeronautics and mechanics at the Technical University at Aachen, Germany, near the Belgian border (von Kármán did not especially like Prandtl as a person, as is evident in his autobiography. Indeed, it was Felix Klein, the president of Göttingen University, who found von Kármán the position at Aachen, not Prandtl, who did little to help advance von Kármán's career). At Aachen, von Kármán was also given the directorship of the Aachen Aerodynamics Institute, an important title and responsibility for someone as young as 31. It was at Aachen that von Kármán met Hugo Junkers, an industrialist and also professor of engines at the university; they collaborated in the aerodynamic design of the Junkers J-1 transport airplane in 1915, the first cantilevered-wing, all-metal airplane. At Aachen, von Kármán also matured as a teacher, delivering animated lectures that were well attended: "Thus, the Aachen Aerodynamics Institute achieved eminence after World War I due as much to Kármán's pedagogical talents as to his scientific achievements and cosmopolitan ways."[157]

In 1926, Robert Millikan, the Nobel Prize experimental physicist and president of the California Institute of Technology (Cal Tech) in Pasadena, invited von Kármán to visit the United States for a series of lectures (and to recruit von Kármán for a new aerodynamics laboratory at Cal Tech under the sponsorship of the Guggenheim Foundation). However, von Kármán was comfortable in the intellectual circles in Europe, and it was not until 1930 that he moved to Pasadena to become director of the Guggenheim Aeronautical Laboratory at the California Institute of Technology (GALCIT). Von Kármán's decision to move to the United States was prompted in part by his alarm at the rising Nazi influence in Germany, but perhaps more by his desire to get away from the influence of Prandtl. Also, he enjoyed the challenge of "bringing aerodynamics to America."[156]

Von Kármán brought to Cal Tech a fresh approach to theoretical aerodynamics and its connections to practical airplane design. He was not the first person to introduce aerodynamics theory into the American aeronautics program: Max Munk, in a rather abortive way, and Theodore Theodorsen had already provided much input, but because of his charismatic personality and his vision for the future of aeronautics, von Kármán soon made GALCIT the intellectual center of aerodynamics thought in America. Figure 9.35 shows von Kármán when he became GALCIT director in 1930.

Figure 9.35 Theodore von Kármán.

World War II propelled von Kármán onto the national scene as General Henry H. ("Hap") Arnold's most trusted scientific advisor, exercising a strong influence on the course of aeronautical research and development. Arnold was the commanding general of the U.S. Army Air Force during the war, and his amicable working relationship with von Kármán set a pattern for the cooperative efforts that have continued in American aeronautics to the present. For example, von Kármán created the Air Force Scientific Advisory Board, a collection of academic, government, and industrial engineers and scientists that still exists today as a major force in determining U.S. Air Force research-and-development policy. Von Kármán was responsible for establishing the Advisory Group for Aeronautical Research and Development (AGARD) as an arm of the North Atlantic Treaty Organization (NATO) after the war; since then, AGARD has played a strong role in technical aeronautical advancements in the NATO countries. Von Kármán was the driving force behind the creation of an educational and research institute in Belgium as part of AGARD. Today that institute is the Von Kármán institute for Fluid Dynamics, one of the leading aerodynamics laboratories in the world, with a long list of distinguished graduates.

After World War II, von Kármán took on the role of an elder statesman, continuing his close policy-making relationship with the U.S. Air Force and traveling and lecturing around the world. Although still director of GALCIT, he became more distant from the classroom and the laboratory, relinquishing the operation to his former students and colleagues. In the mid-1950s he began to spend most of his time in Europe, using Paris as a base. His health began to deteriorate, and he became progressively weaker. On May 6, 1963, von Kármán died in a hospital in Aachen. His body was returned to Pasadena, where he was buried at Hollywood Cemetery. At his funeral, a statement from President John Kennedy was read by a clergyman:

> It is with regret that I have learned of the death of Dr. Theodore von Kármán, to whom only last February I awarded the first National Medal of Science. Dr. von Kármán was known to the world's scientific community as the chairman of the Aeronautical Research and Development Group to NATO which he organized ten years ago. I know that his friends and associates will mourn his loss and join me in paying tribute to a great scientist and humanitarian.

High tribute indeed from a president of the United States to one of the great figures in the field of aerodynamics.

Of relevance to our discussion here is von Kármán's work in supersonic aerodynamics. His first major contribution in that area was a 1932 development in the calculation of wave drag for bodies of revolution – doing for projectiles what Ackeret's theory had done for two-dimensional airfoils. Using the same assumptions as in Ackeret's work (i.e., slender bodies that cause only small perturbations to the flow), von Kármán reduced the governing equations for axisymmetric supersonic flow to a single linear equation in the form of the wave equation from classic physics. He then proposed a novel solution to that equation, namely, a representation of the flow over the projectile by a continuous distribution of sources along the centerline. The idea of distributing sources and sinks in an otherwise uniform flow to construct the incompressible flows around bodies had been conceived by the Scottish engineer W. J. M. Rankine in the nineteenth century; that approach is still used today as a standard analytical technique for calculating incompressible flows around nonlifting bodies. Von Kármán's work extended that approach to the supersonic regime. The governing wave equation was satisfied by a continuous distribution of sources along the centerline of the projectile. The solution to the problem lay in finding the precise distribution of sources that when added to the supersonic free stream would result in the correct flow around the supersonic projectile. That Kármán-Moore theory[158] was the beginning of a new branch of supersonic aerodynamics theory that is still used today in linear-panel solutions for supersonic flows over slender bodies and wings.

Von Kármán had the ability to organize and explain complex technical material in an effective and understandable manner, and therefore he was an excellent spokesman for the aerodynamics community. There was no better example of that than the 10th Wright Brothers Lecture to the Institute of the Aeronautical Sciences in 1946, a milestone lecture that effectively summarized the early development of supersonic aerodynamics theory.[159] What von Kármán laid out in his lecture was a clear tapestry of the fundamentals of supersonic flows, as well as various means for calculating the supersonic aerodynamic properties of bodies and wings, including swept wings, and it was clear to the audience that the state of the art in supersonic aerodynamics had advanced to a quite mature stage, even though it would be almost 10 months before the first supersonic flight of the X-1. Eleven years prior to von Kármán's lecture, the state of the art in supersonic flow had been reviewed by Taylor and Maccoll[160] as part of a definitive six-volume survey of aerodynamics edited by William F. Durand at Stanford University: *Aerodynamic Theory: A General Review of Progress,* which is still a classic record of the state of the art in aerodynamics in the mid-1930s. The lack of knowledge of compressible flows, especially supersonic flows, in 1935 was indicated by the amount of space devoted to the subject: 40 pages out of almost 2,000 for the complete series. Taylor and Maccoll[160] discussed the basic aspects of shock-wave and expansion-wave theory, subsonic-supersonic nozzle flows, and wave propagation, and that was about it. In contrast, in 1946 von Kármán covered the physical and analytical aspects of supersonic wave drag and its calculation. Much attention was given to finite wings, including major emphasis on swept wings. In the article by Taylor and Maccoll in 1935, there was a bibliography of only 12 prior publications. Eleven years later, von Kármán had 186 entries in his bibliography, reflecting the explosion of research on compressible flow between 1935 and 1946. Von Kármán's Wright Brothers Lecture provides a convenient milestone, for in 1947 the state of the art in supersonic aerodynamics theory was on a firm foundation. Of course, it would continue to mature during the last half of the twentieth century, but that later growth would be incremental compared with the outstanding achievements prior to 1947. In the history of aerodynamics, 1947 marked the end of an era, the era of the early history of supersonic aerodynamics, and von Kármán's Wright Brothers Lecture was a fitting capstone for that era.

The Swept Wing: An Aerodynamic Breakthrough in High-Speed Flight

The novel idea of using swept wings for more efficient high-speed flight belongs to the era of early supersonic theory, as just discussed, but it is useful to isolate our discussion of swept-wing theory from the earlier main discussion in order to highlight its special nature.

The concept of the swept wing for high-speed flight was introduced by Adolf Busemann at the 1935 Volta conference in the presence of the world's leading high-speed aerodynamicists. It should have startled the conference delegates like an electric shock. Instead, it was virtually ignored by the audience. Even von Kármán and Eastman Jacobs did not mention the idea on their return to the United States. Indeed, 10 years later, when World War II was reaching its conclusion and jet airplanes were beginning to revolutionize aviation, the idea of swept wings was suggested independently by Robert T. Jones, an ingenious aerodynamicist at the NACA Langley Memorial Laboratory. When Jones presented his proposal to Jacobs and von Kármán in 1945, neither man seemed to remember Busemann's idea from the Volta conference. Von Kármán mentioned that oversight in his autobiography: "I must admit that I did not give this suggestion much attention until years later" (p. 219).[67] Being human, von Kármán offered an excuse: "My direction of effort at this time was not in design, but in developing supersonic theory" (p. 219).[67] However, Busemann's idea was not wasted on the German Luftwaffe, which recognized its military significance and classified the concept in 1936, one year after the conference. That set into motion a research program on swept wings in Germany that had produced a mass of technical data by the end of the war, to the great surprise and concern of the Allied technical teams that swooped into the German research laboratories at Peenemünde and Braunschweig in early 1945.

At the time of the Volta conference, Adolf Busemann was a relatively young (age 34) but accomplished aerodynamicist. Born in Lübeck, Germany, in 1901, he completed high school in his home town and received his engineering diploma and doctorate in engineering in 1924 and 1925, respectively, from the Technische Hochschule at Braunschweig. Busemann was one of the few important German aerodynamicists of that era who did not begin as one of Prandtl's students, but in 1925 Busemann began his professional career at the Kaiser Wilhelm Institute in Göttingen and soon entered Prandtl's sphere. From 1931 to 1935, Busemann broke away from that sphere to teach in the Engine Laboratory of the Technische Hochschule in Dresden. In 1935 he went to Braunschweig as chief of the Gas Dynamics Division of the Aeronautical Research Laboratory. When the Allied technical teams moved into German laboratories at the end of the war, they not only scooped up masses of technical aerodynamic data but also effectively scooped up Busemann, who accepted an invitation to join the NACA Langley Memorial Laboratory under Operation Paperclip in 1947. Busemann continued his research on high-speed aerodynamics for the NACA after joining Langley. He subsequently became chairman of the advanced-study committee at Langley and among other responsibilities supervised the preparation of science lectures used for training the early groups of astronauts in the manned space program. Later, Busemann became a professor in the Department of Aerospace Engineering Sciences at the University of Colorado in Boulder. After retirement, he remained in Boulder, leading an active life until his death in 1986.

Busemann's paper at the 1935 Volta conference was based on the simple idea that the aerodynamic characteristics of a wing are governed mainly by the component of the flow velocity perpendicular to the leading edge. Figure 9.21 shows the sketch used by Busemann to illustrate that normal component of the velocity. The angle of sweep is the angle ϕ. As the wing sweep is increased (as ϕ is increased) for a fixed free-stream velocity, the component of velocity normal to the leading edge will decrease. Because the wing essentially "sees" the normal component rather than the full free-stream velocity, the onset of high-speed compressibility effects on the wing will be delayed to a higher free-stream Mach number. The meaning of that for transonic flight is that as the wing is swept, the critical Mach number for the wing will be increased, and hence the free-stream Mach number at which the large rise in drag is encountered will be increased. Its meaning for supersonic flight is that the onset of wave drag will be delayed, and its magnitude will be reduced. Although Busemann was not aware of it, Max Munk had described the effect of the normal component of velocity in regard to wings in low-speed, incompressible flow in 1924. In studying the effects of sweepback and dihedral angle on airplane stability, Munk pointed out that only the component of flow velocity perpendicular to the leading edge was "effective for the creation of lift," though Munk's work had nothing to do with high-speed flow.[161] Much later, Robert T. Jones, a student of Munk, would remember Munk's work, and that provided the basis for Jones's independent discovery in 1945 of the advantages of a swept wing for transonic and supersonic flight.

The swept-wing concept in Busemann's 1935 Volta conference paper was, for everybody outside of Germany, an idea before its time. It is difficult to understand how von Kármán and other attendees failed to appreciate the significance of Busemann's idea, even forgetting it entirely, for that very evening Busemann went to dinner with von Kármán, Dryden, and General Arturo Crocco, the organizer of the conference. During dinner, Crocco sketched on the back of a menu card an airplane with swept wings, swept tail, and swept propeller, calling it, facetiously, "Busemann's airplane of the future."[162,163]

There was no such facetiousness in Germany. Under the Nazi government, the Luftwaffe was expanding rapidly. With Busemann in charge of aerodynamics research at Braunschweig, high-speed wind-tunnel testing of swept wings began. By 1939, the data confirmed the aerodynamic advantage of swept wings, as originally theorized by Busemann. For examples of the German experimental swept-wing data, see the paper by Blair.[164] In 1942, the senior airplane designer at Messerschmitt, Woldemar Voigt, used Busemann's idea during the paper design for an advanced experimental jet fighter. Designated Projekt 1101, the airplane had sharply swept wings, in contrast to the mild sweep of the Me-262 twin-jet fighter that Voigt was also designing. Because of the high priority placed on the Me-262, Voigt was not able to spend enough time on Projekt 1101. By the end of the war, there had been only wind-tunnel tests on models of the highly swept jet. The data, however, were most promising.[163]

In May 1945 von Kármán led one of the teams of U.S. scientists and engineers that swept into a crumbling Germany to search for information on German research and development. Because of his education in Germany and familiarity with the leading German scientists, von Kármán was a particularly effective member of the team. On May 7, the day before the surrender was signed, the team arrived at Braunschweig and was amazed to find numerous swept-wing wind-tunnel models and a mass of swept-wing data.

One member of that team was George Schairer, now a retired vice-president of Boeing, but at that time a young Boeing aeronautical engineer working on a preliminary design for a new generation of jet-powered bombers. After studying the German data on swept wings, Schairer quickly wrote a letter to his colleague, Ben Cohn, at Boeing, alerting the design team to the interesting features of such wings. Moreover, Schairer asked Cohn to distribute copies of his letter to all the major aircraft manufacturers so that the entire aeronautical community would know about the benefits of swept wings for high-speed airplanes.[165] In the short run, only two companies would take advantage of that information: Boeing and North American.

It is unlikely that the swept wing would have revolutionized airplane design so soon after the war if it had not been for the independent discovery of its advantages by Robert T. Jones, an aerodynamicist at the NACA Langley Memorial Laboratory. Jones was a self-made person in much the same mold as nineteenth-century aerodynamicists like Cayley, Wenham, and Phillips. Born in Macon, Missouri, in 1910, Jones was totally captivated by aeronautics at an early age:

> All during the late twenties the weekly magazine *Aviation* appeared on the local newsstand in my hometown, Macon, Missouri. *Aviation* carried technical articles by eminent aeronautical

engineers such as B. V. Korvin-Krovkovsky, Alexander Klemin, and others. Included in both *Aero Digest* and *Aviation* were notices of forthcoming *NACA Technical Reports* and *Notes*. These could be procured from the Government Printing Office usually for ten cents and sometimes even free simply by writing NACA Headquarters in Washington. The contents of these reports seemed much more interesting to me than the regular high school and college curricula, and I suspect that my English teachers may have been quite perplexed by the essays I wrote for them on aeronautical subjects.[117]

Jones attended the University of Missouri for one year, but left to take a series of aeronautics-related jobs, first as a crew member with the Marie Meyer Flying Circus, and then with the Nicholas-Beazley Airplane Company in Marshall, Missouri, which was just starting to produce a single-engine, low-wing monoplane designed by the noted British aeronautical engineer Walter H. Barling. At one time, Nicholas-Beazley was producing and selling one of those aircraft each day. However, the company became a victim of the Depression, and in 1933 Jones found himself working as an elevator operator in Washington, D.C., and taking night classes in aeronautics at Catholic University, taught by Max Munk. That contact began a lifelong friendship between Jones and Munk. In 1934 the Public Works Administration created a number of temporary scientific positions in the federal government. On the recommendation of Congressman David J. Lewis, from Jones's hometown, Jones received a nine-month appointment at the NACA Langley Memorial Laboratory. That was the beginning of a lifetime career for Jones at the NACA/NASA. Through a passionate interest and self-study in aeronautics, Jones had become exceptionally knowledgeable in aerodynamics theory. His talents were recognized at Langley, and he was kept on at the laboratory through a series of temporary and emergency reappointments for the next two years. Unable to promote him into the lowest professional engineering grade because of civil-service regulations that required a college degree, in 1936 the laboratory management was finally able to hire Jones permanently via a loophole: It hired Jones at the next grade above the lowest, for which the requirement of a college degree was not specifically stated (although presumed).

By 1944, Jones was one of the most respected aerodynamicists in the NACA. At that time he was working on the design of an experimental air-to-air missile for the Army Air Force and was also studying the aerodynamics of a proposed glide bomb having a low-aspect-ratio delta wing.[84] The Ludington-Griswold Company of Saybrook, Connecticut, had carried out wind-tunnel tests on a dart-shaped missile of their design, and Roger Griswold, president of the company, showed the data to Jones in 1944. Griswold had compared the lift data for the missile's low-aspect-ratio delta wing with calculations made from Prandtl's tried-and-proved lifting-line theory. Jones realized that Prandtl's lifting-line theory was not valid for low-aspect-ratio wings, and he began to construct a more appropriate theory for the delta-wing planform. Jones obtained rather simple analytical equations for the low-speed, incompressible flow over delta wings, but considered the theory to be "so crude" that "nobody would be interested in it." He placed his analysis in his desk and went on with other matters.

In early 1945 Jones (Figure 9.36) began to look at the mathematical theory of supersonic potential (irrotational) flows. When applied to delta wings, Jones found that he was obtaining equations similar to those he had found for incompressible flow using the crude theory that was now buried in his desk. Searching for an explanation, he recalled the statement by Max Munk in 1924 that the aerodynamic characteristics of a wing were governed mainly by the component of the free-stream velocity perpendicular to the leading edge.[161] The answer suddenly was quite simple: For the delta wing, the reason his supersonic findings were the

Figure 9.36 Robert T. Jones.

same as his earlier low-speed findings was that the leading edge of the delta wing was swept far enough that the component of the supersonic free-stream Mach number perpendicular to the leading edge was subsonic, and hence the supersonic swept wing acted as if it were in a subsonic flow. With that revelation, Jones had independently discovered the high-speed aerodynamic advantage of swept wings, albeit 10 years after Busemann's paper at the Volta conference.

Jones began to discuss his swept-wing theory with colleagues at NACA Langley. In mid-February 1945 he outlined his thoughts to Jean Roche and Ezra Kotcher of the Army Air Force at Wright Field. On March 5, 1945, he sent a memo to Gus Crowley, chief of research at Langley, stating that he had "recently made a theoretical analysis which indicates that a V-shaped wing traveling point foremost would be less affected by compressibility than other planforms. In fact, if the angle of the V is kept small relative to the Mach angle, the lift and center of pressure remain the same at speeds both above and below the speed of sound." In the same memo, Jones asked Crowley to approve experimental work on swept wings. Such work was quickly initiated by the Flight Research Section of Langley, under the direction of Robert Gilruth, beginning with a series of free-flight tests using bodies with swept wings dropped from high altitude.

Jones finished a formal report on his low-aspect-ratio wing theory in late April 1945, including the effects of compressibility and the concept of a swept wing. However, during the in-house editorial review of that report, Theodore Theodorsen raised some serious objections. Theodorsen did not like the heavily intuitive nature of Jones's theory, and he asked Jones to clarify the "hocus-pocus" with some "real mathematics." Furthermore, because supersonic flow was so different physically and mathematically from subsonic flow, Theodorsen could not accept the "subsonic" behavior of Jones's highly swept wings at supersonic speeds. Criticizing Jones's entire swept-wing concept, calling it "a snare and a delusion," Theodorsen insisted that Jones take out the part about swept wings.[84]

Theodorsen's insistence prevailed, and publication of Jones's report was delayed. However, at the end of May 1945, Gilruth's free-flight tests dramatically verified Jones's predictions, showing a factor-of-4 reduction in drag due to sweeping the wings.[166] Quickly following those data, wind-tunnel tests carried out in a small supersonic wind tunnel at

Langley showed a large reduction in drag on a section of wire in the test section when the wire was placed at a substantial angle of sweep relative to the flow in the test section.[84] With that experimental proof of the validity of the swept-wing concept, Langley forwarded Jones's report to NACA Headquarters in Washington for publication. But Theodorsen would not give up: The transmittal letter to NACA Headquarters contained the statement that "Dr. Theodore Theodorsen (still) does not agree with the arguments presented and the conclusions reached and accordingly declined to participate in editing the paper." Such recalcitrance on the part of Theodorsen is reminiscent of his refusal to believe that the shock waves seen in John Stack's schlieren photographs of the transonic flow over an airfoil 11 years earlier were real. Theodorsen certainly made important contributions to airfoil theory in the 1930s, but he was also capable of errors in judgment (i.e., he was human).

On June 21, 1945, the NACA issued Jones's report as a confidential memorandum, chiefly for the army and navy. Three weeks later, the report was reissued as an advance confidential report, sent by registered mail to those people in industry with a "need to know." Entitled "Wing Plan Forms for High-Speed Flight," Jones's report quickly spread the idea of the swept wing to selected members of the aeronautical community in the United States, but by that time, information about the German swept-wing research was beginning to reach the same aeronautical community. Jones's work appeared in the open literature about a year later, as NACA TR 863, a technical report only five pages long, but a classic explanation of how a swept wing works aerodynamically.

Credit for the idea of the swept wing for high-speed flight is shared between Busemann and Jones. Separated by a time interval of 10 years, and the closed shops of military security in both Germany and the United States, each independently developed the concept, not knowing of the other's work. The full impact of the swept-wing concept on the aeronautical industry came directly after the end of World War II, with almost simultaneous release of similar information from both sides of the ocean, thus promoting confidence in the validity of the concept.

The Swept Wing: Its Impact on Flying Machines

Boeing's George Schairer was aware of Jones's swept-wing concept before the intelligence team led by von Kármán left for Europe in April 1945, and he was aware that Jones's report was being held up in the editorial process. Indeed, Schairer reported that the swept wing was the main topic of conversation as the intelligence team lumbered across the Atlantic Ocean, taking 26 hours to fly from Washington to Paris in an army C-54 (the military version of the DC-4). Schairer stated that "I concluded during this flight that sweepback was a valid concept."[162] When the team saw the German swept-wing data at Braunschweig on May 7, Schairer needed no convincing of its value. The letter he sent to Boeing on May 10 set into motion a revolution in high-speed airplane design, a revolution that we see reflected in almost every commercial and military jet airplane today.

At that time, Boeing was working on the design of a jet-powered bomber with conventional straight wings. The design team was frustrated because none of their airframe designs seemed, on paper, capable of achieving the high Mach numbers that were potentially available with jet engines. John Steiner, an M.I.T. graduate and Boeing engineer since 1941, stated that "the Boeing bomber team at Seattle was redirected by George's letter sent May 10, 1945, which told us to investigate sweepback, which we did, although not without the usual resistance from within the organization."[167] Boeing had built a high-

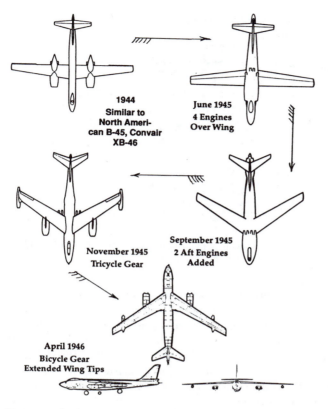

Figure 9.37 Design evolution of the Boeing B-47. (From Cook,[168] with permission.)

speed wind tunnel in 1944 (with an 8- × 12-ft test section and airspeeds as high as Mach 0.975). After the redirection, that tunnel was essentially dedicated to the examination of new bomber designs with swept wings. The result was the B-47, the first operational swept-wing aircraft.[168] Figure 9.37 shows the design evolution of the B-47, with the final configuration at the bottom. The B-47 was the progenitor of the famous Boeing 707, the first successful jet transport, and the direct ancestor of almost all the modern civil jet transports. When addressing the question of the credit for the swept wing, Schairer gave credit to the 1935 Volta conference paper by Busemann, but went on to say that "Jones separately invented sweepback and first brought it to the attention of U.S. aerodynamicists. I give him full credit for the XB-47 being built with swept wings."[162]

About 1,000 miles south of Seattle, similar activity was taking place in the North American aircraft plant in Los Angeles. In 1944, North American's confidential Design Group was working on a series of jet-propelled fighter airplanes, including a jet version of their highly successful P-51 Mustang. One design was a straight-wing jet fighter for the navy, designed the XFJ-1, and a very similar design was prepared for the air force, designated the XP-86 (Figure 9.38). When the team headed by von Kármán brought back German technical reports on swept-wing aerodynamics, North American's head of design aerodynamics, Larry Green, who read technical German, studied those reports carefully. He knew that the straight-wing design for the XP-86 would not be markedly better than the existing Lockheed

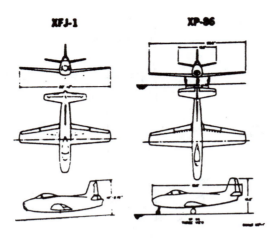

Figure 9.38 Original straight-wing designs for the North American XFJ-1 and XP-86. (From Blair,[164] copyright © 1980 AIAA, with permission.)

P-80 Shooting Star or the Republic P-84, both of which were already on order by the Army Air Force. However, if the XP-86 were to be redesigned with a swept wing, its top speed would be at least 10% higher than that of the competition. The Aerodynamics Group at North American was soon converted to the swept-wing design, and in August 1945 the top management approved the swept-wing design for the XP-86. Blair stated that the in-house proposal to change to a swept wing met some internal resistance, with claims that it was "Germanizing our design,"[164] but in the end there was no real contest. North American initiated a battery of wind-tunnel experiments on swept wings, resulting in a wing design for the XP-86 with a 35° sweep, aspect ratio of 4.8, and taper ratio (root chord to tip chord) of 0.51. The airfoil section was a very conservative NACA 0009 symmetric airfoil of 9% thickness (the use of symmetric airfoils for airplane wing sections was not unheard of, but it was certainly unusual). Figure 9.39 shows the North American data for drag divergence for various wings. Note that the straight-wing XP-86 was not much of an improvement over a P-51 Mustang, but the swept-wing F-86 had a much higher drag-divergence Mach number. (In 1948 the U.S. Air Force replaced the "P" designation, for pursuit, with the "F" designation, for fighter.)

On October 1, 1947, exactly two weeks before Chuck Yeager broke the sound barrier in the X-1, the XP-86 made its first flight at Muroc in California, with North American test pilot George Welch at the controls. Less than a year later, on April 26, 1948, Welch put the XP-86 into a shallow dive and went supersonic. It was the first time that an airplane designed for combat had exceeded the speed of sound. A three-view of the F-86 is shown in Figure 9.40.

By 1948 the swept wing had become an accepted design feature. It had done for the high-speed jet airplane what streamlining had done in the 1930s for the advanced propeller-driven airplane, namely, provided the aerodynamic means for efficient flight in the desired flight regime. Virtually all high-speed, jet-propelled airplanes today have wings with highly swept leading edges, and these contemporary aircraft can trace their ancestry directly back to the B-47 and the F-86 and to the innovative ideas and genius of Adolf Busemann and Robert Jones.

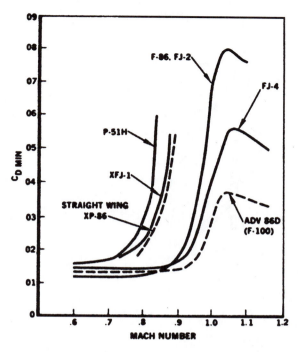

Figure 9.39 Drag-divergence characteristics for several aircraft, straight and swept wings. (From Blair,[164] copyright © 1980, with permission.)

Engineering Science and Aircraft Design: Technology Transfer in the Jet Age

This is a good place to pause for a moment and reflect on a rather dramatic paradigm shift that took place at the end of the era of the advanced propeller-driven airplane and the beginning of the age of the jet-propelled airplane in regard to the speed and means by which new aerodynamic knowledge was transferred to the practical world of airplane design. In the nineteenth century, mathematical and experimental advances were being made in basic fluid dynamics, primarily by highly educated scientists at universities in Britain, France, and Germany. The work of Navier, Stokes, Helmholtz, Reynolds, Kelvin, and Rayleigh, to name just a few, had advanced the state of the art of fluid mechanics to an impressive degree by the end of the nineteenth century, but there was a schism between those well-respected academics and the class of self-educated engineers who were struggling to solve the mechanical problems of heavier-than-air flight. There was virtually no transfer of technology from the academic knowledge base in fluid dynamics to the engineering designs of Cayley, Wenham, Phillips, Lilienthal, and even the Wright brothers.

That situation began to change in the early twentieth century, when Prandtl and his group at Göttingen began to bridge the gap between academic research and the airplane designer. They provided the first conduit by which the body of knowledge in basic fluid dynamics accumulated in the nineteenth century, augmented by Prandtl's extensions and applications of that knowledge for an understanding of aerodynamics problems, began to be made readily available to those working in the practical world of airplane design. The flow of information

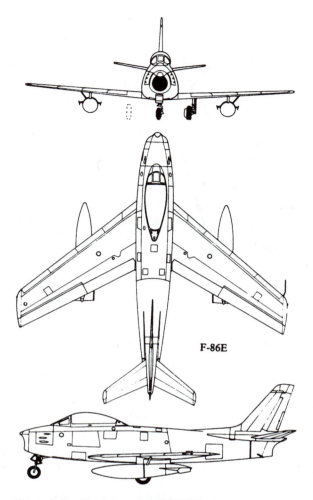

Figure 9.40   North American F-86E Sabre.

was slow at first: It took 20 years for Prandtl's boundary-layer theory to become widely known and appreciated by aircraft designers; ultimately, the wider understanding of the boundary layer and its association with skin-friction drag and flow separation set the stage for the massive efforts toward aerodynamic streamlining during the 1920s and 1930s. It took less time for the designers to appreciate Prandtl's lifting-line theory for finite wings and its importance for an understanding of induced drag and a comprehension of the advantages of high-aspect-ratio wings, but there was still a lag of about 10 years from the beginning of Prandtl's work to its impact on flying machines. Munk's thin-airfoil theory in the 1920s and Theodorsen's more generalized airfoil analysis in the early 1930s reached aircraft designers sooner, mainly because of the widespread dissemination of NACA technical reports and the respect those reports enjoyed in the aircraft industry, but even in those cases there was some lag time – a sort of intellectual divide, such that the typical airplane designer did not have sufficient depth of mathematical knowledge to fully understand the theories, though that did not prevent the designers from using the important practical findings and data from the more complex theories.

The type of work carried out by Prandtl, Munk, and Theodorsen can be classified as *engineering science,* because its purpose was to use scientific knowledge to synthesize a type of *engineering predictive capability.* Finally, in the 1920s and 1930s, for the first time in the history of aerodynamics, we see the beginnings of a real technology transfer of academic science to the design of flying machines.

The speed at which that technology transfer was occurring took a quantum leap in the age of the jet-propelled airplane. A progressively better understanding of the physics of airflows over airfoils and wings near the speed of sound, and the attendant creation of shock waves in those transonic flows, was provided by Briggs, Dryden, Stack, and Jacobs during the 1920s and early 1930s. Although that work did not immediately lead to a direct predictive capability, it still can be classified as engineering science, because it was new scientific knowledge that was applied by the engineer in a qualitative sense to enhance the design of high-speed aircraft. I contend that engineering science can be defined as one or both of the following: the use of scientific knowledge to obtain an engineering predictive capability and/or to obtain a fundamental physical understanding that can qualitatively and intuitively influence engineering design. In that sense, the early research on compressibility effects was engineering science, and that engineering science was quickly transferred to the designers.

It is obvious that the speed at which technology transfer takes place from engineering science to the airplane designer will to a considerable extent be determined by the needs of the designer. If the only way that the designer can meet the stipulated design specifications is by digging into the latest scientific research findings, the speed at which technology transfer will take place will be rapid. That is a twentieth-century development. In the nineteenth century, the designers of flying machines certainly needed the findings from academic research in fluid dynamics, but there was such a large schism between the two cultures – the university-educated academician and the self-educated engineer – that there simply was no one to bridge the gap to bring about the transfer. Even the creation of the Aeronautical Society of Great Britain in 1866 did not effectively close the gap, although that was one of the stated purposes for its creation. In the twentieth century, most of the necessary bridges have been built, but the amount of information transferred and the speed of its flow across those bridges still depend on the needs of the designer.

A case in point concerns the compressibility research in the 1930s. When the Lockheed P-38 began to suffer from the tuck-under problem in dives, and it was recognized that the cause of that tendency was the transonic flow over the wing, the transfer of the qualitative knowledge from 1930s research (engineering science) to fix the problem (design) was very rapid. In that case, the needs of the designer were determinative. The necessary bridge already existed (partly because of the work of John Stack and his people and their accumulated knowledge base), and the needs of the designer elicited the flow of information across the bridge.

The design of the Bell X-1 provides another case in point. Although it was to be a research airplane whose purpose would be to generate new knowledge about the aerodynamics of high-speed flight (i.e., to generate new engineering science), its design process required the use of the existing engineering-science knowledge of transonic flows. The Bell designers of the X-1 explored every conduit for technology transfer that they could find.

Application of the swept-wing concept provided a more graphic example of technology transfer from engineering science to the design process in aerodynamics. The engineering science had been worked out by Busemann, his German colleagues, and Robert Jones. When the development of the jet engine made transonic and supersonic airplanes possible, the next great problem was to reduce wave drag. Because of the simultaneous discovery of the German swept-wing research and Robert Jones's independent development of the concept in 1945, the industry had an answer already in hand. The transfer of technology was almost immediate.

So it is important to recognize that at the beginning of the age of the jet-propelled airplane there was a definite paradigm shift in regard to the use of state-of-the-art research findings in the design of flying machines. In the nineteenth century, the state of the art in aerodynamics was not reflected in flying machines. That situation had been completely reversed by the 1940s, when the design process for new jet aircraft used, indeed demanded, the latest aerodynamic knowledge. That situation has continued to the present. Indeed, in a sense, the designs for the new stealth airplanes have gone beyond the state of the art in aerodynamics; the configurations for the F-117 stealth fighter and the B-2 stealth bomber have been influenced more by the state of the art in electrical engineering than by that in aerodynamics.

Supersonic Wind Tunnels: Early Development

In the twentieth century, the primary laboratory device for aerodynamic testing has been the wind tunnel. We have described the development of the wind tunnel during the various periods in the history of aerodynamics, ranging from the low-speed tunnels of Wenham, Phillips, the Wright brothers, Eiffel, and Prandtl to the early high-speed subsonic and transonic tunnels used for compressibility research in the 1920s and 1930s. Therefore, to parallel our discussion of the evolution of supersonic aerodynamics, we shall briefly examine the early development of supersonic wind tunnels.

The first supersonic nozzle was developed by de Laval in the late nineteenth century, for use with steam turbines, and Stodola studied the physics of the flows in such nozzles at the turn of the century. In 1905, Prandtl built a small Mach-1.5 tunnel at Göttingen to study steam-turbine flows and the movements of sawdust around sawmills. The first practical supersonic wind tunnel for aerodynamic testing was built by Busemann at Braunschweig in the mid-1930s. Using the method-of-characteristics technique, which he had developed in 1929, Busemann designed the first smooth supersonic nozzle contour that could produce shock-free flow (Figure 9.41). He had a diffuser with a second throat downstream to decelerate the flow and to ensure efficient operation of the tunnel. All supersonic tunnels today look essentially the same. During the same time period, Ackeret helped to design and construct supersonic wind tunnels in many European locations, most notably at Guidonia, Italy.

Participants at the 1935 Volta conference toured the Guidonia aerodynamics laboratory and saw the supersonic wind tunnel designed by Ackeret, and when von Kármán returned to the United States, he immediately proposed that the NACA construct a large supersonic wind tunnel:

> I proposed a large modern supersonic wind tunnel to augment the small high-speed wind tunnels, which had been built by NACA aerodynamicists some years earlier for testing models in winds up to 650 miles per hour. I felt that as a start we should increase our effort in supersonic research if we were to keep up with the rest of the world. Dr. George Lewis, the director of NACA, told me that he did not see why anybody would need a major wind tunnel that developed a speed greater than the speed of the propeller blade tips – about 500

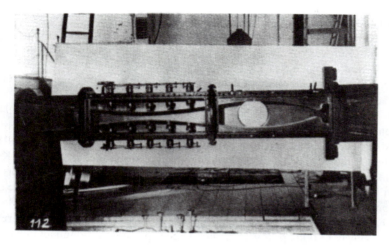

Figure 9.41   The first practical supersonic wind tunnel, built by Busemann in the mid-1930s.

to 600 miles per hour, the point of maximum propeller efficiency. So he turned down my proposal. In the light of subsequent events such as the development of jet propulsion, such decisions appear very shortsighted indeed.

I went home to Pasadena, discouraged. I could only recall an old axiom – good judgment comes from experience, but experience comes from bad judgment [p. 224].[67]

In 1937, after another trip to Europe, von Kármán tried again:

I made another strong attempt to convince the U.S. government to build a modern supersonic wind tunnel. Not only was I turned down again, but a few years later in the face of the growing need for supersonic data for missile research, the Bureau of the Budget went so far as to order hearings with the idea of restricting the building of all new research facilities to "avoid duplication," implying I suppose that all wind tunnels were alike [pp. 224–5].[67]

However, unknown to von Kármán and to most others, Eastman Jacobs also returned from the Volta conference convinced that the NACA should study supersonic flows. He asked Arthur Kantrowitz, a young physicist with a recent M.A. from Columbia, to examine the nature of supersonic flow: "Jacobs made this decision on his own, in defiance of a cautious NACA management stance against supersonics."[84] In 1939, after both Jacobs and Kantrowitz had become convinced that supersonic flight was inevitable, Jacobs gave Kantrowitz a free hand to design a supersonic wind tunnel. In July 1942 that tunnel became operational. It had a square test section 9 × 9 in., with a Mach number of 2.5. It was the first meaningful supersonic tunnel in the United States, the result of a farsighted but unauthorized order from Eastman Jacobs. Although it had water-condensation problems (the water vapor in the airflow condensed as the airstream became cold in the supersonic expansion), that tunnel was important for two reasons: (1) It provided a pioneering educational experience for NACA aerodynamicists in the fundamentals of supersonic flows. (2) It served as a model for a larger supersonic tunnel that had just been authorized by NACA Headquarters to be constructed at the new NACA Ames Laboratory in Mountain View, California, near San Francisco. Thus, in retrospect, the Langley model supersonic tunnel finally received the official blessing.

But the United States was still well behind Germany in the use of supersonic wind tunnels at that time. By 1936, Dr. Rudolph Hermann, at the Technical University of Aachen, had constructed, with Luftwaffe funding, a Mach-3.3 tunnel with a square test section 10×10 cm. Later, Hermann joined the V-2 rocket-development center at Peenemünde, where several supersonic wind tunnels were put into operation. The definitive study of the German supersonic aerodynamics program in support of V-2 development is by Neufeld.[169]

Von Kármán finally had his chance to build a supersonic wind tunnel when, in 1942, the army suddenly realized that it needed one. After General G. M. Barnes, chief of ordnance research and engineering, returned from a trip to England, where he was shown a small model of a supersonic tunnel the British were building for guided-missile research, von Kármán was asked to prepare a preliminary design. Assisted by one of his graduate students at Cal Tech, Allen Puckett, von Kármán designed and supervised the construction of the army's first supersonic tunnel, which went into operation at Aberdeen Proving Ground in Maryland in 1944. It had a test section that was 15×20 in. in cross section, and the power required for its operation was 13,000 hp. It was the first large supersonic wind tunnel in the United States, and it was the first to be used for practical testing of supersonic bodies.

That first introduction to supersonic wind tunnels gave aerodynamicists in the United States some badly needed experience in dealing with supersonic flow phenomena and the aerodynamic characteristics of supersonic configurations, and that increasing maturity was reflected in the rapid construction of a very large (6×6 ft) supersonic tunnel at the NACA Ames Laboratory beginning in 1945. When it was finished, that tunnel occupied an entire building at Ames (Figure 9.42).

By 1947, when Chuck Yeager proved that the sound barrier was no barrier, testing with supersonic wind tunnels by NACA and army aerodynamicists had finally begun in earnest. The early development of supersonic wind tunnels had a parallel in the development of swept wings: The supersonic wind tunnel, which was first developed and fostered in Germany (as

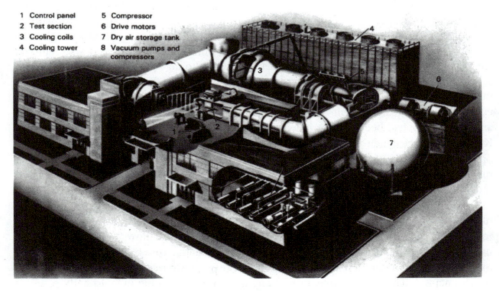

Figure 9.42   The NACA Ames Complex for the 6- $\times$ 6-feet supersonic wind tunnel.

was the swept wing), reached its highest expression in the United States (as did the swept wing). We are reminded again that much of the advanced aerodynamics work in Germany during World War II did more for the victors after the war than it did for Germany during the war.

A final note relevant to this section is that the accelerated program for design and use of supersonic wind tunnels in the United States carried the state of the art in aerodynamics well beyond the supersonic regime. In November 1947, only a month after the Bell X-1 had broken the sound barrier, aerodynamicists at NACA Langley successfully operated an 11-in. tunnel at Mach 6.9 – the first operation of a hypersonic tunnel in the United States. That was not, however, the world's first hypersonic tunnel: When the Allies captured the German V-2 test facilities at Peenemünde, they discovered a 1.2-ft wind tunnel that could generate Mach 5, the threshold of the hypersonic regime.

Modern Developments in Aerodynamics: Hypersonics and Computational Fluid Dynamics

The field of aerodynamics in the age of the jet-propelled airplane is still evolving. The progression of the theory and the empirical knowledge in aerodynamics in the early part of that age has been discussed, leading us to the modern period of that age, the period from 1950 to the present. Looking back to ancient Greek science, where we began our review of the history of aerodynamics, and considering the evolution of aerodynamics over the centuries, clearly the short period from 1950 to the present is "modern" by comparison. There has been a virtual explosion of research and development in aerodynamics during this modern period, and that story would fill another book. Also, the events of the modern period are still in progress; in light of the broad survey and vast span of time covered herein, it is perhaps premature to look at the modern period through the same eyes, from the same perspective. However, there have already been two spectacular advances in aerodynamics during this period that stand out above all others, two advances that future historians surely will rank as pivotal developments in modern aerodynamics: (1) hypersonic aerodynamics, driven by the evolution of ballistic missiles and the space program during the last half of the century, and (2) computational fluid dynamics, made possible by the development of the high-speed digital computer.

Hypersonic Flight

On February 24, 1949, the pens on automatic plotting boards at South Station were busy tracking the altitude and course of a rocket that just moments before had been launched from a site three miles away on the test range of the White Sands Proving Ground. The rocket was a V-2, one of many brought to the United States from Germany after World War II. By that time, launching V-2s had become almost routine for the crews at White Sands, but on that day neither the launch nor the rocket was routine. Mounted on top of that V-2 was a slender, needle-like rocket called the WAC Corporal, which would function as a second stage following the burnout of the V-2. That test firing of the combination V-2/WAC Corporal was the first meaningful attempt to implement a multistage rocket for achieving high velocities and high altitudes, as part of a larger program labeled "Bumper" by the U.S. Army. All previous rocket launchings of any importance, both in the United States and in Europe, had used the single-stage V-2 by itself. Figure 9.43 shows the "Bumper" rocket as

Figure 9.43   V-2/WAC Corporal lift-off on February 24, 1949, the first object of human invention to achieve hypersonic flight.

it lifted off the New Mexico desert. The pen plotters tracked the V-2 to an altitude of 100 miles at a velocity of 3,500 mph, at which point the WAC Corporal was ignited. The slender upper stage accelerated to a maximum velocity of 5,150 mph and reached an altitude of 244 miles, exceeding by a healthy 130 miles the previous record set by a V-2 alone. After reaching that peak, the WAC Corporal nosed over and careered back into the atmosphere at more than 5,000 mph. Thus it became the first object of human invention to achieve hypersonic flight – the first time that any vehicle has flown faster than five times the speed of sound. In spite of the pen plotters charting its course, the WAC Corporal could not be found in the desert after the test. The only remnants to be recovered later were a charred electric switch and part of the tail section, found in April 1950.

Near the small village of Smelooka in the Ternov district, Saratov region, Russia, at 10:55 a.m. (Moscow time) on April 12, 1961, a strange spherical object landed under the canopy of a parachute. Its surface was charred black, and it had three small viewing ports covered with heat-resistant glass. Inside that capsule was Flight Major Yuri Gagarin, who only 108 min earlier had been sitting on top of a rocket at the Russian comsodrome at Baikonur, near the Aral Sea. Part of what had happened during those 108 min was announced to the world by a broadcast from the Soviet news agency Tass:

> The world's first spaceship, Vostok [East], with a man on board was launched into orbit from the Soviet Union on April 12, 1961. The pilot space-navigator of the satellite-spaceship Vostok is a citizen of the U.S.S.R., Flight Major Yuri Gagarin.
>
> The launching of the multistage space rocket was successful and, after attaining the first escape velocity and the separation of the last stage of the carrier rocket, the spaceship went into free flight on around-the-earth orbit. According to preliminary data, the period of the revolution of the satellite spaceship around the earth is 89.1 min. The minimum distance from the earth at perigee is 175 km [108.7 miles] and the maximum at apogee is 302 km

Figure 9.44   Vostok I, in which Russian Major Yuri Gagarin became the first human to fly at hypersonic speed, during the world's first manned orbital flight, April 12, 1961.

[187.6 miles], and the angle of inclination of the orbit plane to the equator is 65°4'. The spaceship with the navigator weighs 4725 kg [10,418.6 lb], excluding the weight of the final stage of the carrier rocket [*New York Times,* April 12, 1961].

After that announcement was made, Gagarin's orbital craft, called Vostok I (Figure 9.44), was slowed at 10:25 a.m. by the firing of a retro-rocket and entered the earth's atmosphere at a speed in excess of 25 times the speed of sound. Thirty minutes later, Gagarin had become the first man to fly in space, to orbit the earth, and to return safely. Also, he had been the first human to experience hypersonic flight.

That year would produce a bumper crop of manned hypersonic flights. On May 4, Alan B. Shepard became the second man in space by virtue of a suborbital flight over the Atlantic Ocean, reaching an altitude of 115.7 miles and reentering the atmosphere at a speed above Mach 5. On June 23, U.S. Air Force test pilot Major Robert White flew the X-15 airplane at Mach 5.3, the first X-15 flight to exceed Mach 5 (in so doing, White accomplished the first mile-per-second flight in an airplane, reaching a maximum velocity of 3,603 mph). That record was extended by White on November 9, flying the X-15 at Mach 6.

Thus hypersonic flight, both unmanned and manned, suddenly burst onto the scene in the middle of the twentieth century without anyone having made much preparation for it.[170] Whereas an understanding of supersonic aerodynamics had been more or less in place in the 1930s, ready to be used for the new supersonic vehicles that appeared after the war, the study of hypersonic aerodynamics was in somewhat of a catch-up mode right from the start. Hypersonic vehicles proliferated rapidly because of the ready availability of powerful rocket engines, and there was little hesitation to send them off at their incredible speeds, in spite of a dearth of knowledge at that time about aerodynamic flows at very high Mach numbers.

Hypersonic flows are characterized by new physical phenomena that become important at high Mach numbers, but are not so important at lower supersonic Mach numbers. For instance, the enormous amounts of aerodynamic heating imparted to a surface in a hypersonic flow (due to the frictional effect of the high-speed flow over the surface, as well as the high gas temperatures behind strong shock waves) become a paramount concern at hypersonic speeds, indeed the limiting factor in aerodynamic designs for hypersonic vehicles. The flow fields around hypersonic vehicles become hot enough to cause the nitrogen and oxygen molecules in air to dissociate, literally tear apart, leading to various chemical reactions. Such massive aerodynamic heating and the accompanying high-temperature chemical reactions in hypersonic flows were uncharted regions to the aerodynamicists responsible for examining the new hypersonic regime. Prior to the sudden appearance of hypersonic vehicles that were already being launched, such phenomena had hardly ever been examined in the field of aerodynamics. Unexpectedly faced with the reality of hypersonic vehicles in the sky, the aerodynamics community found a brand new challenge – a challenge that was eagerly taken up in the 1950s.

One of the most dramatic examples to illustrate the differences between supersonic aerodynamics and hypersonic aerodynamics involved a finding that completely changed the design configurations for hypersonic vehicles, as compared with the more conventional supersonic bodies: It was found that a blunt-nose hypersonic body experienced far less aerodynamic heating than a sharp-nose body, which seemed contrary to conventional wisdom and intuition based on supersonic experience. That blunt-body finding was one of the most important research breakthroughs in hypersonic aerodynamics. But what was the nature of blunt-body flows that allowed this breakthrough?

High-speed, supersonic flight had become the dominant topic in aerodynamics by the end of World War II. By that time, aerodynamicists appreciated the advantages of using slender, pointed body shapes to reduce the drag on supersonic vehicles. The more pointed and slender the body, the weaker the shock wave attached to the nose, and hence the lower the wave drag. The German V-2 rocket used during the last stages of World War II had a pointed nose, and all short-range rocket vehicles flown during the next decade followed suit. Then, in 1953, the first hydrogen bomb was exploded by the United States. That immediately spurred the race for long-range intercontinental ballistic missiles (ICBMs) to deliver such bombs. Such vehicles were designed to fly outside of the earth's atmosphere for distances of 5,000 miles or more and to reenter the atmosphere at suborbital speeds of 20,000–22,000 ft/s. At such high velocities, the aerodynamic heating of a reentering vehicle became severe, and that heating problem dominated the thinking in high-speed aerodynamics. The logical first approach was to stay conventional: Design a sharp-pointed, slender body for reentry. Efforts to minimize aerodynamic heating centered on maintaining a laminar boundary-layer flow on the vehicle's surface; such laminar flow produces far less heating than turbulent flow. However, Nature much prefers turbulent flow, and reentry vehicles are no exception. Therefore, the pointed-nose reentry body was doomed to failure, because it would burn up in the atmosphere before reaching the earth's surface.

However, in 1951, one of those major breakthroughs that occur very infrequently in engineering was achieved by H. Julian ("Harvey") Allen at the NACA Ames Aeronautical Laboratory: He hit upon the idea of a blunt reentry body. His thinking was influenced by the following concepts: At the beginning of reentry, near the outer edge of the atmosphere, the vehicle would have a large amount of kinetic energy because of its high velocity, and a large amount of potential energy because of its high altitude. But by the time the vehicle reached the surface of the earth, its velocity would be relatively slow, and its altitude would be zero; thus it would have lost virtually all of its kinetic and potential energy. Where had all the energy gone? The answer was that it had gone into (1) heating the body and (2) heating the airflow around the body, as illustrated in Figure 9.45. The shock wave from the nose of the vehicle will heat the airflow around the vehicle; at the same time, the vehicle will be heated by the intense frictional dissipation within the boundary layer on the surface. Allen reasoned that if more of the total reentry energy could be dumped into the airflow, then less would be available to be transferred to the vehicle itself in the form of heating. The way to increase the heating of the airflow was to create a stronger shock wave at the nose (i.e., to use a blunt-nose

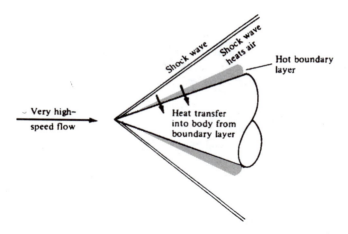

Figure 9.45 The energy of reentry goes into heating both the body and the air around the body.

body). The contrast between slender and blunt reentry bodies is illustrated in Figure 9.46. That was a stunning conclusion – to minimize aerodynamic heating, use a blunt body, rather than a slender body. The finding was so important that it was bottled up in a classified government document. Moreover, because it was so foreign to contemporary intuition, the blunt-body concept for reentry vehicles was accepted only gradually by the technical community. Over the next few years, additional aerodynamic analyses and experiments confirmed the superiority of blunt reentry bodies. By 1955, Allen was publicly recognized for his work, receiving the Sylvanus Albert Reed Award of the Institute of the Aeronautical Sciences (now the American Institute of Aeronautics and Astronautics). Finally, in 1958, his work was made available to the public in NACA Report 1381, "A Study of the Motion and Aerodynamic Heating of Ballistic Missiles Entering the Earth's Atmosphere at High Supersonic Speeds." Since Harvey Allen's early work, all successful reentry bodies, from the first Atlas ICBM to the manned Apollo lunar capsule, have been blunt. Allen went on to distinguish himself in many other areas, becoming the director of the NASA Ames Research Center in 1965, and retiring in 1970. His work on the blunt reentry body is an excellent example of the special importance and somewhat different behavior of hypersonic aerodynamics – a stand-alone discipline within the spectrum of flight aerodynamics.

Today, hypersonic aerodynamics remains a field of great importance, with many applications expected in the future. In the 1980s and 1990s, progress in hypersonic aerodynamics became intimately entwined with progress in another discipline: computational fluid dynamics.

Computational Fluid Dynamics

By the 1970s, a revolution was taking place in aerodynamics, a revolution of such fundamental impact that it would forever change the nature of aerodynamic predictions: the development of computational fluid dynamics (CFD).

It has been a consistent theme in the history of aerodynamics during the past two centuries that although the basic equations of motion – the Euler equations for an inviscid flow (Appendix A) and the Navier-Stokes equations for a viscous flow (Appendix B) – were known and well established by the middle of the nineteenth century, they could not be

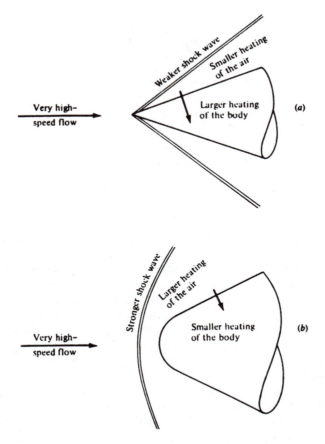

Figure 9.46   Contrast in degrees of aerodynamic heating for slender and blunt reentry vehicles: (a) slender reentry body, (b) blunt reentry body.

solved. They are systems of nonlinear partial differential equations for which there are no known general analytical solutions. Therefore, aerodynamicists over the past 150 years have simplified those equations by making certain approximate assumptions about the flow fields of interest, thus allowing some simplified analytical solutions, albeit at the cost of ignoring some of the physics. Indeed, that has been the world in which aerodynamics theory has functioned. However, because of the development of high-speed digital computers during the past half-century, it is now possible to obtain *numerical solutions* for the full nonlinear equations shown in Appendixes A and B. We can now obtain "exact" solutions for those equations for virtually any aerodynamic configuration ("exact" meaning that the equations themselves are not simplified, but are solved in their full form). However, the numerical answers are *not* "exact" in the purest meaning of the term; they are tainted by numerical round-off and truncation errors, and sometimes the numerical algorithms suffer from numerical stability problems that cause the attempted solutions to "blow up" on the computer. In spite of those problems, CFD has matured to the extent that quite accurate predictive solutions for complex flow problems have been obtained, which would be impossible with any other approach.

CFD can be defined as follows, referring to the partial differential equations for continuity, momentum, and energy in Appendixes A and B:

> Computational fluid dynamics is the art of replacing the ... partial derivatives ... in these equations with discretized algebraic forms, which in turn are solved to obtain *numbers* for the flow field values at discrete points in time and/or space. The end product of CFD is indeed a collection of numbers, in contrast to a closed-form analytical solution. However, in the long run, the objective of most engineering analyses, closed form or otherwise, is a quantitative description of the problem, i.e., *numbers* [pp. 24–5].[171]

CFD – a modern development – is totally consistent with the philosophy advanced by James Clerk Maxwell in 1856: "All the mathematical sciences are founded on relations between physical laws and laws of numbers, so that the aim of exact science is to reduce the problems of nature to the determination of quantities by operations with numbers."

Mathematical techniques to achieve numerical solutions for partial differential equations began to appear about the turn of the century. The first definitive work was carried out by Richardson,[172] who in a paper delivered to the Royal Society in London in 1910 introduced a finite-difference technique for numerical solution of Laplace's equation. Called a "relaxation technique," that approach is still used today to obtain numerical solutions for so-called elliptic partial differential equations (the equations that govern inviscid subsonic flows are such equations). However, modern numerical analysis is usually considered to have begun in 1928, when Courant, Friedrichs, and Lewy[173] published a definitive paper on the numerical solution of so-called hyperbolic partial differential equations (the equations that govern inviscid compressible flow are such equations).

Those early researchers had no idea that they were helping to lay the foundation for a revolution in aerodynamics predictions in the 1970s. They were mathematicians developing intellectually interesting numerical methods that in the early twentieth century were mainly of academic interest. The arithmetic operations necessary for such a solution for a practical flow field were so numerous that to carry them out by hand (the only way available at that time) was simply out of the question. However, in the 1950s, International Business Machines (IBM) introduced the first practical high-speed digital computers: the IBM 650 and 704. Those were followed in the 1960s by the IBM 7090 and 7094 mainframes and by the CDC 6400 and 6600 computers. The computer age had begun, and aerodynamicists quickly realized that those were computing tools that could make numerical solutions for the governing flow equations feasible. The result: CFD was born.

CFD provides a new dimension in the study and development of the whole discipline of fluid dynamics. The experimental tradition in fluid dynamics began in France during the middle of the seventeenth century, and rational theoretical analysis was begun by Newton at the end of the seventeenth century. For the next 250 years, the study of fluid dynamics took place in the binary world of pure experiment on one hand and pure theory on the other. Anyone learning fluid dynamics as recently as 1960 was operating in that binary, somewhat Janus-faced world of theory and experiment. However, CFD has added a new dimension, a third instrument for the study and practice of fluid dynamics. As sketched in Figure 9.47, CFD is today an equal partner with pure theory and pure experiment. And CFD is no flash in the pan – it will continue to fill that role for as long as our civilization endures.

An example of the power of CFD is shown in Figure 9.48, which illustrates the three-dimensional variation of the pressure over the surface of a modern generic fighter aircraft, obtained with a CFD solution for the Euler equations for inviscid flow.[174] The lines are pressure contour lines on the surface of the airplane (a pressure contour line is a curve along

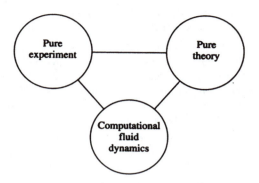

Figure 9.47   The three dimensions of modern fluid dynamics.

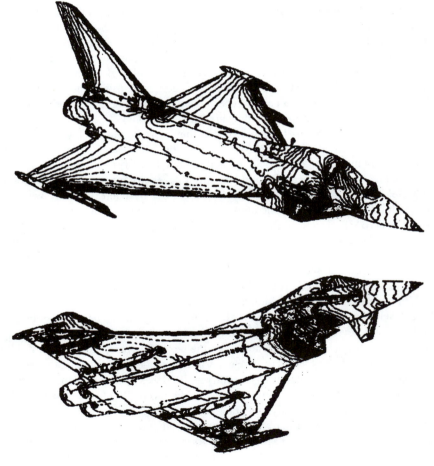

Figure 9.48   Three-dimensional pressure-coefficient contours over the surface of a generic fighter aircraft: $M_\infty = 0.85$, angle of attack $= 10°$, angle of yaw $= 30°$. (From Selmin et al.,[174] with kind permission from Elsevier Science, Amsterdam, The Netherlands.)

which the pressure is constant; different contour lines correspond to different pressures). The flow is transonic, with a free-stream Mach number of 0.85, and the airplane is at a 10° angle of attack and a 30° angle of yaw. This complex figure is an appropriate image with which to end our study of the evolution of high-speed, compressible-flow aerodynamics. We have come a long way since the pioneering efforts in the 1920s and 1930s at the NACA and the British Royal Aircraft Establishment. Looking at Figure 9.48, John Stack would be delighted.

Epilogue

So here we are at the end of the twentieth century and at the end of our discussion of the history of aerodynamics and its impact on flying machines. Some final comments may add to our perspective on that history.

There are three themes whose threads have been running through this history of aerodynamics. First, the discipline of aerodynamics – the fundamental understanding of the physical nature of aerodynamic flows and the gradual evolution of the basic governing equations – developed quite independently of any drive toward practical applications such as flying machines. The period from early Greek science through the Middle Ages generated a few basic thoughts, such as the concept of a continuum, the recognition that an object moving through the air experienced some type of aerodynamic "resistance," da Vinci's quantitative understanding of the continuity equation, and his careful observations of separated-flow patterns in water. Major advances came during the seventeenth century with the rise of serious experimental work in western Europe, particularly the development of rational mechanics by Newton. Human understanding of the physical sciences began to flower, as did the beginnings of a comprehension of aerodynamics. With Newtonian mechanics as a springboard, the intellectual giants of the eighteenth and nineteenth centuries, men such as Bernoulli, Euler, d'Alembert, Lagrange, Laplace, and Helmholtz, built an intellectual framework for aerodynamics that led to stunning quantitative breakthroughs: the Euler equations for an inviscid flow and the Navier-Stokes equations for a viscous flow. By the middle of the nineteenth century those equations embodied a full, rational understanding of the fundamental physics of aerodynamic flows. The only problem was that those equations were too complicated to be solved! All of that intellectual progress took place quite independently of any desire to design flying machines. The men involved in that progress were, for the most part, intellectuals simply pursuing a basic understanding of nature. In the eighteenth and nineteenth centuries, most of them were university-educated scientists who had only disdain for any ideas about machines that could fly. A subtheme for that period was the evolution of an understanding of how the aerodynamic force on a body would vary with velocity, density, and body area. Da Vinci found that the force was directly proportional to the body area, Galileo found that the force was proportional to the density, and, most important, Mariotte and Huygens proved that the force varied as the square of the velocity, rather than the first power as had previously been thought. Such understanding would be critical for the design of flying machines, but not until more than a century later.

Second, the group of inventors who were avidly trying to design, build, and take off in flying machines were the first practitioners of applied aerodynamics. They were mostly self-educated, well-read inventors who were also laying the foundation for the beginnings of the profession of engineering. Those men, such as Cayley, Wenham, Phillips, Lilienthal (who had an engineering degree), Langley, and, most important, the Wright brothers, carried out investigations that gathered the data that became the essence of applied aerodynamics in the nineteenth and at the beginning of the twentieth century. Those findings in applied aerodynamics were used to design flying machines. It is remarkable that the flying machine

447

was developed and advanced well into the beginning of the twentieth century without direct recourse to the state of the art of theoretical aerodynamics that existed in academic circles.

Third, all of that changed in the second decade of the twentieth century. Suddenly, with the airplane already a reality, academic scientists embraced a new cause. It became fashionable to carry out research in basic aerodynamics for the purpose of understanding the flight of airplanes. Indeed, the airplane became a primary justification for basic research in aerodynamics, as illustrated by the important work carried out by Prandtl in Germany. From that time forward, the enormous expansion of the engineering sciences – the understanding of nature and the development of predictive capabilities for engineering design purposes – has been driven by the needs of aeronautics and, more recently, astronautics. Today, state-of-the-art aerodynamic data are instantly available to and are understood by airplane designers. The former gap between academic researchers and aeronautical engineers is today virtually nonexistent.

Where is all this going? Because of the opening of new frontiers in hypersonic flight and computational fluid dynamics, aerodynamics continues as a vital discipline. One wonders what the story of aerodynamics will be like in the far future, when the developments discussed in this book will have become ancient history. Experience has shown us that an accurate answer to that question is impossible. We can only dream. I hope that this book will provide the reader more substance for such dreams.

The Euler Equations

Consider any general flow in a three-dimensional space, such as a Cartesian x–y–z space. The nature of that flow is governed by three fundamental physical principles:

(1) Mass is conserved. It is a basic principle of classical mechanics that mass can be neither created nor destroyed. That principle gives rise to the *continuity equation* of aerodynamics.

(2) Newton's second law states that the force on a moving object is equal to its mass times its acceleration:

$$F = ma$$

That principle, when applied to a moving fluid, gives rise to the *momentum equation*.

(3) Energy is conserved. There are various forms of energy (kinetic, potential, internal, etc.), and as one form of energy decreases, it reappears as an increase in other forms. Also, the first law of thermodynamics states that the heat added to a system and the work done on the system both contribute to an increase in the energy of the system. Those concepts, when applied to a moving fluid, give rise to the *energy equation*.

The three equations for continuity, momentum, and energy, when derived for an unsteady, three-dimensional flow, appear in a mathematical form called partial differential equations. When the flow is *inviscid*, those partial differential equations take a form called the *Euler equations*. For the reader familiar with such mathematical nomenclature, these equations will look reasonable. For the reader who is not, these equations should simply be viewed as an example of the mathematics necessary to work in theoretical aerodynamics.

Continuity equation

$$\frac{\partial \rho}{\partial t} + \boldsymbol{\nabla} \cdot (\rho V) = 0$$

Momentum equation

$$\rho \frac{Du}{Dt} = -\frac{\partial p}{\partial x}$$

$$\rho \frac{Dv}{Dt} = -\frac{\partial p}{\partial y}$$

$$\rho \frac{Dw}{Dt} = -\frac{\partial p}{\partial z}$$

Energy equation

$$\rho \frac{D(e + V^2/2)}{Dt} = \rho \dot{q} - \boldsymbol{\nabla} \cdot (pV)$$

449

where

$$\frac{D}{Dt} \equiv \underbrace{\frac{\partial}{\partial t}}_{\substack{\text{local} \\ \text{derivative}}} + \underbrace{(V \cdot \nabla)}_{\substack{\text{convective} \\ \text{derivative}}}$$

APPENDIX B

The Navier-Stokes Equations

Consider any general flow, such as that introduced in Appendix A. Now assume that the flow is *viscous* (i.e., is influenced by friction and thermal conduction). The three fundamental physical principles discussed in Appendix A also apply here; however, because friction and thermal conduction are added to the physical picture, the momentum and energy equations are more elaborate. The governing equations for a *viscous* flow are called the *Navier-Stokes equations*, and they are longer and involve more quantities than the simpler Euler equations shown in Appendix A. The Navier-Stokes equations are displayed below. The qualifying statement regarding mathematics in Appendix A also applies here.

Continuity equation

$$\frac{\partial \rho}{\partial t} + \nabla \cdot (\rho V) = 0$$

x-momentum equation

$$\rho \frac{Du}{Dt} = -\frac{\partial p}{\partial x} + \frac{\partial \tau_{xx}}{\partial x} + \frac{\partial \tau_{yx}}{\partial y} + \frac{\partial \tau_{zx}}{\partial z}$$

y-momentum equation

$$\rho \frac{Dv}{Dt} = -\frac{\partial p}{\partial y} + \frac{\partial \tau_{xy}}{\partial x} + \frac{\partial \tau_{yy}}{\partial y} + \frac{\partial \tau_{zy}}{\partial z}$$

z-momentum equation

$$\rho \frac{Dw}{Dt} = -\frac{\partial p}{\partial z} + \frac{\partial \tau_{xz}}{\partial x} + \frac{\partial \tau_{yz}}{\partial y} + \frac{\partial \tau_{zz}}{\partial z}$$

Energy equation

$$\rho \frac{D(e + V^2/2)}{Dt} = \rho \dot{q} + \frac{\partial}{\partial x}\left(k\frac{\partial T}{\partial x}\right) + \frac{\partial}{\partial y}\left(k\frac{\partial T}{\partial y}\right) + \frac{\partial}{\partial z}\left(k\frac{\partial T}{\partial z}\right) - \nabla \cdot pV$$

$$+ \frac{\partial(u\tau_{xx})}{\partial x} + \frac{\partial(u\tau_{yx})}{\partial y} + \frac{\partial(u\tau_{zx})}{\partial z} + \frac{\partial(v\tau_{xy})}{\partial x} + \frac{\partial(v\tau_{yy})}{\partial y}$$

$$+ \frac{\partial(v\tau_{zy})}{\partial z} + \frac{\partial(w\tau_{xz})}{\partial x} + \frac{\partial(w\tau_{yz})}{\partial y} + \frac{\partial(w\tau_{zz})}{\partial z}$$

where

$$\tau_{xy} = \tau_{yx} = \mu\left(\frac{\partial v}{\partial x} + \frac{\partial u}{\partial y}\right)$$

$$\tau_{yz} = \tau_{zy} = \mu\left(\frac{\partial w}{\partial y} + \frac{\partial v}{\partial z}\right)$$

451

$$\tau_{zx} = \tau_{xz} = \mu\left(\frac{\partial u}{\partial z} + \frac{\partial w}{\partial x}\right)$$

$$\tau_{xx} = \lambda(\boldsymbol{\nabla} \cdot \boldsymbol{V}) + 2\mu\frac{\partial u}{\partial x}$$

$$\tau_{yy} = \lambda(\boldsymbol{\nabla} \cdot \boldsymbol{V}) + 2\mu\frac{\partial v}{\partial y}$$

$$\tau_{zz} = \lambda(\boldsymbol{\nabla} \cdot \boldsymbol{V}) + 2\mu\frac{\partial w}{\partial z}$$

Whirling-Arm Calculation Showing the Benefit of a Large-Radius Arm

Refer to Figure 4.48. We have

$$V_1^2 = R_1^2 \omega^2 \tag{C.1}$$

and

$$V_2^2 = R_2^2 \omega^2 \tag{C.2}$$

Assuming low-speed, incompressible flow, from Bernoulli's equation the difference in stagnation pressure between the outer and inner edges is

$$p_{0_2} - p_{0_1} = \tfrac{1}{2}\rho \left[V_2^2 - V_1^2 \right] \tag{C.3}$$

Denoting $p_{0_2} - p_{0_1}$ by Δp_0, and substituting equations (C.1) and (C.2) into (C.3), we have

$$\Delta p_0 = \tfrac{1}{2}\rho \left[R_2^2 - R_1^2 \right] \omega^2 \tag{C.4}$$

The dynamic pressure at the inner edge, by definition, is

$$q_1 = \tfrac{1}{2}\rho V_1^2 = \tfrac{1}{2}\rho R_1^2 \omega^2 \tag{C.5}$$

Divide equation (C.4) by (C.5):

$$\frac{\Delta p_0}{q_1} = \left[\frac{R_2}{R_1} \right]^2 - 1 \tag{C.6}$$

In equation (C.6), as both R_1 and R_2 become very large, then

$$\frac{R_2}{R_1} \to 1$$

and hence, from equation (C.6),

$$\frac{\Delta p_0}{q_1} \to 0 \tag{C.7}$$

Therefore, the larger the value of R for a whirling arm, the smaller the difference in the total pressure across the lifting surface compared to the dynamic pressure, and hence the smaller the effect of such a nonuniformity on the aerodynamic measurements obtained with a whirling arm.

Calculation of Drag for Langley's Flat Plate at a Zero Angle of Attack

Samuel Langley's data for flat-plate drag obtained from his soaring experiments are shown in Figure 4.52. Here we use modern aerodynamic procedures to calculate the drag on Langley's flat-plate model at a zero angle of attack.

For low-speed incompressible flow, the laminar friction drag coefficient for a flat plate, including both the top and bottom surfaces of the plate, is

$$C_{D_f} = \frac{2.656}{\sqrt{Re}} \tag{D.1}$$

where the Reynolds number is defined as

$$Re = \frac{\rho_\infty V_\infty c}{\mu_\infty} \tag{D.2}$$

For the conditions of Langley's experiment (Figure 4.52), the sea-level density and viscosity coefficient are $\rho_\infty = 1.23 \, \text{kg/m}^3$ and $\mu_\infty = 1.7894 \times 10^{-5} \, \text{kg} \cdot \text{m}^{-1} \cdot \text{s}^{-1}$. The chord length of the plate is $c = 4.8 \, \text{in.} = 0.1219 \, \text{m}$. Hence, from equations (D.1) and (D.2), we have

$$C_{D_f} = \frac{2.656}{\left(\dfrac{(1.23)(0.1219)V_\infty}{1.7894 \times 10^{-5}} \right)^{1/2}} = \frac{0.029}{\sqrt{V_\infty}} \tag{D.3}$$

In equation (D.3), V_∞ is in meters per second. For the experimental situation in Figure 4.52, Langley measured a soaring velocity at a low angle of attack (i.e., at $\alpha = 2°$) of 20 m/s. For that velocity, the value of C_{D_f} from equation (D.3) is

$$C_{D_f} = \frac{0.029}{\sqrt{20}} = 0.00648 \tag{D.4}$$

The drag force due to skin friction is given by

$$D_f = \tfrac{1}{2}\rho_\infty V_\infty^2 S C_{D_f} \tag{D.5}$$

In equation (D.5), S is the planform area of the plate, equal to $0.929 \, \text{m}^2$. Hence, from equation (D.5),

$$D_f = 0.148 \, \text{N} = \boxed{15 \, \text{g (grams force)}}$$

[We note that the Reynolds number for these conditions is 167,580, certainly low enough to justify the assumption of laminar flow, and hence the use of equation (D.1).]

The frontal cross-sectional dimensions of the plate are 30 in. $\times \frac{1}{8}$ in., giving an area of $2.4384 \times 10^{-3} \, \text{m}^2$. Assuming a vertical-flat-plate drag coefficient of 1.0, the combined pressure drag due to the front and back edges of the plate perpendicular to the flow is

$$D_p = \frac{1}{2}\rho_\infty V_\infty^2 S C_D$$
$$= \frac{1}{2}(1.23)(20)^2(2.4384 \times 10^{-3})(1)$$
$$= 0.6\,\text{N} = \boxed{61\,\text{g (grams force)}}$$

Hence, the net predicted drag on the plate at a zero angle of attack is

$$D = D_f + D_p = 15 + 61 = \boxed{76\,\text{g (grams force)}}$$

This is the value shown as the black square in Figure 4.52.

Calculation of the Power-Required Curve for Langley's Flat-Plate Models

This appendix gives the calculations for the power-required curve plotted in Figure 4.54.

The Reynolds numbers associated with Langley's 30 × 4.8 in. plates (chord length = 4.8 in.) for the velocity range from 10 to 20 m/s are 84,000–168,000, low enough to safely assume that Langley's data were obtained for laminar flow. For laminar flow, the flat-plate skin-friction coefficient is given by equation (D.3) in Appendix D, accounting for skin friction on both the top and bottom surfaces of the plate. However, at even small angles of attack, the flow over a flat plate will readily separate from the top surface, creating a low-energy, dead-air region over the top surface. Therefore, at an angle of attack, only the bottom of the plate will experience an attached flow, and that is where the major effect of skin friction will be found. Hence, for the present calculation, we use, for the skin-friction drag coefficient, half the value given by equation (D.3):

$$C_{D_f} = \frac{0.0145}{\sqrt{V_\infty}} \tag{E.1}$$

The pressure acts perpendicular to the plate surface. Ignoring the thickness of the plate (i.e., ignoring the pressures acting on the front and rear edges, which have very small surface areas compared with the planform area), at an angle of attack the resultant *pressure* force will be essentially perpendicular to the plate. Hence, from the geometry of Figure 4.53a, the pressure drag will be related to the lift via

$$D_p = L \tan \alpha \tag{E.2}$$

or, in coefficient form,

$$C_{D_p} = C_L \tan \alpha \tag{E.3}$$

The total drag coefficient due to both the shear-stress distribution and the pressure distribution exerted over the plate is given by the sum of equations (E.1) and (E.3):

$$C_D = \frac{0.0145}{\sqrt{V_\infty}} + C_L \tan \alpha \tag{E.4}$$

This is the *drag polar* for Langley's flat-plate model.

The procedure for calculating the power required for steady, level flight, given the drag polar, is described elsewhere.[17] For the present calculation, it is as follows:

(1) Specify V_∞.
(2) Calculate C_L from

$$L = W = \tfrac{1}{2}\rho_\infty V_\infty^2 S C_L$$

or

$$C_L = \frac{2W}{\rho_\infty V_\infty^2 S} = \frac{(2)(0.5)(9.8)}{(1.23)(0.0929)V_\infty^2} = \frac{85.76}{V_\infty^2} \tag{E.5}$$

In equation (E.5), the mass of the plate is 0.5 kg, the standard density is 1.23 kg/s, the planform area is 0.0929 m^2, and 9.8 m · s^{-2} is the acceleration due to gravity necessary to convert the mass in kilograms to weight in newtons.

(3)   For the value of C_L calculated, obtain the corresponding angle of attack α from the flat-plate data in Figure 3.24, with an aspect-ratio correction to the lift slope. Using the standard correction to the lift slope due to a finite aspect ratio, from Prandtl's lifting-line theory,[1] we have

$$\frac{dC_L}{d\alpha} \equiv a = \frac{a_0}{1 + (57.3 a_0/\pi e \text{AR})} \tag{E.6}$$

where a is the lift slope of the finite wing, a_0 is the infinite wing lift slope taken as 0.1 per degree, e is a span efficiency factor taken as 0.943, and AR is the aspect ratio, given as 6.25. Hence, from equation (E.6) we have $a = 0.076$ per degree, and the angle of attack pertaining to the lift coefficient calculated in step 2 is

$$\alpha = \frac{C_L}{a} = \frac{C_L}{0.076} \quad \text{(in degrees)} \tag{E.7}$$

(4)   Calculate C_D from equation (E.4).
(5)   Calculate the power required from

$$P_R = D V_\infty = \tfrac{1}{2} \rho_\infty V_\infty^2 S C_D V_\infty \tag{E.8}$$

For the values of ρ_∞ and S pertaining here, equation (E.8) is written as

$$P_R = 0.057 V_\infty^3 C_D \tag{E.9}$$

where P_R is in watts.

Some tabulated values from this calculation are as follows:

V_∞ (m/s)	C_L [Eq. (E.5)]	α (degrees) [Eq. (E.7)]	C_D [Eq. (E.4)]	P_R (watts) [Eq. (E.9)]
12	0.596	7.84	0.0863	8.5
14	0.4376	5.76	0.0480	7.5
16	0.335	4.41	0.0295	6.89
18	0.265	3.49	0.0194	6.45
20	0.2144	2.82	0.0138	6.29
22	0.177	2.33	0.0103	6.25
24	0.149	1.96	0.00806	6.35
26	0.127	1.67	0.0065	6.51
28	0.109	1.43	0.00546	6.83
30	0.095	1.25	0.00472	7.26

The power required (P_R) from this tabulation is plotted versus V_∞ in Figure 4.54, yielding the power-required curve for Langley's flat-plate model.

Calculation of Lift and Drag for a Glider Flown as a Kite

Referring to Figure F.1, consider a lifting surface A tethered to the ground by a string B. The lifting surface has weight W. The net force on the lifting surface in the vertical direction is $L - W$, where L is the lift. The pulling force R' on the string is the resultant of the drag (the component of the force parallel to the wind) and the net vertical force, $L - W$. From the force diagram in Figure F.1b, we have

$$D = R' \cos \theta \qquad \text{(F.1)}$$

and

$$L - W = R' \sin \theta \qquad \text{(F.2)}$$

Hence, from (F.2),

$$L = R' \sin \theta + W \qquad \text{(F.3)}$$

From a measurement of R' (the pulling force on the string) and the angle θ, the corresponding measured drag and lift for a known weight W are obtained from equations (F.1) and (F.3), respectively.

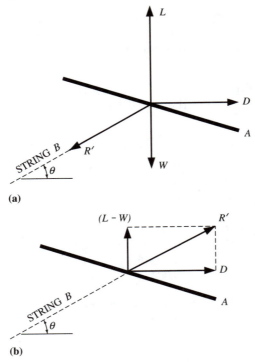

Figure F.1 Resolution of aerodynamic forces acting on a kite: (a) forces acting on a kite of weight W; (b) resolution of forces.

Aspect-Ratio Effect on the Wrights' 1900 Glider

For the Wrights' 1900 glider, the aspect ratio $AR_1 = 3.5$; for Lilienthal's wing, the aspect ratio $AR_2 = 6.48$. From Prandtl's lifting-line theory, the lift slope for a finite wing, $dC_L/d\alpha \equiv a$, is given in terms of the lift slope for an infinite wing, a_0, as

$$a = \frac{a_0}{1 + (a_0/\pi e AR)} \tag{G.1}$$

were e is a span efficiency factor. For both the Wrights' rectangular wing and Lilienthal's pointed-tip wing, $e = 0.84$. Hence, the *ratio* of the lift slopes for the Wrights' wing, a_1, and Lilienthal's wing, a_2, is obtained from equation (G.1) as

$$\frac{a_1}{a_2} = \frac{1 + (a_0/\pi e AR_2)}{1 + (a_0/\pi e AR_1)} \tag{G.2}$$

Using the finding from thin-airfoil theory that the lift slope for an infinite wing is 2π per radian (i.e., $a_0 = 2\pi$), equation (G.2) yields

$$\frac{a_1}{a_2} = \frac{1 + \{2\pi/[\pi(0.84)(6.48)]\}}{1 + \{2\pi/[\pi(0.84)(3.5)]\}} = 0.814$$

Hence, for the Wrights' conditions, the aspect-ratio effect dictates that the lift coefficients obtained from the Lilienthal table be reduced by the factor 0.814, or by approximately 19%.

Lift Coefficient According to Kutta

In 1902, Wilhelm Kutta published the following equation for the lift on a circular-arc airfoil at a zero angle of attack:

$$L = 4\pi a \rho V^2 \sin^2(\theta/2) \tag{H.1}$$

where a is the radius of the circular arc, and 2θ is the angle subtended by the arc at the center of the circle. This geometric construction is shown in Figure H.1. In terms of the airfoil lift coefficient c_l, defined as

$$c_l = \frac{L}{\frac{1}{2}\rho V^2 c}$$

where c is the chord of the airfoil, Kutta's formula, equation (H.1), becomes

$$c_l = \frac{8\pi a \sin^2(\theta/2)}{c} \tag{H.2}$$

Let us apply equation (H.2) to Lilienthal's circular-arc airfoil with a camber ratio of $\frac{1}{12}$. Referring to Figure H.1, the radius a can be found from the equation for a circle:

$$x^2 + y^2 = a^2$$

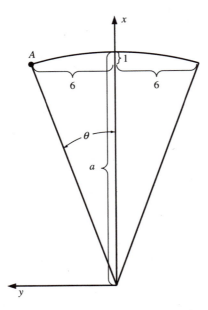

Figure H.1 Geometry for calculating the radius and subtended angle of a circular-arc airfoil of given camber ratio, in this case equal to $\frac{1}{12}$.

Applying that equation at point A in Figure H.1, we have

$$(a - 1)^2 + 6^2 = a^2 \tag{H.3}$$

Solving equation (H.3) for a, we find that $a = 18.5$.

To find θ, note from Figure H.1 that

$$\tan \theta = \frac{6}{17.5} = 0.3428$$

Hence $\theta = 18.92°$. From that,

$$\sin^2(\theta/2) = \sin^2(9.46°) = 0.027$$

Returning to equation (H.2), we have

$$c_l = \frac{8\pi(18.5)(0.027)}{12} = 1.047$$

Hence, for Lilienthal's circular-arc airfoil, Kutta, in 1902, predicted a lift-coefficient value of 1.047. That was the first meaningful theoretical prediction of airfoil lift.

Compressibility Error in the Caldwell and Fales Data Reduction (1920)

In Chapter 9, where we discussed the historic significance of Caldwell and Fales's early data on the compressibility effects on airfoils,[128] it was noted that they made a mistake in the data reduction. This appendix explains the nature and magnitude of that mistake.

In studying Caldwell and Fales's detailed data reduction, I have found that although they recognized that the density of the airflow inside the wind tunnel changed at the higher speeds, their attempt to account for that in calculating their lift and drag coefficients from their measured lift and drag forces was done incorrectly. They thought that they had carried out their data reduction so that density would not enter into the calculation. Rather, they expressed their lift and drag coefficients in terms of the impact pressure – the difference between total pressure and static pressure. That was why they said that "density does not enter into the calculation." But they incorrectly and rather naively used the incompressible Bernoulli equation to replace the velocity-squared term in the definition of the lift coefficient with the impact pressure. That resulted in errors of about 10% in the values of their reported lift and drag coefficients at high speeds.

Caldwell and Fales defined the lift coefficient as

$$K_y = \frac{L}{\rho A V^2} \tag{I.1}$$

which, as discussed in Chapter 8, is one-half the value used by convention today. (Hence, a value of $K_y = 0.5$ in Figure 9.11 is equivalent to a value of $C_L = 1.0$ used today.) Caldwell and Fales made an error when they used the incompressible Bernoulli equation

$$p_0 - p = \tfrac{1}{2}\rho V^2 \tag{I.2}$$

to write equation (I.1) as

$$K_y = \frac{L}{2A(p_0 - p)} \tag{I.3}$$

The difference $p_0 - p$ is the impact pressure, which was measured in the wind tunnel. Because equation (I.3) does not show the density explicitly, they concluded that density did not enter into the calculation, thus accounting for the effect of compressibility. The values for K_y reported by Caldwell and Fales[128] and shown in Figure 9.11 were obtained from equation (I.3).

To introduce the impact pressure into equation (I.1), to properly account for compressibility, we note that instead of equation (I.2), which assumes incompressible flow, the difference $p_0 - p$ is given for compressible flow as

$$p_0 - p = \frac{1}{2}\rho V^2 \left[1 + \frac{1}{4}M^2 + \frac{1}{40}M^4 + \frac{1}{1,600}M^6 + \cdots \right] \tag{I.4}$$

or, solving for ρV^2,

$$\rho V^2 = \frac{2(p_0 - p)}{1 + \dfrac{1}{4}M^2 + \dfrac{1}{40}M^4 + \dfrac{1}{1,600}M^6 + \cdots} \tag{I.5}$$

462

Hence, equation (I.1) can be written as

$$(K_y)_{\text{true}} = \frac{L}{2A(p_0 - p)} \left[1 + \frac{1}{4}M^2 + \frac{1}{40}M^4 + \frac{1}{1,600}M^6 + \cdots \right] \quad (I.6)$$

The value for K_y given by equation (I.6) is the true value for the lift coefficient for subsonic compressible flow, expressed in terms of the impact pressure $p_0 - p$. Substituting equation (I.3) into (I.6), we have

$$(K_y)_{\text{true}} = K_y \left[1 + \frac{1}{4}M^2 + \frac{1}{40}M^4 + \frac{1}{1,600}M^6 + \cdots \right] \quad (I.7)$$

where, in equation (I.7), K_y is the Caldwell and Fales value.

To assess the magnitude of the error, assume a free-stream Mach number of 0.63. From equation (I.7),

$$(K_y)_{\text{true}} = K_y(1.103)$$

Hence, the true lift coefficient is about 10% higher than the value reported by Caldwell and Fales. At lower Mach numbers, the differences are smaller; at higher Mach numbers, the differences are larger. If Caldwell and Fales had carried out their calculation properly, the values for the lift coefficient in the intermediate-velocity regime shown in the middle of Figure 9.11 would have been slightly increasing (the proper trend), instead of slightly decreasing as shown.

References

1. Anderson, John D., Jr. 1991. *Fundamentals of Aerodynamics*, 2nd ed. New York: McGraw-Hill.
2. Rouse, Hunter, and Ince, Simon. 1957. *History of Hydraulics*. Iowa City: Iowa Institute of Hydraulic Research.
3. Clagett, Marshall. 1978. *Archimedes in the Middle Ages. Vol. 3: The Fate of the Medieval Archimedes, 1300 to 1565*. Philadelphia: American Philosophical Society.
4. Hart, Ivor B. 1961. *The World of Leonardo da Vinci*. London: MacDonald.
5. Giacomelli, R. 1930. The Aerodynamics of Leonardo da Vinci. *Journal of the Royal Aeronautical Society* 34(240):1016–38.
6. Gibbs-Smith, C. H. 1962. *Sir George Cayley's Aeronautics, 1796–1855*. London: HMSO.
7. Mach, Ernst. 1942. *The Science of Mechanics* (trans. T. J. McCormack). London: Open Court Publishing. (Originally published 1893.)
8. Tokaty, G. A. 1971. *A History and Philosophy of Fluid Mechanics*. Henley-on-Thames: G. T. Foulis & Co.
9. Mahoney, Michael S. 1974. Edme Mariotte. In *Dictionary of Scientific Biography*, vol. 9, ed. C. C. Gillispie, pp. 114–22. New York: Scribner.
10. Giacomelli, R., and Pistolesi, E. 1934. Historical Sketch. In *Aerodynamic Theory*, vol. 1, ed. W. F. Durand. Berlin: Springer.
11. Bos, H. J. M. 1972. Christiaan Huygens. In *Dictionary of Scientific Biography*, vol. 6, ed. C. C. Gillispie, pp. 597–613. New York: Scribner.
12. Newton, Isaac. 1947. *Mathematical Principles of Natural Philosophy* (rev. trans. Florian Cajori). Berkeley: University of California Press.
13. Mason, Stephen F. 1962. *A History of the Sciences*. New York: Macmillan.
14. Bernoulli, Daniel. 1968. *Hydrodynamics* (trans. Thomas Carmody and Helmut Kobus). New York: Dover.
15. Salas, M. D. 1988. Leonhard Euler and His Contributions to Fluid Mechanics. American Institute of Aeronautics and Astronautics, AIAA paper 88-3566-CP.
16. Gillispie, Charles Coulston. 1978. Laplace. In *Dictionary of Scientific Biography*, vol. 15, suppl. 1, ed. C. C. Gillispie, pp. 273–303. New York: Scribner.
17. Anderson, John D., Jr. 1989. *Introduction to Flight*, 3rd ed. New York: McGraw-Hill.
18. Airey, John. 1913. Notes on the Pitot Tube. *Engineering News* 69(16):783.
19. Mayr, Otto. 1975. Henri Pitot. In *Dictionary of Scientific Biography*, vol. 11, ed. C. C. Gillispie, pp. 4–5. New York: Scribner.
20. Pritchard, J. L. 1957. The Dawn of Aerodynamics. *Journal of the Royal Aeronautical Society* 61:149–80.
21. Gillmor, C. S. 1970. Jean-Charles Borda. In *Dictionary of Scientific Biography*, vol. 2, ed. C. C. Gillispie, pp. 299–300. New York: Scribner.
22. Anderson, John D., Jr. 1987. Sir George Cayley. In *Great Lives from History: British and Commonwealth Series*. Englewood Cliffs, NJ: Salem Press.

23. Pritchard, J. Lawrence. 1961. *Sir George Cayley: The Inventor of the Aeroplane*. London: Max Parrish & Co.

24. Jakab, Peter L. 1990. *Visions of a Flying Machine*. Washington, DC: Smithsonian Institution Press.

25. Gibbs-Smith, Charles H. 1970. *Aviation, An Historical Survey*. London: HMSO.

26. Navier, L. M. H. 1822. Mémoire sur les lois du mouvement des fluides. *Mémoires de l'Académie Royale des Sciences* 6:389.

27. Saint-Venant, B. 1843. Note à joindre au mémoire sur la dynamique des fluids. *Comptes Rendus des Seances de l'Académie des Sciences* 17:1240.

28. Stokes, G. G. 1845. On the Theories of the Internal Friction of Fluids in Motion, and of the Equilibrium and Motion of Elastic Solids. *Transactions of the Cambridge Philosophical Society* 8:287.

29. Cauchy, A. L. 1827. Mémoire sur la theorie des ondes. *Mémoires de l'Académie Royale des Sciences* 11.

30. Helmholtz, Hermann L. 1858. On the Integrals of the Hydrodynamical Equations Corresponding to Vortex Motions. *Crelles Journal für die reine und angewandte Mathematik* 60:23–55.

31. Helmholtz, Hermann L. 1868. On the Discontinuous Motions of a Fluid. *Monatsberichte d. Kon. Akademie der Wissenschaften zu Berlin*, pp. 215–28.

32. Randers-Pehrson, N. H. 1935. Pioneer Wind Tunnels. *Smithsonian Miscellaneous Collections* 93(4).

33. Marey, Etienne. 1900. *Comptes Rendus des Séances de l'Académie des Sciences* 131:160–3.

34. Kirchhoff, G. R. 1869. Zur Theorie freier Flüssigkeitsstrahlen. *Crelles Journal Für die reine und angewandte Mathematik* 70:289.

35. Strutt, J. W. (Lord Rayleigh). 1876. On the Resistance of Fluids. *Philosophical Magazine*, ser. 5 2:430–41.

36. Vincenti, Walter G. 1990. *What Engineers Know and How They Know It*. Baltimore: Johns Hopkins University Press.

37. Hoerner, S. F., and Borst, H. V. 1975. *Fluid-Dynamic Lift*. Brick Town, NJ: published by authors.

38. Hoerner, S. F. 1958. *Fluid-Dynamic Drag*. Brick Town, NJ: published by author.

39. Hagen, G. H. L. 1839. Ueber die Bewegung des Wassers in engen cylindrischen Röhren. *Poggendorfs Annalen der Physik und Chemie* 16.

40. Hagen, G. H. L. 1855. Ueber den Einfluss der Temperatur auf die Bewegung des Wasser in Röhren. *Mathematisch Abhandlungen der Akademie der Wissenschaften zu Berlin*.

41. Reynolds, Osborne. 1883. An Experimental Investigation of the Circumstances which Determine whether the Motion of Water in Parallel Channels Shall be Direct or Sinuous, and of the Law of Resistance in Parallel Channels. *Philosophical Transactions of the Royal Society* 174.

42. Reynolds, Osborne. 1894. On the Dynamical Theory of Incompressible Fluids and the Determination of the Criterion. *Philosophical Transactions of the Royal Society* 186.

43. Reynolds, Osborne. 1874–5. On the Extent and Action of the Heating of Steam Boilers. *Proceedings of the Manchester Literary and Philosophical Society* 14:8.

44. Pritchard, J. Laurence. 1958. Francis Herbert Wenham, Honorary Member, 1824–1908; An Appreciation of the First Lecturer to the Aeronautical Society. *Journal of the Royal Aeronautical Society* 62:571–96.

45. Chanute, Octave. 1976. *Progress in Flying Machines*. Long Beach, CA: Lorenz & Herweg. (Originally published 1894.)

46. Lilienthal, Otto. 1889. *Der Vogelflug als Grundlage der Fliegekunst*. Berlin: R. Gaertners Verlagsbuchhandlung. (Translated by I. W. Isenthal and published in 1911 as *Birdflight as the Basis of Aviation*, London: Longmans, Green.)

47. Moedebeck, Hermann W. L. 1895. *Taschenbuch zum praktischen Gebrauch für Flugtechniker und Liftschiffer*. Berlin: Verlag von W. H. Kuhl.

48. Schwipps, Werner. 1979. *Lilienthal; Die Biographie des ersten Fliegers*. Grafelling, Germany: Aviatic Verlag. (English translation available in the library of the National Air and Space Museum, Smithsonian Institution.)

49. Langley, S. P. 1891. *Experiments in Aerodynamics*. Smithsonian Contributions to Knowledge no. 801. Washington, DC: Smithsonian Institution.

50. Langley, S. P., and Manly, C. M. 1911. *Langley Memoir on Mechanical Flight*. Smithsonian Contributions to Knowledge, vol. 27, no. 3. Washington, DC: Smithsonian Institution.

51. Crouch, Tom D. 1981. *A Dream of Wings*. New York: Norton.

52. Biddle, Wayne. 1991. *Barons of the Sky*. New York: Simon & Schuster.

53. Crouch, Tom. 1989. *The Bishop's Boys*. New York: Norton.

54. Lamb, H. 1924. *Hydrodynamics*, 5th ed. Cambridge University Press.

55. McFarland, Marvin W. (ed.) 1953. *The Papers of Wilbur and Orville Wright*. New York: McGraw-Hill.

56. Wright, Wilbur. 1901. Some Aeronautical Experiments. *Journal of the Western Society of Engineers* 6:489–510.

57. Culick, F. E. C., and Jex, H. R. 1987. Aerodynamics, Stability, and Control of the 1903 Wright Flyer. In *The Wright Flyer: An Engineering Perspective*, ed. Howard Wolko. Washington, DC: Smithsonian Institution.

58. Jacobs, Eastman N., Ward, Kenneth E., and Pinkerton, Robert. 1933. *The Characteristics of 78 Related Airfoil Sections from Tests in the Variable-Density Wind Tunnel*. NACA TR 460.

59. Lanchester, F. W. 1926. Sustentation in Flight. *Journal of the Royal Aeronautical Society* 30:587–606.

60. Lanchester, F. W. 1907. *Aerodynamics*. London: A. Constable & Co.

61. Prandtl, L. 1905. Ueber Flüssigkeitsbewegung bei sehr kleiner Reibung. In *Proceedings of the 3rd International Mathematical Congress, Heidelberg, 1904*. Leipzig.

62. Goldstein, Sydney. 1969. Fluid Mechanics in the First Half of This Century. In *Annual Review of Fluid Mechanics*, vol. 1, ed. W. R. Sears and M. Van Dyke, pp. 1–28. Palo Alto, CA: Annual Reviews, Inc.

63. Blasius, H. 1908. Boundary Layers in Fluids with Small Friction. *Zeitschrift für Mathematik und Physik* 56:1.

64. Schlichting, Hermann. 1979. *Boundary-Layer Theory*, 7th ed. New York: McGraw-Hill.

65. Oswatitsch, Klaus, and Wieghardt, K. 1987. Ludwig Prandtl and His Kaiser-Wilhelm-Institut. In *Annual Review of Fluid Mechanics*, vol. 19, ed. J. L. Lumley, M. Van Dyke, and H. L. Reed, pp. 1–25. Palo Alto, CA: Annual Reviews Inc.

66. Flügge-Lotz, Irmgard, and Flügge, Wilhelm. 1973. Ludwig Prandtl in the Nineteen-Thirties: Reminiscences. In *Annual Review of Fluid Mechanics*, vol. 5, ed. M. Van Dyke, W. G. Vincenti, and J. V. Wehausen, pp. 1–8. Palo Alto, CA: Annual Reviews, Inc.

67. von Kármán, Theodore (with Lee Edson). 1967. *The Wind and Beyond*. Boston: Little, Brown.

68. von Kármán, Theodore. 1954. *Aerodynamics*. Ithaca, NY: Cornell University Press.

69. Loftin, Laurence K. 1985. *Quest for Performance: The Evolution of Modern Aircraft*. NASA SP-468. Washington, DC: National Aeronautics and Space Administration.

70. Eiffel, Gustave. 1907. *Recherches expérimentales sur la résistance de l'air exécutées à la tour*. Paris: Maretheux.

71. Eiffel, Gustave. 1910. *The Resistance of the Air and Aviation: Experiments Conducted at the Champ-de-Mars Laboratory.* Paris: Dunot & Pinat. (English trans., Jerome C. Hunsaker, Houghton Mifflin, Boston, 1913.)

72. Black, Joseph. 1990. Gustave Eiffel – Pioneer of Experimental Aerodynamics. *Aeronautical Journal* 94:231–44.

73. Rotta, Julius C. 1990. *Die aerodynamische Versuchsanstalt in Göttingen, ein Werk Ludwing Prandtls,* Göttingen: Vandenhoeck & Ruprecht.

74. Prandtl, Ludwig. 1913. Flüssigkeitsbewegung. In *Handworterbuch der Naturwissenschaften.* Jena: Verlag von Gustav Fischer.

75. Prandtl, Ludwig. 1927. The Generation of Vortices in Fluids of Small Viscosity. *Journal of the Royal Aeronautical Society* 31:720–41.

76. Betz, A. 1914. Untersuchungen von Tragflächen mit Verwundenen and nach Ruckwarts gerichteten Enden. *Zeitschrift für Flugtechnik und Motorluftschiffahrt* 16–17.

77. Prandtl, Ludwig. 1918. *Tragflächentheorie. I.* Mitteilung, Nachrichten der K. Gesellschaft der Wissenschaften zu Göttingen, Math-phys. Klasse.

78. Prandtl, Ludwig. 1919. *Tragflächentheorie. II.* Mitteilung, Nachrichten der K. Gesellschaft der Wissenschaften zu Göttingen, Math-phys. Klasse.

79. Prandtl, Ludwig. 1921. *Applications of Modern Hydrodynamics to Aeronautics.* NACA technical report 116.

80. Glauert, Hermann. 1926. *The Elements of Aerofoil and Airscrew Theory.* Cambridge University Press.

81. Houghton, E. L., and Carruthers, N. B. 1982. *Aerodynamics for Engineering Students,* 3rd ed. London: Edward Arnold.

82. Betz, A. 1913. Auftrieb und Widerstand eines Doppeldeckers. *Zeitschrift für Flugtechnik und Motorluftschiffahrt* 1.

83. Theodorsen, Theodore. 1931. *Theory of Wing Sections of Arbitrary Shape.* NACA report 411.

84. Hansen, James R. 1987. *Engineer in Charge.* NASA SP-4305.

85. Munk, Max. 1922. *General Theory of Thin Wing Sections.* NACA report 142.

86. Glauert, H. 1925–4. Experimental Tests of the Vortex Theory of Aerofoils. In *Technical Report of the Aeronautical Research Committee. Vol. I. Reports and Memoranda,* no. 889. London: HMSO.

87. Fage, A., and Nixon, H. L. 1923–4. The Prediction on the Prandtl Theory of the Lift and Drag for Infinite Span from Measurements on Aerofoils on Finite Span. In *Technical Report of the Aeronautical Research Committee. Vol. I. Reports and Memoranda,* no. 903. London: HMSO.

88. Bairstow, Leonard. 1920. *Applied Aerodynamics.* London: Longmans, Green & Co.

89. Diehl, Walter S. 1928. *Engineering Aerodynamics.* New York: Ronald Press.

90. Millikan, Clark B. 1941. *Aerodynamics of the Airplane.* New York: Wiley.

91. Vander Meulen, Jacob. 1991. *The Politics of Aircraft.* Lawrence: University Press of Kansas.

92. Roseberry, C. R. 1972. *Glenn Curtiss: Pioneer of Flight.* New York: Doubleday.

93. Baals, Donald D., and Corliss, William R. 1981. *Wind Tunnels of NASA.* NASA SP-440. Washington, DC: National Aeronautics and Space Administration.

94. Hunsaker, Jerome C. 1956. *Forty Years of Aeronautical Research.* Smithsonian Institution publication 4247. Washington, DC: Smithsonian Institution Press.

95. Gorrell, Edgar S., and Martin, H. S. 1918. *Aerofoils and Aerofoil Structural Combinations.* NACA report 18.

96. Munk, Max M., and Miller, Elton W. 1925. *Model Tests with a Systematic Series of 27 Wing Sections at Full Reynolds Number.* NACA report 221.

97. Carter, C. C. 1929. *Simple Aerodynamics and the Airplane*. New York: Ronald Press.

98. Higgins, George J., Diehl, Walter S., and DeFoe, George L. 1927. *Tests on Models of Three British Airplanes in the Variable Density Wind Tunnel*. NACA report 279.

99. Breguet, Louis. 1922. Aerodynamical Efficiency and the Reduction of Air Transport Costs. *Aeronautical Journal* 26:307–13.

100. Jones, B. Melvill. 1929. The Streamline Airplane. *Aeronautical Journal* 33(221):358–85.

101. Miller, Ronald, and Sawers, David. 1970. *The Technical Development of Modern Aviation*. New York: Praeger.

102. Binnie, A. M. 1978. Some Notes on the Study of Fluid Mechanics in Cambridge, England. In *Annual Review of Fluid Mechanics*, vol, 10, ed. M. Van Dyke, J. V. Wehausen, and J. L. Lumley, pp. 1–10. Palo Alto, CA: Annual Reviews, Inc.

103. Weick, Fred E., and Hansen James R. 1988. *From the Ground Up: The Autobiography of an Aeronautical Engineer*. Washington, DC: Smithsonian Institution Press.

104. Weick, Fred E. 1929. *Drag and Cooling with Various Forms of Cowling for a "Whirlwind" Radial Air-Cooled Engine – I*. NACA TR 313.

105. Weick, Fred E. 1928. The New NACA Low Drag Cowling. *Aviation* 25:1556–7, 1586–90.

106. Townend, H. C. H. 1929. *Reduction of Drag of Radial Engines by the Attachment of Rings of Airfoil Section, Including Interference Experiments of an Allied Nature, with Some Further Applications*. R&M no. 1267. British A.R.C.

107. Wood, Donald H. 1932. *Tests of Nacelle-Propeller Combinations in Various Positions with Reference to Wings. II. Thick Wing–Various Radial-Engine Cowlings – Tractor Propeller*. NACA TR 436.

108. Wood, Donald H. 1932. *Tests of Nacelle-Propeller Combinations in Various Positions with Reference to Wings. I. Thick – NACA Cowled Nacelle – Tractor Propeller*. NACA TR 415.

109. Garrick, I. E. 1992. Sharing His Insights and Innovations. In *A Modern View of Theodore Theodorsen*, ed. E. H. Dowell, pp. 21–6. Washington, DC: American Institute of Aeronautics and Astronautics.

110. Theodorsen, Theodore. 1938. The Fundamental Principles of the NACA Cowling. *Journal of the Aeronautical Sciences* 5(3).

111. Theodorsen, Theodore, Brevoort, M. J., and Stickle, George W. 1937. *Full Scale Tests of N.A.C.A. Cowlings*. NACA TR 592.

112. Stickle, George. 1939. *Design of N.A.C.A. Cowlings for Radial Air-cooled Engines*. NACA TR 662.

113. Theodorsen, Theodore, and Garrick, I. E. 1933. *General Potential Theory of Arbitrary Wing Sections*. NACA TR 452.

114. Garrick, I. E. 1933. *Determination of the Theoretical Pressure Distribution for Twenty Airfoils*. NACA TR 465.

115. Jacobs, Eastman N., and Pinkerton, Robert M. 1935. *Tests in the Variable-Density Wind Tunnel of Related Airfoils Having the Maximum Camber Unusually Far Forward*. NACA TR 537.

116. Jacobs, Eastman N., Pinkerton, Robert M., and Greenberg, Harry. 1937. *Tests of Related Forward-Camber Airfoils in the Variable-Density Wind Tunnel*. NACA TR 610.

117. Jones, R. T. 1977. Recollections from an Earlier Period in American Aeronautics. In *Annual Review of Fluid Mechanics*, vol. 9, ed. M. Van Dyke, J. V. Wehausen, and J. L. Lumley, pp. 1–11. Palo Alto, CA: Annual Reviews, Inc.

118. Abbott, Ira H. 1980. Airfoils: Significance and Early Development. In *The Evolution of Aircraft Wing Design* (AIAA Symposium), pp. 21–48. Washington, DC: American Institute of Aeronautics and Astronautics.

119. Jones, B. Melvill. 1938. Flight Experiments on the Boundary Layer. *Journal of the Aeronautical Sciences* 5(3):81–101.

120. Farren, W. S. 1944. Research for Aeronautics – Its Planning and Applications. *Journal of the Aeronautical Sciences* 11(2):95–105.

121. van der Linden, F. Robert. 1991. *The Boeing 247: The First Modern Airliner.* Seattle: University of Washington Press.

122. Kindleburger, J. H. 1953. The Design of Military Aircraft. *Aeronautical Engineering Review* 12:44.

123. Minutes of a meeting of the Committee on Aerodynamics, October 12, 1943, p. 9, John Stack files, NASA Langley Research Center archives.

124. Stodola, A. B. 1905. *Steam Turbines.* Trans. L. C. Loewenstein. New York: Van Nostrand. (Originally published 1904 as *Die Dampfturbinen.* Berlin: Verlag von Julius Springer.)

125. Meyer, T. 1908. Ueber zweidimensionale Bewegungsvorgange in einen Gas, Das mit Ueberschallgeschwindigkeit Strömt. Ph.D. dissertation, Göttingen University.

126. Bryan, G. H. 1918. *The Effect of Compressibility on Streamline Motions.* R&M no. 555, vol. 1. London: Advisory Committee for Aeronautics.

127. Bryan, G. H. 1919. *The Effect of Compressibility on Streamline Motions, Part II.* R&M no. 640. London: Advisory Committee for Aeronautics.

128. Caldwell, F. W., and Fales, F. M. 1920. *Wind Tunnel Studies in Aerodynamic Phenomena at High Speed.* NACA TR 83.

129. Douglas, G. P., and Wood, R. M. 1923. *The Effects of Tip Speed on Airscrew Performance. An Experimental Investigation of the Performance of an Airscrew Over a Range of Speeds of Revolution from "Model" Speeds up to Tip Speeds in Excess of the Velocity of Sound in Air.* R&M no. 884. London: Advisory Committee for Aeronautics.

130. Bensberg, H., and Cranz, C. 1910. Ueber eine photographische Methode zur Messung von Geschwindigkeiten und Geschwindigkeitsverlusten bei Infanteriegeschossen. *Artillerische Monatshefte (Berlin),* no. 41.

131. Briggs, L. J., Hull, G. F., and Dryden, H. L. 1924. *Aerodynamic Characteristics of Airfoils at High Speeds.* NACA TR 207.

132. Liepmann, Hans W., and Puckett, Allen E. 1947. *Introduction to Aerodynamics of a Compressible Fluid.* New York: Wiley.

133. Briggs, L. J., and Dryden, H. L. 1926. *Pressure Distribution Over Airfoils at High Speeds.* NACA TR 255.

134. Briggs, L. J., and Dryden, H. L. 1929. *Aerodynamic Characteristics of Twenty-Four Airfoils at High Speeds.* NACA TR 319.

135. Glauert, H. 1927. The Effect of Compressibility on the Lift of an Airfoil. R&M no. 1135. London: Advisory Committee for Aeronautics. Reprinted 1928 in *Proceedings of the Royal Society* 118:113.

136. Douglas, G. P., and Perring, W. G. A. 1927. *Wind Tunnel Tests with High Tip Speed Airscrews. The Characteristics of the Aerofoil Section R.A.F. 31a at High Speeds.* R&M no. 1086. London: Advisory Committee for Aeronautics.

137. Douglas, G. P., and Perring, W. G. A. 1927. *Wind Tunnel Tests with High Speed Airscrews. The Characteristics of a Bi-Convex Aerofoil at High Speeds.* R&M no. 1091. London: Advisory Committee for Aeronautics.

138. Douglas, G. P., and Perring, W. G. A. 1927. *Wind Tunnel Tests with High Speed Airscrews. The Characteristics of Bi-Convex No. 2 Aerofoil Section at High Speeds.* R&M no. 1123. London: Advisory Committee for Aeronautics.

139. Taylor, G. I., and Sharman, C. F. 1928. *A Mechanical Method for Solving Problems of Flow in Compressible Fluids*. R&M no. 1195. London: Advisory Committee for Aeronautics. *Proceedings of the Royal Society*, A 121:194.

140. Davis, Lou. 1963. No Time for Soft Talk. *National Aeronautics* (January), pp. 9–12.

141. Stack, John. 1933. *The N.A.C.A. High-Speed Wind Tunnel and Tests of Six Propeller Sections*. NACA TR 463.

142. Stack, John, 1934. Effects of Compressibility on High Speed Flight. *Journal of the Aeronautical Sciences* 1(1):40–3.

143. John Stack Files, NASA Langley Research Center historical archives, Hampton, VA.

144. Becker, John V. 1980. *The High-Speed Frontier*. NASA SP-445. Washington, DC: National Aeronautics and Space Administration.

145. Ferri, Antonio. 1949. *Elements of Aerodynamics of Supersonic Flows*. New York: Macmillan.

146. Jacobs, Eastman. 1936. *Methods Employed in America for the Experimental Investigation of Aerodynamic Phenomena at High Speeds*. NACA miscellaneous paper 42.

147. Tsien, H. S. 1939. Two-Dimensional Subsonic Flow of Compressible Fluids. *Journal of the Aeronautical Sciences* 6(10):399.

148. von Kármán, T. H. 1941. Compressibility Effects in Aerodynamics. *Journal of the Aeronautical Sciences* 8(9):337.

149. Stack, John, Lindsey, W. F., and Littell, Robert E. 1938. *The Compressibility Burble and the Effect of Compressibility on Pressures and Forces Acting on an Airfoil*. NACA TR 646.

150. Ackeret, Von J. 1925. Luftkrafte auf Flugel, die mit Grosserer als Schallgeschwindigkeit Bewegt werden. *Zeitschrift für Flugtechnik und Motorluftschiffahrt* 16:72–4.

151. Foss, R. L. 1978. From Propellers to Jets in Fighter Aircraft Design. *Diamond Jubilee of Powered Flight: The Evolution of Aircraft Design*, ed. J. D. Pinson, pp. 51–64. Washington, DC: American Institute of Aeronautics and Astronautics.

152. Hallion, Richard P. 1972. *Supersonic Flight*. New York: Macmillan.

153. Hilton, W. F. 1966. British Aeronautical Research Facilities. *Journal of the Royal Aeronautical Society* 70:103–4.

154. Young, James O. 1990. *Supersonic Symposium: The Men of Mach 1*. Air Force Flight Test Center History Office.

155. Whitcomb, R. T., and Clark, L. R. 1965. *An Airfoil Shape for Efficient Flight at Supercritical Mach Numbers*. NASA TMX-1109.

156. Hanle, Paul A. 1982. *Bringing Aerodynamics to America*. Cambridge, MA: M.I.T. Press.

157. Gorn, Michael H. 1992. *The Universal Man: Theodore von Kármán's Life in Aeronautics*. Washington, DC: Smithsonian Institution Press.

158. von Kármán, Theodore, and Moore, Norton B. 1932. Resistance of Slender Bodies Moving with Supersonic Velocities, with Special Reference to Projectiles. *Transactions of the American Society of Mechanical Engineers* 34:303–10.

159. von Kármán, Theodore. 1947. Supersonic Aerodynamics – Principles and Applications. *Journal of the Aeronautical Sciences* 14(7):373–402.

160. Taylor, S. I., and Maccoll, J. W. 1935. The Mechanics of Compressible Fluids. In *Aerodynamic Theory*, vol. 3, ed. W. F. Durand, pp. 209–49. Berlin: Springer.

161. Munk, Max M. 1924. *Note on the Relative Effect of Dihedral and the Sweep Back of Airplane Wings*. NACA technical note 177.

162. Schairer, George S. 1980. Evolution of Modern Air Transport Wings. In *Evolution of Wing Design*, pp. 61–5. Washington, DC: American Institute of Aeronautics and Astronautics.

163. Hallion, Richard P. 1983. *Designers and Test Pilots*. Alexandria, VA: Time-Life Books.

164. Blair, Morgan W. 1980. Evolution of the F-86. In *Evolution of Aircraft Wing Design*, pp. 75–89. Washington, DC: American Institute of Aeronautics and Astronautics.

165. Copley, Steve. 1995. A Look Back at Swept-Back Wings. *AIAA Student Journal*, 33(3):2–3.

166. Matthews, C. W., and Thompson, J. R. 1945. *Comparative Drag Measurements at Transonic Speeds of Straight and Sweptback NACA 65-009 Airfoils Mounted on a Freely-Falling Body*. NACA memorandum report L5G23a. Released in 1949 as TR 988.

167. Steiner, John E. 1979. Jet Aviation Development: A Company Perspective. In *The Jet Age*, ed. W. Boyne and D. Lopez, pp. 140–83. Washington, DC: Smithsonian Institution Press.

168. Cook, William H. 1991. *The Road to the 707*. Bellevue, WA: TYC Publishing.

169. Neufeld, Michael J. 1995. *The Rocket and the Reich*. New York: Free Press.

170. Anderson, John D., Jr. 1989. *Hypersonic and High-Temperature Gas Dynamics*. New York: McGraw-Hill.

171. Anderson, John D., Jr. 1995. *Computational Fluid Dynamics: The Basics with Applications*. New York: McGraw-Hill.

172. Richardson, L. F. 1910. The Approximate Arithmetical Solution of Finite Differences of Physical Problems Involving Differential Equations, with an Application to the Stresses in a Masonry Dam. *Philosophical Transactions of the Royal Society*, A 210:304–57.

173. Courant, R., Friedrichs, K. O., and Lewy, H. 1928. Ueber die partiellen Differenzengleichungen der mathematischen Physik. *Mathematische Annalen* 100:32–74.

174. Selmin, V., Hettena, E., and Formaggia, L. 1992. An Unstructured Node Centered Scheme for the Simulation of 3-D Inviscid Flows. In *Computational Fluid Dynamics '92*, vol. 2, ed. C. Hirsch, J. Periaux, and W. Kordulla, pp. 823–8. Amsterdam: Elsevier.

175. Angelucci, Enzo. 1973. *Airplanes*. New York: McGraw-Hill.

Index

473